1997

NATIONAL PAINTING COST ESTIMATOR

by Dennis D. Gleason, CPE

Includes inside the back cover:
A 3½" high density disk with all the cost estimates in the book plus an estimating program for *Windows*™.

Craftsman Book Company
6058 Corte del Cedro / P.O. Box 6500 / Carlsbad, CA 92018

The author thanks the following individuals and organizations for furnishing materials and information used in the preparation of various portions of this book.

Howard Shahan, *American Design Painting & Drywall,* Poway, CA
American Society of Professional Estimators (ASPE), Wheaton, MD
Ameritone Paint, San Diego, CA
Benjamin Moore Paints, San Diego, CA
Gordon H. Brevcort, *Brevcort Consulting Associates,* Ridgewood, NJ
Paul Ross, *Deen Pierce Paint Company,* Pleasant Hill, CA
John San Marcos, *Devoe Coatings,* San Diego Marine Hardware, San Diego, CA
Gene Christopher, *Dunn-Edwards Paints & Wallcovering,* Concord, CA
Randy Martin, *Dunn-Edwards Paints & Wallcovering,* San Diego, CA
Expo Builders Supply, San Diego, CA
Larry Henderson, *Frazee Paint & Wallcovering,* San Diego, CA
John Christiansen, *Fuller O'Brien Paints,* South San Francisco, CA
Mike Crowley, *Fuller O'Brien Paints,* Cupertino, CA
The Glidden Company, San Diego, CA
Tim Boland, *Kelly Moore Paint Company,* Concord, CA
Hugh Champeny, *Kelly Moore Paint Company,* San Carlos, CA
Dennis Cripe, *R.W. Little Co, Inc., Sandblasting,* San Diego, CA
Bruce McMullan, *McMullan & Son Painting,* San Diego, CA
Joe Garrigan, *Mr. Paints,* San Diego, CA
John Horne, *Old Quaker Paint Company,* Escondido, CA
PPG Industries, Inc., *Pittsburgh Paints,* Torrance, CA
Bruce Asbill, *Rich Paint Company,* Oakland, CA
Richardson Engineering Services, Inc., Mesa, AZ
Rust-Oleum Protective Coatings, Los Angeles, CA
Keith Braswell, *Scotty Rents,* Pleasant Hill, CA
Al Chavez, *Sinclair Paints,* San Diego, CA
Squires-Belt Material Co., San Diego, CA
Steel Structures Painting Council, Pittsburgh, PA
The Sherwin-Williams Company, Dublin, CA
The Sherwin-Williams Company, San Diego, CA
John Meyer, *U.S. Government, Department of the Navy, Public Works,* San Diego, CA
Vista Paint Centers, Fullerton, CA
Vista Paint Centers, San Diego, CA
Jerry Rittgarn, *Waco-Arise Scaffolding & Equipment,* San Diego, CA
Ron Colton, *Wall Dimensions Wallcovering,* San Diego, CA

Cover photos by Pete Rintye, Hi-Country Photography
Painting contractor: Bob Petritz, Custom Painting, the High Desert

Sheila M. Scott, Calligraphy

ISBN 1-57218-033-1

Contents

List of Figures

How to Use This Book

Paint estimating is more of an art than a science. There's no price that's exactly right for every job and for every bidder. That's because every painting job is unique. No single material cost, no labor estimate, no pricing system fits all types of work. And just as every job varies, so do painting companies. No two painting contractors have the same productivity rates, the same labor burden, the same overhead expense and the same profit requirements.

The best paint estimates are always custom-made for a particular job. They're based on the contractor's actual productivity rate, material cost, labor cost, overhead percentage and profit expectations. No estimating book, no computerized estimating system, no estimating service can possibly account for all the variables that make every job and every painting company different. Only a skilled estimator using professional judgment and a proven estimating system can produce consistently reliable estimates on a wide variety of painting jobs.

So Why Buy This Book?

That's easy. This is the most complete, authoritative and reliable unit cost guide ever made available to paint estimators. No matter what types of work you estimate, no matter what your costs are, this book will help produce consistently accurate painting cost estimates in dollars and cents. But it isn't a substitute for expertise. It's not a simple way to do in minutes what an experienced paint estimator might not be able to do in hours. Instead, this unit cost guide will aid you in developing a good estimate of costs for any painting operation on any project. Think of this manual as *one* good estimating tool. But it's not (or at least shouldn't be) the only estimating tool you'll use.

For most jobs, I expect that the figures you see here will prove to be good estimates. But anyone who understands paint estimating will understand why judgment is needed when applying figures from this manual – or any other paint estimating guide. It's your responsibility to decide which conditions on the job you're bidding are like conditions assumed in this manual, and which conditions are different. Where conditions are different, you'll need good professional judgment to arrive at a realistic estimated cost.

National Estimator '97 Inside the back cover of this book you'll find an envelope with a 3½" floppy disk. This disk has the entire 1997 *National Painting Cost Estimator* in a form that can be installed on the hard disk of a computer. Once

	Manhour productivity	Labor cost per hour	Labor burden percent	Labor burden dollars	Labor cost plus burden	Material price discount	Overhead percent	Profit
Slow (1P)	Low	$10.50	28.45%	$2.99	$13.49	20%	31.0%	15%
Medium (2P)	Average	16.75	34.85%	5.84	22.59	30%	24.5%	11%
Fast (3P)	High	22.50	41.75%	9.39	31.89	40%	18.5%	6%

Note: These rates are for painters. Hourly rates for wallcovering are different. See page 29. Slow, Medium and Fast jobs are defined on page 13. National Estimator computer software program uses hourly rates in the Labor cost plus burden column. National Estimator shows productivity rates (Slow, Medium and Fast) and copies the words Slow, Medium or Fast to your estimate. It also copies the crew productivity code, either 1P (Slow), 2P (Medium), or 3P (Fast) to your estimating form. National Estimator allows you to enter any percentage you select for overhead and profit.

Figure 1
The basis for cost estimates in this book

installed, it's easy to use costs in this book to compile estimates for your jobs. Instructions for using National Estimator begin on page 433.

How to Use the Tables

The estimating tables in this book show typical costs and bid prices for every painting operation you're likely to encounter, whether paint is applied by brush, roller, mitt or spray. Selecting the right cost table and the correct application method is easy. Tables are divided into four parts:

Part I: General Painting Costs

Part II: Preparation Costs

Part III: Industrial, Institutional and Heavy Commercial Painting Costs

Part IV: Wallcovering Costs

Each section is arranged alphabetically by operation. If you have trouble finding the tables you need, use the Table of Contents at the front of the book or the Index at the back of the book.

Once you've found the right table and the appropriate application method, you have to select the correct application rate. For each of the application methods (brush, roll, mitt or spray), the tables show three application rates: "Slow," "Medium," or "Fast." That's a very important decision when using this book, because each application rate assumes different manhour productivity, material coverage, material cost per gallon, hourly labor cost, labor burden, overhead and profit. Your decision on the application rate to use (or which combination of rates to use) has to be based on your evaluation of the job, your painters and your company. That's where good common sense is needed.

Figure 1 shows crew codes, labor costs, labor burdens, material discounts, and profit for each of the three production rates for painting.

The "Slow" application rate in Figure 1 assumes lower productivity (less area covered per manhour), a lower labor cost (due to a less skilled crew), a lower labor burden (due to lower fringe benefits), a lower discount on materials (because of low volume), higher overhead (due to lower volume) and a higher profit margin (typical on small repaint or custom jobs). Figures in this "Slow" application row will apply where painters with lower skill levels are working on smaller or more difficult repaint jobs with minimum supervision.

Look at the "Fast" row in Figure 1. These estimates will apply where a skilled crew (higher hourly rate and larger fringe benefits) is working under good supervision and good conditions (more area covered per manhour) on larger (volume discount on materials) and more competitive jobs (lower profit margin). Figures in the "Fast" application row assume high productivity and lower material coverage, like that of a residential tract job.

Each of the three application rates is described more completely later in this section.

	Pricing variables			Unit cost estimate					
	1	2	3	4	5	6	7	8	9
	Labor SF per manhour	Material coverage SF/gallon	Material cost per gallon	Labor cost per 100 SF	Labor burden 100 SF	Material cost per 100 SF	Overhead per 100 SF	Profit per 100 SF	Total cost per 100 SF
Walls, gypsum drywall, orange peel or knock-down, roll, per 100 sq ft of wall									
Flat latex, water base (material #5)									
Roll 1st coat									
Slow	400	300	17.10	2.63	.74	5.70	2.81	1.78	13.66
Medium	538	(275)	(15.00)	3.11	1.09	(5.45)	2.36	1.32	13.33
Fast	(675)	250	12.90	3.33	1.39	5.16	1.83	.70	12.41
				2.81	.80	5.45	2.22	1.24	12.52

Figure 2
Customize the tables

The Easy Case: No Adjustments

Let's suppose the "Slow" application rate fits the job you're estimating almost perfectly. Your crew's productivity is expected to be low. The labor cost will be $10.50 per hour. Labor burden (fringes, taxes and insurance) will be 28.45 percent. Discount on materials will be 20 percent. Overhead will be 31 percent and profit will be 15 percent. Then your task is easy. All of your costs match the costs in the "Slow" row. No modifications are needed. The same is true if your costs fit the "Medium" or "Fast" rows.

But that's not always going to happen. More often, the job, your crew and your company won't fit exactly into any of the three rows. What then? More evaluation is required. You'll combine costs from several application rate rows to reach an accurate bid price. I call that *customizing your costs* and it's nearly always required for an accurate estimate.

Customizing Your Costs

Every company has a different combination of worker speed and experience, taxes, benefits, spread rates, equipment needs, percentage for overhead, and profit margin. These are the cost variables in paint estimating. This book is designed so you can quickly and easily adjust estimates to reflect actual costs on the job you're estimating. It's important that you *read the rest of this section before using the cost tables in this book.* That's the only way to get from this manual all the accuracy and flexibility that's built into it.

In the remainder of this section I'll describe the assumptions I've made and the methods I used to compile the cost tables in this manual. Once you understand them, you'll be able to combine and modify costs in the estimating tables so your bids fit the job, your crew and your company as closely as possible.

When you start using the cost tables in this book, I suggest you circle numbers in the "Slow," "Medium," or "Fast" application rate rows that best fit your company and your jobs. To improve accuracy even more, write your own figures in the blank row below the "Fast" row in each table, like I've done in Figure 2.

A Practical Example

Figure 2 is part of an estimating table from Part I of this book, General Painting Costs. I'm going to use it to show how to customize estimates to match your actual costs. In Figure 2 I've circled some of the costs I plan to use in a sample estimate and calculated others.

In column 1, *Labor SF per manhour*, I've circled 675 because I feel the journeyman painter assigned to this job can paint walls at the "Fast" rate of 675 square feet per hour. That's the number I plan to use for my estimate.

For column 2, *Material coverage SF/gallon*, I've reviewed my past performance and I expect coverage will be about 275 square feet per gallon of paint. So I've circled that figure.

In column 3, *Material cost per gallon*, I've circled 15.00 for my cost per gallon for flat water base latex (including tax and an allowance for consumable supplies), based on a 30 percent discount from the retail price.

So far, so good. That completes the first three columns, what I call the *pricing variables*. Now we can begin on the *unit cost estimate*, columns 4 through 9. Each of these columns show a price per 100 square feet of wall.

We'll start with column 4, *Labor cost per 100 SF*. Notice that I've entered 2.81 for this column. Here's why. Look back at Figure 1. Throughout this book the painting labor rate for "Fast" work is assumed to be $22.50 per hour. See page 29 for the wall covering application rate. I can't use the labor cost per 100 SF for "Fast" work because the journeymen on my job earn $19.00 per hour. That pay rate is between the "Medium" and "Fast" pay rates as shown in Figure 1. To calculate the labor cost per 100 SF, divide $19.00 by 675 and multiply by 100: 19/675 = .0281 x 100 = 2.81.

In column 5, *Labor burden 100 SF*, I've entered .80. This figure is a result of my labor cost at $2.81 x 28.45 percent, my labor burden (taxes, insurance and benefits) from the "Slow" row of Figure 1. Even though the labor rate is "Fast" and the labor cost is between "Fast" and "Medium," for this example labor burden will be most like work done at the "Slow" rate because this company doesn't offer many benefits.

In column 6, *Material cost per 100 SF*, I've circled 5.45, the number in the "Medium" row. Since I've used numbers in the "Medium" row in both columns 2 and 3, I can take the figure in column 6 for material costs directly from the table, without any calculations.

In column 7, *Overhead per 100 SF*, I've calculated the overhead dollar value by adding the labor cost, labor burden and material cost then multiplying that sum by the "Medium" overhead at 24.5 percent: $2.81 + $.80 + $5.45 = $9.06 x .245 = $2.22.

In column 8, *Profit per 100 SF*, I've calculated the profit dollar value by adding the labor cost, labor burden, material cost and overhead then multiplying that sum by the "Medium" profit at 11 percent from Figure 1. The result is $2.81 + $.80 + $5.45 + $2.22 = $11.28 x .11 = $1.24.

Column 9, *Total cost per 100 SF*, is the bid price – it's the sum of columns 4 through 8 for each row. Because I've circled costs that fall in more than one row, I can't use any figure in column 9. Instead, I simply add the circled or calculated figures in columns 4 through 8: $2.81 + $.80 + $5.45 + $2.22 + $1.24 = $12.52. That's my bid price per 100 square feet on this job. It's the combination of costs that fit my company, my painters and the job.

Using Your Good Judgment

Of course, judgment is required when using these tables, as it is when making any estimate. For example, if your journeymen painters earn the top rate of $22.50 but work at the "Medium" production rate or slower, your labor cost per unit will be higher than the highest cost listed in column 4. An adjustment will be required.

Because figures in columns 7 and 8 are percentages of figures in columns 4, 5 and 6, you have to be careful when you blend costs from different rows. Let's look at an extreme (and unlikely) example.

Suppose you use costs from the "Slow" application row for columns 4 (2.63), 5 (.74) and 6 (5.70) of Figure 2. The total of those three costs is 9.07. Then you decide to use overhead from the "Fast" row because your overhead is about 18.5 percent of cost, not 31 percent of cost as in the "Slow" row. "Fast" overhead is listed as 1.83 in Figure 2. The correct overhead figure is $1.68, 18.5 percent of the sum of "Slow" costs in columns 4, 5 and 6. Be aware of this small discrepancy and calculate figures for all the categories yourself if extreme accuracy is essential.

Converting Unit Prices

The last column in Figure 2 shows the total cost per 100 square feet of wall. Some estimating tables in this book show a total cost per 100 linear feet (such as for baseboard) or total costs per unit (such as for doors). To convert a cost per 100 square feet to a cost per square foot, move the decimal point two places to the left. Thus the cost per 100 square feet for the "Fast" rate in Figure 2 is $12.41 or about 12-2/5 cents per square foot.

General Qualifications

It's important that you understand the conditions the tables are based upon. I call these conditions the job *qualifications*. A qualifications statement follows each estimating table to help you understand what's included and what's excluded. Please read those qualifications before using costs from this manual in your estimates. The following points apply to *all* tables in this book:

Included Costs

- Minor preparation, both time and material. Normal preparation for new residential construction is included in the "Fast" row and for new commercial jobs in the "Medium" row. Minimal preparation is included for repaint jobs in the "Slow" row.
- Minimum setup and cleanup
- Equipment such as ladders, spray rigs and brushes are included in overhead for the "Fast" rate (residential tracts) or "Medium" (commercial) work. Add equipment costs at their rental rate for "Slow" (repaint) jobs.

Excluded Costs

- Equipment costs such as ladders, spray rigs, etc. for "Slow" (repaint) jobs. Add these at their rental rate whether or not you own the equipment.
- Extensive surface preparation. Add the cost of time and materials needed for more than "normal" preparation work. Also add time to remove and replace hardware and accessories, protect adjacent surfaces, and do any extensive setup, cleanup, or touchup. (See the discussion of SURRPTUCU on the next page.)
- Mobilization or demobilization
- Supervision
- Material handling, delivery, or storage
- Sample preparation
- Mixing coatings
- Excessive material waste or spillage
- Equipment rental or placement costs
- Scaffolding rental and erection costs
- Subcontract costs
- Contingency allowance
- Owner allowances
- Commissions, bonuses, overtime, premium pay for shift adjustments (evening work), travel time or per diem.
- Bonds, fees, or permits
- Additional insurance to meet owner requirements
- Work at heights above 8 feet or beyond the reach of a wand or extension pole. (See the table for High Time Difficulty Factors on page 137.)

Surface Preparation

The Preparation estimating tables that follow Part I: General Painting Costs, apply to both interior and exterior surfaces.

Surface preparation is one of the hardest parts of the job to estimate accurately. Any experienced painter can make a reasonably good estimate of the quantity of paint and time needed for application. But the amount of prep work needed will vary widely – especially for repaint jobs. Some will need very little work. Others will take more time for prep than for painting.

Preparation work for new construction jobs is relatively standard and consistent. You'll have to mask cabinets before spraying sealer on wet area walls, caulk at the baseboards, putty the nail holes in wood trim, and occasionally use a wire brush to smooth and clean a surface. The time required for this work is fairly predictable.

Labor cost for normal preparation of unpainted surfaces in new residential construction is included in the "Fast" *labor* costs and for new commercial construction in the "Medium" *labor* cost. The cost of *materials* for normal surface preparation on unpainted surfaces is included in the sundries allowance that's part of the "Fast" or "Medium" material cost.

But if more than normal surface prep work is needed, estimate the extra manhours and materials required and add these costs to your estimate.

Add for Repaint Preparation

The "Slow" unit costs include no surface preparation other than a quick wipedown. Preparation on a repaint job may take longer than the painting itself. That's why you have to estimate surface prep as a separate item and add that cost to your estimate.

A misjudgment in estimating preparation work can be very expensive. That's why I recommend that you bid surface preparation by the hour, using your shop rate for "time and material" jobs, or some other specified hourly rate. That protects you against cost overruns if the preparation takes longer than anticipated. But there's a danger here. Owners may be angry about the cost because they don't understand what's involved in preparation and why it takes so long. You can avoid this with a "not to exceed" bid that contains a maximum price for the prep work. Your bid should define the scope of preparation work in detail and list exactly what's included and excluded. Be sure to consider all the labor, material, and equipment costs involved.

If you have to bid repaint work, be sure to include all the miscellaneous costs. The acronym I use to identify these miscellaneous costs is SURRPTUCU: Setup (SU), Remove and Replace (RR), Protection (P), Touchup (TU) and Cleanup (CU). Add these costs to your repaint estimate if they require anything beyond minimum attention.

1) *Setup* includes unloading the vehicle, spreading the tarp and setting up the tools – everything that has to be done before prep or painting can begin.

2) *Remove and replace* everything that will interfere with painting, including door and cabinet hardware, the contents of cabinets, light fixtures, bathroom accessories, switch covers and outlet plates, among others.

3) *Protection* for furniture and adjacent surfaces such as floors, cabinets, plumbing or electrical fixtures, windows, and doors. Protection methods include masking, applying visqueen, laying drop cloths and applying a protective coating on windows.

4) *Touchup* time varies with the speed and quality of the painting job and how fussy the owner is. The more careful your painters are, the less touchup time needed. You can estimate touchup time accurately only if you know how well your crews perform. The Touchup table in this book is based on a percentage of total job cost.

5) *Cleanup* time is usually about the same as setup time, about 20 to 30 minutes each day for repaint jobs. Cleanup time begins when work stops for the day and ends when the crew is back in the truck and ready to go home. It includes cleaning tools, dismantling the paint shop and loading the vehicle.

Subcontractors

Painting contractors don't hire many subcontractors. But once in a while you'll need a specialist for sandblasting, waterblasting, wallcovering, scaffolding or pavement marking. Subcontract costs are not included in the estimating tables. Add the cost of any subcontract work that will be required.

Figure 3 shows some typical rates quoted by sandblasting subcontractors. Of course, prices in your area will probably be different. You could also figure sandblasting unit costs from the sandblasting estimating tables included in Part II, Preparation Costs, in this book.

Figure 4 shows typical subcontract bids for pavement marking. Again, prices in your area may be different.

Item	Price	Item	Price
Minimum charges: $500.00, scaffolding not included		Epoxy coated - add	.95 to 1.05/SF
Additional insurance: May be required to cover adjacent personal and real property which may not be protected.		With portable equipment - add	.55 to .80/SF
Sandblasting water soluble paints	$.80 to .90/SF	**Commercial blast** - 67% white stage	
Sandblasting oil paints	.85 to .95/SF	Flat welded, new, uncoated	
Sandblasting heavy mastic		ground runs	.85 to 1.00/SF
(depends on coating thickness)	1.10 to 1.20/SF	above ground	1.05 to 1.65/SF
Sandblasting brick - light blast	.80 to .90/SF	Previously painted surfaces - add	.50 to .90/SF
Sandblasting masonry block walls		Epoxy coated - add	.90 to 1.05/SF
Clean up & remove grime - light	.75 to .80/SF	With portable equipment - add	.65 to .80/SF
- heavy	1.15 to 1.25/SF	**Near white blast** - 95% white stage	
Sandblasting structural steel		Field welded, new, uncoated	
Pricing rules of thumb:		ground runs	1.00 to 1.15/SF
Pipe up to 12" O.D.	1.15 to 1.70/SF	above ground	1.15 to 1.75/SF
Structural steel up to 2 SF/LF	1.05 to 1.15/SF	Previously painted surfaces - add	.50 to .90/SF
Structural steel from 2 to 5 SF/LF	1.25 to 1.40/SF	Epoxy coated - add	.90 to 1.05/SF
Structural steel over 5 SF/LF	(depends on shape)	With portable equipment - add	.65 to .80/SF
Tanks and vessels up to 12'0" O.D.	1.65 to 1.90/SF	**White blast** - 100% uniform white stage	
Tanks and vessels over 12'0" O.D.	1.65 to 1.90/SF	Field welded, new, uncoated	
Brush off blast - light blast (loose mill scale)		ground runs	1.50 to 1.75/SF
Field welded, new, uncoated		above ground	1.65 to 1.95/SF
ground runs	.50 to .65/SF	Previously painted surfaces - add	.50 to .85/SF
above ground	.75 to 1.40/SF	Epoxy coated - add	.90 to 1.05/SF
Previously painted surfaces - add	.50 to .90/SF	With portable equipment - add	.50 to .75/SF

Figure 3
Sandblasting pricing table

If you do much repainting, you'll probably want to buy a waterblasting rig. Even if you own the blaster, include a charge in each estimate for the equipment as though you rented it from a rental yard just for that job. Figure the unit costs for waterblasting from Part II of this book, Preparation Costs.

Consider using a waterblasting subcontractor if you don't need the service often. Figure 5 shows some typical rates for waterblasting. Make up a table like this based on quotes from subcontractors in your area. For a more detailed table, see Sandblasting in the Preparation section, page 301.

When you hire a subcontractor, make sure the quoted price includes everything that contractor has to do – all labor, material (with tax, if applicable), equipment, overhead and profit. Add your overhead and profit percentage to the subcontractor's bid price when you enter that item on the estimate.

Contingencies

Occasionally you'll add a contingency allowance on bids for repaint projects where there are unknowns that can't be forecast before work actually begins. Contingency allowances are rarely needed when estimating new construction. When necessary, the contingency amount is usually from 3 to 5 percent. It can go higher, however, if there are unusual conditions or unknowns that make it hard to produce an accurate estimate. Include a contingency allowance in your estimates only if you have reason to expect:

- An uncertain scope of work (unknown job conditions)
- An inexperienced owner or general contractor
- Incomplete drawings
- Delays in beginning the project
- Owner involvement in supervision
- Below-standard working conditions

Don't use contingency allowances as a substitute for complete estimates. Include contingency only to cover what can't be estimated, not what you don't have time to estimate accurately.

Pricing rule of thumb:	
Number of parking spaces: Figure on one space per 300 SF of pavement	
Single line striping with light graphics apply	$ 7.50 per space
Single line striping with heavy graphics apply	13.00 per space
Single striping, light graphics and 3' wheel stop	18.50 per space
Single striping, heavy graphics and 3' wheel stop	24.00 per space
Equipment pricing:	
Simple "inverted spray can" approximate cost	$185.00
Professional striping machine cost range	$3,800 to 4,200
Professional road/highway striper	$209,000
Subcontractor pricing:	
Move on	$125.00 to 150.00
Striping prices:	
Single line striping	$.40 to .50 per lineal foot
Bike lane striping	.50 to .60 per lineal foot
Fire lane, red curb	.40 to .50 per lineal foot
Symbol pricing:	
Templates - 8'0" template	$145.00 to 170.00 each
Arrows	32.00 to 37.00 each
Handicap symbol, one color	13.00 to 18.00 each
two color	24.00 to 29.00 each
No parking fire lane stencil	2.55 to 3.10 each
Wheel stops:	
3'0" stops	$18.00 to 23.00 each if pinned on asphalt
	24.00 to 29.00 each if glued and pinned
6'0" stops	29.00 to 34.00 each if pinned on asphalt
	34.00 to 39.00 each if glued and pinned
	(add for wheel stops pinned to concrete)
Signs and posts:	
Sign only 12" x 18"	$40.00 to 56.00
Post mounted 12" x 18"	105.00 to 145.00
Pavement markers:	
One way pavement markers	$8.50 each
Two way pavement markers	11.00 each

Figure 4
Pavement marking pricing table

Minimum charges: $500.00, scaffolding not included	
Additional insurance: May be required to cover adjacent personal and real property	
Pricing rules of thumb:	
Up to 5,000 PSI blast	4 hour minimum $100.00/hour
5,000 to 10,000 PSI blast	8 hour minimum $145.00/hour
Over 10,000 PSI blast	8 hour minimum $180.00/hour
Wet sandblasting	4 hour minimum $115.00/hour

Figure 5
Waterblasting pricing table

Column Headings Defined

Take another look at Figure 2. The heading describes the surface to be coated: the type, texture, and often, condition. Sections within each surface heading are divided according to coating material, then by application method, and further into the "Slow," "Medium," and "Fast" application rates.

Column 1: Labor Productivity

This column shows units of work completed per manhour. My estimates assume that painters are experienced and motivated professionals. The labor productivity categories are shown in Figure 6.

My experience is that a painting company that can handle larger projects will have more skilled, better qualified and more productive painters. The estimating tables also assume that repainting a surface usually takes about 35 percent more time than painting newly constructed surfaces. Much of this extra time is spent protecting adjacent areas.

To establish your company's production levels, ask your field superintendent to monitor the time needed to complete each task and to keep records of crew productivity. Your best guide to productivity on future jobs is productivity on jobs already completed. The more you know about your painters' performance, the more accurate your estimates will be. But don't expect your estimates and actual production to match exactly. Painters are human beings, not robots. You can't expect them to work at the same rate at all times.

Slow	**Medium**	**Fast**
Repaint jobs	New commercial projects	New residential production
Custom painting	Industrial painting	Repetitious painting
Tenant improvements	—	—
Small jobs	Medium-sized jobs	Large projects
Single units	Two to four units	Five or more units
Low production	Average production	High production
High difficulty	Average difficulty	Low difficulty
Poor conditions	Average conditions	Good conditions
High quality	Average quality	Minimum quality
Unskilled crew	Semi-skilled crew	Skilled crew
No supervision	Some supervision	Good supervision

Figure 6
Labor productivity categories

Reduced Productivity

The tables assume no overtime work. Excessive overtime puts a strain on your craftsmen and reduces productivity. A few consecutive days of overtime can drag productivity down to well below average. It's good practice not to assign overtime work on more than two consecutive days.

Work efficiency is also lower when men, materials and equipment are confined in a small area or required to work in cluttered, poorly lit or dirty rooms. Painters need elbow room to work efficiently and get maximum productivity. They're also more productive in a clean environment where they can see what they're doing. It's easier – and safer – to work in a well-lighted area that's relatively clear of debris. If the work area is confined or dirty, reduce estimated productivity accordingly.

Supervision

Supervision expense is *not* included in the cost tables. Add the cost of supervision to your estimates.

Most supervision is done by foremen. Every crew should have a project foreman designated, usually the most experienced and reliable painter on the job. When not supervising, project foremen should be painting. Thus the project foreman is a working supervisor. Part of the foreman's time will be productive (applying coatings) and part will be nonproductive (directing the work).

If you have more than three or four jobs going at one time, you need a field superintendent. The field superintendent is the foreman's supervisor. His or her primary responsibility is to be sure that each foreman has the manpower, materials and equipment needed to get the job done. The field superintendent should monitor job progress to be sure man-hour productivity and materials used are in line with estimates. Field superintendents usually are not working supervisors; all their time is nonproductive. Figure the field superintendent's salary as overhead expense, because you can't charge his salary to a specific job.

Your project foremen and field superintendent can make or break a job. The better they are, the more work will be done. You want a field superintendent who assigns the right painters to the right foreman, and a foremen who puts the right painters on the right tasks. The most experienced tradesmen should work on tasks that require more skill. Other painters should be used where less skill is needed. The project foreman is also responsible for job safety and quality control.

Your estimates will be more competitive if you can assume high productivity. That's only possible when you have good supervision, from both foremen and superintendent and motivated crews.

Allowances for Supervision

Supervision isn't considered productive labor. A foreman isn't painting when he's scheduling, organizing a job and instructing his workers. Here are my rule-of-thumb allowances for nonproductive labor on painting jobs.

Custom homes. Allow 2.5 hours of nonproductive supervision for a home up to 1,500 square feet, 3 hours on a home between 1,500 and 2,000 square feet, 4 hours on a custom home between 2,000 and 2,500 SF, and 5 hours on a larger home.

Model homes in a tract. One hour of nonproductive supervision for each day your crew will be on the job.

Most tract homes. One hour per house.

Higher-quality tract homes. Two hours per house.

Apartments and condos. Allow 1 hour per unit if there are 10 units or less. For 11 to 30 units, allow 0.75 hour of nonproductive time per unit. If there are more than 30 units, allow 0.5 hour per unit.

Nonproductive labor on commercial, industrial, institutional and government projects varies considerably. More complex jobs will require proportionately more nonproductive labor. Use your knowledge based on past experience to estimate supervision either as a percentage of job cost or by the square foot of floor.

Slow application and light coverage (Repaint jobs)	Medium application and medium coverage (Commercial projects)	Fast application and heavy coverage (Residential tracts)
Repaint jobs	Commercial projects	Residential production
Light usage	Moderate usage	Heavy usage
Low absorption	Moderate absorption	High absorption
Light application	Medium application	Heavy application
Low waste	Moderate waste	High waste
Quality paint	Standard paint	Production paint
Skilled painters	Semi-skilled crew	Unskilled crew

Figure 7
Material coverage rates

Column 2: Material Coverage

The second column in the cost tables shows the estimated material coverage in units (usually square feet or linear feet) per gallon. Figure 7 shows the conditions likely to apply for each of the three material coverage rates. Every condition listed in each of these categories won't necessarily occur on every painting operation. For example, it's possible to have high waste and use low quality paint on a repaint job. But it's more likely that waste will be low and paint quality high on jobs like that.

The "Slow" (repaint) application rate assumes light coverage, "Medium" (commercial project) application rate assumes medium coverage and "Fast" (residential tract) application rate assumes heavy coverage. Light coverage is typical on "Slow" (repaint) jobs because previously painted surfaces usually absorb 10 to 15 percent less paint than an unpainted surface. All coverage rates are based on paint that's been thinned according to the manufacturer's recommendations.

Of course, coverage varies with the paint you're using and the surface you're painting. Paint manufacturers usually list the recommended coverage rate on the container label. I've listed estimated coverage rates in the tables throughout this book.

Calculating Film Thickness

Many project specifications for commercial, industrial and government jobs identify the coating (film) thickness you have to apply to each surface. The thickness is given in *mils*, or thousandths of an inch. One mil is 0.001 inch.

The thickness of the dry paint film depends on the percentage of solids in the paint. If you apply a gallon of paint containing 100 percent solids over 1,600 square feet, the dry film will be 1 mil thick – that is, if 100 percent of the paint adheres to the wall. But if there's 10 percent waste (because of paint that's left in the can, on brushes, or spilled), only 90 percent of the material ends up on the surface.

Here's a formula for coverage rates that makes it easy to calculate mil thickness, including the waste factor. Coverage rate equals:

$$\frac{\text{\% of solids x 1600}}{\text{mil thickness}} \text{ x (1.00 - waste factor)}$$

Here's an example. Assume you're applying paint with 40 percent solids (by volume), using a roller. The waste factor is 10 percent. You need a thickness of 5 mils.

Slow application	Medium application	Fast application
Repaint jobs	Commercial projects	Residential tracts
Low volume	Medium volume	High volume
20% discount	30% discount	40% discount

Figure 8
Material price discounts

Here's the calculation for the coverage rate:

$$\frac{.40 \times 1600}{5} \times (1.00 - .10) = 115.2 \text{ SF per gallon}$$

You may have to apply several coats to get a thickness of 5 mils. In any case, you'll have to use one gallon of paint for each 115.2 square feet of surface.

Waste Factors

Be sure to consider waste and spillage when you figure coverage rates. Professional painters waste very little paint. They rarely kick over a five-gallon paint bucket. But there's always some waste. My material coverage formulas include a typical waste allowance for each application method, whether it's brush, roller or spray. Of course, actual waste depends on the skill of your painters no matter what application method they use.

These are the waste factors I've built into the tables:

Brush . . . . .3 to 5%

Roll . . . . .5 to 10%

Airless spray . . . . .20 to 25%

Conventional spray . . . . .25 to 35%

Changes in Paint Formulation

In the late 1970s, the California State Air Resources Board established a "model rule" for lowering the solvent in oil-based paints. They mandated replacing solvent-based paint with water-based formulas. The objective was to lower the amount of solvents escaping into the air. This change in the formulation of oil-based paints is being adopted by other states, including Arizona, Colorado, New York and New Jersey, and probably will be adopted nationwide eventually.

Changes in paint formulation will affect coverage rates and the cost for non-flat paints. Review actual coverage rates and paint prices and make adjustments where necessary before using the estimates in this book.

Column 3: Material Pricing

The third column in the cost tables shows the cost of materials. The "Slow," "Medium," and "Fast" prices in each table are based on the discounts usually offered by suppliers for volume purchases by contractor customers. The material discounts used in this book are defined in Figure 8.

The more paint a contractor buys over a given period, the greater the discount that contractor can expect. Most paint contractors get a discount of at least 20 percent off retail. Contractors buying in heavy volume usually get discounts that approach 40 percent off retail.

Material Pricing Tables

Figures 9, 10 and 11 show the material prices I've used for each of three application rates throughout this book. In the cost estimating tables each coating is identified by a material number. To find out more about the cost of any of these coatings, refer to the material number listed in Figure 9, 10 or 11.

Figure 9 shows prices at a 20 percent discount off retail. It applies to "Slow" work and assumes light coverage on a previously painted surface. These costs would be typical for a lower-volume company handling mostly repaint or custom work.

Material prices at 20 percent discount

All pricing is based on production grade material purchased in 5 gallon quantities.

		Retail price guide	Contractor price at a 20% discount	Add 15% sundries & escalation	Price with sales tax at 7%	Estimating prices with tax
Interior:						
	Sealer, off white (wet area walls & ceilings)					
#1 -	Water base	16.60	13.28	15.27	16.34	16.30
#2 -	Oil base	23.30	18.64	21.44	22.94	22.90
	Undercoat (doors, casings and other paint grade wood)					
#3 -	Water base	21.45	17.16	19.73	21.11	21.10
#4 -	Oil base	21.90	17.52	20.15	21.56	21.60
	Flat latex (walls, ceilings & paint grade baseboard)					
#5 -	Water base latex paint	17.40	13.92	16.01	17.13	17.10
	Acoustic spray-on texture					
#6 -	Primer	18.35	14.68	16.88	18.06	18.10
#7 -	Finish	15.75	12.60	14.49	15.50	15.50
#8 -	Dripowder mixed (pound)	.42	.34	.39	.42	.42
	Enamel (wet area walls & ceilings and openings)					
#9 -	Water base	24.90	19.92	22.91	24.51	24.50
#10 -	Oil base	27.05	21.64	24.89	26.63	26.60
	System estimate (cabinets, bookshelves, molding, interior windows)					
#11a -	Wiping stain, oil base	31.35	25.08	28.84	30.86	30.90
#11b -	Sanding sealer, lacquer	17.45	13.96	16.05	17.17	17.20
#11c -	Lacquer, semi gloss	19.55	15.64	17.99	19.25	19.30
#11 -	Stain, seal & 2 coat lacquer system					
	Average cost [11a + b + (2 x c)]		17.58	20.22	21.64	21.60
#12 -	Shellac, clear	25.65	20.52	23.60	25.25	25.30
#13 -	Penetrating oil stain	29.50	23.60	27.14	29.04	29.00
#14 -	Penetrating stain wax (molding)	27.65	22.12	25.44	27.22	27.20
#15 -	Wax, per pound (floors)	8.10	6.48	7.45	7.97	8.00
#16 -	Glazing (mottling over enamel)	31.65	25.32	29.12	31.16	31.20
#17 -	Spray can, each (HVAC registers)	4.25	3.40	3.91	4.18	4.20

Figure 9
Material prices at 20% discount

Material prices at 20 percent discount (cont.)

	Retail price guide	Contractor price at a 20% discount	Add 15% sundries & escalation	Price with sales tax at 7%	Estimating prices with tax
Exterior:					
Solid body/color stain (beams, light valance, fascia, overhang, siding, plant-on trim, wood shelves)					
#18 - Water base stain	21.80	17.44	20.06	21.46	21.50
#19 - Oil base	24.25	19.40	22.31	23.87	23.90
Semi-transparent stain (beams, siding, T & G ceiling)					
#20 - Water base stain	21.70	17.36	19.96	21.36	21.40
#21 - Oil base stain	26.55	21.24	24.43	26.14	26.10
#22 - Polyurethane (exterior doors)	34.80	27.84	32.02	34.26	34.30
#23 - Marine spar varnish, flat or gloss (exterior doors)					
Interior or exterior	40.00	32.00	36.80	39.38	39.40
Exterior enamel (exterior doors & trim)					
#24 - Water base	27.75	22.20	25.53	27.32	27.30
#25 - Oil base	31.15	24.92	28.66	30.67	30.70
Porch & deck enamel - interior or exterior					
#26 - Water base	28.00	22.40	25.76	27.56	27.60
#27 - Oil base	30.60	24.48	28.15	30.12	30.10
#28 - " Epoxy, 1 part, water base	37.85	30.28	34.82	37.26	37.30
#29 - " Epoxy, 2 part system	74.35	59.48	68.40	73.19	73.20
System estimate (exterior windows)					
#30a - Wiping stain, oil base	31.05	24.84	28.57	30.57	30.60
#30b - Sanding sealer, varnish	33.95	27.16	31.23	33.42	33.40
#30c - Varnish, flat or gloss	36.00	28.80	33.12	35.44	35.40
#30 - Stain, seal & 1 coat varnish system					
Average cost (30a + b + c)		26.93	30.97	33.14	33.10
Masonry paint (masonry, concrete, plaster)					
#31 - Water base, flat or gloss	21.20	16.96	19.50	20.87	20.90
#32 - Oil base paint	28.75	23.00	26.45	28.30	28.30
#33 - Masonry block filler	15.85	12.68	14.58	15.60	15.60
#34 - Waterproofing, clear hydro seal	18.05	14.44	16.61	17.77	17.80
Metal primer, rust inhibitor					
#35 - Clean metal	30.35	24.28	27.92	29.87	29.90
#36 - Rusty metal	37.20	29.76	34.22	36.62	36.60
Metal finish, synthetic enamel, gloss, interior or exterior					
#37 - Off white	28.30	22.64	26.04	27.86	27.90
#38 - Colors (except orange/red)	30.50	24.40	28.06	30.02	30.00
Anti-graffiti stain eliminator					
#39 - Water base primer & sealer	25.85	20.68	23.78	25.44	25.40
#40 - Oil base primer & sealer	26.10	20.88	24.01	25.69	25.70
#41 - Polyurethane 2 part system	92.85	74.28	85.42	91.40	91.40

Figure 9 (continued)
Material prices at 20% discount

Material pricing at 20% discount (cont.)

	Retail price guide	Contractor price at a 20% discount	Add 15% sundries & escalation	Price with sales tax at 7%	Estimating prices with tax
Preparation:					
#42 - Caulking, per fluid ounce	.21	.17	.20	.21	.21
Paint remover, per gallon					
#43 - Light duty	19.45	15.56	17.89	19.14	19.10
#44 - Heavy duty	24.30	19.44	22.36	23.93	23.90
#45 - Putty, per pound	7.10	5.68	6.53	6.99	7.00
#46 - Silica sand, per pound	.04	.03	.03	.03	.03
#47 - Window protective coating (wax)	15.20	12.16	13.98	14.96	15.00
#48 - Wood filler, per gallon	33.40	26.72	30.73	32.88	32.90
Industrial:					
#49 - Acid wash (muriatic acid)	5.45	4.36	5.01	5.36	5.40
#50 - Aluminum base paint	40.75	32.60	37.49	40.11	40.10
Epoxy coating, 2 part system					
#51 - Clear	59.25	47.40	54.51	58.33	58.30
#52 - White	62.65	50.12	57.64	61.67	61.70
Heat resistant enamel					
#53 - 500 to 1200 degree range	72.40	57.92	66.61	71.27	71.30
#54 - 300 to 800 degree range	54.90	43.92	50.51	54.05	54.10
#55 - Industrial bonding & penetrating oil paint	28.20	22.56	25.94	27.76	27.80
Industrial enamel, oil base, high gloss					
#56 - Light colors	30.10	24.08	27.69	29.63	29.60
#57 - Dark (OSHA) colors	37.55	30.04	34.55	36.97	37.00
#58 - Industrial waterproofing	25.20	20.16	23.18	24.80	24.80
#59 - Vinyl coating (tanks)	60.00	48.00	55.20	59.06	59.10
Wallcovering:					
Ready-mix					
#60 - Light-weight vinyl (gal)	7.25	5.80	6.67	7.14	7.10
#61 - Heavy weight vinyl (gal)	7.45	5.96	6.85	7.33	7.30
#62 - Cellulose, clear (gal)	7.95	6.36	7.31	7.82	7.80
#63 - Vinyl to vinyl (gal)	17.10	13.68	15.73	16.83	16.80
#64 - Powdered cellulose, 2 - 4 ounces	2.75	2.20	2.53	2.71	2.70
#65 - Powdered vinyl, 2 - 4 ounces	3.25	2.60	2.99	3.20	3.20
#66 - Powdered wheat paste, 2-4 ounces	2.75	2.20	2.53	2.71	2.70

Note: Typically, powdered paste is in 2 to 4 ounce packages which will adhere 6 to 12 rolls of wallcovering.

Figure 9 (continued)
Material prices at 20% discount

Material pricing at 30% discount

All pricing is based on production grade material purchased in 5 gallon quantities.

		Retail price guide	Contractor price at a 30% discount	Add 15% sundries & escalation	Price with sales tax at 7%	Estimating prices with tax
Interior:						
	Sealer, off white (wet area walls & ceilings)					
#1 -	Water base	16.60	11.62	13.36	14.30	14.30
#2 -	Oil base	23.30	16.31	18.76	20.07	20.10
	Undercoat (doors, casings and other paint grade wood)					
#3 -	Water base	21.45	15.02	17.27	18.48	18.50
#4 -	Oil base	21.90	15.33	17.63	18.86	18.90
	Flat latex (walls, ceilings & paint grade baseboard)					
#5 -	Water base latex paint	17.40	12.18	14.01	14.99	15.00
	Acoustic spray-on texture					
#6 -	Primer	18.35	12.85	14.78	15.81	15.80
#7 -	Finish	15.75	11.03	12.68	13.57	13.60
#8 -	Dripowder mixed (pound)	.42	.29	.33	.35	.40
	Enamel (wet area walls & ceilings and openings)					
#9 -	Water base enamel	24.90	17.43	20.04	21.44	21.40
#10 -	Oil base enamel	27.05	18.94	21.78	23.30	23.30
	System estimate (cabinets, bookshelves, molding, interior windows)					
#11a -	Wiping stain, oil base	31.35	21.95	25.24	27.01	27.00
#11b -	Sanding sealer, lacquer	17.45	12.22	14.05	15.03	15.00
#11c -	Lacquer, semi gloss	19.55	13.69	15.74	16.84	16.80
#11 -	Stain, seal & 2 coat lacquer system					
	Average cost [11a + b + (2 x c)]		15.95	18.34	19.62	19.60
#12 -	Shellac, clear	25.65	17.96	20.65	22.10	22.10
#13 -	Penetrating oil stain	29.50	20.65	23.75	25.41	25.40
#14 -	Penetrating stain wax (molding)	27.65	19.36	22.26	23.82	23.80
#15 -	Wax, per pound (floors)	8.10	5.67	6.52	6.98	7.00
#16 -	Glazing (mottling over enamel)	31.65	22.16	25.48	27.26	27.30
#17 -	Spray can, each (HVAC registers)	4.25	2.98	3.43	3.67	3.70

Figure 10
Material prices at 30% discount

Material pricing at 30% discount (cont.)

		Retail price guide	Contractor price at a 30% discount	Add 15% sundries & escalation	Price with sales tax at 7%	Estimating prices with tax
Exterior:						
	Solid body/color stain (beams, light valance, fascia, overhang, siding, plant-on trim, wood shelves)					
#18 -	Water base stain	21.80	15.26	17.55	18.78	18.80
#19 -	Oil base stain	24.25	16.98	19.53	20.90	20.90
	Semi-transparent stain (beams, siding, T & G ceiling)					
#20 -	Water base stain	21.70	15.19	17.47	18.69	18.70
#21 -	Oil base stain	26.55	18.59	21.38	22.88	22.90
#22 -	Polyurethane (exterior doors)	34.80	24.36	28.01	29.97	30.00
#23 -	Marine spar varnish, flat or gloss (exterior doors)					
	Interior or exterior	40.00	28.00	32.20	34.45	34.50
	Exterior enamel (exterior doors & trim)					
#24 -	Water base	27.75	19.43	22.34	23.90	23.90
#25 -	Oil base	31.15	21.81	25.08	26.84	26.80
	Porch & deck enamel - interior or exterior					
#26 -	Water base	28.00	19.60	22.54	24.12	24.10
#27 -	Oil base	30.60	21.42	24.63	26.35	26.40
#28 -	Epoxy, 1 part, water base	37.85	26.50	30.48	32.61	32.60
#29 -	Epoxy, 2 part system	74.35	52.05	59.86	64.05	64.10
	System estimate (exterior windows)					
#30a -	Wiping stain, oil base	31.05	21.74	25.00	26.75	26.80
#30b -	Sanding sealer, varnish	33.95	23.77	27.34	29.25	29.30
#30c -	Varnish, flat or gloss	36.00	25.20	28.98	31.01	31.00
#30 -	Stain, seal & 1 coat varnish system					
	Average cost (30a + b + c)		23.57	27.11	29.01	29.00
	Masonry paint (masonry, concrete, plaster)					
#31 -	Water base, flat or gloss	21.20	14.84	17.07	18.26	18.30
#32 -	Oil base paint	28.75	20.13	23.15	24.77	24.80
#33 -	Block filler	15.85	11.10	12.77	13.66	13.70
#34 -	Waterproofing, clear hydro seal	18.05	12.64	14.54	15.56	15.60
	Metal primer, rust inhibitor					
#35 -	Clean metal	30.35	21.25	24.44	26.15	26.20
#36 -	Rusty metal	37.20	26.04	29.95	32.05	32.10
	Metal finish, synthetic enamel, gloss, interior or exterior					
#37 -	Off white	28.30	19.81	22.78	24.37	24.40
#38 -	Colors (except orange/red)	30.50	21.35	24.55	26.27	26.30
	Anti-graffiti stain eliminator					
#39 -	Water base primer & sealer	25.85	18.10	20.82	22.28	22.30
#40 -	Oil base primer & sealer	26.10	18.27	21.01	22.48	22.50
#41 -	Polyurethane 2 part system	92.85	65.00	74.75	79.98	80.00

Figure 10 (continued)
Material prices at 30% discount

Material pricing at 30% discount (cont.)

	Retail price guide	Contractor price at a 30% discount	Add 15% sundries & escalation	Price with sales tax at 7%	Estimating prices with tax
Preparation:					
#42 - Caulking, per fluid ounce	.21	.15	.17	.18	.20
Paint remover, per gallon					
#43 - Light duty	19.45	13.62	15.66	16.76	16.80
#44 - Heavy duty	24.30	17.01	19.56	20.93	20.90
#45 - Putty, per pound	7.10	4.97	5.72	6.12	6.10
#46 - Silica sand, per pound	.04	.03	.03	.03	.03
#47 - Window protective coating (wax)	15.20	10.64	12.24	13.10	13.10
#48 - Wood filler, per gallon	33.40	23.38	26.89	28.77	28.80
Industrial:					
#49 - Acid wash (muriatic acid)	5.45	3.82	4.39	4.70	4.70
#50 - Aluminum base paint	40.75	28.53	32.81	35.11	35.10
Epoxy coating, 2 part system					
#51 - Clear	59.25	41.48	47.70	51.04	51.00
#52 - White	62.65	43.86	50.44	53.97	54.00
Heat resistant enamel					
#53 - 500 to 1200 degree range	72.40	50.68	58.28	62.36	62.40
#54 - 300 to 800 degree range	54.90	38.43	44.19	47.28	47.30
#55 - Industrial bonding & penetrating oil paint	28.20	19.74	22.70	24.29	24.30
Industrial enamel, oil base, high gloss					
#56 - Light colors	30.10	21.07	24.23	25.93	25.90
#57 - Dark (OSHA) colors	37.55	26.29	30.23	32.35	32.40
#58 - Industrial waterproofing	25.20	17.64	20.29	21.71	21.70
#59 - Vinyl coating (tanks)	60.00	42.00	48.30	51.68	51.70
Wallcovering:					
Ready-mix:					
#60 - Light-weight vinyl (gal)	7.25	5.08	5.84	6.25	6.30
#61 - Heavy weight vinyl (gal)	7.45	5.22	6.00	6.42	6.40
#62 - Cellulose, clear (gal)	7.95	5.57	6.41	6.86	6.90
#63 - Vinyl to vinyl (gal)	17.10	11.97	13.77	14.73	14.70
#64 - Powdered cellulose, 2 - 4 ounces	2.75	1.93	2.22	2.38	2.40
#65 - Powdered vinyl, 2 - 4 ounces	3.25	2.28	2.62	2.80	2.80
#66 - Powdered wheat paste, 2-4 ounces	2.75	1.93	2.22	2.38	2.40

Note: Typically, powdered paste is in 2 to 4 ounce packages which will adhere 6 to 12 rolls of wallcovering.

Figure 10 (continued)
Material prices at 30% discount

Material pricing at 40% discount

All pricing is based on production grade material purchased in 5 gallon quantities.

		Retail price guide	Contractor price at a 40% discount	Add 15% sundries & escalation	Price with sales tax at 7%	Estimating prices with tax
Interior:						
	Sealer, off white (wet area walls & ceilings)					
#1 -	Water base	16.60	9.96	11.45	12.25	12.30
#2 -	Oil base	23.30	13.98	16.08	17.21	17.20
	Undercoat (doors, casings and other paint grade wood)					
#3 -	Water base	21.45	12.87	14.80	15.84	15.80
#4 -	Oil base	21.90	13.14	15.11	16.17	16.20
	Flat latex (walls, ceilings & paint grade baseboard)					
#5 -	Water base latex paint	17.40	10.44	12.01	12.85	12.90
	Acoustic spray-on texture					
#6 -	Primer	18.35	11.01	12.66	13.55	13.60
#7 -	Finish	15.75	9.45	10.87	11.63	11.60
#8 -	Dripowder mixed (pound)	.42	.25	.29	.31	.30
	Enamel (wet area walls & ceilings and openings)					
#9 -	Water base enamel	24.90	14.94	17.18	18.38	18.40
#10 -	Oil base enamel	27.05	16.23	18.66	19.97	20.00
	System estimate (cabinets, bookshelves, molding, interior windows)					
#11a -	Wiping stain, oil base	31.35	18.81	21.63	23.14	23.10
#11b -	Sanding sealer, lacquer	17.45	10.47	12.04	12.88	12.90
#11c -	Lacquer, semi gloss	19.55	11.73	13.49	14.43	14.40
#11 -	Stain, seal & 2 coat lacquer system					
	Average cost [11a + b + (2 x c)]		13.19	15.17	16.23	16.20
#12 -	Shellac, clear	25.65	15.39	17.70	18.94	18.90
#13 -	Penetrating oil stain	29.50	17.70	20.36	21.79	21.80
#14 -	Penetrating stain wax (molding)	27.65	16.59	19.08	20.42	20.40
#15 -	Wax, per pound (floors)	8.10	4.86	5.59	5.98	6.00
#16 -	Glazing (mottling over enamel)	31.65	18.99	21.84	23.37	23.40
#17 -	Spray can, each (HVAC registers)	4.25	2.55	2.93	3.14	3.10

Figure 11
Material prices at 40% discount

Material pricing at 40% discount (cont.)

	Retail price guide	Contractor price at a 40% discount	Add 15% sundries & escalation	Price with sales tax at 7%	Estimating prices with tax
Exterior:					
Solid body/color stain (beams, light valance, fascia, overhang, siding, plant-on trim, wood shelves)					
#18 - Water base stain	21.80	13.08	15.04	16.09	16.10
#19 - Oil base stain	24.25	14.55	16.73	17.90	17.90
Semi-transparent stain (beams, siding, T & G ceiling)					
#20 - Water base stain	21.70	13.02	14.97	16.02	16.00
#21 - Oil base stain	26.55	15.93	18.32	19.60	19.60
#22 - Polyurethane (exterior doors)	34.80	20.88	24.01	25.69	25.70
#23 - Marine spar varnish, flat or gloss (exterior doors)					
Interior or exterior	40.00	24.00	27.60	29.53	29.50
Exterior enamel (exterior doors & trim)					
#24 - Water base	27.75	16.65	19.15	20.49	20.50
#25 - Oil base	31.15	18.69	21.49	22.99	23.00
Porch & deck enamel - interior or exterior					
#26 - Water base enamel	28.00	16.80	19.32	20.67	20.70
#27 - Oil base enamel	30.60	18.36	21.11	22.59	22.60
#28 - Epoxy, 1 part, water base	37.85	22.71	26.12	27.95	28.00
#29 - Epoxy, 2 part system	74.35	44.61	51.30	54.89	54.90
System estimate (exterior windows)					
#30a - Wiping stain, oil base	31.05	18.63	21.42	22.92	22.90
#30b - Sanding sealer, varnish	33.95	20.37	23.43	25.07	25.10
#30c - Varnish, flat or gloss	36.00	21.60	24.84	26.58	26.60
#30 - Stain, seal & 1 coat varnish system					
Average cost (30a + b + c)		20.20	23.23	24.86	24.90
Masonry paint (masonry, concrete, plaster)					
#31 - Water base, flat or gloss	21.20	12.72	14.63	15.65	15.70
#32 - Oil base paint	28.75	17.25	19.84	21.23	21.20
#33 - Block filler	15.85	9.51	10.94	11.71	11.70
#34 - Waterproofing, clear hydro seal	18.05	10.83	12.45	13.32	13.30
Metal primer, rust inhibitor					
#35 - Clean metal	30.35	18.21	20.94	22.41	22.40
#36 - Rusty metal	37.20	22.32	25.67	27.47	27.50
Metal finish, synthetic enamel, gloss, interior or exterior					
#37 - Off white	28.30	16.98	19.53	20.90	20.90
#38 - Colors (except orange/red)	30.50	18.30	21.05	22.52	22.50
Anti-graffiti stain eliminator					
#39 - Water base primer & sealer	25.85	15.51	17.84	19.09	19.10
#40 - Oil base primer & sealer	26.10	15.66	18.01	19.27	19.30
#41 - Polyurethane 2 part system	92.85	55.71	64.07	68.55	68.60

Figure 11 (continued)
Material prices at 40% discount

Material pricing at 40% discount (cont.)

	Retail price guide	Contractor price at a 40% discount	Add 15% sundries & escalation	Price with sales tax at 7%	Estimating prices with tax
Preparation:					
#42 - Caulking, per fluid ounce	.21	.13	.15	.16	.20
Paint remover, per gallon					
#43 - Light duty	19.45	11.67	13.42	14.36	14.40
#44 - Heavy duty	24.30	14.58	16.77	17.94	17.90
#45 - Putty, per pound	7.10	4.26	4.90	5.24	5.20
#46 - Silica sand, per pound	.04	.02	.02	.02	.02
#47 - Window protective coating (wax)	15.20	9.12	10.49	11.22	11.20
#48 - Wood filler, per gallon	33.40	20.04	23.05	24.66	24.70
Industrial:					
#49 - Acid wash (muriatic acid)	5.45	3.27	3.76	4.02	4.00
#50 - Aluminum base paint	40.75	24.45	28.12	30.09	30.10
Epoxy coating, 2 part system					
#51 - Clear	59.25	35.55	40.88	43.74	43.70
#52 - White	62.65	37.59	43.23	46.26	46.30
Heat resistant enamel					
#53 - 500 to 1200 degree range	72.40	43.44	49.96	53.46	53.50
#54 - 300 to 800 degree range	54.90	32.94	37.88	40.53	40.50
#55 - Industrial bonding & penetrating oil paint	28.20	16.92	19.46	20.82	20.80
Industrial enamel, oil base, high gloss					
#56 - Light colors	30.10	18.06	20.77	22.22	22.20
#57 - Dark (OSHA) colors	37.55	22.53	25.91	27.72	27.70
#58 - Industrial waterproofing	25.20	15.12	17.39	18.61	18.60
#59 - Vinyl coating (tanks)	60.00	36.00	41.40	44.30	44.30
Wallcovering:					
Ready-mix:					
#60 - Light-weight vinyl (gal)	7.25	4.35	5.00	5.35	5.40
#61 - Heavy weight vinyl (gal)	7.45	4.47	5.14	5.50	5.50
#62 - Cellulose, clear (gal)	7.95	4.77	5.49	5.87	5.90
#63 - Vinyl to vinyl (gal)	17.10	10.26	11.80	12.63	12.60
#64 - Powdered cellulose, 2 - 4 ounces	2.75	1.65	1.90	2.03	2.00
#65 - Powdered vinyl, 2 - 4 ounces	3.25	1.95	2.24	2.40	2.40
#66 - Powdered wheat paste, 2-4 ounces	2.75	1.65	1.90	2.03	2.00

Note: Typically, powdered paste is in 2 to 4 ounce packages which will adhere 6 to 12 rolls of wallcovering

Figure 11 (continued)
Material prices at 40% discount

Figure 10 reflects a 30 percent discount. It applies to "Medium" work and assumes medium coverage, as in commercial work.

Figure 11 is the 40 percent discount table. It applies to "Fast" work and assumes heavier coverage typically required on unpainted surfaces in new construction. This discount is usually available only to large, high-volume painting companies that purchase materials in large quantities.

Here's an explanation of the columns in Figures 9, 10 and 11:

Retail price guide: This is an average based on a survey of eleven paint manufacturers or distributors, for standard grade, construction-quality paint, purchased in five gallon quantities.

Material pricing and discount percentages will vary from supplier to supplier and from area to area. Always keep your supplier's current price list handy. It should show your current cost for all the coatings and supplies you use. Also post a list of all suppliers, their phone numbers, and the salesperson's name beside your phone.

Prices change frequently. Paint quality, your supplier's discount programs, their marketing strategy and competition from other paint manufacturers will influence the price you pay. Never guess about paint prices – especially about less commonly used coatings. Don't assume that a product you haven't used before costs about the same as similar products. It might not. A heavy-duty urethane finish, for example, will cost about twice as much as a heavy-duty vinyl coating. If you don't know that, your profit for the job can disappear very quickly.

Prices at discount: The retail price, less the appropriate discount.

Allowance for sundries: It's not practical to figure the cost of every sheet of sandpaper and every rag you'll use on a job. And there's no way to accurately predict how many jobs you'll get out of each brush or roller pole, roller handle, ladder, or drop cloth. But don't let that keep you from including an allowance for these important costs in your estimates. If you leave them out, it's the same as estimating the cost of those items as *zero*. That's a 100 percent miss. Too many of those, and you're out of the painting business. It's better to estimate any amount than to omit some costs entirely.

Figure 12 is a sundries inventory checklist. Use it to keep track of the actual cost of expendable tools and equipment.

I've added 15 percent to the paint cost to cover expendable tools and supplies. This is enough for sundries on most jobs. There is one exception, however. On repaint jobs where there's extensive prep work, the cost of sundries may be more than 15 percent of the paint cost. When preparation work is extensive, figure the actual cost of supplies. Then add to the estimate that portion of the sundries cost that exceeds 15 percent of the paint cost. You might have to double the normal sundries allowance. When it comes to prep work, make sure your estimate covers *all* your supplies.

Price with sales tax at 7 percent: This column increases the material cost, including sundries, by 7 percent to cover sales tax. If sales tax in your area is more or less than 7 percent, you can adjust the material cost, or use the price that's closest to your actual cost.

In most cases contractors have to pay sales tax. If you don't pay the tax yourself, you may have to collect it from the building owner or general contractor and remit it to the state taxing authority. In either case, include sales tax in your estimate.

Estimating prices with tax: The figures in the last column of Figures 9 through 11 are rounded to the nearest dime unless the total is under a dollar. Those prices are rounded to the nearest penny.

This system for pricing materials isn't exact. But it's quick, easy and flexible. Compare your current material costs with costs in Figures 9, 10 and 11. If your costs are more than a few percent higher or lower than my costs, make a note on the blank line below "Fast" in the estimating tables.

Price Escalation

Escalation is the change in prices between the time you bid a job and the time you pay for labor and materials. Painting contractors seldom include escalation clauses in their bids because they don't expect lengthy delays. That's why escalation isn't included as a separate item in the estimating forms, Figures 18 and 19.

Sundry Inventory Checklist

Suppliers: D-Dumphy Paints
F-Fisher Paints
S-Superior Paints
P-Pioneer Paints

Supplier	Product number	Product	Inventory quantity	Unit	Cost	7/22	7/29	8/5	8/12
D	# —	Bender paint pads	3	Each	$3.50				
D	#792	Brush - 3" nylon Peacock	2	Each	$18.85		1		
D	#783	Brush - 4" nylon Scooter	2	Each	$27.85			1	
D	#115	Brush - 5" nylon Pacer	2	Each	$44.60			1	
D	#784	Brush - 3" bristle	2	Each	$17.55			1	
D	#2170	Caulking bags	2	Each	$3.75				
D	Latex	Caulking-DAP Acrylic latex	12	Each	$2.05		12		
D	#2172	Caulking gun (Newborn)	2	Each	$ 7.15		1		
P	# —	Hydraulic fluid	2	Qt	$8.05				
P	# —	Lemon oil	2	Pint	$4.05		1		
F	# —	Masking paper 18" wide	3	Roll	$20.60				
F	Anchor	Masking tape 1½"	24	Roll	$3.05		12		12
P	#2176	Lacquer	2	5's	$24.90			1	
P	#2173	Sanding sealer	2	5's	$20.35		1		
P	#9850	Resin sealer	2	5's	$17.75				
P	#131	PVA sealer (clear)	2	5's	$18.80		1		
F	#8500	Particle masks 100/box	1	Box	$13.45			1	
P	# —	Putty (Crawfords)	3	Qt	$9.20		2		
F	#R-10	Respirators	1	Each	$37.65				1
F	#R-49	Respirator cartridges 20/box	2	Box	$43.90				
F	#R-51	Respirator filters 20/box	2	Box	$31.35			1	
P	# —	Rags	10	Lb	$2.25				
F	#AR 691	Roller covers 9" x 3/4"	6	Each	$4.20		2		
F	#AR 692	Roller covers 9" x 3/8"	6	Each	$3.50	3			2
F	#AR 671	Roller covers 7" x 3/4"	3	Each	$3.45			1	
F	#AR 672	Roller covers 7" x 3/8"	3	Each	$3.75		1		

Figure 12
Sundry inventory checklist

Supplier	Product number	Product	Inventory quantity	Unit	Cost	7/22	7/29	8/5	8/12
F	#AR 611	Roller covers mini	3	Each	$2.90			1	
F	#95	Roller frames 9"	6	Each	$5.40	1	2		
F	#75	Roller frames 7"	5	Each	$5.20	3		3	
F	#TSR	Roller frames mini	2	Each	$2.95				
D	#40	Roller poles 4' wood tip	3	Each	$2.60		1		
D	#10	Roller poles 6' wood tip	10	Each	$4.05				
P	# 1	Roller pole tips metal	2	Each	$3.30			2	
P	# —	Sandpaper (120C production)	2	Slve	$48.65			2	
P	# —	Sandpaper (220A trimite)	2	Slve	$37.85				1
P	# —	Sandpaper (220A garnet)	1	Slve	$34.45		1		
D	# —	Spackle (Synkloid)	3	Qt	$5.30	1		1	
D	#42/61	Spray bombs (black[B]/white[W])	12	Each	$3.10	[B]12			[W]12
F	# —	Spray gun tips #3 or #4	10	Each	$7.65			3	
F	#2762	Spray gun couplers	10	Each	$2.10			5	
F	#S-27	Spray socks 48/Box	1	Box	$16.70				
D	#5271	Stip fill	1	Gal	$9.00			1	
D	#5927	Strainer bags	2	Each	$1.45	1			
D	#JT-21	Staples - 5/16"	2	Box	$2.35				
P	50 Gal	Thinner, lacquer	1	Drum	$420.00				
P	50 Gal	Thinner, paint	1	Drum	$210.00				1
P	# —	Thinner, shellac (alcohol)	1	Gal	$9.80				
D	#5775	Work pots (2 gal. plastic)	3	Each	$2.80		1		2
	#				$				
	#				$				
	#				$				
	#				$				
		Order date:				7/21	7/27	8/2	8/10
		Ordered by: (initials)				jj	jj	jj	jj
		Purchase order no.				0352	0356	0361	0371

Figure 12 (continued)
Sundry inventory checklist

Production Rate	Residential Wallcovering: Computer Program Crew Code	Labor Cost per Hour	Labor Burden per Hour	Labor Cost + Burden	Commercial Wallcovering: Computer Program Crew Code	Labor Cost per Hour	Labor Burden per Hour	Labor Cost + Burden	Flexible Wood Wallcovering: Computer Program Crew Code	Labor Cost per Hour	Labor Burden per Hour	Labor Cost + Burden
Slow	1W	$10.00	$2.85	$12.85	4W	$9.00	$2.56	$11.56	7W	$9.50	$2.70	$12.20
Medium	2W	15.75	5.49	21.24	5W	14.25	4.97	19.22	8W	15.00	5.23	20.23
Fast	3W	21.00	8.77	29.77	6W	19.00	7.93	26.93	9W	20.00	8.35	28.35

Figure 13
Hourly wage rates for wallcovering application

Any minor price escalation will be covered by the 15 percent added to material prices for sundries. But don't rely on that small cushion to absorb major inflationary cost increases. Plan ahead if prices are rising. In that case, add escalation as a separate item in the estimate.

Many formal construction contracts include an escalator clause that allows the contractor to recover for cost increases during the time of construction – especially if there was an unreasonable delay through no fault of the subcontractor. This clause may give you the right to collect for increases in both labor and material costs.

If work is delayed after you've been awarded the contract, you may be able to recover for cost increases under the escalator clause. This is more likely on public projects than on private jobs. Also, if there's a significant delay due to weather, you may have a good argument for adjusting the contract amount.

You can protect yourself against escalation if you include an expiration date on your bids. If the contract award is delayed beyond your expiration date, you can review your costs and make necessary adjustments. But be careful here. Increase the bid too much and you'll probably lose the contract. So raise your bid only if necessary, and then only by the amount of the actual cost increases. Don't try to make a killing on the job just because the bid prices have expired.

Column 4: Labor Cost

Column 4 in Figure 2 on page 7 shows the labor cost per unit. This figure is based on the productivity rate in column 1 and the wage rate in Figure 1. The wage rate for "Slow" (repaint) work is assumed to be $10.50 per hour. The wage rate for "Medium" (commercial) work is $16.75 per hour. The wage rate for "Fast" (residential tract) work is $22.50 per hour. Wage rates for wallcovering are different.

Wage Rates Vary

Wages vary from city to city. Recently I saw a survey of hourly union rates for painters in U.S. cities. The lowest rate shown was $8.40 an hour for painters in El Paso, Texas. The highest rate was $27.75 for painters in Anchorage, Alaska. You might ask, "Why don't all the painters in El Paso move to Anchorage?"

I don't know the answer, except to suggest that painters aren't starving in El Paso. Nor are they getting rich in Anchorage. Working conditions and the cost of living are very different in those two cities. However, on private jobs using non-union tradesmen, wage rates usually don't vary as much from city to city. The wage you pay depends on the demand for painting and how many painters are available for work.

Wages also change over time. For example, wage rates increased very quickly between 1979 and 1986. The national average union wage (including fringes) for painters in large cities went from $12.37 in 1979 to $19.50 per hour in 1986. Since then, wage rates haven't changed as quickly. Always base your estimates on the actual wages you'll pay your experienced painters.

	Fixed burden					Fringe benefits					
	FICA	FUTA	SUI	WCI	Liab. Ins.	Vac	Med	Life	Pension	Training	Total
Slow	7.65%	.8%	3.0%	12.5%	4.5%	0	0	0	0	0	28.45%
Medium	7.65%	.8%	4.5%	14.5%	4.9%	1.0%	1.0%	.25%	.25%	0	34.85%
Fast	7.65%	.8%	6.0%	16.5%	5.3%	2.5%	2.0%	.25%	.5%	.25%	41.75%

Figure 14
Labor burden percentages

Wages for Higher Skilled Specialists

Wages also vary with a workers' skill, dependability and with job difficulty. Generally higher paid painters are more productive than lower paid painters. Here's a chart I use to determine how much more per hour I should estimate for painting and surface preparation specialists. These figures are in addition to the basic journeyman rate.

Foremen ....$.50 to 2.00

Field superintendents ....3.00 to 4.50

Swing stage brush painters,

spray painters, or paperhangers 25

Iron, steel and bridge painters

(ground work) 50

Sandblasters, iron, steel, or

bridge painters (swing stage) 1.00

Steeplejacks 1.50

Most government and defense painting contracts require compliance with the Davis Bacon Act, which specifies that contractors pay at least the prevailing wage for each trade in the area where the job is located.

Calculate Your Labor Rate

Use the wage rate in Figure 1 ($10.50, $16.75 or $22.50 for "Slow," "Medium," or "Fast") that's appropriate for your company. Or, use a rate somewhere in between the rates listed. If you use your own wage rate, divide the hourly wage by the labor productivity (such as square feet per manhour in column 1). That's your labor cost per unit. Multiply by 100 if the units used are 100 linear feet or 100 square feet. ($10 ÷ 400 x 100 = $2.50.)

Column 5: Labor Burden

For each dollar of wages your company pays, at least another 28 cents has to be paid in payroll tax and for insurance. That's part of your labor burden. The rest is fringe benefits such as vacation pay, health benefits and pension plans.

Federal taxes are the same for all employers. State taxes vary from state to state. Fringe benefits vary the most. Generally, larger companies with more skilled painters offer considerably more fringe benefits than smaller companies.

In the estimating tables, the labor burden percentage varies with the application rate. For "Slow" (repaint) work, it's assumed to be 28.45 percent of $10.50 or $2.99 per hour. For "Medium" (commercial) work, the estimating tables use 34.85 percent of $16.75 or $5.84 per hour. For "Fast" (residential tract) work, the labor burden is 41.75 percent of $22.50 or $9.39 per hour.

Figure 14 shows how the labor burden percentages were compiled for each application rate.

FICA - Social Security tax: This is the portion paid by employers and is set by federal law. A similar amount is withheld from each employee's wage and deposited with a Federal Reserve bank by the employer.

FUTA - Federal Unemployment Insurance tax: Paid entirely by the employer and set by federal law. No portion is deducted from employee wages.

SUI - State Unemployment Insurance: Varies from state to state.

WCI - Workers' Compensation Insurance: Provides benefits for employees in case of injury on the job. Workers' comp is required by state law. Rates vary by state, job description and the loss experience of the employer.

Liab. Ins. - Liability Insurance: Covers injury or damage done to the public by employees. Comprehensive contractor's liability insurance includes current operations, completed operations, bodily injury, property damage, protective and contractual coverages with a $1,000,000 policy limit.

Fringe benefits: *Vac* is vacation pay. *Med* is medical insurance. *Life* is life insurance contribution. *Pension* is a pension plan contribution. *Training* is an apprentice training fund.

Vacation, life, pension and training payments depend on the agreement between employers and employees. These are voluntary contributions if not required by a collective bargaining agreement. Smaller companies are less likely to provide these benefits. The cost of fringe benefits in a painting company can range from zero to more than 10 percent of wages.

Column 6: Material Cost per Unit

This column is the result of dividing column 3 (material cost) by column 2 (material coverage) for each application rate. For example, in Figure 2 in the "Medium" row, a material cost of 15.00 is divided by material coverage of 275, then multiplied by 100 to arrive at $5.45 per 100 square feet. That's the figure listed for "Medium" work in column 6.

Column 7: Overhead

The overhead rate for "Slow" (repaint) jobs is assumed to be 31 percent. For "Medium" (commercial projects), overhead is 24.5 percent. For "Fast" (residential tracts), overhead is 18.5 percent. The overhead cost per unit in each row is calculated by adding the labor cost per unit, labor burden per unit, and material cost per unit and then multiplying by the appropriate overhead percentage.

There are two types of overhead, direct overhead and indirect overhead. Only indirect overhead is included in the "Overhead" column of the estimating cost tables. Enter your direct overhead costs on a separate line on your take-off sheet.

Direct overhead is job site overhead, expenses you charge to a specific job. Examples include performance bonds, special insurance premiums, or rental of a job site storage trailer. These expenses are not included in the estimating tables and have to be added to your estimates. On many jobs, there may be little or no direct overhead.

Indirect overhead is office overhead, expenses that aren't related to any particular job and that tend to continue whether the volume of work increases or decreases. Examples are non-trade salaries, office rent, vehicles, sales and financial expenses, insurance, taxes and licenses.

The percentage of income spent on overhead is assumed to be lower for high volume companies and higher for low volume companies. A large company working many projects at the same time can spread overhead costs over many projects – charging a smaller percentage of overhead to each job. The more jobs, the lower the overhead per job – assuming overhead doesn't increase faster than business volume.

On the other hand, a small business may have to absorb all overhead on a single job. Even painting contractors who work out of their homes have overhead expenses.

Here's one overhead expense every paint contractor has and that you might overlook: the cost of estimating jobs. That's part of the salary cost of the employee who does the estimating.

Figure Overhead Carefully

Estimating indirect (office) overhead isn't as easy as estimating labor and material. There aren't as many clear-cut answers. That's why indirect overhead is often underestimated. Don't make that mistake in your estimates. Underestimating overhead is the same as giving up part of your profit. After all, indirect overhead expenses are real costs, just like paint, labor and taxes.

In large painting companies, management accumulates indirect overhead costs and translates them into a percentage the estimator should add to the costs of each job. In smaller companies, the estimator should keep a record of indirect overhead expenses. With a good record of overhead expense, you can calculate your overhead percentage for future periods very accurately. Then it's easy to add a percentage for indirect overhead costs into your estimate.

Computing Your Overhead Percentage

Here's how to decide which overhead rate to use in the cost estimating tables:

1) List all your overhead expenses for at least the last six months; a year would be better. You need overhead cost information that goes back far enough to eliminate the effect of seasonal changes in business volume.

   If your company is new, estimate your annual overhead by projecting overhead costs for the first full year. For example, if you've been in business for five months and overhead has been $5,500 so far, you can expect annual overhead to be about $13,200 ($5,500 divided by 5 and multiplied by 12).

2) Here's how to calculate your indirect overhead percentage:

$$\frac{\text{Annual indirect overhead}}{\text{Annual job expenses}} = \text{Overhead percentage}$$

Calculate your indirect overhead by adding together your real (or anticipated) annual expenses for the following:

Salaries. Include what you pay for all employees except trade workers, plus payroll-related expenses for all employees.

Office and shop expense. Rent or mortgage, utilities, furniture and equipment, maintenance, office supplies and postage, storage sheds, warehouses, fences or yard maintenance.

Vehicles. Lease or purchase payments, maintenance, repairs and fuel.

Sales promotion. Advertising, entertainment and sales-related travel.

Taxes. Property tax and income tax, and sales tax (if not included in your material prices).

Licenses. Contractor's and business licenses.

Insurance. General liability, property and vehicle policies.

Interest expense. Loan interest and bank charges. Also consider loss of interest on payments retained by the general contractor until the job is finished.

Miscellaneous expenses. This could include depreciation and amortization on building and vehicles, bad debts, legal and accounting fees, and educational expenses.

Direct overhead is easier to figure. It's all job expenses except tradesman labor, payroll taxes and insurance, materials, equipment, subcontracts, and contingency expenses. Permits, bonds, fees and special insurance policies for property owners are also examples of direct overhead. Add the direct overhead expense on the appropriate lines in your estimate. Direct overhead is not included in the estimating tables in this manual.

Field Equipment May Be Part of Overhead

As you may have noticed, there's no equipment cost column in the estimating tables. Instead, field equipment expense is included in the overhead percentage for "Fast" and "Medium" work but not "Slow" work.

New Construction and Commercial Work: The overhead percentage for "Fast" (residential tract) work and "Medium" (commercial) projects *includes* equipment costs such as ladders, spray

Equipment Rental Rates

Use the following rates only as a guide. They may not be accurate for your area.
Verify equipment rental rates at your local yard.

	Rental		
	$/day	$/week	$/month
Acoustical sprayer	45.00	135.00	337.50
Air compressors			
Electric or gasoline, wheel mounted			
5 CFM, 1.5 HP, electric	28.00	84.00	210.00
8 CFM, 1.5 HP, electric	32.00	96.00	240.00
10 CFM, 5.5 HP, gasoline	37.00	111.00	277.50
15 CFM, shop type, electric	42.00	126.00	315.00
50 CFM, shop type, electric	55.00	165.00	412.50
100 CFM, gasoline	75.00	225.00	562.50
125 CFM, gasoline	85.00	255.00	637.50
150 CFM, gasoline	95.00	285.00	712.50
175 CFM, gasoline	105.00	315.00	787.50
190 CFM, gasoline	115.00	345.00	862.50
Diesel, wheel mounted			
to 159 CFM	85.00	255.00	765.00
160 to 249 CFM	105.00	315.00	945.00
250 to 449 CFM	155.00	465.00	1,395.00
450 to 749 CFM	230.00	690.00	2,070.00
750 to 1199 CFM	315.00	945.00	2,835.00
1200 CFM & over	460.00	1,380.00	4,140.00
Air hose - with coupling, 50' lengths			
1/4" I.D.	6.00	18.00	45.00
3/8" I.D.	7.00	21.00	52.50
1/2" I.D.	8.00	24.00	60.00
5/8" I.D.	9.00	27.00	67.50
3/4" I.D.	10.00	30.00	75.00
1" I.D.	11.00	33.00	82.50
1 1/2" I.D.	16.00	48.00	120.00
Boomlifts			
3' x 4' to 3' x 8' basket			
20' two wheel drive	140.00	420.00	1,260.00
30' two wheel drive	170.00	510.00	1,530.00
40' four wheel drive	195.00	585.00	1,755.00
50' - 1000 lb.	320.00	960.00	2,880.00
Telescoping and articulating booms, self propelled, gas or diesel powered, 2-wheel drive			
21' to 30' high	200.00	600.00	1,800.00
31' to 40' high	250.00	750.00	2,250.00
41' to 50' high	325.00	975.00	2,925.00
51' to 60' high	400.00	1,200.00	3,600.00
Burner, paint	12.00	36.00	90.00
Dehumidifier - 5000 Btu, 89 lb, 8.7 amp	55.00	165.00	412.50
Ladders			
Aluminum extension			
16' to 36'	30.00	90.00	225.00
40' to 60'	45.00	135.00	337.50

	Rental		
	$/day	$/week	$/month
Ladders (Continued)			
Step - fiberglass or wood			
6'	8.00	24.00	60.00
8'	10.00	30.00	75.00
10'	12.00	36.00	90.00
12'	14.00	42.00	105.00
14'	16.00	48.00	120.00
16'	20.00	60.00	150.00
20'	26.00	78.00	195.00
Ladder jacks - No guardrail.	8.00	20.00	50.00
Masking paper dispenser	20.00	50.00	125.00
Painter's pic (walkboards); No guardrail. (Also known as airplane planks, toothpicks and banana boards)			
16' long	8.00	24.00	60.00
20' long	16.00	48.00	120.00
24' long	20.00	60.00	150.00
28' long	24.00	72.00	180.00
32' long	28.00	84.00	210.00
Planks - plain end microlam scaffold plank			
9" wide	10.00	30.00	75.00
10" wide	12.00	36.00	90.00
12" wide	14.00	42.00	105.00
Pressure washers (See Water pressure washers)			
Sandblast compressor and hopper			
To 250 PSI	60.00	180.00	450.00
Over 250 to 300 PSI	85.00	255.00	637.50
Over 600 to 1000 PSI	110.00	330.00	825.00
Sandblast machines			
150 lb pot with hood, 175 CFM compressor	230.00	690.00	1,725.00
300 lb pot with hood, 325 CFM compressor	410.00	1,230.00	3,075.00
600 lb pot with hood, 600 CFM compressor	745.00	2,235.00	5,587.50
Sandblast hoses - 50' lengths, coupled			
3/8" I.D.	10.00	30.00	75.00
3/4" I.D.	14.00	42.00	105.00
1" I.D.	18.00	54.00	135.00
1 1/4" I.D.	20.00	60.00	150.00
1 1/2" I.D.	22.00	66.00	165.00
Sandblast accessories			
Nozzles, all types	18.00	54.00	135.00
Hood, air-fed	28.00	84.00	210.00
Valves, remote control (deadman, all sizes)	30.00	90.00	225.00

Figure 15
Typical equipment purchase and rental prices

	Rental		
	$/day	$/week	$/month
Sanders			
Belt - 3"	14.00	42.00	105.00
Belt - 4" x 24"	17.00	51.00	127.50
Disc - 7"	22.00	66.00	165.00
Finish sander, 6"	12.00	36.00	90.00
Floor edger, 7" disk, 29#, 15 amp.	20.00	60.00	150.00
Floor sander, 8" drum, 118#, 14 amp	45.00	135.00	337.50
Palm sander, 4" x 4"	10.00	30.00	75.00
Palm sander, 4 1/2" x 9 1/4"	12.00	36.00	90.00
Scaffolding: rolling stage, caster mounted, 30" wide by 7' or 10' long			
4' to 6' reach	40.00	80.00	160.00
7' to 11' reach	50.00	100.00	200.00
12' to 16' reach	70.00	140.00	280.00
17' to 21' reach	95.00	190.00	380.00
22' to 26' reach	105.00	210.00	420.00
27' to 30' reach	115.00	230.00	460.00
Casters - each	10.00	20.00	30.00
Scissor lifts			
Electric powered, rolling with 2' x 3' platform, 650 lb capacity			
30' high	75.00	225.00	675.00
40' high	130.00	390.00	1,170.00
50' high	150.00	450.00	1,350.00
Rolling, self-propelled, hydraulic, electric powered			
to 20' high	110.00	330.00	990.00
21' to 30' high	135.00	405.00	1,215.00
31' to 40' high	170.00	510.00	1,530.00
Rolling, self-propelled, hydraulic, diesel powered			
to 20' high	125.00	375.00	1,125.00
21' to 30' high	155.00	465.00	1,395.00
31' to 40' high	200.00	600.00	1,800.00
Spray rigs			
Airless pumps, complete with gun and 50' of line			
Titan 447, 7/8 HP, electric	70.00	210.00	630.00
Titan 660, 1 HP, electric	80.00	240.00	720.00
Gasoline, .75 gpm	85.00	255.00	765.00
Emulsion pumps			
65 gal, 5 HP engine	70.00	210.00	630.00
200 gal, 5 HP engine	80.00	240.00	720.00
Emulsion airless, 1.25 gpm, gasoline	85.00	255.00	765.00

	Rental		
	$/day	$/week	$/month
Spray rigs (Continued)			
Conventional pumps, gas, portable			
High pressure, low volume (HVLP)	45.00	135.00	405.00
8 CFM complete	60.00	180.00	540.00
17 CFM complete	65.00	195.00	585.00
85 CFM complete	75.00	225.00	675.00
150 CFM complete	110.00	330.00	990.00
Spray rig accessories: 6' wand	7.00	21.00	52.50
Striper, paint (parking lot striping)			
Aerosol	20.00	60.00	150.00
Pressure regulated	28.00	84.00	210.00
Swing stage			
Any length drop, motor operated, excluding safety gear and installation or dismantling. Note: Must be set up by a professional to ensure safety.			
Swing stage	100.00	300.00	900.00
Basket	50.00	150.00	450.00
Bosun's chair	50.00	150.00	450.00
Swing stage safety gear, purchase only			
Safety harness (90.00)			
4' lanyard with locking snap at each end (65.00)			
DBI rope grab for 5/8" safety line (70.00)			
Komet rope grab for 3/4" safety line (100.00)			
Texturing equipment			
Texturing gun - w/ hopper, no compressor	5.00	15.00	45.00
Texturing mud paddle mixer	7.00	21.00	63.00
Texturing outfit - 1 HP w/ gun, 50' hose, 75 PSI	11.00	33.00	99.00
Wallpaper hanging kit	17.00	51.00	153.00
Wallpaper steamer			
Electric, small, 10 amp	20.00	60.00	180.00
Electric, 15 amp	30.00	90.00	270.00
Pressurized, electric	38.00	114.00	342.00
Water pressure washer (pressure washer, water blaster, power washer)			
1000 PSI, electric, 15 amp	45.00	135.00	405.00
2000 PSI, gas	75.00	225.00	675.00
2500 PSI, gas	80.00	240.00	720.00
3500 PSI, gas	88.00	264.00	792.00

Figure 15 (continued)
Typical equipment purchase and rental prices

equipment, and masking paper holders. Those items are used on many jobs, not just one specific job. The overhead allowance covers equipment purchase payments, along with maintenance, repairs and fuel. If you have to rent equipment for a specific new construction project, *add* that rental expense as a separate cost item in your estimate.

Repaint Jobs: Overhead rates for "Slow" (repaint) work do *not* include equipment costs. When you estimate a repaint job, any small or short-term job, or a job that uses only a small quantity of materials, *add* the cost of equipment at the rental rate – even if the equipment is owned by your company.

Rental yards quote daily, weekly and monthly equipment rental rates. Figure 15 shows typical rental costs for painting equipment. Your actual equipment costs may be different. Here's a suggestion that can save you more than a few minutes on the telephone collecting rental rates. Make up a blank form like Figure 15 and give it to your favorite rental equipment suppliers. Ask each supplier to fill in current rental costs. Use the completed forms until you notice that rates have changed. Then ask for a new set of rental rates.

Commissions and Bonuses

Any commissions or bonuses you have to pay on a job aren't included in the estimating tables. You must add these expenses to your bid.

Painting contractors rarely have a sales staff, so there won't be sales commissions to pay on most jobs. There's one exception, however. Most room addition and remodeling contractors have salespeople. And many of their remodeling projects exclude painting. In fact, their contract may specify that the owner is responsible for the painting. These jobs may be a good source of leads for a painting contractor. Develop a relationship with the remodeling contractor's sales staff (with the remodeling contractor's approval, of course). If you have to pay a sales commission for the referral, this is direct overhead and has to be added to the estimate.

Some painting contractors pay their estimators a bonus of 1 to 3 percent per job in addition to their salary. If you offer an incentive like this, add the cost to your estimate, again as a direct overhead item.

An Example of Overhead

Here's an example of how overhead is added into an estimate. A painting company completed 20 new housing projects in the last year. Average revenue per project was $50,000. Gross receipts were $1,000.000 and the company made a 5 percent profit.

Gross income	$1,000,000
Less the profit earned (5%)	50,000
Gross expenses	950,000
Less total direct job cost	825,000
Indirect overhead expense	125,000

$$\frac{125{,}000 \text{ (overhead cost)}}{825{,}000 \text{ (direct job cost)}} = 0.1515 \text{ or } 15.15\%$$

When you've calculated indirect overhead as a percentage of direct job cost, add that percentage to your estimates. If you leave indirect overhead out of your estimates, you've left out some very significant costs.

Column 8: Profit

The estimating tables assume that profit on "Slow" (repaint) jobs is 15 percent, profit on "Medium" (commercial) projects is 11 percent and profit on "Fast" (residential tract) jobs is 6 percent. Calculate the profit per unit by first adding together the costs in columns 4 (labor cost per unit), column 5 (labor burden per unit), column 6 (material costs per unit), and column 7 (overhead per unit). Then multiply the total by the appropriate profit percentage to find the profit per unit.

It's my experience that larger companies with larger projects can survive with a smaller profit percentage. Stiff competition for high volume tract work forces bidders to trim their profit margin. Many smaller companies doing custom work earn a higher profit margin because they produce better quality work, have fewer jobs, and face less competition.

Risk factor	Normal profit (assume 10%)		Difficulty factor		Proposed profit range
High risk	10%	x	1.5 to 3.5	=	15% to 35%
Average risk	10%	x	1.3 to 1.4	=	13% to 14%
Moderate risk	10%	x	1.0 to 1.2	=	10% to 12%
Low risk	10%	x	0.5 to 0.9	=	5% to 9%

Figure 16
Risk factors and profit margin

Profit and Risk

Profit is usually proportionate to risk. The more risk, the greater the potential profit has to be to attract bidders. Smaller companies handling custom or repaint work have more risk of a major cost overrun because there are many more variables in that type of work. It's usually safe to estimate a smaller profit on new work because new work tends to be more predictable. The risk of loss smaller.

How do you define risk? Here's my definition: Risk is the *headache factor*, the number and size of potential problems you could face in completing the project. Repaint jobs have more unknowns, so they're a greater risk. And dealing with an indecisive or picky homeowner can be the greatest headache of all. You may need to use a profit margin even higher than the 15 to 35 percent range indicated for high-risk work in Figure 16.

Tailoring Your Profit Margin

Of course, your profit margin has to be based on the job, your company and the competition. But don't cut your profit to the bone just to get more work. Instead, review your bid to see if there are reasons why the standard costs wouldn't apply.

I use the term *standard base* bid to refer to my usual charge for all the estimated costs, including my standard profit. Before submitting any bid, spend a minute or two deciding whether your standard base bid will apply.

Risk Factors

Your assessment of the difficulty of the job may favor assigning a risk factor that could be used to modify your profit percentage. The higher the risk, the higher potential profit should be. My suggestions are in Figure 16.

As you might expect, opinions on difficulty factors can vary greatly. There's a lot of knowledge involved. You need experience and good judgment to apply these factors effectively.

Bidding Variables

Of course, your profit may be affected by an error in evaluating the job risk factor. You can greatly reduce the risk by accurately evaluating the bidding variables in Figure 17. Make adjustments to your standard base bid for example, if you expect your crews to be more or less efficient on this project, or if you expect competition to be intense. If there are logical reasons to modify your standard base bid, make those changes.

But remember, if you adjust your standard base bid, you're not changing your profit margin. You're only allowing for cost variables in the job. Adjust your standard base costs for unusual labor productivity, material or equipment cost changes, or because of unusual overhead conditions. Review the following bidding variables when deciding how to adjust your standard base bid.

The Bottom Line

The profit margin you include in estimates depends on the way you do business, the kind of work you do, and your competition. Only you can decide what percentage is right for your bids. Don't take another paint estimator's advice on the "correct" profit margin. There's no single correct answer. Use your own judgment. But here are some typical profit margins for the kinds of work most painting contractors do.

Repaints:	Custom	20 to 35%
	Average	15 to 20%
Commercial or industrial		10 to 15%
New residential:	1 to 4 units	10 to 12%
	5 or more	5 to 7%
Government work		5 to 7%

Column 9: Total Cost

The costs in Column 9 of Figure 2, and all the estimating tables in this book, are the totals per unit for each application rate in columns 4, 5, 6, 7, and 8. That includes labor, labor burden, material cost, overhead and profit.

Sample Estimate

Figure 18 is a sample repaint estimate for a small house with many amenities. The final bid total is the bid price. Figure 19 is a blank estimating form for your use.

Reputations and attitudes

- Owner
- Architect
- General Contractor
- Lender
- Inspector

The project

- Building type
- Project size
- Your financial limits
- Start date
- Weather conditions
- Manpower availability and capability

The site

- Location (distance from shop and suppliers)
- Accessibility
- Working conditions
- Security requirements
- Safety considerations

Competition

- Number bidding
- Their strength, size and competence

Desire for the work

Figure 17
Bidding variables

Date 6/7/97
Customer Dan Gleason
Address 3333 A Street
City/State/Zip Yourtown, USA 77777
Phone (619)555-1212
Estimated by CHS

Due date 6/15/97
Job name Gleason Repaint
Job location 3333 A Street
Estimate # 97-051
Total square feet 1,020 SF (5 rooms)
Checked by Jack

Interior Costs

	Operation	Material	Application method	Dimensions	Quantity SF/LF/Each		Unit cost		Total cost
1	Ceiling-T&G	Semi-Trans	R + B	17.5 x 15.3 x 1.3	348 SF	x	.21	= $	73.00
2	Beams to 13'H	Solid	R + B	17.5 x 7	122.5 LF	x	1.23	= $	151.00
3	Ceiling-Drywall	Flat	R	127 + 127	254 SF	x	.14	= $	36.00
4	Ceiling-Drywall	Enamel	R	75 + 15 + 40	130 SF	x	.16	= $	21.00
5	Walls-Drywall	Flat	R	675 + 392 + 392	1,459 SF	x	.15	= $	219.00
6	Walls-Above 8' (clip)	Flat	R + B	70 + 85 = 155 x 1.3	201.5 SF	x	.15	= $	30.00
7	Walls-Drywall	Enamel	R	280 + 128 + 208	616 SF	x	.20	= $	123.00
8	Doors-Finish	Enamel	R + B	Opening Count	10 Ea	x	93.61	= $	94.00
9	Baseboard-Prime	Flat w/ walls	R + B	64 + 49 + 49	162 LF	x	.05	= $	8.00
10	Baseboard-Finish	Enamel (cut-in)	B	11 + 16 + 35	62 LF	x	.26	= $	16.00
11	Railing-W.I.	Enamel	B	42" High	15 LF	x	1.40	= $	21.00
12	Light Valance	Solid	B	2 x 8	10 LF	x	1.03	= $	10.00
13	Registers	Can	Spray	1,020 SF Home	1,020 SF	x	.03	= $	31.00
14						x		= $	
15						x		= $	
16						x		= $	
17						x		= $	
18						x		= $	
				Total Interior Costs (includes overhead and profit)				= $	833.00

Exterior Costs

	Operation	Material	Application method	Dimensions	Quantity SF/LF/Each		Unit cost		Total cost
1	S.M. Diverter-3'W	Finish	B	14	14 LF	x	.13	= $	2.00
2	Roof Jacks	Finish	B	1 Story	1,020 SF	x	.01	= $	10.00
3	S.M. Vents & Flashings	Finish	B	1 Story	1 Ea	x	31.04	= $	31.00
4	Fascia-2 x 4	Solid	R + B	66 + 59	125 LF	x	.46	= $	58.00
5	Overhang-24"	Solid	R + B	(132 + 76) x 1.5	312 SF	x	.39	= $	122.00
6	Siding-R.S. Wood	Solid	R + B	(½ x 24 x 4.5) x 2	108 SF	x	.33	= $	36.00
7	Plaster/Stucco	Masonry	Spray	255+255+204+204	918 SF	x	.19	= $	174.00
8	Door-Panel (Entry)	Enamel 2 coats	R + B	Entry	1 Ea	x	38.72	= $	39.00
9	Door - Flush	Enamel 2 coats	R + B	Exterior	1 Ea	x	16.03	= $	16.00
10	Plant-On Trim-2x4	Solid	R + B	66 + 62 + 52	180 LF	x	.36	= $	65.00
11	Pot Shelf	Solid	R + B	27	27 LF	x	1.11	= $	30.00
12	Pass Through	Finish	B	10	10 LF	x	1.04	= $	10.00
13						x		= $	
14						x		= $	
15						x		= $	
16						x		= $	
17						x		= $	
18						x		= $	
				Total Exterior Costs (includes overhead and profit)				= $	593.00

Figure 18
Sample painting estimate

Preparation Costs

	Operation	Dimensions	Quantity SF/LF/Each		Unit cost		Total cost
1	Sand Wood Ceiling	17.5 x 15.3 x 1.3	348 SF	x	.11	=	$ 38.00
2	Sand and Putty Int. Wall	675 + 392 + 392	1,459 SF	x	.11	=	$ 160.00
3	Sand Openings	14 Ea x 42 SF	588 SF	x	.08	=	$ 47.00
4	Wash Int. Walls / Ceil-Enam	280 + 128 + 208	616 SF	x	.11	=	$ 68.00
5	Waterblast	125 + 210 + 108 + 918	1,361 SF	x	.05	=	$ 68.00
6	Sand and Putty Ext. Trim	125 + 210 + 108	443 SF	x	.20	=	$ 89.00
7	Caulk Ext. Windows-1/8" gap	20 + 15 + 10 + 20 + 12	77 SF	x	.36	=	$ 28.00
8	Window Protective Coating	34 + 25 + 14 + 5 + 48 + 18	144 SF	x	.29	=	$ 42.00
9				x		=	$
10				x		=	$
	Total Preparation Costs (includes overhead and profit)					=	$ 540.00

SURRPTUCU Costs

Operation	Description	Labor hours	Labor cost (at $ 16.75)	Material cost	Totals
SetUp	2 Days @.5 hr/day	1.0	16.75	—	= $ 17.00
Remove/Replace	Hardware & SW Plates	1.25	20.94	—	= $ 21.00
Protection	Furniture & Floors	1.0	16.75	15.00	= $ 32.00
TouchUp is applied as a percentage of the total costs. See *Extensions*					
CleanUp	2 Days @.5 hr/day	1.0	16.75	—	= $ 17.00

Equipment Costs

Equipment description	Rental days	Daily cost	Total cost
Waterblast	1	50.00	$ 50.00
Ladders, 2 Ea.	1	5.00	$ 10.00
Sander Belt	1	14.00	$ 14.00
			$
			$
			$
		Total Equipment Costs	$ 74.00

Subcontractor Costs

Trade	Bid amount
Pavement marking	$ 0
Sandblasting	$ 0
Scaffolding	$ 0
Wallcovering	$ 0
Waterblasting	$ 0
Other ___	$ 0
Other ___	$ 0
Other ___	$ 0
Total Subcontractor Costs	$ —

Extensions

Supervision 1Hr.	$ 18.00
Setup	17.00
Remove/replace	21.00
Protection	32.00
Cleanup	17.00
Equipment	74.00
Subcontracts	0
Commissions	0
Other costs	0
Subtotal	179.00
Overhead (24.5%)	44.00
Profit (11%)	20.00
Subtotal	243.00
Preparation	540.00
Interior total	838.00
Exterior total	595.00
Subtotal	2,216.00
Touchup (10%)	222.00
Contingency (0%)	0
Total base bid	2,438.00
Adjustment (-2%)	49.00
Final bid total	2,389.00
Price per SF	2.30
Price per room	478.00

Figure 18 (continued)
Sample painting estimate

Date ______	Due date ______
Customer ______	Job name ______
Address ______	Job location ______
City/State/Zip ______	Estimate # ______
Phone ______	Total square feet ______
Estimated by ______	Checked by ______

Interior Costs

Operation	Material	Application method	Dimensions	Quantity SF/LF/Each		Unit cost	Total cost
1					x		= $
2					x		= $
3					x		= $
4					x		= $
5					x		= $
6					x		= $
7					x		= $
8					x		= $
9					x		= $
10					x		= $
11					x		= $
12					x		= $
13					x		= $
14					x		= $
15					x		= $
16					x		= $
17					x		= $
18					x		= $
			Total Interior Costs (includes overhead and profit)				= $

Exterior Costs

Operation	Material	Application method	Dimensions	Quantity SF/LF/Each		Unit cost	Total cost
1					x		= $
2					x		= $
3					x		= $
4					x		= $
5					x		= $
6					x		= $
7					x		= $
8					x		= $
9					x		= $
10					x		= $
11					x		= $
12					x		= $
13					x		= $
14					x		= $
15					x		= $
16					x		= $
17					x		= $
18					x		= $
			Total Exterior Costs (includes overhead and profit)				= $

Figure 19
Blank painting estimate

Preparation Costs

	Operation	Dimensions	Quantity SF/LF/Each		Unit cost		Total cost
1				x		=	$
2				x		=	$
3				x		=	$
4				x		=	$
5				x		=	$
6				x		=	$
7				x		=	$
8				x		=	$
9				x		=	$
10				x		=	$

Total Preparation Costs (includes overhead and profit) = $ ______

SURRPTUCU Costs

Operation	Description	Labor hours	Labor cost (at $ ______)	Material cost	Totals
Set**U**p					= $
Remove/**R**eplace					= $
Protection					= $

Touch**U**p is applied as a percentage of the total costs. See *Extensions*

Operation	Description	Labor hours	Labor cost	Material cost	Totals
Clean**U**p					= $

Equipment Costs

Equipment description	Rental days	Daily cost	Total cost
			$
			$
			$
			$
			$
			$
		Total Equipment Costs	$

Subcontractor Costs

Trade	Bid amount
Pavement marking	$
Sandblasting	$
Scaffolding	$
Wallcovering	$
Waterblasting	$
Other ___	$
Other ___	$
Other ___	$
Total Subcontractor Costs	$

Extensions

Supervision ___	$
Setup	
Remove/replace	
Protection	
Cleanup	
Equipment	
Subcontracts	
Commissions	
Other costs	
Subtotal	
Overhead (__%)	
Profit (__%)	
Subtotal	
Preparation	
Interior total	
Exterior total	
Subtotal	
Touchup (___%)	
Contingency (__%)	
Total base bid	
Adjustment (__%)	
Final bid total	
Price per SF	
Price per room	

Figure 19 (continued)
Sample painting estimate

Part I

General Painting Costs

	Labor LF per manhour	Material coverage LF/gallon	Material cost per gallon	Labor cost per 100 LF	Labor burden 100 LF	Material cost per 100 LF	Overhead per 100 LF	Profit per 100 LF	Total price per 100 LF
Baseboard, per linear foot									
Roll 1 coat with walls, brush touchup, paint grade base									
Flat latex, water base, (material #5)									
Slow	900	800	17.10	1.17	.33	2.14	1.13	.72	5.49
Medium	1200	750	15.00	1.40	.47	2.00	.95	.53	5.35
Fast	1500	700	12.90	1.50	.64	1.84	.73	.28	4.99
Enamel, water base (material #9)									
Slow	600	750	24.50	1.75	.50	3.27	1.71	1.08	8.31
Medium	800	725	21.40	2.09	.73	2.95	1.41	.79	7.97
Fast	1000	700	18.40	2.25	.94	2.63	1.08	.41	7.31
Enamel, oil base (material #10)									
Slow	600	750	26.60	1.75	.50	3.55	1.80	1.14	8.74
Medium	800	725	23.30	2.09	.73	3.21	1.48	.83	8.34
Fast	1000	700	20.00	2.25	.94	2.86	1.12	.43	7.60
Brush 1 coat, cut-in, paint grade base									
Enamel, water base (material #9)									
Slow	100	700	24.50	10.50	2.98	3.50	5.27	3.34	25.59
Medium	120	675	21.40	13.96	4.86	3.17	5.39	3.01	30.39
Fast	140	650	18.40	16.07	6.70	2.83	4.74	1.82	32.16
Enamel, oil base (material #10)									
Slow	100	700	26.60	10.50	2.98	3.80	5.36	3.40	26.04
Medium	120	675	23.30	13.96	4.86	3.45	5.46	3.05	30.78
Fast	140	650	20.00	16.07	6.70	3.08	4.78	1.84	32.47
Spray 1 coat, stain in boneyard, stain grade base									
Wiping stain (material #11a)									
Slow	--	--	--	--	--	--	--	--	--
Medium	1500	1750	27.00	1.12	.39	1.54	.75	.42	4.22
Fast	2000	1500	23.10	1.13	.46	1.54	.58	.22	3.93

Use these figures for 1-1/2 inch to 3 inch baseboard stock, painted or stained on one side. Measurements are based on linear feet of baseboard. Paint grade base is painted after it is installed but stain grade base is usually stained in a boneyard. Typically, finger joint stock is paint grade and butt joint stock is stain grade. These figures include minimal preparation time and material. Add for extensive preparation. "Slow" work is based on a $10.50 hourly wage, "Medium" work on a $16.75 hourly wage, and "Fast" work on a $22.50 hourly wage. Other qualifications that apply to this table are on page 9.

	Labor SF per manhour	Material coverage SF/gallon	Material cost per gallon	Labor cost per 100 SF	Labor burden 100 SF	Material cost per 100 SF	Overhead per 100 SF	Profit per 100 SF	Total price per 100 SF
Baseboard, per square foot of floor area									
Roll 1 coat with walls, brush touchup, paint grade base									
Flat latex, water base, (material #5)									
Slow	2500	1500	17.10	.42	.12	1.14	.52	.33	2.53
Medium	2750	1250	15.00	.61	.20	1.20	.49	.28	2.78
Fast	3000	1000	12.90	.75	.30	1.29	.43	.17	2.94
Enamel, water base (material #9)									
Slow	2000	1000	24.50	.53	.14	2.45	.97	.62	4.71
Medium	2200	900	21.40	.76	.26	2.38	.83	.47	4.70
Fast	2400	800	18.40	.94	.40	2.30	.67	.26	4.57
Enamel, oil base (material #10)									
Slow	2000	1000	26.60	.53	.14	2.66	1.04	.66	5.03
Medium	2200	900	23.30	.76	.26	2.59	.88	.49	4.98
Fast	2400	800	20.00	.94	.40	2.50	.71	.27	4.82
Brush 1 coat, cut-in, paint grade base									
Enamel, water base (material #9)									
Slow	500	1500	24.50	2.10	.60	1.63	1.34	.85	6.52
Medium	550	1350	21.40	3.05	1.06	1.59	1.40	.78	7.88
Fast	600	1200	18.40	3.75	1.58	1.53	1.27	.49	8.62
Enamel, oil base (material #10)									
Slow	500	1500	26.60	2.10	.60	1.77	1.39	.88	6.74
Medium	550	1350	23.30	3.05	1.06	1.73	1.43	.80	8.07
Fast	600	1200	20.00	3.75	1.58	1.67	1.29	.50	8.79
Spray 1 coat, stain in boneyard, stain grade base									
Wiping stain (material #11a)									
Slow	--	--	--	--	--	--	--	--	--
Medium	4000	1350	27.00	.42	.14	2.00	.63	.35	3.54
Fast	5000	1200	23.10	.45	.19	1.93	.48	.18	3.23

Baseboard measurements are based on square feet of floor area. Use these figures for 1-1/2 inch to 3 inch stock, painted or stained on one side. Stain grade base is to be stained in a boneyard. Typically, finger joint stock is paint grade and butt joint stock is stain grade. These figures include minimal preparation time and material. Add for extensive preparation. "Slow" work is based on a $10.50 hourly wage, "Medium" work on a $16.75 hourly wage, and "Fast" work on a $22.50 hourly wage. Other qualifications that apply to this table are on page 9.

	Labor LF per manhour	Material coverage LF/gallon	Material cost per gallon	Labor cost per 100 LF	Labor burden 100 LF	Material cost per 100 LF	Overhead per 100 LF	Profit per 100 LF	Total price per 100 LF
Beams, per linear foot, heights to 13 feet									
Solid body stain, water base (material #18)									
Roll & brush each coat									
Slow	35	50	21.50	30.00	8.51	43.00	25.28	16.02	122.81
Medium	40	45	18.80	41.88	14.60	41.78	24.07	13.46	135.79
Fast	45	40	16.10	50.00	20.86	40.25	20.56	7.90	139.57
Solid body stain, oil base (material #19)									
Roll & brush each coat									
Slow	35	50	23.90	30.00	8.51	47.80	26.77	16.97	130.05
Medium	40	45	20.90	41.88	14.60	46.44	25.22	14.10	142.24
Fast	45	40	17.90	50.00	20.86	44.75	21.39	8.22	145.22
Semi-transparent stain, water base (material #20)									
Roll & brush each coat									
Slow	40	55	21.40	26.25	7.45	38.91	22.52	14.27	109.40
Medium	45	50	18.70	37.22	12.97	37.40	21.46	12.00	121.05
Fast	50	45	16.00	45.00	18.78	35.56	18.38	7.06	124.78
Semi-transparent stain, oil base (material #21)									
Roll & brush each coat									
Slow	40	55	26.10	26.25	7.45	47.45	25.16	15.95	122.26
Medium	45	50	22.90	37.22	12.97	45.80	23.52	13.15	132.66
Fast	50	45	19.60	45.00	18.78	43.56	19.86	7.63	134.83

Beam measurements are based on linear feet of installed 4" x 6" to 8" x 14" beams. High time difficulty factors are already figured into the formula. "Slow" work is based on a $10.50 hourly wage, "Medium" work on a $16.75 hourly wage, and "Fast" work on a $22.50 hourly wage. Other qualifications that apply to this table are on page 9.

	Labor LF per manhour	Material coverage LF/gallon	Material cost per gallon	Labor cost per 100 LF	Labor burden 100 LF	Material cost per 100 LF	Overhead per 100 LF	Profit per 100 LF	Total price per 100 LF
Beams, per linear foot, heights from 13 to 17 feet									
Solid body stain, water base (material #18)									
Roll & brush each coat									
Slow	24	50	21.50	43.75	12.42	43.00	30.75	19.49	149.41
Medium	27	45	18.80	62.04	21.63	41.78	30.73	17.18	173.36
Fast	30	40	16.10	75.00	31.29	40.25	27.11	10.42	184.07
Solid body stain, oil base (material #19)									
Roll & brush each coat									
Slow	24	50	23.90	43.75	12.42	47.80	32.24	20.44	156.65
Medium	27	45	20.90	62.04	21.63	46.44	31.87	17.82	179.80
Fast	30	40	17.90	75.00	31.29	44.75	27.95	10.74	189.73
Semi-transparent stain, water base (material #20)									
Roll & brush each coat									
Slow	28	55	21.40	37.50	10.64	38.91	26.99	17.11	131.15
Medium	31	50	18.70	54.03	18.85	37.40	27.01	15.10	152.39
Fast	34	45	16.00	66.18	27.61	35.56	23.93	9.20	162.48
Semi-transparent stain, oil base (material #21)									
Roll & brush each coat									
Slow	28	55	26.10	37.50	10.64	47.45	29.64	18.79	144.02
Medium	31	50	22.90	54.03	18.85	45.80	29.07	16.25	164.00
Fast	34	45	19.60	66.18	27.61	43.56	25.41	9.77	172.53

Beam measurements are based on linear feet of installed 4" x 6" to 8" x 14" beams. High time difficulty factors are already figured into the formula. "Slow" work is based on a $10.50 hourly wage, "Medium" work on a $16.75 hourly wage, and "Fast" work on a $22.50 hourly wage. Other qualifications that apply to this table are on page 9.

	Labor LF per manhour	Material coverage LF/gallon	Material cost per gallon	Labor cost per 100 LF	Labor burden 100 LF	Material cost per 100 LF	Overhead per 100 LF	Profit per 100 LF	Total price per 100 LF
Beams, per linear foot, heights above 18 feet									
Solid body stain, water base (material #18)									
Roll & brush each coat									
Slow	16	50	21.50	65.63	18.62	43.00	39.46	25.01	191.72
Medium	18	45	18.80	93.06	32.45	41.78	40.98	22.91	231.18
Fast	20	40	16.10	112.50	46.95	40.25	36.95	14.20	250.85
Solid body stain, oil base (material #19)									
Roll & brush each coat									
Slow	16	50	23.90	65.63	18.62	47.80	40.95	25.96	198.96
Medium	18	45	20.90	93.06	32.45	46.44	42.12	23.55	237.62
Fast	20	40	17.90	112.50	46.95	44.75	37.78	14.52	256.50
Semi-transparent stain, water base (material #20)									
Roll & brush each coat									
Slow	19	55	21.40	55.26	15.69	38.91	34.07	21.59	165.52
Medium	21	50	18.70	79.76	27.81	37.40	35.52	19.85	200.34
Fast	23	45	16.00	97.83	40.83	35.56	32.23	12.39	218.84
Semi-transparent stain, oil base (material #21)									
Roll & brush each coat									
Slow	19	55	26.10	55.26	15.69	47.45	36.71	23.27	178.38
Medium	21	50	22.90	79.76	27.81	45.80	37.57	21.00	211.94
Fast	23	45	19.60	97.83	40.83	43.56	33.71	12.96	228.89

Beam measurements are based on linear feet of installed 4" x 6" to 8" x 14" beams. High time difficulty factors are already figured into the formula. "Slow" work is based on a $10.50 hourly wage, "Medium" work on a $16.75 hourly wage, and "Fast" work on a $22.50 hourly wage. Other qualifications that apply to this table are on page 9.

	Labor SF per manhour	Material coverage SF/gallon	Material cost per gallon	Labor cost per 100 SF	Labor burden 100 SF	Material cost per 100 SF	Overhead per 100 SF	Profit per 100 SF	Total price per 100 SF

Bookcases and shelves, paint grade, brush application

	Labor SF per manhour	Material coverage SF/gallon	Material cost per gallon	Labor cost per 100 SF	Labor burden 100 SF	Material cost per 100 SF	Overhead per 100 SF	Profit per 100 SF	Total price per 100 SF
Undercoat, water base (material #3)									
Brush 1 coat									
Slow	25	300	21.10	42.00	11.92	7.03	18.90	11.98	91.83
Medium	30	280	18.50	55.83	19.46	6.61	20.07	11.22	113.19
Fast	35	260	15.80	64.29	26.82	6.08	17.98	6.91	122.08
Undercoat, oil base (material #4)									
Brush 1 coat									
Slow	25	340	21.60	42.00	11.92	6.35	18.69	11.85	90.81
Medium	30	318	18.90	55.83	19.46	5.94	19.90	11.12	112.25
Fast	35	295	16.20	64.29	26.82	5.49	17.87	6.87	121.34
Split coat (1/2 undercoat + 1/2 enamel), water base (material #3 + #9)									
Brush each coat									
Slow	40	350	22.80	26.25	7.45	6.51	12.47	7.91	60.59
Medium	45	328	19.95	37.22	12.97	6.08	13.79	7.71	77.77
Fast	50	305	17.10	45.00	18.78	5.61	12.84	4.93	87.16
Split coat (1/2 undercoat + 1/2 enamel), oil base (material #4 + #10)									
Brush each coat									
Slow	40	350	24.10	26.25	7.45	6.89	12.59	7.98	61.16
Medium	45	328	19.95	37.22	12.97	6.08	13.79	7.71	77.77
Fast	50	305	17.10	45.00	18.78	5.61	12.84	4.93	87.16
Enamel, water base (material #9)									
Brush 1st finish coat									
Slow	35	340	24.50	30.00	8.51	7.21	14.18	8.99	68.89
Medium	40	318	21.40	41.88	14.60	6.73	15.49	8.66	87.36
Fast	45	295	18.40	50.00	20.86	6.24	14.27	5.48	96.85
Brush 2nd or additional finish coats									
Slow	40	350	24.50	26.25	7.45	7.00	12.62	8.00	61.32
Medium	45	328	21.40	37.22	12.97	6.52	13.89	7.77	78.37
Fast	50	305	18.40	45.00	18.78	6.03	12.92	4.96	87.69
Enamel, oil base (material #10)									
Brush 1st finish coat									
Slow	35	340	26.60	30.00	8.51	7.82	14.37	9.11	69.81
Medium	40	318	23.30	41.88	14.60	7.33	15.63	8.74	88.18
Fast	45	295	20.00	50.00	20.86	6.78	14.37	5.52	97.53
Brush 2nd or additional finish coats									
Slow	40	350	26.60	26.25	7.45	7.60	12.81	8.12	62.23
Medium	45	318	23.30	37.22	12.97	7.33	14.09	7.88	79.49
Fast	50	305	20.00	45.00	18.78	6.56	13.01	5.00	88.35

Bookcase and shelf estimates are based on overall dimensions (length times width) to 8 feet high and include painting all exposed surfaces (including stiles, interior shelves and backs). For heights above 8 feet, use the High Time Difficulty Factors on page 137. "Slow" work is based on a $10.50 hourly wage, "Medium" work on a $16.75 hourly wage, and "Fast" work on a $22.50 hourly wage. Other qualifications that apply to this table are on page 9.

	Labor SF per manhour	Material coverage SF/gallon	Material cost per gallon	Labor cost per 100 SF	Labor burden 100 SF	Material cost per 100 SF	Overhead per 100 SF	Profit per 100 SF	Total price per 100 SF
Bookcases and shelves, paint grade, spray application									
Undercoat, water base (material #3)									
Spray 1 coat									
Slow	150	145	21.10	7.00	1.99	14.55	7.30	4.63	35.47
Medium	165	133	18.50	10.15	3.54	13.91	6.76	3.78	38.14
Fast	175	120	15.80	12.86	5.35	13.17	5.81	2.23	39.42
Undercoat, oil base (material #4)									
Spray 1 coat									
Slow	150	145	21.60	7.00	1.99	14.90	7.41	4.70	36.00
Medium	165	133	18.90	10.15	3.54	14.21	6.84	3.82	38.56
Fast	175	120	16.20	12.86	5.35	13.50	5.87	2.26	39.84
Split coat (1/2 undercoat + 1/2 enamel), water base (material #3 + #9)									
Spray each coat									
Slow	245	195	22.80	4.29	1.21	11.69	5.33	3.38	25.90
Medium	270	183	19.95	6.20	2.16	10.90	4.72	2.64	26.62
Fast	295	170	17.10	7.63	3.18	10.06	3.86	1.48	26.21
Split coat (1/2 undercoat + 1/2 enamel), oil base (material #4 + #10)									
Spray each coat									
Slow	245	195	24.10	4.29	1.21	12.36	5.54	3.51	26.91
Medium	270	183	21.10	6.20	2.16	11.53	4.87	2.72	27.48
Fast	295	170	18.10	7.63	3.18	10.65	3.97	1.53	26.96
Enamel, water base (material #9)									
Spray 1st finish coat									
Slow	225	170	24.50	4.67	1.32	14.41	6.33	4.01	30.74
Medium	250	158	21.40	6.70	2.34	13.54	5.53	3.09	31.20
Fast	275	145	18.40	8.18	3.43	12.69	4.49	1.73	30.52
Spray 2nd or additional finish coats									
Slow	245	195	24.50	4.29	1.21	12.56	5.60	3.55	27.21
Medium	270	183	21.40	6.20	2.16	11.69	4.91	2.75	27.71
Fast	295	170	18.40	7.63	3.18	10.82	4.00	1.54	27.17
Enamel, oil base (material #10)									
Spray 1st finish coat									
Slow	225	170	26.60	4.67	1.32	15.65	6.71	4.25	32.60
Medium	250	158	23.30	6.70	2.34	14.75	5.83	3.26	32.88
Fast	275	145	20.00	8.18	3.43	13.79	4.70	1.81	31.91
Spray 2nd or additional finish coats									
Slow	245	195	26.60	4.29	1.21	13.64	5.94	3.76	28.84
Medium	270	183	23.30	6.20	2.16	12.73	5.17	2.89	29.15
Fast	295	170	20.00	7.63	3.18	11.76	4.18	1.61	28.36

Bookcase and shelf estimates are based on overall dimensions (length times width) to 8 feet high and include painting all exposed surfaces (including stiles, interior shelves and backs). For heights above 8 feet, use the High Time Difficulty Factors on page 137. "Slow" work is based on a $10.50 hourly wage, "Medium" work on a $16.75 hourly wage, and "Fast" work on a $22.50 hourly wage. Other qualifications that apply to this table are on page 9.

	Labor SF per manhour	Material coverage SF/gallon	Material cost per gallon	Labor cost per 100 SF	Labor burden 100 SF	Material cost per 100 SF	Overhead per 100 SF	Profit per 100 SF	Total price per 100 SF
Bookcases and shelves, stain grade									
Stain, seal & lacquer (7 step process)									
STEP 1: Sand & putty;									
Slow	100	--	--	10.50	2.98	--	4.18	2.65	20.31
Medium	125	--	--	13.40	4.67	--	4.43	2.48	24.98
Fast	150	--	--	15.00	6.27	--	3.93	1.51	26.71
STEP 2 & 3: Stain (material #11a) & wipe									
Brush 1 coat & wipe									
Slow	75	500	30.90	14.00	3.97	6.18	7.49	4.75	36.39
Medium	85	475	27.00	19.71	6.86	5.68	7.90	4.42	44.57
Fast	95	450	23.10	23.68	9.90	5.13	7.16	2.75	48.62
Spray 1 coat & wipe									
Slow	300	175	30.90	3.50	.99	17.66	6.87	4.35	33.37
Medium	400	138	27.00	4.19	1.46	19.57	6.18	3.45	34.85
Fast	500	100	23.10	4.50	1.88	23.10	5.45	2.10	37.03
STEP 4: Sanding sealer (material #11b)									
Brush 1 coat									
Slow	130	550	17.20	8.08	2.29	3.13	4.19	2.66	20.35
Medium	140	525	15.00	11.96	4.17	2.86	4.65	2.60	26.24
Fast	150	500	12.90	15.00	6.27	2.58	4.41	1.70	29.96
Spray 1 coat									
Slow	375	175	17.20	2.80	.80	9.83	4.16	2.64	20.23
Medium	475	138	15.00	3.53	1.24	10.87	3.83	2.14	21.61
Fast	575	100	12.90	3.91	1.64	12.90	3.41	1.31	23.17
STEP 5: Sand lightly									
Slow	175	--	--	6.00	1.70	--	2.39	1.52	11.61
Medium	225	--	--	7.44	2.59	--	2.46	1.37	13.86
Fast	275	--	--	8.18	3.43	--	2.15	.83	14.59
STEP 6 & 7: Lacquer (material #11c), 2 coats									
Brush 1st coat									
Slow	140	400	19.30	7.50	2.12	4.83	4.48	2.84	21.77
Medium	185	375	16.80	9.05	3.17	4.48	4.09	2.28	23.07
Fast	245	350	14.40	9.18	3.83	4.11	3.17	1.22	21.51
Brush 2nd coat									
Slow	155	425	19.30	6.77	1.92	4.54	4.10	2.60	19.93
Medium	208	413	16.80	8.05	2.82	4.07	3.66	2.04	20.64
Fast	260	400	14.40	8.65	3.63	3.60	2.93	1.13	19.94
Spray 1st coat									
Slow	340	175	19.30	3.09	.87	11.03	4.65	2.95	22.59
Medium	458	138	16.80	3.66	1.26	12.17	4.19	2.34	23.62
Fast	575	100	14.40	3.91	1.64	14.40	3.69	1.42	25.06

	Labor SF per manhour	Material coverage SF/gallon	Material cost per gallon	Labor cost per 100 SF	Labor burden 100 SF	Material cost per 100 SF	Overhead per 100 SF	Profit per 100 SF	Total price per 100 SF
Spray 2nd coat									
Slow	430	200	19.30	2.44	.70	9.65	3.96	2.51	19.26
Medium	530	163	16.80	3.16	1.11	10.31	3.57	2.00	20.15
Fast	630	125	14.40	3.57	1.50	11.52	5.14	3.26	24.99
Complete 7 step stain, seal & lacquer process (material #11)									
Brush all coats									
Slow	30	160	21.60	35.00	9.93	13.50	18.12	11.49	88.04
Medium	35	150	19.60	47.86	16.68	13.07	24.06	15.25	116.92
Fast	40	140	16.20	56.25	23.48	11.57	22.37	12.50	126.17
Spray all coats									
Slow	65	60	21.60	16.15	4.58	36.00	13.90	7.77	78.40
Medium	83	48	19.60	20.18	7.04	40.83	21.09	13.37	102.51
Fast	100	35	16.20	22.50	9.39	46.29	19.15	10.71	108.04
Shellac, clear (material #12)									
Brush each coat									
Slow	205	570	25.30	5.12	1.46	4.44	3.42	2.17	16.61
Medium	230	545	22.10	7.28	2.55	4.06	3.40	1.90	19.19
Fast	255	520	18.90	8.82	3.68	3.63	2.98	1.15	20.26
Varnish, flat or gloss (material #30c)									
Brush each coat									
Slow	175	450	35.40	6.00	1.70	7.87	4.83	3.06	23.46
Medium	200	438	31.00	8.38	2.92	7.08	4.50	2.52	25.40
Fast	225	425	26.60	10.00	4.16	6.26	3.78	1.45	25.65
Penetrating stain wax (material #14) & polish									
Brush 1st coat									
Slow	150	595	27.20	7.00	1.99	4.57	4.20	2.66	20.42
Medium	175	558	23.80	9.57	3.33	4.27	4.21	2.35	23.73
Fast	200	520	20.40	11.25	4.70	3.92	3.68	1.41	24.96
Brush 2nd or additional coats									
Slow	175	600	27.20	6.00	1.70	4.53	3.00	1.68	16.91
Medium	200	575	23.80	8.38	2.92	4.14	4.79	3.03	23.26
Fast	225	550	20.40	10.00	4.16	3.71	4.38	2.45	24.70

Bookcase and shelf estimates are based on overall dimensions (length times width) to 8 feet high and include painting all exposed surfaces (including stiles, interior shelves and backs). For heights above 8 feet, use the High Time Difficulty Factors on page 137. "Slow" work is based on a $10.50 hourly wage, "Medium" work on a $16.75 hourly wage, and "Fast" work on a $22.50 hourly wage. Other qualifications that apply to this table are on page 9.

	Labor SF per manhour	Material coverage SF/gallon	Material cost per gallon	Labor cost per 100 SF	Labor burden 100 SF	Material cost per 100 SF	Overhead per 100 SF	Profit per 100 SF	Total price per 100 SF
Cabinet backs, paint grade, brush									
Flat latex, water base (material #5)									
Brush each coat									
Slow	150	300	17.10	7.00	1.99	5.70	4.55	2.89	22.13
Medium	175	275	15.00	9.57	3.33	5.45	4.50	2.51	25.36
Fast	200	250	12.90	11.25	4.70	5.16	3.91	1.50	26.52
Enamel, water base (material #9)									
Brush each coat									
Slow	125	275	24.50	8.40	2.38	8.91	6.11	3.87	29.67
Medium	150	250	21.40	11.17	3.90	8.56	5.79	3.24	32.66
Fast	175	225	18.40	12.86	5.35	8.18	4.89	1.88	33.16
Enamel, oil base (material #10)									
Brush each coat									
Slow	125	275	26.60	8.40	2.38	9.67	6.34	4.02	30.81
Medium	150	250	23.30	11.17	3.90	9.32	5.97	3.34	33.70
Fast	175	225	20.00	12.86	5.35	8.89	5.02	1.93	34.05

Cabinet back estimates are based on overall dimensions (length times width) to 8 feet high and include painting the inside back wall of paint grade or stain grade cabinets. For heights above 8 feet, use the High Time Difficulty Factors on page 137. Measurements are based on total area of cabinet faces. "Slow" work is based on a $10.50 hourly wage, "Medium" work on a $16.75 hourly wage, and "Fast" work on a $22.50 hourly wage. Other qualifications that apply to this table are on page 9.

	Labor SF per manhour	Material coverage SF/gallon	Material cost per gallon	Labor cost per 100 SF	Labor burden 100 SF	Material cost per 100 SF	Overhead per 100 SF	Profit per 100 SF	Total price per 100 SF
Cabinet faces, stain grade									
Complete 7 step stain, seal & 2 coat lacquer system (material #11)									
Brush all coats									
Slow	30	190	21.60	35.00	9.93	11.37	17.46	11.07	84.83
Medium	40	178	19.60	41.88	14.60	11.01	16.54	9.24	93.27
Fast	50	165	16.20	45.00	18.78	9.82	13.62	5.23	92.45
Spray all coats									
Slow	85	67	21.60	12.35	3.50	32.24	14.91	9.45	72.45
Medium	110	51	19.60	15.23	5.30	38.43	14.45	8.08	81.49
Fast	135	35	16.20	16.67	6.96	46.29	12.94	4.97	87.83

Cabinet face estimates are based on overall dimensions (length times width) to 8 feet high. Use these figures to estimate finishing the faces of stain grade kitchen, bar, linen, pullman or vanity cabinets. For the stain, seal and lacquer process, the figures include finishing the cabinet faces only. Preparation is not included. For heights above 8 feet, use the High Time Difficulty Factors on page 137. Measurements are based on total area of cabinet faces. "Slow" work is based on a $10.50 hourly wage, "Medium" work on a $16.75 hourly wage, and "Fast" work on a $22.50 hourly wage. Other qualifications that apply to this table are on page 9.

	Labor SF per manhour	Material coverage SF/gallon	Material cost per gallon	Labor cost per 100 SF	Labor burden 100 SF	Material cost per 100 SF	Overhead per 100 SF	Profit per 100 SF	Total price per 100 SF
Cabinets, paint grade, roll and brush									
Undercoat, water base (material #3)									
Roll & brush, 1 coat									
Slow	75	260	21.10	14.00	3.97	8.12	8.09	5.13	39.31
Medium	93	250	18.50	18.01	6.27	7.40	7.76	4.34	43.78
Fast	110	240	15.80	20.45	8.54	6.58	6.58	2.53	44.68
Undercoat, oil base (material #4)									
Roll & brush, 1 coat									
Slow	75	275	21.60	14.00	3.97	7.85	8.01	5.08	38.91
Medium	93	268	18.90	18.01	6.27	7.05	7.68	4.29	43.30
Fast	110	250	16.20	20.45	8.54	6.48	6.56	2.52	44.55
Split coat (1/2 undercoat + 1/2 enamel), water base (material #3 + #9)									
Roll & brush each coat									
Slow	95	310	22.80	11.05	3.14	7.35	6.68	4.23	32.45
Medium	113	298	19.95	14.82	5.17	6.69	6.53	3.65	36.86
Fast	130	285	17.10	17.31	7.21	6.00	5.65	2.17	38.34
Split coat (1/2 undercoat + 1/2 enamel), oil base (material #4 + #10)									
Roll & brush each coat									
Slow	95	310	24.10	11.05	3.14	7.77	6.81	4.32	33.09
Medium	113	298	21.10	14.82	5.17	7.08	6.63	3.71	37.41
Fast	130	285	18.10	17.31	7.21	6.35	5.71	2.20	38.78
Enamel, water base (material #9)									
Roll & brush 1st finish coat									
Slow	85	300	24.50	12.35	3.50	8.17	7.45	4.72	36.19
Medium	103	288	21.40	16.26	5.67	7.43	7.19	4.02	40.57
Fast	120	275	18.40	18.75	7.81	6.69	6.15	2.37	41.77
Roll & brush 2nd or additional finish coats									
Slow	95	310	24.50	11.05	3.14	7.90	6.85	4.34	33.28
Medium	113	298	21.40	14.82	5.17	7.18	6.65	3.72	37.54
Fast	130	285	18.40	17.31	7.21	6.46	5.74	2.20	38.92
Enamel, oil base (material #10)									
Roll & brush 1st finish coat									
Slow	85	300	26.60	12.35	3.50	8.87	7.67	4.86	37.25
Medium	103	288	23.30	16.26	5.67	8.09	7.35	4.11	41.48
Fast	120	275	20.00	18.75	7.81	7.27	6.26	2.41	42.50
Roll & brush 2nd or additional finish coats									
Slow	95	310	26.60	11.05	3.14	8.58	7.06	4.47	34.30
Medium	113	298	23.30	14.82	5.17	7.82	6.81	3.81	38.43
Fast	130	285	20.00	17.31	7.21	7.02	5.84	2.24	39.62

Cabinet estimates are based on overall dimensions (length times width) to 8 feet high and include painting the cabinet face, back of doors, stiles and rails. See Cabinet backs for painting the inside back wall of the cabinets. Use these figures to estimate paint grade kitchen cabinets. Use the opening count method to estimate paint grade pullmans, vanities, bars or linen cabinets. For heights above 8 feet, apply the High time difficulty factor to labor costs and the labor burden cost categories and add these figures to the total cost. Measurements are based on total area of cabinet faces. "Slow" work is based on a $10.50 hourly wage, "Medium" work on a $16.75 hourly wage, and "Fast" work on a $22.50 hourly wage. Other qualifications that apply to this table are on page 9.

	Labor SF per manhour	Material coverage SF/gallon	Material cost per gallon	Labor cost per 100 SF	Labor burden 100 SF	Material cost per 100 SF	Overhead per 100 SF	Profit per 100 SF	Total price per 100 SF
Cabinets, paint grade, spray application									
Undercoat, water base (material #3)									
Spray 1 coat									
Slow	125	125	21.10	8.40	2.38	16.88	8.58	5.44	41.68
Medium	140	113	18.50	11.96	4.17	16.37	7.96	4.45	44.91
Fast	155	100	15.80	14.52	6.05	15.80	6.73	2.59	45.69
Undercoat, oil base (material #4)									
Spray 1 coat									
Slow	125	135	21.60	8.40	2.38	16.00	8.30	5.26	40.34
Medium	140	123	18.90	11.96	4.17	15.37	7.72	4.31	43.53
Fast	155	110	16.20	14.52	6.05	14.73	6.53	2.51	44.34
Split coat (1/2 undercoat + 1/2 enamel), water base (material #3 + #9)									
Spray each coat									
Slow	200	175	22.80	5.25	1.49	13.03	6.13	3.89	29.79
Medium	225	163	19.95	7.44	2.59	12.24	5.46	3.05	30.78
Fast	250	150	17.10	9.00	3.76	11.40	4.47	1.72	30.35
Split coat (1/2 undercoat + 1/2 enamel), oil base (material #4 + #10)									
Spray each coat									
Slow	200	175	24.10	5.25	1.49	13.77	6.36	4.03	30.90
Medium	225	163	21.10	7.44	2.59	12.94	5.63	3.15	31.75
Fast	250	150	18.10	9.00	3.76	12.07	4.59	1.77	31.19
Enamel, water base (material #9)									
Spray 1st finish coat									
Slow	185	150	24.50	5.68	1.61	16.33	7.33	4.64	35.59
Medium	210	138	21.40	7.98	2.77	15.51	6.44	3.60	36.30
Fast	235	125	18.40	9.57	4.02	14.72	5.23	2.01	35.55
Spray 2nd or additional finish coats									
Slow	200	175	24.50	5.25	1.49	14.00	6.43	4.08	31.25
Medium	225	163	21.40	7.44	2.59	13.13	5.67	3.17	32.00
Fast	250	150	18.40	9.00	3.76	12.27	4.63	1.78	31.44
Enamel, oil base (material #10)									
Spray 1st finish coat									
Slow	185	160	26.60	5.68	1.61	16.63	7.42	4.70	36.04
Medium	210	148	23.30	7.98	2.77	15.74	6.49	3.63	36.61
Fast	235	135	20.00	9.57	4.02	14.81	5.25	2.02	35.67
Spray 2nd or additional finish coats									
Slow	200	185	26.60	5.25	1.49	14.38	6.55	4.15	31.82
Medium	225	173	23.30	7.44	2.59	13.47	5.76	3.22	32.48
Fast	250	160	20.00	9.00	3.76	12.50	4.67	1.80	31.73

Cabinet estimates are based on overall dimensions (length times width) to 8 feet high and include painting the cabinet face, back of doors, stiles and rails. See Cabinet backs for painting the inside back wall of the cabinets. Use these figures to estimate paint grade kitchen cabinets. Use the opening count method to estimate paint grade pullmans, vanities, bars or linen cabinets. For heights above 8 feet, apply the High time difficulty factor to labor costs and the labor burden cost categories and add these figures to the total cost. Measurements are based on total area of cabinet faces. "Slow" work is based on a $10.50 hourly wage, "Medium" work on a $16.75 hourly wage, and "Fast" work on a $22.50 hourly wage. Other qualifications that apply to this table are on page 9.

	Labor SF per manhour	Material coverage SF/gallon	Material cost per gallon	Labor cost per 100 SF	Labor burden 100 SF	Material cost per 100 SF	Overhead per 100 SF	Profit per 100 SF	Total price per 100 SF
Cabinets, stain grade									
Stain, seal & 2 coats lacquer system (7 step process)									
STEP 1: Sand & putty;									
Slow	125	--	--	8.40	2.38	--	3.34	2.12	16.24
Medium	150	--	--	11.17	3.90	--	3.69	2.06	20.82
Fast	175	--	--	12.86	5.35	--	3.37	1.30	22.88
STEP 2 & 3: Stain (material #11a) & wipe									
Brush 1 coat & wipe									
Slow	65	450	30.90	16.15	4.58	6.87	8.56	5.43	41.59
Medium	75	400	27.00	22.33	7.78	6.75	9.03	5.05	50.94
Fast	85	350	23.10	26.47	11.03	6.60	8.16	3.14	55.40
Spray 1 coat & wipe									
Slow	250	175	30.90	4.20	1.19	17.66	7.15	4.53	34.73
Medium	350	138	27.00	4.79	1.67	19.57	6.38	3.57	35.98
Fast	450	100	23.10	5.00	2.08	23.10	5.59	2.15	37.92
STEP 4: Sanding sealer (material #11b)									
Brush 1 coat									
Slow	110	450	17.20	9.55	2.70	3.82	4.99	3.16	24.22
Medium	120	425	15.00	13.96	4.86	3.53	5.48	3.06	30.89
Fast	130	400	12.90	17.31	7.21	3.23	5.14	1.97	34.86
Spray 1 coat									
Slow	330	175	17.20	3.18	.90	9.83	4.31	2.73	20.95
Medium	430	138	15.00	3.90	1.36	10.87	3.95	2.21	22.29
Fast	530	100	12.90	4.25	1.78	12.90	3.50	1.35	23.78
STEP 5: Sand lightly									
Slow	200	--	--	5.25	1.49	--	2.09	1.32	10.15
Medium	250	--	--	6.70	2.34	--	2.21	1.24	12.49
Fast	300	--	--	7.50	3.12	--	1.97	.76	13.35
STEP 6 & 7: Lacquer (material #11c), 2 coats									
Brush 1st coat									
Slow	120	375	19.30	8.75	2.48	5.15	5.08	3.22	24.68
Medium	165	350	16.80	10.15	3.54	4.80	4.53	2.53	25.55
Fast	215	325	14.40	10.47	4.36	4.43	3.56	1.37	24.19
Brush 2nd coat									
Slow	130	400	19.30	8.08	2.29	4.83	4.72	2.99	22.91
Medium	173	388	16.80	9.68	3.38	4.33	4.26	2.38	24.03
Fast	225	375	14.40	10.00	4.16	3.84	3.33	1.28	22.61
Spray 1st coat									
Slow	275	150	19.30	3.82	1.09	12.87	5.51	3.49	26.78
Medium	388	100	16.80	4.32	1.51	16.80	5.54	3.10	31.27
Fast	500	75	14.40	4.50	1.88	19.20	4.73	1.82	32.13

	Labor SF per manhour	Material coverage SF/gallon	Material cost per gallon	Labor cost per 100 SF	Labor burden 100 SF	Material cost per 100 SF	Overhead per 100 SF	Profit per 100 SF	Total price per 100 SF
Spray 2nd coat									
Slow	350	200	19.30	3.00	.86	9.65	4.19	2.65	20.35
Medium	475	163	16.80	3.53	1.24	10.31	3.69	2.06	20.83
Fast	600	125	14.40	3.75	1.58	11.52	3.12	1.20	21.17
Complete 7 step stain, seal & 2 coat lacquer system (material #11)									
Brush all coats									
Slow	20	125	21.60	52.50	14.90	17.28	26.26	16.65	127.59
Medium	25	113	19.60	67.00	23.36	17.35	26.39	14.75	148.85
Fast	30	100	16.20	75.00	31.29	16.20	22.66	8.71	153.86
Spray all coats									
Slow	40	40	21.60	26.25	7.45	54.00	27.19	17.24	132.13
Medium	50	30	19.60	33.50	11.68	65.33	27.07	15.13	152.71
Fast	60	21	16.20	37.50	15.66	77.14	24.11	9.26	163.67
Shellac, clear (material #12)									
Brush each coat									
Slow	175	525	25.30	6.00	1.70	4.82	3.07	1.72	17.31
Medium	200	513	22.10	8.38	2.92	4.31	2.89	1.11	19.61
Fast	225	500	18.90	10.00	4.16	3.78	3.32	1.28	22.54
Varnish, flat or gloss (material #30c)									
Brush each coat									
Slow	155	475	35.40	6.77	1.92	7.45	2.99	1.15	20.28
Medium	180	463	31.00	9.31	3.25	6.70	5.97	3.78	29.01
Fast	205	450	26.60	10.98	4.58	5.91	5.26	2.94	29.67
Penetrating stain wax (material #14) & polish									
Brush 1st coat									
Slow	125	575	27.20	8.40	2.38	4.73	2.87	1.10	19.48
Medium	150	538	23.80	11.17	3.90	4.42	6.04	3.83	29.36
Fast	175	500	20.40	12.86	5.35	4.08	5.47	3.06	30.82
Brush 2nd or additional coats									
Slow	150	600	27.20	7.00	1.99	4.53	2.50	.96	16.98
Medium	175	575	23.80	9.57	3.33	4.14	3.15	1.21	21.40
Fast	200	550	20.40	11.25	4.70	3.71	6.09	3.86	29.61

Cabinet estimates are based on overall dimensions (length times width) to 8 feet high. Use these figures to estimate stain grade kitchen, bar, linen, pullman or vanity cabinets. For the stain, seal and lacquer process, the figures include finishing both sides of cabinet doors, stiles and rails with a fog coat of stain on shelves and the wall behind the cabinet (cabinet back). See Cabinet backs for painting the inside back wall of the cabinets. For heights above 8 feet, use the High Time Difficulty Factors on page 137. Measurements are based on total area of cabinet faces. "Slow" work is based on a $10.50 hourly wage, "Medium" work on a $16.75 hourly wage, and "Fast" work on a $22.50 hourly wage. Other qualifications that apply to this table are on page 9.

	Labor SF per manhour	Material coverage SF/gallon	Material cost per gallon	Labor cost per 100 SF	Labor burden 100 SF	Material cost per 100 SF	Overhead per 100 SF	Profit per 100 SF	Total price per 100 SF
Ceiling panels, suspended, fiber panels in T-bar frames, brush application									
Flat latex, water base (material #5)									
Brush 1st coat									
Slow	80	260	17.10	13.13	3.72	6.58	7.27	4.61	35.31
Medium	110	230	15.00	15.23	5.30	6.52	6.63	3.71	37.39
Fast	140	200	12.90	16.07	6.70	6.45	5.41	2.08	36.71
Brush 2nd or additional coats									
Slow	130	300	17.10	8.08	2.29	5.70	4.98	3.16	24.21
Medium	150	275	15.00	11.17	3.90	5.45	5.02	2.81	28.35
Fast	170	250	12.90	13.24	5.51	5.16	4.43	1.70	30.04
Enamel, water base (material #9)									
Brush 1st coat									
Slow	65	260	24.50	16.15	4.58	9.42	9.35	5.93	45.43
Medium	100	230	21.40	16.75	5.84	9.30	7.81	4.37	44.07
Fast	125	200	18.40	18.00	7.51	9.20	6.42	2.47	43.60
Brush 2nd or additional coats									
Slow	115	300	24.50	9.13	2.60	8.17	6.17	3.91	29.98
Medium	135	275	21.40	12.41	4.33	7.78	6.00	3.36	33.88
Fast	155	250	18.40	14.52	6.05	7.36	5.17	1.99	35.09
Enamel, oil base (material #10)									
Brush 1st coat									
Slow	65	250	26.60	16.15	4.58	10.64	9.73	6.17	47.27
Medium	95	213	23.30	17.63	6.16	10.94	8.50	4.75	47.98
Fast	125	175	20.00	18.00	7.51	11.43	6.84	2.63	46.41
Brush 2nd or additional coats									
Slow	115	275	26.60	9.13	2.60	9.67	6.63	4.20	32.23
Medium	135	260	23.30	12.41	4.33	8.96	6.29	3.52	35.51
Fast	155	240	20.00	14.52	6.05	8.33	5.35	2.06	36.31

Ceiling panel estimates are based on overall dimensions (length times width) to 8 feet high. For heights above 8 feet, use the High Time Difficulty Factors on page 137. "Slow" work is based on a $10.50 hourly wage, "Medium" work on a $16.75 hourly wage, and "Fast" work on a $22.50 hourly wage. Other qualifications that apply to this table are on page 9.

	Labor SF per manhour	Material coverage SF/gallon	Material cost per gallon	Labor cost per 100 SF	Labor burden 100 SF	Material cost per 100 SF	Overhead per 100 SF	Profit per 100 SF	Total price per 100 SF
Ceiling panels, suspended, fiber panels in T-bar frame, roll application									
Flat latex, water base (material #5)									
Roll 1st coat									
Slow	150	270	17.10	7.00	1.99	6.33	4.75	3.01	23.08
Medium	215	235	15.00	7.79	2.71	6.38	4.14	2.31	23.33
Fast	280	200	12.90	8.04	3.34	6.45	3.30	1.27	22.40
Roll 2nd or additional coats									
Slow	225	280	17.10	4.67	1.32	6.11	3.75	2.38	18.23
Medium	288	260	15.00	5.82	2.02	5.77	3.34	1.87	18.82
Fast	350	240	12.90	6.43	2.69	5.38	2.68	1.03	18.21
Enamel, water base (material #9)									
Roll 1st coat									
Slow	135	250	24.50	7.78	2.21	9.80	6.13	3.89	29.81
Medium	200	220	21.40	8.38	2.92	9.73	5.15	2.88	29.06
Fast	265	190	18.40	8.49	3.53	9.68	4.02	1.54	27.26
Roll 2nd or additional finish coats									
Slow	210	280	24.50	5.00	1.42	8.75	4.70	2.98	22.85
Medium	273	260	21.40	6.14	2.13	8.23	4.04	2.26	22.80
Fast	335	240	18.40	6.72	2.82	7.67	3.18	1.22	21.61
Enamel, oil base (material #10)									
Roll 1st coat									
Slow	135	240	26.60	7.78	2.21	11.08	6.53	4.14	31.74
Medium	200	230	23.30	8.38	2.92	10.13	5.25	2.93	29.61
Fast	265	210	20.00	8.49	3.53	9.52	3.99	1.53	27.06
Roll 2nd or additional finish coats									
Slow	210	275	26.60	5.00	1.42	9.67	4.99	3.16	24.24
Medium	273	250	23.30	6.14	2.13	9.32	4.31	2.41	24.31
Fast	335	230	20.00	6.72	2.82	8.70	3.37	1.30	22.91

Ceiling panel estimates are based on overall dimensions (length times width) to 8 feet high. For heights above 8 feet, use the High Time Difficulty Factors on page 137. "Slow" work is based on a $10.50 hourly wage, "Medium" work on a $16.75 hourly wage, and "Fast" work on a $22.50 hourly wage. Other qualifications that apply to this table are on page 9.

Ceiling panels, suspended, fiber panels in T-bar frame, spray application

	Labor SF per manhour	Material coverage SF/gallon	Material cost per gallon	Labor cost per 100 SF	Labor burden 100 SF	Material cost per 100 SF	Overhead per 100 SF	Profit per 100 SF	Total price per 100 SF
Flat latex, water base (material #5)									
Spray 1st coat									
Slow	300	250	17.10	3.50	.99	6.84	3.52	2.23	17.08
Medium	345	238	15.00	4.86	1.69	6.30	3.15	1.76	17.76
Fast	390	225	12.90	5.77	2.39	5.73	2.57	.99	17.45
Spray 2nd or additional coats									
Slow	500	300	17.10	2.10	.60	5.70	2.60	1.65	12.65
Medium	545	288	15.00	3.07	1.06	5.21	2.29	1.28	12.91
Fast	590	275	12.90	3.81	1.58	4.69	1.87	.72	12.67
Enamel, water base (material #9)									
Spray 1st coat									
Slow	275	250	24.50	3.82	1.09	9.80	4.56	2.89	22.16
Medium	325	238	21.40	5.15	1.81	8.99	3.90	2.18	22.03
Fast	375	225	18.40	6.00	2.51	8.18	3.09	1.19	20.97
Spray 2nd or additional coats									
Slow	450	325	24.50	2.33	.66	7.54	3.26	2.07	15.86
Medium	500	313	21.40	3.35	1.17	6.84	2.78	1.56	15.70
Fast	550	300	18.40	4.09	1.71	6.13	2.21	.85	14.99
Enamel, oil base (material #10)									
Spray 1st coat									
Slow	275	240	26.60	3.82	1.09	11.08	4.96	3.14	24.09
Medium	325	220	23.30	5.15	1.81	10.59	4.29	2.40	24.24
Fast	375	200	20.00	6.00	2.51	10.00	3.42	1.32	23.25
Spray 2nd or additional coats									
Slow	450	300	26.60	2.33	.66	8.87	3.68	2.33	17.87
Medium	500	288	23.30	3.35	1.17	8.09	3.09	1.73	17.43
Fast	550	275	20.00	4.09	1.71	7.27	2.42	.93	16.42

Ceiling panel estimates are based on overall dimensions (length times width) to 8 feet high. For heights above 8 feet, use the High Time Difficulty Factors on page 137. "Slow" work is based on a $10.50 hourly wage, "Medium" work on a $16.75 hourly wage, and "Fast" work on a $22.50 hourly wage. Other qualifications that apply to this table are on page 9.

	Labor SF per manhour	Material coverage SF/gallon	Material cost per gallon	Labor cost per 100 SF	Labor burden 100 SF	Material cost per 100 SF	Overhead per 100 SF	Profit per 100 SF	Total price per 100 SF
Ceiling pans, metal, exterior enamel finish									
Enamel, water base (material #24)									
Brush each coat									
Slow	80	450	27.30	13.13	3.72	6.07	7.11	4.51	34.54
Medium	100	388	23.90	16.75	5.84	6.16	7.04	3.94	39.73
Fast	125	325	20.50	18.00	7.51	6.31	5.89	2.26	39.97
Enamel, oil base (material #25)									
Brush each coat									
Slow	80	400	30.70	13.13	3.72	7.68	7.61	4.82	36.96
Medium	103	338	26.80	16.26	5.67	7.93	7.32	4.09	41.27
Fast	125	275	23.00	18.00	7.51	8.36	6.27	2.41	42.55
Enamel, water base (material #24)									
Roll each coat									
Slow	175	425	27.30	6.00	1.70	6.42	4.38	2.78	21.28
Medium	200	368	23.90	8.38	2.92	6.49	4.36	2.44	24.59
Fast	225	300	20.50	10.00	4.16	6.83	3.89	1.49	26.37
Enamel, oil base (material #25)									
Roll each coat									
Slow	175	375	30.70	6.00	1.70	8.19	4.93	3.12	23.94
Medium	200	313	26.80	8.38	2.92	8.56	4.87	2.72	27.45
Fast	225	250	23.00	10.00	4.16	9.20	4.33	1.66	29.35
Enamel, water base (material #24)									
Spray each coat									
Slow	550	380	27.30	1.91	.54	7.18	2.99	1.89	14.51
Medium	600	370	23.90	2.79	.98	6.46	2.50	1.40	14.13
Fast	650	260	20.50	3.46	1.45	7.88	2.36	.91	16.06
Enamel, oil base (material #25)									
Spray each coat									
Slow	550	330	30.70	1.91	.54	9.30	3.64	2.31	17.70
Medium	600	270	26.80	2.79	.98	9.93	3.35	1.87	18.92
Fast	650	210	23.00	3.46	1.45	10.95	2.93	1.13	19.92

Ceiling panel estimates are based on overall dimensions (length times width) to 8 feet high. For heights above 8 feet, use the High Time Difficulty Factors on page 137. "Slow" work is based on a $10.50 hourly wage, "Medium" work on a $16.75 hourly wage, and "Fast" work on a $22.50 hourly wage. Other qualifications that apply to this table are on page 9.

	Labor SF per manhour	Material coverage SF/gallon	Material cost per gallon	Labor cost per 100 SF	Labor burden 100 SF	Material cost per 100 SF	Overhead per 100 SF	Profit per 100 SF	Total price per 100 SF
Ceilings, acoustic on drywall, spray-on texture									
Acoustic spray-on texture, primer (material #6)									
Spray prime coat									
Slow	250	100	18.10	4.20	1.19	18.10	7.28	4.62	35.39
Medium	300	90	15.80	5.58	1.94	17.56	6.14	3.43	34.65
Fast	350	80	13.60	6.43	2.69	17.00	4.83	1.86	32.81
Acoustic spray-on texture, finish (material #7)									
Spray 1st finish coat									
Slow	400	180	15.50	2.63	.74	8.61	3.72	2.36	18.06
Medium	450	170	13.60	3.72	1.29	8.00	3.19	1.78	17.98
Fast	500	160	11.60	4.50	1.88	7.25	2.52	.97	17.12
Spray 2nd or additional finish coats									
Slow	500	200	15.50	2.10	.60	7.75	3.24	2.05	15.74
Medium	550	188	13.60	3.05	1.06	7.23	2.78	1.55	15.67
Fast	600	175	11.60	3.75	1.58	6.63	2.21	.85	15.02

Ceiling texture estimates are based on overall dimensions (length times width) to 8 feet high. For heights above 8 feet, use the High Time Difficulty Factors on page 137. "Slow" work is based on a $10.50 hourly wage, "Medium" work on a $16.75 hourly wage, and "Fast" work on a $22.50 hourly wage. Other qualifications that apply to this table are on page 9.

	Labor SF per manhour	Material coverage SF/pound	Material cost per pound	Labor cost per 100 SF	Labor burden 100 SF	Material cost per 100 SF	Overhead per 100 SF	Profit per 100 SF	Total price per 100 SF
Ceilings, drywall, stipple finish, spray									
Stipple finish texture paint, Dripowder mix (material #8)									
Spray each coat									
Slow	225	10.0	.42	4.67	1.32	4.20	3.16	2.00	15.35
Medium	250	7.5	.40	6.70	2.34	5.33	3.52	1.97	19.86
Fast	275	5.0	.30	8.18	3.43	6.00	3.26	1.25	22.12

Ceiling texture estimates are based on overall dimensions (length times width) to 8 feet high. For heights above 8 feet, use the High Time Difficulty Factors on page 137. "Slow" work is based on a $10.50 hourly wage, "Medium" work on a $16.75 hourly wage, and "Fast" work on a $22.50 hourly wage. Other qualifications that apply to this table are on page 9.

	Labor SF per manhour	Material coverage SF/gallon	Material cost per gallon	Labor cost per 100 SF	Labor burden 100 SF	Material cost per 100 SF	Overhead per 100 SF	Profit per 100 SF	Total price per 100 SF

Ceilings, drywall, smooth finish, brush

	Labor SF per manhour	Material coverage SF/gallon	Material cost per gallon	Labor cost per 100 SF	Labor burden 100 SF	Material cost per 100 SF	Overhead per 100 SF	Profit per 100 SF	Total price per 100 SF
Flat latex, water base (material #5)									
Brush 1st coat									
Slow	175	325	17.10	6.00	1.70	5.26	4.02	2.55	19.53
Medium	200	313	15.00	8.38	2.92	4.79	3.94	2.20	22.23
Fast	225	300	12.90	10.00	4.16	4.30	3.42	1.31	23.19
Brush 2nd coat									
Slow	225	400	17.10	4.67	1.32	4.28	3.19	2.02	15.48
Medium	250	375	15.00	6.70	2.34	4.00	3.19	1.78	18.01
Fast	275	350	12.90	8.18	3.43	3.69	2.83	1.09	19.22
Brush 3rd or additional coats									
Slow	250	425	17.10	4.20	1.19	4.02	2.92	1.85	14.18
Medium	275	400	15.00	6.09	2.13	3.75	2.93	1.64	16.54
Fast	300	375	12.90	7.50	3.12	3.44	2.60	1.00	17.66
Sealer, water base (material #1)									
Brush prime coat									
Slow	200	325	16.30	5.25	1.49	5.02	3.65	2.31	17.72
Medium	225	313	14.30	7.44	2.59	4.57	3.58	2.00	20.18
Fast	250	300	12.30	9.00	3.76	4.10	3.12	1.20	21.18
Sealer, oil base (material #2)									
Brush prime coat									
Slow	200	350	22.90	5.25	1.49	6.54	4.12	2.61	20.01
Medium	225	338	20.10	7.44	2.59	5.95	3.92	2.19	22.09
Fast	250	325	17.20	9.00	3.76	5.29	3.34	1.28	22.67
Enamel, water base (material #9)									
Brush 1st finish coat									
Slow	200	400	24.50	5.25	1.49	6.13	3.99	2.53	19.39
Medium	225	375	21.40	7.44	2.59	5.71	3.86	2.16	21.76
Fast	250	350	18.40	9.00	3.76	5.26	3.33	1.28	22.63
Brush 2nd and additional finish coats									
Slow	225	425	24.50	4.67	1.32	5.76	3.65	2.31	17.71
Medium	250	400	21.40	6.70	2.34	5.35	3.52	1.97	19.88
Fast	275	375	18.40	8.18	3.43	4.91	3.05	1.17	20.74
Enamel, oil base (material #10)									
Brush 1st finish coat									
Slow	200	400	26.60	5.25	1.49	6.65	4.15	2.63	20.17
Medium	225	388	23.30	7.44	2.59	6.01	3.93	2.20	22.17
Fast	250	375	20.00	9.00	3.76	5.33	3.35	1.29	22.73

	Labor SF per manhour	Material coverage SF/gallon	Material cost per gallon	Labor cost per 100 SF	Labor burden 100 SF	Material cost per 100 SF	Overhead per 100 SF	Profit per 100 SF	Total price per 100 SF
Brush 2nd or additional finish coats									
Slow	225	425	26.60	4.67	1.32	6.26	3.80	2.41	18.46
Medium	250	413	23.30	6.70	2.34	5.64	3.59	2.01	20.28
Fast	275	400	20.00	8.18	3.43	5.00	3.07	1.18	20.86
Epoxy coating, white (material #52)									
Brush 1st coat									
Slow	175	425	61.70	6.00	1.70	14.52	6.89	4.37	33.48
Medium	200	400	54.00	8.38	2.92	13.50	6.08	3.40	34.28
Fast	225	375	46.30	10.00	4.16	12.35	4.91	1.89	33.31
Brush 2nd or additional coats									
Slow	200	450	61.70	5.25	1.49	13.71	6.34	4.02	30.81
Medium	225	425	54.00	7.44	2.59	12.71	5.57	3.11	31.42
Fast	250	400	46.30	9.00	3.76	11.58	4.50	1.73	30.57
Stipple finish									
Slow	200	--	--	5.25	1.49	--	2.09	1.32	10.15
Medium	225	--	--	7.44	2.59	--	2.46	1.37	13.86
Fast	250	--	--	9.00	3.76	--	2.36	.91	16.03

Ceiling estimates are based on overall dimensions (length times width) to 8 feet high. For heights above 8 feet, use the High Time Difficulty Factors on page 137. "Slow" work is based on a $10.50 hourly wage, "Medium" work on a $16.75 hourly wage, and "Fast" work on a $22.50 hourly wage. Other qualifications that apply to this table are on page 9.

	Labor SF per manhour	Material coverage SF/gallon	Material cost per gallon	Labor cost per 100 SF	Labor burden 100 SF	Material cost per 100 SF	Overhead per 100 SF	Profit per 100 SF	Total price per 100 SF
Ceilings, drywall, smooth finish, roll									
Flat latex, water base (material #5)									
Roll 1st coat									
Slow	325	350	17.10	3.23	.92	4.89	2.80	1.78	13.62
Medium	375	325	15.00	4.47	1.56	4.62	2.61	1.46	14.72
Fast	425	300	12.90	5.29	2.20	4.30	2.18	.84	14.81
Roll 2nd coat									
Slow	375	375	17.10	2.80	.80	4.56	2.53	1.60	12.29
Medium	413	363	15.00	4.06	1.41	4.13	2.35	1.31	13.26
Fast	450	350	12.90	5.00	2.08	3.69	1.99	.77	13.53
Roll 3rd or additional coats									
Slow	425	400	17.10	2.47	.70	4.28	2.31	1.46	11.22
Medium	450	388	15.00	3.72	1.29	3.87	2.18	1.22	12.28
Fast	475	375	12.90	4.74	1.99	3.44	1.88	.72	12.77
Sealer, water base (material #1)									
Roll prime coat									
Slow	350	350	16.30	3.00	.86	4.66	2.64	1.67	12.83
Medium	400	325	14.30	4.19	1.46	4.40	2.46	1.38	13.89
Fast	450	300	12.30	5.00	2.08	4.10	2.07	.80	14.05
Sealer, oil base (material #2)									
Roll prime coat									
Slow	350	300	22.90	3.00	.86	7.63	3.56	2.26	17.31
Medium	400	288	20.10	4.19	1.46	6.98	3.09	1.73	17.45
Fast	450	275	17.20	5.00	2.08	6.25	2.47	.95	16.75
Enamel, water base, (material #9)									
Roll 1st finish coat									
Slow	350	375	24.50	3.00	.86	6.53	3.22	2.04	15.65
Medium	400	363	21.40	4.19	1.46	5.90	2.83	1.58	15.96
Fast	450	350	18.40	5.00	2.08	5.26	2.28	.88	15.50
Roll 2nd or additional finish coats									
Slow	425	400	24.50	2.47	.70	6.13	2.88	1.83	14.01
Medium	450	388	21.40	3.72	1.29	5.52	2.58	1.44	14.55
Fast	475	375	18.40	4.74	1.99	4.91	2.15	.83	14.62
Enamel, oil base (material #10)									
Roll 1st finish coat									
Slow	350	350	26.60	3.00	.86	7.60	3.55	2.25	17.26
Medium	400	338	23.30	4.19	1.46	6.89	3.07	1.72	17.33
Fast	450	325	20.00	5.00	2.08	6.15	2.45	.94	16.62

	Labor SF per manhour	Material coverage SF/gallon	Material cost per gallon	Labor cost per 100 SF	Labor burden 100 SF	Material cost per 100 SF	Overhead per 100 SF	Profit per 100 SF	Total price per 100 SF
Roll 2nd or additional finish coats									
Slow	425	375	26.60	2.47	.70	7.09	3.18	2.02	15.46
Medium	450	363	23.30	3.72	1.29	6.42	2.80	1.57	15.80
Fast	475	350	20.00	4.74	1.99	5.71	2.30	.88	15.62
Epoxy coating, white (material #52)									
Roll 1st finish coat									
Slow	325	400	61.70	3.23	.92	15.43	6.07	3.85	29.50
Medium	363	375	54.00	4.61	1.60	14.40	5.05	2.82	28.48
Fast	400	350	46.30	5.63	2.34	13.23	3.92	1.51	26.63
Roll 2nd or additional finish coats									
Slow	400	425	61.70	2.63	.74	14.52	5.55	3.52	26.96
Medium	425	400	54.00	3.94	1.37	13.50	4.61	2.58	26.00
Fast	450	375	46.30	5.00	2.08	12.35	3.60	1.38	24.41

Ceiling estimates are based on overall dimensions (length times width) to 8 feet high. For heights above 8 feet, use the High Time Difficulty Factors on page 137. "Slow" work is based on a $10.50 hourly wage, "Medium" work on a $16.75 hourly wage, and "Fast" work on a $22.50 hourly wage. Other qualifications that apply to this table are on page 9.

	Labor SF per manhour	Material coverage SF/gallon	Material cost per gallon	Labor cost per 100 SF	Labor burden 100 SF	Material cost per 100 SF	Overhead per 100 SF	Profit per 100 SF	Total price per 100 SF
Ceilings, drywall, smooth finish, spray									
Flat latex, water base (material #5)									
Spray 1st coat									
Slow	750	300	17.10	1.40	.39	5.70	2.33	1.47	11.29
Medium	850	275	15.00	1.97	.70	5.45	1.99	1.11	11.22
Fast	950	250	12.90	2.37	.98	5.16	1.58	.61	10.70
Spray 2nd coat									
Slow	850	350	17.10	1.24	.35	4.89	2.01	1.27	9.76
Medium	950	325	15.00	1.76	.61	4.62	1.71	.96	9.66
Fast	1050	300	12.90	2.14	.89	4.30	1.36	.52	9.21
Spray 3rd or additional coats									
Slow	900	350	17.10	1.17	.33	4.89	1.98	1.26	9.63
Medium	1000	338	15.00	1.68	.58	4.44	1.64	.92	9.26
Fast	1100	325	12.90	2.05	.85	3.97	1.27	.49	8.63
Sealer, water base (material #1)									
Spray prime coat									
Slow	800	300	16.30	1.31	.38	5.43	2.20	1.40	10.72
Medium	900	275	14.30	1.86	.65	5.20	1.89	1.06	10.66
Fast	1000	250	12.30	2.25	.94	4.92	1.50	.58	10.19
Sealer, oil base (material #2)									
Spray prime coat									
Slow	800	250	22.90	1.31	.38	9.16	3.36	2.13	16.34
Medium	900	238	20.10	1.86	.65	8.45	2.69	1.50	15.15
Fast	1000	225	17.20	2.25	.94	7.64	2.00	.77	13.60
Enamel, water base (material #9)									
Spray 1st finish coat									
Slow	800	350	24.50	1.31	.38	7.00	2.69	1.71	13.09
Medium	900	325	21.40	1.86	.65	6.58	2.23	1.25	12.57
Fast	1000	300	18.40	2.25	.94	6.13	1.72	.66	11.70
Spray 2nd or additional finish coats									
Slow	850	350	24.50	1.24	.35	7.00	2.66	1.69	12.94
Medium	950	338	21.40	1.76	.61	6.33	2.13	1.19	12.02
Fast	1050	325	18.40	2.14	.89	5.66	1.61	.62	10.92
Enamel, oil base (material #10)									
Spray 1st finish coat									
Slow	800	300	26.60	1.31	.38	8.87	3.27	2.07	15.90
Medium	900	280	23.30	1.86	.65	8.32	2.65	1.48	14.96
Fast	1000	260	20.00	2.25	.94	7.69	2.01	.77	13.66
Spray 2nd or additional finish coats									
Slow	850	325	26.60	1.24	.35	8.18	3.03	1.92	14.72
Medium	950	313	23.30	1.76	.61	7.44	2.40	1.34	13.55
Fast	1050	300	20.00	2.14	.89	6.67	1.79	.69	12.18

Ceiling estimates are based on overall dimensions (length times width) to 8 feet high. For heights above 8 feet, use the High Time Difficulty Factors on page 137. "Slow" work is based on a $10.50 hourly wage, "Medium" work on a $16.75 hourly wage, and "Fast" work on a $22.50 hourly wage. Other qualifications that apply to this table are on page 9.

	Labor SF per manhour	Material coverage SF/gallon	Material cost per gallon	Labor cost per 100 SF	Labor burden 100 SF	Material cost per 100 SF	Overhead per 100 SF	Profit per 100 SF	Total price per 100 SF
Ceilings, gypsum drywall, anti-graffiti stain eliminator									
Water base primer and pigmented sealer (material #39)									
Roll & brush each coat									
Slow	350	450	25.40	3.00	.86	5.64	2.94	1.86	14.30
Medium	375	425	22.30	4.47	1.56	5.25	2.76	1.54	15.58
Fast	400	400	19.10	5.63	2.34	4.78	2.36	.91	16.02
Oil base primer and pigmented sealer (material #40)									
Roll & brush each coat									
Slow	350	400	25.70	3.00	.86	6.43	3.19	2.02	15.50
Medium	375	388	22.50	4.47	1.56	5.80	2.90	1.62	16.35
Fast	400	375	19.30	5.63	2.34	5.15	2.43	.93	16.48
Polyurethane 2 part system (material #41)									
Roll & brush each coat									
Slow	300	400	91.40	3.50	.99	22.85	8.48	5.37	41.19
Medium	325	375	80.00	5.15	1.81	21.33	6.93	3.87	39.09
Fast	350	350	68.60	6.43	2.69	19.60	5.31	2.04	36.07

Ceiling estimates are based on overall dimensions (length times width) to 8 feet high. For heights above 8 feet, use the High Time Difficulty Factors on page 137. "Slow" work is based on a $10.50 hourly wage, "Medium" work on a $16.75 hourly wage, and "Fast" work on a $22.50 hourly wage. Other qualifications that apply to this table are on page 9.

	Labor SF per manhour	Material coverage SF/gallon	Material cost per gallon	Labor cost per 100 SF	Labor burden 100 SF	Material cost per 100 SF	Overhead per 100 SF	Profit per 100 SF	Total price per 100 SF
Ceilings, gypsum drywall, orange peel texture, brush									
Flat latex, water base (material #5)									
Brush 1st coat									
Slow	150	300	17.10	7.00	1.99	5.70	4.55	2.89	22.13
Medium	175	288	15.00	9.57	3.33	5.21	4.44	2.48	25.03
Fast	200	275	12.90	11.25	4.70	4.69	3.82	1.47	25.93
Brush 2nd coat									
Slow	175	350	17.10	6.00	1.70	4.89	3.91	2.48	18.98
Medium	200	338	15.00	8.38	2.92	4.44	3.86	2.16	21.76
Fast	225	325	12.90	10.00	4.16	3.97	3.36	1.29	22.78
Brush 3rd or additional coats									
Slow	200	400	17.10	5.25	1.49	4.28	3.42	2.17	16.61
Medium	225	375	15.00	7.44	2.59	4.00	3.44	1.92	19.39
Fast	250	350	12.90	9.00	3.76	3.69	3.04	1.17	20.66
Sealer, water base (material #1)									
Brush prime coat									
Slow	175	300	16.30	6.00	1.70	5.43	4.07	2.58	19.78
Medium	200	288	14.30	8.38	2.92	4.97	3.99	2.23	22.49
Fast	225	275	12.30	10.00	4.16	4.47	3.45	1.33	23.41
Sealer, oil base (material #2)									
Brush prime coat									
Slow	175	250	22.90	6.00	1.70	9.16	5.23	3.32	25.41
Medium	200	238	20.10	8.38	2.92	8.45	4.84	2.70	27.29
Fast	225	225	17.20	10.00	4.16	7.64	4.04	1.55	27.39
Enamel, water base (material #9)									
Brush 1st finish coat									
Slow	150	350	24.50	7.00	1.99	7.00	4.96	3.14	24.09
Medium	175	338	21.40	9.57	3.33	6.33	4.71	2.63	26.57
Fast	200	325	18.40	11.25	4.70	5.66	4.00	1.54	27.15
Brush 2nd or additional finish coats									
Slow	175	400	24.50	6.00	1.70	6.13	4.29	2.72	20.84
Medium	200	375	21.40	8.38	2.92	5.71	4.17	2.33	23.51
Fast	225	350	18.40	10.00	4.16	5.26	3.60	1.38	24.40

	Labor SF per manhour	Material coverage SF/gallon	Material cost per gallon	Labor cost per 100 SF	Labor burden 100 SF	Material cost per 100 SF	Overhead per 100 SF	Profit per 100 SF	Total price per 100 SF
Enamel, oil base (material #10)									
Brush 1st finish coat									
Slow	150	325	26.60	7.00	1.99	8.18	5.32	3.37	25.86
Medium	175	313	23.30	9.57	3.33	7.44	4.99	2.79	28.12
Fast	200	300	20.00	11.25	4.70	6.67	4.18	1.61	28.41
Brush 2nd or additional finish coats									
Slow	150	400	26.60	7.00	1.99	6.65	4.85	3.07	23.56
Medium	175	375	23.30	9.57	3.33	6.21	4.68	2.62	26.41
Fast	200	350	20.00	11.25	4.70	5.71	4.01	1.54	27.21
Epoxy coating, white (material #52)									
Brush 1st coat									
Slow	125	350	61.70	8.40	2.38	17.63	8.81	5.58	42.80
Medium	150	325	54.00	11.17	3.90	16.62	7.76	4.34	43.79
Fast	175	300	46.30	12.86	5.35	15.43	6.23	2.39	42.26
Brush 2nd or additional coats									
Slow	175	375	61.70	6.00	1.70	16.45	7.49	4.75	36.39
Medium	200	350	54.00	8.38	2.92	15.43	6.55	3.66	36.94
Fast	225	325	46.30	10.00	4.16	14.25	5.26	2.02	35.69

Ceiling estimates are based on overall dimensions (length times width) to 8 feet high. For heights above 8 feet, use the High Time Difficulty Factors on page 137. "Slow" work is based on a $10.50 hourly wage, "Medium" work on a $16.75 hourly wage, and "Fast" work on a $22.50 hourly wage. Other qualifications that apply to this table are on page 9.

	Labor SF per manhour	Material coverage SF/gallon	Material cost per gallon	Labor cost per 100 SF	Labor burden 100 SF	Material cost per 100 SF	Overhead per 100 SF	Profit per 100 SF	Total price per 100 SF
Ceilings, gypsum drywall, orange peel texture, roll									
Flat latex, water base (material #5)									
Roll 1st coat									
Slow	325	300	17.10	3.23	.92	5.70	3.05	1.94	14.84
Medium	350	275	15.00	4.79	1.67	5.45	2.92	1.63	16.46
Fast	375	250	12.90	6.00	2.51	5.16	2.53	.97	17.17
Roll 2nd coat									
Slow	350	325	17.10	3.00	.86	5.26	2.82	1.79	13.73
Medium	375	313	15.00	4.47	1.56	4.79	2.65	1.48	14.95
Fast	400	300	12.90	5.63	2.34	4.30	2.27	.87	15.41
Roll 3rd or additional coats									
Slow	400	350	17.10	2.63	.74	4.89	2.56	1.62	12.44
Medium	425	338	15.00	3.94	1.37	4.44	2.39	1.34	13.48
Fast	450	325	12.90	5.00	2.08	3.97	2.05	.79	13.89
Sealer, water base (material #1)									
Roll prime coat									
Slow	350	300	16.30	3.00	.86	5.43	2.88	1.82	13.99
Medium	375	275	14.30	4.47	1.56	5.20	2.75	1.54	15.52
Fast	400	250	12.30	5.63	2.34	4.92	2.39	.92	16.20
Sealer, oil base (material #2)									
Roll prime coat									
Slow	350	275	22.90	3.00	.86	8.33	3.78	2.39	18.36
Medium	375	250	20.10	4.47	1.56	8.04	3.45	1.93	19.45
Fast	400	225	17.20	5.63	2.34	7.64	2.89	1.11	19.61
Enamel, water base (material #9)									
Roll 1st finish coat									
Slow	325	325	24.50	3.23	.92	7.54	3.62	2.30	17.61
Medium	350	313	21.40	4.79	1.67	6.84	3.26	1.82	18.38
Fast	375	300	18.40	6.00	2.51	6.13	2.71	1.04	18.39
Roll 2nd or additional finish coats									
Slow	375	350	24.50	2.80	.80	7.00	3.29	2.08	15.97
Medium	400	338	21.40	4.19	1.46	6.33	2.94	1.64	16.56
Fast	425	325	18.40	5.29	2.20	5.66	2.43	.94	16.52

	Labor SF per manhour	Material coverage SF/gallon	Material cost per gallon	Labor cost per 100 SF	Labor burden 100 SF	Material cost per 100 SF	Overhead per 100 SF	Profit per 100 SF	Total price per 100 SF
Enamel, oil base (material #10)									
Roll 1st finish coat									
Slow	325	300	26.60	3.23	.92	8.87	4.04	2.56	19.62
Medium	350	275	23.30	4.79	1.67	8.47	3.66	2.04	20.63
Fast	375	250	20.00	6.00	2.51	8.00	3.05	1.17	20.73
Roll 2nd or additional finish coats									
Slow	375	300	26.60	2.80	.80	8.87	3.87	2.45	18.79
Medium	400	288	23.30	4.19	1.46	8.09	3.37	1.88	18.99
Fast	425	275	20.00	5.29	2.20	7.27	2.73	1.05	18.54
Epoxy coating, white (material #52)									
Roll 1st coat									
Slow	300	300	61.70	3.50	.99	20.57	7.77	4.93	37.76
Medium	325	288	54.00	5.15	1.81	18.75	6.29	3.52	35.52
Fast	350	275	46.30	6.43	2.69	16.84	4.80	1.85	32.61
Roll 2nd or additional coats									
Slow	350	300	61.70	3.00	.86	20.57	7.57	4.80	36.80
Medium	375	288	54.00	4.47	1.56	18.75	6.07	3.39	34.24
Fast	400	275	46.30	5.63	2.34	16.84	4.59	1.76	31.16

Ceiling estimates are based on overall dimensions (length times width) to 8 feet high. For heights above 8 feet, use the High Time Difficulty Factors on page 137. "Slow" work is based on a $10.50 hourly wage, "Medium" work on a $16.75 hourly wage, and "Fast" work on a $22.50 hourly wage. Other qualifications that apply to this table are on page 9.

	Labor SF per manhour	Material coverage SF/gallon	Material cost per gallon	Labor cost per 100 SF	Labor burden 100 SF	Material cost per 100 SF	Overhead per 100 SF	Profit per 100 SF	Total price per 100 SF

Ceilings, gypsum drywall, orange peel texture, spray

	Labor SF per manhour	Material coverage SF/gallon	Material cost per gallon	Labor cost per 100 SF	Labor burden 100 SF	Material cost per 100 SF	Overhead per 100 SF	Profit per 100 SF	Total price per 100 SF
Flat latex, water base (material #5)									
Spray 1st coat									
Slow	650	225	17.10	1.62	.46	7.60	3.00	1.90	14.58
Medium	750	200	15.00	2.23	.77	7.50	2.57	1.44	14.51
Fast	850	175	12.90	2.65	1.11	7.37	2.06	.79	13.98
Spray 2nd coat									
Slow	775	250	17.10	1.35	.39	6.84	2.66	1.68	12.92
Medium	875	225	15.00	1.91	.67	6.67	2.27	1.27	12.79
Fast	975	200	12.90	2.31	.97	6.45	1.80	.69	12.22
Spray 3rd or additional coats									
Slow	825	275	17.10	1.27	.36	6.22	2.43	1.54	11.82
Medium	925	250	15.00	1.81	.63	6.00	2.07	1.16	11.67
Fast	1025	225	12.90	2.20	.93	5.73	1.64	.63	11.13
Sealer, water base (material #1)									
Spray prime coat									
Slow	700	225	16.30	1.50	.43	7.24	2.84	1.80	13.81
Medium	800	200	14.30	2.09	.73	7.15	2.44	1.37	13.78
Fast	900	175	12.30	2.50	1.04	7.03	1.96	.75	13.28
Sealer, oil base (material #2)									
Spray prime coat									
Slow	700	200	22.90	1.50	.43	11.45	4.15	2.63	20.16
Medium	800	188	20.10	2.09	.73	10.69	3.31	1.85	18.67
Fast	900	175	17.20	2.50	1.04	9.83	2.47	.95	16.79
Enamel, water base (material #9)									
Spray 1st finish coat									
Slow	725	250	24.50	1.45	.41	9.80	3.61	2.29	17.56
Medium	825	225	21.40	2.03	.70	9.51	3.00	1.68	16.92
Fast	925	200	18.40	2.43	1.01	9.20	2.34	.90	15.88
Spray 2nd or additional finish coat									
Slow	775	275	24.50	1.35	.39	8.91	3.30	2.09	16.04
Medium	875	250	21.40	1.91	.67	8.56	2.73	1.53	15.40
Fast	975	225	18.40	2.31	.97	8.18	2.12	.81	14.39

	Labor SF per manhour	Material coverage SF/gallon	Material cost per gallon	Labor cost per 100 SF	Labor burden 100 SF	Material cost per 100 SF	Overhead per 100 SF	Profit per 100 SF	Total price per 100 SF
Enamel, oil base (material #10)									
Spray 1st finish coat									
Slow	725	225	26.60	1.45	.41	11.82	4.24	2.69	20.61
Medium	825	213	23.30	2.03	.70	10.94	3.35	1.87	18.89
Fast	925	200	20.00	2.43	1.01	10.00	2.49	.96	16.89
Spray 2nd or additional finish coat									
Slow	775	250	26.60	1.35	.39	10.64	3.83	2.43	18.64
Medium	875	238	23.30	1.91	.67	9.79	3.03	1.69	17.09
Fast	975	225	20.00	2.31	.97	8.89	2.25	.86	15.28

Ceiling estimates are based on overall dimensions (length times width) to 8 feet high. For heights above 8 feet, use the High Time Difficulty Factors on page 137. "Slow" work is based on a $10.50 hourly wage, "Medium" work on a $16.75 hourly wage, and "Fast" work on a $22.50 hourly wage. Other qualifications that apply to this table are on page 9.

	Labor SF per manhour	Material coverage SF/gallon	Material cost per gallon	Labor cost per 100 SF	Labor burden 100 SF	Material cost per 100 SF	Overhead per 100 SF	Profit per 100 SF	Total price per 100 SF

Ceilings, gypsum drywall, sand finish texture, brush

	Labor SF per manhour	Material coverage SF/gallon	Material cost per gallon	Labor cost per 100 SF	Labor burden 100 SF	Material cost per 100 SF	Overhead per 100 SF	Profit per 100 SF	Total price per 100 SF
Flat latex, water base (material #5)									
Brush 1st coat									
Slow	175	325	17.10	6.00	1.70	5.26	4.02	2.55	19.53
Medium	200	313	15.00	8.38	2.92	4.79	3.94	2.20	22.23
Fast	225	300	12.90	10.00	4.16	4.30	3.42	1.31	23.19
Brush 2nd coat									
Slow	200	400	17.10	5.25	1.49	4.28	3.42	2.17	16.61
Medium	238	375	15.00	7.04	2.45	4.00	3.31	1.85	18.65
Fast	275	350	12.90	8.18	3.43	3.69	2.83	1.09	19.22
Brush 3rd or additional coats									
Slow	225	425	17.10	4.67	1.32	4.02	3.11	1.97	15.09
Medium	263	400	15.00	6.37	2.21	3.75	3.02	1.69	17.04
Fast	300	375	12.90	7.50	3.12	3.44	2.60	1.00	17.66
Sealer, water base (material #1)									
Brush prime coat									
Slow	200	325	16.30	5.25	1.49	5.02	3.65	2.31	17.72
Medium	225	313	14.30	7.44	2.59	4.57	3.58	2.00	20.18
Fast	250	300	12.30	9.00	3.76	4.10	3.12	1.20	21.18
Sealer, oil base (material #2)									
Brush prime coat									
Slow	200	325	22.90	5.25	1.49	7.05	4.27	2.71	20.77
Medium	225	313	20.10	7.44	2.59	6.42	4.03	2.25	22.73
Fast	250	300	17.20	9.00	3.76	5.73	3.42	1.31	23.22
Enamel, water base (material #9)									
Brush 1st finish coat									
Slow	200	400	24.50	5.25	1.49	6.13	3.99	2.53	19.39
Medium	225	375	21.40	7.44	2.59	5.71	3.86	2.16	21.76
Fast	250	350	18.40	9.00	3.76	5.26	3.33	1.28	22.63
Brush 2nd or additional finish coats									
Slow	225	425	24.50	4.67	1.32	5.76	3.65	2.31	17.71
Medium	263	400	21.40	6.37	2.21	5.35	3.42	1.91	19.26
Fast	300	375	18.40	7.50	3.12	4.91	2.87	1.10	19.50

	Labor SF per manhour	Material coverage SF/gallon	Material cost per gallon	Labor cost per 100 SF	Labor burden 100 SF	Material cost per 100 SF	Overhead per 100 SF	Profit per 100 SF	Total price per 100 SF
Enamel, oil base (material #10)									
Brush 1st finish coat									
Slow	200	375	26.60	5.25	1.49	7.09	4.29	2.72	20.84
Medium	225	350	23.30	7.44	2.59	6.66	4.09	2.29	23.07
Fast	250	325	20.00	9.00	3.76	6.15	3.50	1.34	23.75
Brush 2nd or additional finish coats									
Slow	225	400	26.60	4.67	1.32	6.65	3.92	2.49	19.05
Medium	263	375	23.30	6.37	2.21	6.21	3.63	2.03	20.45
Fast	300	350	20.00	7.50	3.12	5.71	3.02	1.16	20.51
Epoxy coating, white (material #52)									
Brush 1st coat									
Slow	150	375	61.70	7.00	1.99	16.45	7.89	5.00	38.33
Medium	175	350	54.00	9.57	3.33	15.43	6.94	3.88	39.15
Fast	225	325	46.30	10.00	4.16	14.25	5.26	2.02	35.69
Brush 2nd or additional coats									
Slow	175	400	61.70	6.00	1.70	15.43	7.17	4.55	34.85
Medium	200	375	54.00	8.38	2.92	14.40	6.30	3.52	35.52
Fast	225	350	46.30	10.00	4.16	13.23	5.07	1.95	34.41

Ceiling estimates are based on overall dimensions (length times width) to 8 feet high. For heights above 8 feet, use the High Time Difficulty Factors on page 137. "Slow" work is based on a $10.50 hourly wage, "Medium" work on a $16.75 hourly wage, and "Fast" work on a $22.50 hourly wage. Other qualifications that apply to this table are on page 9.

	Labor SF per manhour	Material coverage SF/gallon	Material cost per gallon	Labor cost per 100 SF	Labor burden 100 SF	Material cost per 100 SF	Overhead per 100 SF	Profit per 100 SF	Total price per 100 SF
Ceilings, gypsum drywall, sand finish texture, roll									
Flat latex, water base (material #5)									
Roll 1st coat									
Slow	300	325	17.10	3.50	.99	5.26	3.03	1.92	14.70
Medium	350	300	15.00	4.79	1.67	5.00	2.81	1.57	15.84
Fast	400	275	12.90	5.63	2.34	4.69	2.34	.90	15.90
Roll 2nd coat									
Slow	350	350	17.10	3.00	.86	4.89	2.71	1.72	13.18
Medium	388	338	15.00	4.32	1.51	4.44	2.52	1.41	14.20
Fast	425	325	12.90	5.29	2.20	3.97	2.12	.82	14.40
Roll 3rd or additional coats									
Slow	425	350	17.10	2.47	.70	4.89	2.50	1.58	12.14
Medium	450	338	15.00	3.72	1.29	4.44	2.32	1.30	13.07
Fast	475	325	12.90	4.74	1.99	3.97	1.98	.76	13.44
Sealer, water base (material #1)									
Roll prime coat									
Slow	325	325	16.30	3.23	.92	5.02	2.84	1.80	13.81
Medium	375	300	14.30	4.47	1.56	4.77	2.65	1.48	14.93
Fast	425	275	12.30	5.29	2.20	4.47	2.21	.85	15.02
Sealer, oil base (material #2)									
Roll prime coat									
Slow	325	300	22.90	3.23	.92	7.63	3.65	2.31	17.74
Medium	375	275	20.10	4.47	1.56	7.31	3.27	1.83	18.44
Fast	425	250	17.20	5.29	2.20	6.88	2.66	1.02	18.05
Enamel, water base (material #9)									
Roll 1st finish coat									
Slow	325	350	24.50	3.23	.92	7.00	3.46	2.19	16.80
Medium	363	338	21.40	4.61	1.60	6.33	3.07	1.72	17.33
Fast	400	325	18.40	5.63	2.34	5.66	2.52	.97	17.12
Roll 2nd or additional finish coats									
Slow	400	350	24.50	2.63	.74	7.00	3.22	2.04	15.63
Medium	425	338	21.40	3.94	1.37	6.33	2.85	1.59	16.08
Fast	450	325	18.40	5.00	2.08	5.66	2.36	.91	16.01
Enamel, oil base (material #10)									
Roll 1st finish coat									
Slow	325	325	26.60	3.23	.92	8.18	3.82	2.42	18.57
Medium	363	313	23.30	4.61	1.60	7.44	3.35	1.87	18.87
Fast	400	300	20.00	5.63	2.34	6.67	2.71	1.04	18.39

	Labor SF per manhour	Material coverage SF/gallon	Material cost per gallon	Labor cost per 100 SF	Labor burden 100 SF	Material cost per 100 SF	Overhead per 100 SF	Profit per 100 SF	Total price per 100 SF
Roll 2nd or additional finish coats									
Slow	400	350	26.60	2.63	.74	7.60	3.40	2.16	16.53
Medium	425	338	23.30	3.94	1.37	6.89	2.99	1.67	16.86
Fast	450	325	20.00	5.00	2.08	6.15	2.45	.94	16.62
Epoxy coating, white (material #52)									
Roll 1st coat									
Slow	300	350	61.70	3.50	.99	17.63	6.86	4.35	33.33
Medium	350	325	54.00	4.79	1.67	16.62	5.65	3.16	31.89
Fast	375	300	46.30	6.00	2.51	15.43	4.43	1.70	30.07
Roll 2nd or additional coats									
Slow	375	375	61.70	2.80	.80	16.45	6.22	3.94	30.21
Medium	400	350	54.00	4.19	1.46	15.43	5.16	2.89	29.13
Fast	425	325	46.30	5.29	2.20	14.25	4.02	1.55	27.31

Ceiling estimates are based on overall dimensions (length times width) to 8 feet high. For heights above 8 feet, use the High Time Difficulty Factors on page 137. "Slow" work is based on a $10.50 hourly wage, "Medium" work on a $16.75 hourly wage, and "Fast" work on a $22.50 hourly wage. Other qualifications that apply to this table are on page 9.

	Labor SF per manhour	Material coverage SF/gallon	Material cost per gallon	Labor cost per 100 SF	Labor burden 100 SF	Material cost per 100 SF	Overhead per 100 SF	Profit per 100 SF	Total price per 100 SF

Ceilings, gypsum drywall, sand finish texture, spray

	Labor SF per manhour	Material coverage SF/gallon	Material cost per gallon	Labor cost per 100 SF	Labor burden 100 SF	Material cost per 100 SF	Overhead per 100 SF	Profit per 100 SF	Total price per 100 SF
Flat latex, water base (material #5)									
Spray 1st coat									
Slow	700	275	17.10	1.50	.43	6.22	2.53	1.60	12.28
Medium	800	250	15.00	2.09	.73	6.00	2.16	1.21	12.19
Fast	900	225	12.90	2.50	1.04	5.73	1.71	.66	11.64
Spray 2nd coat									
Slow	800	325	17.10	1.31	.38	5.26	2.15	1.36	10.46
Medium	900	300	15.00	1.86	.65	5.00	1.84	1.03	10.38
Fast	1000	275	12.90	2.25	.94	4.69	1.46	.56	9.90
Spray 3rd or additional coats									
Slow	850	325	17.10	1.24	.35	5.26	2.12	1.35	10.32
Medium	950	313	15.00	1.76	.61	4.79	1.75	.98	9.89
Fast	1050	300	12.90	2.14	.89	4.30	1.36	.52	9.21
Sealer, water base (material #1)									
Spray prime coat									
Slow	750	275	16.30	1.40	.39	5.93	2.40	1.52	11.64
Medium	850	250	14.30	1.97	.70	5.72	2.05	1.15	11.59
Fast	950	225	12.30	2.37	.98	5.47	1.63	.63	11.08
Sealer, oil base (material #2)									
Spray prime coat									
Slow	750	225	22.90	1.40	.39	10.18	3.71	2.35	18.03
Medium	850	213	20.10	1.97	.70	9.44	2.96	1.66	16.73
Fast	950	200	17.20	2.37	.98	8.60	2.21	.85	15.01
Enamel, water base (material #9)									
Spray 1st finish coat									
Slow	750	325	24.50	1.40	.39	7.54	2.90	1.84	14.07
Medium	850	300	21.40	1.97	.70	7.13	2.40	1.34	13.54
Fast	950	275	18.40	2.37	.98	6.69	1.86	.71	12.61
Spray 2nd or additional finish coat									
Slow	800	325	24.50	1.31	.38	7.54	2.86	1.81	13.90
Medium	900	313	21.40	1.86	.65	6.84	2.29	1.28	12.92
Fast	1000	300	18.40	2.25	.94	6.13	1.72	.66	11.70

	Labor SF per manhour	Material coverage SF/gallon	Material cost per gallon	Labor cost per 100 SF	Labor burden 100 SF	Material cost per 100 SF	Overhead per 100 SF	Profit per 100 SF	Total price per 100 SF
Enamel, oil base (material #10)									
Spray 1st finish coat									
Slow	750	300	26.60	1.40	.39	8.87	3.31	2.10	16.07
Medium	850	288	23.30	1.97	.70	8.09	2.63	1.47	14.86
Fast	950	275	20.00	2.37	.98	7.27	1.97	.76	13.35
Spray 2nd or additional finish coat									
Slow	800	325	26.60	1.31	.38	8.18	3.06	1.94	14.87
Medium	900	313	23.30	1.86	.65	7.44	2.44	1.36	13.75
Fast	1000	300	20.00	2.25	.94	6.67	1.82	.70	12.38

Ceiling estimates are based on overall dimensions (length times width) to 8 feet high. For heights above 8 feet, use the High Time Difficulty Factors on page 137. "Slow" work is based on a $10.50 hourly wage, "Medium" work on a $16.75 hourly wage, and "Fast" work on a $22.50 hourly wage. Other qualifications that apply to this table are on page 9.

	Labor SF per manhour	Material coverage SF/gallon	Material cost per gallon	Labor cost per 100 SF	Labor burden 100 SF	Material cost per 100 SF	Overhead per 100 SF	Profit per 100 SF	Total price per 100 SF

Ceilings, tongue & groove, paint grade, brush

	Labor SF per manhour	Material coverage SF/gallon	Material cost per gallon	Labor cost per 100 SF	Labor burden 100 SF	Material cost per 100 SF	Overhead per 100 SF	Profit per 100 SF	Total price per 100 SF
Flat latex, water base (material #5)									
Brush 1st coat									
Slow	55	300	17.10	19.09	5.42	5.70	9.37	5.94	45.52
Medium	65	288	15.00	25.77	8.97	5.21	9.79	5.47	55.21
Fast	75	275	12.90	30.00	12.51	4.69	8.74	3.36	59.30
Brush 2nd coat									
Slow	65	350	17.10	16.15	4.58	4.89	7.95	5.04	38.61
Medium	75	338	15.00	22.33	7.78	4.44	8.46	4.73	47.74
Fast	85	325	12.90	26.47	11.03	3.97	7.68	2.95	52.10
Brush 3rd or additional coats									
Slow	80	375	17.10	13.13	3.72	4.56	6.64	4.21	32.26
Medium	90	363	15.00	18.61	6.49	4.13	7.16	4.00	40.39
Fast	100	350	12.90	22.50	9.39	3.69	6.58	2.53	44.69
Sealer, water base (material #1)									
Brush prime coat									
Slow	60	300	16.30	17.50	4.97	5.43	8.65	5.48	42.03
Medium	70	288	14.30	23.93	8.35	4.97	9.12	5.10	51.47
Fast	80	275	12.30	28.13	11.73	4.47	8.20	3.15	55.68
Sealer, oil base (material #2)									
Brush prime coat									
Slow	60	375	22.90	17.50	4.97	6.11	8.86	5.62	43.06
Medium	70	363	20.10	23.93	8.35	5.54	9.26	5.18	52.26
Fast	80	350	17.20	28.13	11.73	4.91	8.28	3.18	56.23
Enamel, water base (material #9)									
Brush 1st finish coat									
Slow	55	300	24.50	19.09	5.42	8.17	10.13	6.42	49.23
Medium	65	288	21.40	25.77	8.97	7.43	10.33	5.78	58.28
Fast	75	275	18.40	30.00	12.51	6.69	9.11	3.50	61.81
Brush 2nd or additional finish coats									
Slow	70	375	24.50	15.00	4.26	6.53	8.00	5.07	38.86
Medium	80	363	21.40	20.94	7.30	5.90	8.36	4.68	47.18
Fast	90	350	18.40	25.00	10.43	5.26	7.53	2.89	51.11

	Labor SF per manhour	Material coverage SF/gallon	Material cost per gallon	Labor cost per 100 SF	Labor burden 100 SF	Material cost per 100 SF	Overhead per 100 SF	Profit per 100 SF	Total price per 100 SF
Enamel, oil base (material #10)									
Brush 1st finish coat									
Slow	55	375	26.60	19.09	5.42	7.09	9.80	6.21	47.61
Medium	65	363	23.30	25.77	8.97	6.42	10.09	5.64	56.89
Fast	75	350	20.00	30.00	12.51	5.71	8.92	3.43	60.57
Brush 2nd or additional finish coats									
Slow	70	425	26.60	15.00	4.26	6.26	7.91	5.02	38.45
Medium	80	413	23.30	20.94	7.30	5.64	8.30	4.64	46.82
Fast	90	400	20.00	25.00	10.43	5.00	7.48	2.88	50.79

Ceiling estimates are based on overall dimensions (length times width) to 8 feet high. For heights above 8 feet, use the High Time Difficulty Factors on page 137. "Slow" work is based on a $10.50 hourly wage, "Medium" work on a $16.75 hourly wage, and "Fast" work on a $22.50 hourly wage. Other qualifications that apply to this table are on page 9.

	Labor SF per manhour	Material coverage SF/gallon	Material cost per gallon	Labor cost per 100 SF	Labor burden 100 SF	Material cost per 100 SF	Overhead per 100 SF	Profit per 100 SF	Total price per 100 SF
Ceilings, tongue & groove, paint grade, roll									
Flat latex, water base (material #5)									
Roll 1st coat									
Slow	90	275	17.10	11.67	3.31	6.22	6.58	4.17	31.95
Medium	110	263	15.00	15.23	5.30	5.70	6.43	3.59	36.25
Fast	130	250	12.90	17.31	7.21	5.16	5.49	2.11	37.28
Roll 2nd coat									
Slow	140	325	17.10	7.50	2.12	5.26	4.62	2.93	22.43
Medium	155	313	15.00	10.81	3.76	4.79	4.75	2.65	26.76
Fast	170	300	12.90	13.24	5.51	4.30	4.27	1.64	28.96
Roll 3rd or additional coats									
Slow	190	350	17.10	5.53	1.56	4.89	3.72	2.36	18.06
Medium	200	338	15.00	8.38	2.92	4.44	3.86	2.16	21.76
Fast	210	325	12.90	10.71	4.47	3.97	3.54	1.36	24.05
Sealer, water base (material #1)									
Roll prime coat									
Slow	100	275	16.30	10.50	2.98	5.93	6.02	3.82	29.25
Medium	120	263	14.30	13.96	4.86	5.44	5.95	3.32	33.53
Fast	150	250	12.30	15.00	6.27	4.92	4.84	1.86	32.89
Sealer, oil base (material #2)									
Roll prime coat									
Slow	100	350	22.90	10.50	2.98	6.54	6.21	3.94	30.17
Medium	120	325	20.10	13.96	4.86	6.18	6.13	3.43	34.56
Fast	150	300	17.20	15.00	6.27	5.73	4.99	1.92	33.91
Enamel, water base (material #9)									
Roll 1st finish coat									
Slow	130	325	24.50	8.08	2.29	7.54	5.56	3.52	26.99
Medium	145	313	21.40	11.55	4.04	6.84	5.49	3.07	30.99
Fast	160	300	18.40	14.06	5.87	6.13	4.82	1.85	32.73
Roll 2nd or additional finish coats									
Slow	180	350	24.50	5.83	1.66	7.00	4.49	2.85	21.83
Medium	190	338	21.40	8.82	3.06	6.33	4.46	2.49	25.16
Fast	200	325	18.40	11.25	4.70	5.66	4.00	1.54	27.15

	Labor SF per manhour	Material coverage SF/gallon	Material cost per gallon	Labor cost per 100 SF	Labor burden 100 SF	Material cost per 100 SF	Overhead per 100 SF	Profit per 100 SF	Total price per 100 SF
Enamel, oil base (material #10)									
Roll 1st finish coat									
Slow	130	375	26.60	8.08	2.29	7.09	5.42	3.43	26.31
Medium	145	363	23.30	11.55	4.04	6.42	5.39	3.01	30.41
Fast	160	350	20.00	14.06	5.87	5.71	4.74	1.82	32.20
Roll 2nd or additional finish coats									
Slow	180	400	26.60	5.83	1.66	6.65	4.38	2.78	21.30
Medium	190	388	23.30	8.82	3.06	6.01	4.39	2.45	24.73
Fast	200	375	20.00	11.25	4.70	5.33	3.94	1.51	26.73

Ceiling estimates are based on overall dimensions (length times width) to 8 feet high. For heights above 8 feet, use the High Time Difficulty Factors on page 137. "Slow" work is based on a $10.50 hourly wage, "Medium" work on a $16.75 hourly wage, and "Fast" work on a $22.50 hourly wage. Other qualifications that apply to this table are on page 9.

	Labor SF per manhour	Material coverage SF/gallon	Material cost per gallon	Labor cost per 100 SF	Labor burden 100 SF	Material cost per 100 SF	Overhead per 100 SF	Profit per 100 SF	Total price per 100 SF
Ceilings, tongue & groove, paint grade, spray									
Flat latex, water base (material #5)									
Spray 1st coat									
Slow	300	180	17.10	3.50	.99	9.50	4.34	2.75	21.08
Medium	360	155	15.00	4.65	1.63	9.68	3.91	2.18	22.05
Fast	420	125	12.90	5.36	2.23	10.32	3.32	1.27	22.50
Spray 2nd coat									
Slow	420	250	17.10	2.50	.71	6.84	3.12	1.98	15.15
Medium	470	225	15.00	3.56	1.25	6.67	2.81	1.57	15.86
Fast	520	200	12.90	4.33	1.79	6.45	2.33	.90	15.80
Spray 3rd or additional coats									
Slow	520	350	17.10	2.02	.57	4.89	2.32	1.47	11.27
Medium	570	338	15.00	2.94	1.01	4.44	2.06	1.15	11.60
Fast	620	325	12.90	3.63	1.50	3.97	1.69	.65	11.44
Sealer, water base (material #1)									
Spray prime coat									
Slow	320	180	16.30	3.28	.94	9.06	4.11	2.61	20.00
Medium	380	155	14.30	4.41	1.53	9.23	3.72	2.08	20.97
Fast	440	125	12.30	5.11	2.13	9.84	3.16	1.21	21.45
Sealer, oil base (material #2)									
Spray prime coat									
Slow	320	200	22.90	3.28	.94	11.45	4.85	3.08	23.60
Medium	380	190	20.10	4.41	1.53	10.58	4.05	2.26	22.83
Fast	440	180	17.20	5.11	2.13	9.56	3.11	1.19	21.10
Enamel, water base (material #9)									
Spray 1st finish coat									
Slow	400	250	24.50	2.63	.74	9.80	4.09	2.59	19.85
Medium	450	225	21.40	3.72	1.29	9.51	3.56	1.99	20.07
Fast	500	200	18.40	4.50	1.88	9.20	2.88	1.11	19.57
Spray 2nd or additional finish coat									
Slow	500	350	24.50	2.10	.60	7.00	3.01	1.91	14.62
Medium	550	325	21.40	3.05	1.06	6.58	2.62	1.46	14.77
Fast	600	300	18.40	3.75	1.58	6.13	2.12	.81	14.39

	Labor SF per manhour	Material coverage SF/gallon	Material cost per gallon	Labor cost per 100 SF	Labor burden 100 SF	Material cost per 100 SF	Overhead per 100 SF	Profit per 100 SF	Total price per 100 SF
Enamel, oil base (material #10)									
Spray 1st finish coat									
Slow	400	270	26.60	2.63	.74	9.85	4.10	2.60	19.92
Medium	450	250	23.30	3.72	1.29	9.32	3.51	1.96	19.80
Fast	500	230	20.00	4.50	1.88	8.70	2.79	1.07	18.94
Spray 2nd or additional finish coat									
Slow	500	325	26.60	2.10	.60	8.18	3.37	2.14	16.39
Medium	550	313	23.30	3.05	1.06	7.44	2.83	1.58	15.96
Fast	600	300	20.00	3.75	1.58	6.67	2.22	.85	15.07

Ceiling estimates are based on overall dimensions (length times width) to 8 feet high. For heights above 8 feet, use the High Time Difficulty Factors on page 137. "Slow" work is based on a $10.50 hourly wage, "Medium" work on a $16.75 hourly wage, and "Fast" work on a $22.50 hourly wage. Other qualifications that apply to this table are on page 9.

	Labor SF per manhour	Material coverage SF/gallon	Material cost per gallon	Labor cost per 100 SF	Labor burden 100 SF	Material cost per 100 SF	Overhead per 100 SF	Profit per 100 SF	Total price per 100 SF
Ceilings, tongue & groove, stain grade									
Semi-transparent stain, water base (material #20)									
Roll & brush each coat									
Slow	200	300	21.40	5.25	1.49	7.13	4.30	2.73	20.90
Medium	240	275	18.70	6.98	2.44	6.80	3.97	2.22	22.41
Fast	280	250	16.00	8.04	3.34	6.40	3.29	1.27	22.34
Semi-transparent stain, oil base (material #21)									
Roll & brush each coat									
Slow	200	280	26.10	5.25	1.49	9.32	4.98	3.16	24.20
Medium	240	260	22.90	6.98	2.44	8.81	4.46	2.49	25.18
Fast	280	240	19.60	8.04	3.34	8.17	3.62	1.39	24.56
Semi-transparent stain, water base (material #20)									
Spray each coat									
Slow	300	220	21.40	3.50	.99	9.73	4.41	2.80	21.43
Medium	350	200	18.70	4.79	1.67	9.35	3.87	2.16	21.84
Fast	400	180	16.00	5.63	2.34	8.89	3.12	1.20	21.18
Semi-transparent stain, oil base (material #21)									
Spray each coat									
Slow	300	200	26.10	3.50	.99	13.05	5.44	3.45	26.43
Medium	350	188	22.90	4.79	1.67	12.18	4.57	2.55	25.76
Fast	400	175	19.60	5.63	2.34	11.20	3.55	1.36	24.08
Stain, seal and 2 coat lacquer system (7 step process)									
STEP 1: Sand & putty									
Slow	100	--	--	10.50	2.98	--	4.18	2.65	20.31
Medium	125	--	--	13.40	4.67	--	4.43	2.48	24.98
Fast	150	--	--	15.00	6.27	--	3.93	1.51	26.71
STEP 2 & 3: Wiping stain, oil base (material #11a) & wipe									
Roll & brush, 1 coat & wipe									
Slow	75	300	30.90	14.00	3.97	10.30	8.77	5.56	42.60
Medium	100	275	27.00	16.75	5.84	9.82	7.94	4.44	44.79
Fast	125	250	23.10	18.00	7.51	9.24	6.43	2.47	43.65
Spray, 1 coat & wipe									
Slow	275	150	30.90	3.82	1.09	20.60	7.91	5.01	38.43
Medium	200	125	27.00	8.38	2.92	21.60	8.06	4.51	45.47
Fast	325	100	23.10	6.92	2.90	23.10	6.09	2.34	41.35

	Labor SF per manhour	Material coverage SF/gallon	Material cost per gallon	Labor cost per 100 SF	Labor burden 100 SF	Material cost per 100 SF	Overhead per 100 SF	Profit per 100 SF	Total price per 100 SF
STEP 4: Sanding sealer (material #11b)									
Brush, 1 coat									
Slow	125	325	17.20	8.40	2.38	5.29	4.98	3.16	24.21
Medium	150	300	15.00	11.17	3.90	5.00	4.91	2.75	27.73
Fast	175	275	12.90	12.86	5.35	4.69	4.24	1.63	28.77
Spray, 1 coat									
Slow	350	150	17.20	3.00	.86	11.47	4.75	3.01	23.09
Medium	400	125	15.00	4.19	1.46	12.00	4.32	2.42	24.39
Fast	450	100	12.90	5.00	2.08	12.90	3.70	1.42	25.10
STEP 5: Sand lightly									
Slow	175	--	--	6.00	1.70	--	2.39	1.52	11.61
Medium	225	--	--	7.44	2.59	--	2.46	1.37	13.86
Fast	275	--	--	8.18	3.43	--	2.15	.83	14.59
STEP 6 & 7: Lacquer, 2 coats (material #11c)									
Brush, 1st coat									
Slow	150	350	19.30	7.00	1.99	5.51	4.50	2.85	21.85
Medium	200	338	16.80	8.38	2.92	4.97	3.99	2.23	22.49
Fast	275	325	14.40	8.18	3.43	4.43	2.97	1.14	20.15
Brush, 2nd coat									
Slow	200	400	19.30	5.25	1.49	4.83	3.59	2.27	17.43
Medium	250	375	16.80	6.70	2.34	4.48	3.31	1.85	18.68
Fast	325	350	14.40	6.92	2.90	4.11	2.58	.99	17.50
Spray, 1st coat									
Slow	425	350	19.30	2.47	.70	5.51	2.69	1.71	13.08
Medium	525	300	16.80	3.19	1.10	5.60	2.43	1.36	13.68
Fast	625	250	14.40	3.60	1.50	5.76	2.01	.77	13.64
Spray, 2nd coat									
Slow	475	425	19.30	2.21	.63	4.54	2.29	1.45	11.12
Medium	588	375	16.80	2.85	.99	4.48	2.04	1.14	11.50
Fast	650	325	14.40	3.46	1.45	4.43	1.73	.66	11.73

	Labor SF per manhour	Material coverage SF/gallon	Material cost per gallon	Labor cost per 100 SF	Labor burden 100 SF	Material cost per 100 SF	Overhead per 100 SF	Profit per 100 SF	Total price per 100 SF
Complete 7 step stain, seal & 2 coat lacquer system (material #11)									
Brush all coats									
Slow	30	165	21.60	35.00	9.93	13.09	18.00	11.41	87.43
Medium	35	155	19.60	47.86	16.68	12.65	18.91	10.57	106.67
Fast	40	145	16.20	56.25	23.48	11.17	16.82	6.46	114.18
Spray all coats									
Slow	60	60	21.60	17.50	4.97	36.00	18.13	11.49	88.09
Medium	70	50	19.60	23.93	8.35	39.20	17.51	9.79	98.78
Fast	80	40	16.20	28.13	11.73	40.50	14.87	5.71	100.94

Ceiling estimates are based on overall dimensions (length times width) to 8 feet high. For heights above 8 feet, use the High Time Difficulty Factors on page 137. "Slow" work is based on a $10.50 hourly wage, "Medium" work on a $16.75 hourly wage, and "Fast" work on a $22.50 hourly wage. Other qualifications that apply to this table are on page 9.

	Labor SF per manhour	Material coverage SF/gallon	Material cost per gallon	Labor cost per 100 SF	Labor burden 100 SF	Material cost per 100 SF	Overhead per 100 SF	Profit per 100 SF	Total price per 100 SF
Closet pole, stain grade									
Penetrating oil stain, (material #13)									
Brush & wipe, 1 coat									
Slow	40	225	29.00	26.25	7.45	12.89	14.45	9.16	70.20
Medium	50	213	25.40	33.50	11.68	11.92	13.99	7.82	78.91
Fast	60	200	21.80	37.50	15.66	10.90	11.85	4.55	80.46

To stain poles in new construction, apply stain before installation. On repaints, remove the pole before staining. When estimating by the Opening count method, count one opening for each 10 linear feet of pole. "Slow" work is based on a $10.50 hourly wage, "Medium" work on a $16.75 hourly wage, and "Fast" work on a $22.50 hourly wage. Other qualifications that apply to this table are on page 9.

	Labor LF per manhour	Material coverage LF/gallon	Material cost per gallon	Labor cost per 100 LF	Labor burden 100 LF	Material cost per 100 LF	Overhead per 100 LF	Profit per 100 LF	Total price per 100 LF
Closet shelf & pole, paint grade									
Undercoat, water base (material #3)									
Brush 1 coat									
Slow	17	80	21.10	61.76	17.53	26.38	32.77	20.77	159.21
Medium	22	70	18.50	76.14	26.53	26.43	31.63	17.68	178.41
Fast	33	60	15.80	68.18	28.45	26.33	22.75	8.74	154.45
Undercoat, oil base (material #4)									
Brush 1 coat									
Slow	17	90	21.60	61.76	17.53	24.00	32.03	20.30	155.62
Medium	22	80	18.90	76.14	26.53	23.63	30.94	17.30	174.54
Fast	33	65	16.20	68.18	28.45	24.92	22.49	8.64	152.68
Split coat (1/2 undercoat + 1/2 enamel), water base (material #3 & #9)									
Brush 1 coat									
Slow	16	80	22.80	65.63	18.62	28.50	34.97	22.17	169.89
Medium	21	70	19.95	79.76	27.81	28.50	33.33	18.63	188.03
Fast	32	60	17.10	70.31	29.35	28.50	23.71	9.11	160.98
Split coat (1/2 undercoat + 1/2 enamel), oil base (material #4 & #10)									
Brush 1 coat									
Slow	16	90	24.10	65.63	18.62	26.78	34.43	21.83	167.29
Medium	21	80	21.10	79.76	27.81	26.38	32.82	18.34	185.11
Fast	32	65	18.10	70.31	29.35	27.85	23.59	9.07	160.17
Enamel, water base (material #9)									
Brush each coat									
Slow	15	80	24.50	70.00	19.87	30.63	37.37	23.69	181.56
Medium	20	70	21.40	83.75	29.20	30.57	35.16	19.65	198.33
Fast	30	60	18.40	75.00	31.29	30.67	25.34	9.74	172.04
Enamel, oil base (material #10)									
Brush each coat									
Slow	15	90	26.60	70.00	19.87	29.56	37.04	23.48	179.95
Medium	20	80	23.30	83.75	29.20	29.13	34.81	19.46	196.35
Fast	30	65	20.00	75.00	31.29	30.77	25.36	9.75	172.17

Use these costs for painting the wardrobe closet shelves and poles with an undercoat and enamel system. If painting wardrobe closet shelves and poles with flat latex paint along with walls, use the Opening count method discribed under Doors, interior openings. Measurements are based on linear feet (LF) of shelves and poles. "Slow" work is based on a $10.50 hourly wage, "Medium" work on a $16.75 hourly wage, and "Fast" work on a $22.50 hourly wage. Other qualifications that apply to this table are on page 9.

	Labor LF per manhour	Material coverage LF/gallon	Material cost per gallon	Labor cost per 100 LF	Labor burden 100 LF	Material cost per 100 LF	Overhead per 100 LF	Profit per 100 LF	Total price per 100 LF
Closet shelves, paint grade									
Undercoat, water base (material #3)									
Brush 1 coat									
Slow	30	100	21.10	35.00	9.93	21.10	20.48	12.98	99.49
Medium	37	90	18.50	45.27	15.79	20.56	19.99	11.18	112.79
Fast	44	80	15.80	51.14	21.35	19.75	17.06	6.56	115.86
Undercoat, oil base (material #4)									
Brush 1 coat									
Slow	30	110	21.60	35.00	9.93	19.64	20.03	12.69	97.29
Medium	37	100	18.90	45.27	15.79	18.90	19.59	10.95	110.50
Fast	44	90	16.20	51.14	21.35	18.00	16.74	6.43	113.66
Split coat (1/2 undercoat + 1/2 enamel), water base (material #3 & #9)									
Brush 1 coat									
Slow	27	100	22.80	38.89	11.04	22.80	22.55	14.30	109.58
Medium	35	90	19.95	47.86	16.68	22.17	21.24	11.87	119.82
Fast	42	80	17.10	53.57	22.36	21.38	18.00	6.92	122.23
Split coat (1/2 undercoat + 1/2 enamel), oil base (material #4 & #10)									
Brush 1 coat									
Slow	27	110	24.10	38.89	11.04	21.91	22.28	14.12	108.24
Medium	35	100	21.10	47.86	16.68	21.10	20.98	11.73	118.35
Fast	42	90	18.10	53.57	22.36	20.11	17.77	6.83	120.64
Enamel, water base (material #9)									
Brush each coat									
Slow	25	100	24.50	42.00	11.92	24.50	24.32	15.42	118.16
Medium	33	90	21.40	50.76	17.69	23.78	22.60	12.63	127.46
Fast	40	80	18.40	56.25	23.48	23.00	19.01	7.30	129.04
Enamel, oil base (material #10)									
Brush each coat									
Slow	25	110	26.60	42.00	11.92	24.18	24.22	15.35	117.67
Medium	33	100	23.30	50.76	17.69	23.30	22.48	12.57	126.80
Fast	40	90	20.00	56.25	23.48	22.22	18.86	7.25	128.06

Use these costs for painting the wardrobe closet shelves with an undercoat and enamel system. If painting wardrobe closet shelves with flat latex paint along with walls, use the Opening count method described under Doors, interior openings. Measurements are based on linear feet (LF) of shelves. "Slow" work is based on a $10.50 hourly wage, "Medium" work on a $16.75 hourly wage, and "Fast" work on a $22.50 hourly wage. Other qualifications that apply to this table are on page 9.

	Labor LF per manhour	Material coverage LF/gallon	Material cost per gallon	Labor cost per 100 LF	Labor burden 100 LF	Material cost per 100 LF	Overhead per 100 LF	Profit per 100 LF	Total price per 100 LF
Closets, molding at perimeter, paint grade									
Undercoat, water base (material #3)									
Brush 1 coat									
Slow	80	225	21.10	13.13	3.72	9.38	8.14	5.16	39.53
Medium	95	213	18.50	17.63	6.16	8.69	7.95	4.45	44.88
Fast	110	200	15.80	20.45	8.54	7.90	6.82	2.62	46.33
Undercoat, oil base (material #4)									
Brush 1 coat									
Slow	80	250	21.60	13.13	3.72	8.64	7.91	5.01	38.41
Medium	95	238	18.90	17.63	6.16	7.94	7.77	4.34	43.84
Fast	110	220	16.20	20.45	8.54	7.36	6.72	2.58	45.65
Split coat (1/2 undercoat + 1/2 enamel), water base (material #3 & #9)									
Brush 1 coat									
Slow	75	225	22.80	14.00	3.97	10.13	8.71	5.52	42.33
Medium	90	213	19.95	18.61	6.49	9.37	8.45	4.72	47.64
Fast	105	200	17.10	21.43	8.93	8.55	7.20	2.77	48.88
Split coat (1/2 undercoat + 1/2 enamel), oil base (material #4 & #10)									
Brush 1 coat									
Slow	75	250	24.10	14.00	3.97	9.64	8.56	5.43	41.60
Medium	90	238	21.10	18.61	6.49	8.87	8.32	4.65	46.94
Fast	105	220	18.10	21.43	8.93	8.23	7.14	2.75	48.48
Enamel, water base (material #9)									
Brush each coat									
Slow	70	225	24.50	15.00	4.26	10.89	9.35	5.93	45.43
Medium	85	213	21.40	19.71	6.86	10.05	8.97	5.02	50.61
Fast	100	200	18.40	22.50	9.39	9.20	7.60	2.92	51.61
Enamel, oil base (material #10)									
Brush each coat									
Slow	70	250	26.60	15.00	4.26	10.64	9.27	5.88	45.05
Medium	85	238	23.30	19.71	6.86	9.79	8.91	4.98	50.25
Fast	100	220	20.00	22.50	9.39	9.09	7.58	2.91	51.47

Use these costs for molding around wardrobe closets. Measurements are based on linear feet (LF) of molding. "Slow" work is based on a $10.50 hourly wage, "Medium" work on a $16.75 hourly wage, and "Fast" work on a $22.50 hourly wage. Other qualifications that apply to this table are on page 9.

	Labor LF per manhour	Material coverage LF/gallon	Material cost per gallon	Labor cost per 100 LF	Labor burden 100 LF	Material cost per 100 LF	Overhead per 100 LF	Profit per 100 LF	Total price per 100 LF
Corbels, wood trim, stain grade, average size 4" x 8"									
Solid body stain, water base (material #18)									
Brush each coat									
Slow	15	50	21.50	70.00	19.87	43.00	41.21	26.12	200.20
Medium	20	48	18.80	83.75	29.20	39.17	37.27	20.83	210.22
Fast	25	45	16.10	90.00	37.56	35.78	30.22	11.61	205.17
Solid body stain, oil base (material #19)									
Brush each coat									
Slow	15	55	23.90	70.00	19.87	43.45	41.34	26.21	200.87
Medium	20	53	20.90	83.75	29.20	39.43	37.33	20.87	210.58
Fast	25	50	17.90	90.00	37.56	35.80	30.23	11.62	205.21
Semi-transparent stain, water base (material #20)									
Brush each coat									
Slow	18	55	21.40	58.33	16.56	38.91	35.29	22.37	171.46
Medium	22	53	18.70	76.14	26.53	35.28	33.80	18.89	190.64
Fast	28	50	16.00	80.36	33.52	32.00	26.99	10.37	183.24
Semi-transparent stain, oil base (material #21)									
Brush each coat									
Slow	18	60	26.10	58.33	16.56	43.50	36.71	23.27	178.37
Medium	22	58	22.90	76.14	26.53	39.48	34.83	19.47	196.45
Fast	28	55	19.60	80.36	33.52	35.64	27.67	10.63	187.82

Use these costs for painting corbels averaging 4" x 8" in size. Measurements are based on linear feet (LF) of corbels that are painted or stained with a different material or color than the surface they extend from. "Slow" work is based on a $10.50 hourly wage, "Medium" work on a $16.75 hourly wage, and "Fast" work on a $22.50 hourly wage. Other qualifications that apply to this table are on page 9.

Deck overhang, wood

Multiply the horizontal surface area by 1.5 to allow for painting floor joists and use the overhang table for areas greater than 2.5 feet to determine pricing.

Deck surfaces, steps, stair treads & porches, wood

Measure the surface area and apply the prices for smooth siding.

	Labor LF per manhour	Material coverage LF/gallon	Material cost per gallon	Labor cost per 100 LF	Labor burden 100 LF	Material cost per 100 LF	Overhead per 100 LF	Profit per 100 LF	Total price per 100 LF

Door frames and trim only, per 100 linear feet

	Labor LF per manhour	Material coverage LF/gallon	Material cost per gallon	Labor cost per 100 LF	Labor burden 100 LF	Material cost per 100 LF	Overhead per 100 LF	Profit per 100 LF	Total price per 100 LF
Undercoat, water base (material #3)									
Brush 1 coat									
Slow	220	510	21.10	4.77	1.36	4.14	3.18	2.02	15.47
Medium	270	465	18.50	6.20	2.16	3.98	3.02	1.69	17.05
Fast	320	425	15.80	7.03	2.95	3.72	2.53	.97	17.20
Undercoat, oil base (material #4)									
Brush 1 coat									
Slow	220	560	21.60	4.77	1.36	3.86	3.10	1.96	15.05
Medium	270	510	18.90	6.20	2.16	3.71	2.96	1.65	16.68
Fast	320	465	16.20	7.03	2.95	3.48	2.49	.96	16.91
Split coat (1/2 undercoat + 1/2 enamel), water base (material #3 & #9)									
Brush 1 coat									
Slow	210	510	22.80	5.00	1.42	4.47	3.38	2.14	16.41
Medium	258	465	19.95	6.49	2.27	4.29	3.19	1.79	18.03
Fast	305	425	17.10	7.38	3.08	4.02	2.68	1.03	18.19
Split coat (1/2 undercoat + 1/2 enamel), oil base (material #4 & #10)									
Brush 1 coat									
Slow	210	560	24.10	5.00	1.42	4.30	3.32	2.11	16.15
Medium	258	510	21.10	6.49	2.27	4.14	3.16	1.77	17.83
Fast	305	465	18.10	7.38	3.08	3.89	2.65	1.02	18.02
Enamel, water base (material #9)									
Brush each coat									
Slow	200	510	24.50	5.25	1.49	4.80	3.58	2.27	17.39
Medium	245	465	21.40	6.84	2.38	4.60	3.39	1.89	19.10
Fast	290	425	18.40	7.76	3.24	4.33	2.84	1.09	19.26
Enamel, oil base (material #10)									
Brush each coat									
Slow	200	560	26.60	5.25	1.49	4.75	3.56	2.26	17.31
Medium	245	510	23.30	6.84	2.38	4.57	3.38	1.89	19.06
Fast	290	465	20.00	7.76	3.24	4.30	2.83	1.09	19.22

Use these costs for painting door frames and wood trim on all sides when doors are not to be painted. When doors are painted along with the frames and trim, use the Opening count method under Doors, interior openings, or the exterior door costs under Doors, exterior. Measurements for door frames and trim are based on linear feet (LF) of one side of the frame. Prices are for about 17 linear feet of frame at each opening. "Slow" work is based on a $10.50 hourly wage, "Medium" work on a $16.75 hourly wage, and "Fast" work on a $22.50 hourly wage. Other qualifications that apply to this table are on page 9.

	Openings per manhour	Openings per gallon	Material cost per gallon	Labor cost per opening	Labor burden opening	Material cost per opening	Overhead per opening	Profit per opening	Total price per opening
Door frames and trim only, per opening									
Undercoat, water base (material #3)									
Brush 1 coat									
Slow	13	30	21.10	.81	.23	.70	.54	.34	2.62
Medium	16	28	18.50	1.05	.37	.66	.51	.28	2.87
Fast	18	25	15.80	1.25	.54	.63	.44	.17	3.03
Undercoat, oil base (material #4)									
Brush 1 coat									
Slow	13	33	21.60	.81	.23	.65	.52	.33	2.54
Medium	16	31	18.90	1.05	.37	.61	.50	.28	2.81
Fast	18	28	16.20	1.25	.54	.58	.43	.17	2.97
Split coat (1/2 undercoat + 1/2 enamel), water base (material #3 & #9)									
Brush 1 coat									
Slow	12	30	22.80	.88	.24	.76	.59	.37	2.84
Medium	14	28	19.95	1.20	.40	.71	.57	.32	3.20
Fast	16	25	17.10	1.41	.60	.68	.50	.19	3.38
Split coat (1/2 undercoat + 1/2 enamel), oil base (material #4 & #10)									
Brush 1 coat									
Slow	12	33	24.10	.88	.24	.73	.58	.37	2.80
Medium	14	31	21.10	1.20	.40	.68	.56	.31	3.15
Fast	16	28	18.10	1.41	.60	.65	.49	.19	3.34
Enamel, water base (material #9)									
Brush each coat									
Slow	11	30	24.50	.95	.28	.82	.63	.40	3.08
Medium	14	28	21.40	1.20	.40	.76	.58	.33	3.27
Fast	16	25	18.40	1.41	.60	.74	.51	.20	3.46
Enamel, oil base (material #10)									
Brush each coat									
Slow	11	33	26.60	.95	.28	.81	.63	.40	3.07
Medium	14	31	23.30	1.20	.40	.75	.58	.32	3.25
Fast	16	28	20.00	1.41	.60	.71	.50	.19	3.41

Use these costs for painting door frames and wood trim on all sides when doors are not to be painted. When doors are painted along with the frames and trim, use the Opening count method under Doors, interior openings, or the exterior door costs under Doors, exterior. These costs are based on a count of the openings requiring paint. Prices are for about 17 linear feet of frame at each opening. "Slow" work is based on a $10.50 hourly wage, "Medium" work on a $16.75 hourly wage, and "Fast" work on a $22.50 hourly wage. Other qualifications that apply to this table are on page 9.

Doors, exterior

The tables that follow include costs for both time and material needed to apply two coats of a high quality finish to all six sides of each door, finish the jamb and trim, and lay-off each door smoothly. These costs are in addition to those shown under Doors, interior openings, for both the opening count method, and the per door method which include one coat of undercoat and one coat of enamel for each exterior door along with the interior doors. New paint grade doors actually receive two coats of exterior enamel. New stain grade doors actually receive a coat of stain, sealer, and then a coat of either marine spar varnish or polyurethane finish. The following example gives total cost to finish a flush exterior door with polyurethane, and includes the cost for the two coats from the interior take-off.

Included in the interior take-off	
Opening count method, Interior undercoat cost	$ 10.66
Opening count method, Interior enamel cost	$ 11.07
Included in the exterior take-off	
Exterior polyurethane cost	$ 24.55
Total to finish the exterior door	$ 46.28

Under this system, almost half the cost to paint the exterior door is included in the interior take-off. Be sure you include all the exterior doors whether you use either the opening count or per door method to estimate doors.

	Manhours per door	Doors per gallon	Material cost per gallon	Labor cost per door	Labor burden door	Material cost per door	Overhead per door	Profit per door	Total price per door
Doors, exterior, flush, two coat system									
Exterior enamel, 2 coats, water base (material #24)									
Roll & brush each coat									
Slow	0.5	7	27.30	5.25	1.49	3.90	3.30	2.09	16.03
Medium	0.4	6	23.90	6.70	2.34	3.98	3.19	1.78	17.99
Fast	0.3	5	20.50	6.75	2.82	4.10	2.53	.97	17.17
Exterior enamel, 2 coats, oil base (material #25)									
Roll & brush each coat									
Slow	0.5	7	30.70	5.25	1.49	4.39	3.45	2.19	16.77
Medium	0.4	6	26.80	6.70	2.34	4.47	3.31	1.85	18.67
Fast	0.3	5	23.00	6.75	2.82	4.60	2.62	1.01	17.80
Polyurethane (material #22)									
Brush 2 coats									
Slow	0.7	5.0	34.30	7.35	2.09	6.86	5.05	3.20	24.55
Medium	0.6	4.5	30.00	10.05	3.50	6.67	4.95	2.77	27.94
Fast	0.5	4.0	25.70	11.25	4.70	6.43	4.14	1.59	28.11
Marine spar varnish, flat or gloss (material #23)									
Brush 2 coats									
Slow	0.5	6	39.40	5.25	1.49	6.57	4.13	2.62	20.06
Medium	0.4	5	34.50	6.70	2.34	6.90	3.90	2.18	22.02
Fast	0.3	4	29.50	6.75	2.82	7.38	3.14	1.21	21.30
ADD - Preparation for spar varnish									
Steel wool buff	0.2	--	--	3.35	1.17	--	1.11	.62	6.25
Wax application	0.2	--	--	3.35	1.17	--	1.11	.62	6.25

Use these figures for painting flush exterior doors, other than entry doors. These doors require a two coat system, and these costs are included with the exterior door take-off. Add minimum preparation time for varnishing as indicated above. These costs are in addition to the costs that are included in the interior take-off as explained by the example in the previous section. "Slow" work is based on a $10.50 hourly wage, "Medium" work on a $16.75 hourly wage, and "Fast" work on a $22.50 hourly wage. Other qualifications that apply to this table are on page 9.

	Manhours per door	Doors per gallon	Material cost per gallon	Labor cost per door	Labor burden door	Material cost per door	Overhead per door	Profit per door	Total price per door
Doors, exterior, French, two coat system									
Exterior enamel, 2 coats, water base (material #24)									
Roll & brush each coat									
Slow	1.0	12	27.30	10.50	2.98	2.28	4.89	3.10	23.75
Medium	0.8	10	23.90	13.40	4.67	2.39	5.01	2.80	28.27
Fast	0.6	8	20.50	13.50	5.63	2.56	4.01	1.54	27.24
Exterior enamel, 2 coats, oil base (material #25)									
Roll & brush each coat									
Slow	1.0	12	30.70	10.50	2.98	2.56	4.98	3.15	24.17
Medium	0.8	10	26.80	13.40	4.67	2.68	5.08	2.84	28.67
Fast	0.6	8	23.00	13.50	5.63	2.88	4.07	1.57	27.65
Polyurethane (material #22)									
Brush 2 coats									
Slow	1.5	8.0	34.30	15.75	4.47	4.29	7.60	4.82	36.93
Medium	1.3	7.5	30.00	21.78	7.59	4.00	8.18	4.57	46.12
Fast	1.0	7.0	25.70	22.50	9.39	3.67	6.58	2.53	44.67
Marine spar varnish, flat or gloss (material #23)									
Brush 2 coats									
Slow	1.0	12	39.40	10.50	2.98	3.28	5.20	3.30	25.26
Medium	0.8	10	34.50	13.40	4.67	3.45	5.27	2.95	29.74
Fast	0.6	8	29.50	13.50	5.63	3.69	4.22	1.62	28.66
ADD - Preparation for spar varnish									
Steel wool buff	0.2	--	--	3.35	1.17	--	1.11	.62	6.25
Wax application	0.2	--	--	3.35	1.17	--	1.11	.62	6.25

Use these figures for painting exterior French doors that have 10 to 15 lites. These doors require a two coat system, and these costs are included with the exterior door take-off. Add minimum preparation time for varnishing as indicated above. These costs are in addition to the costs that are included in the interior take-off as explained by the example in the previous section. "Slow" work is based on a $10.50 hourly wage, "Medium" work on a $16.75 hourly wage, and "Fast" work on a $22.50 hourly wage. Other qualifications that apply to this table are on page 9.

	Manhours per door	Doors per gallon	Material cost per gallon	Labor cost per door	Labor burden door	Material cost per door	Overhead per door	Profit per door	Total price per door
Doors, exterior, louvered, two coat system									
Exterior enamel, 2 coats, water base (material #24)									
Roll & brush each coat									
Slow	1.4	7	27.30	14.70	4.17	3.90	7.06	4.48	34.31
Medium	1.1	6	23.90	18.43	6.42	3.98	7.06	3.95	39.84
Fast	0.7	5	20.50	15.75	6.57	4.10	4.89	1.88	33.19
Exterior enamel, 2 coats, oil base (material #25)									
Roll & brush each coat									
Slow	1.4	7	30.70	14.70	4.17	4.39	7.21	4.57	35.04
Medium	1.1	6	26.80	18.43	6.42	4.47	7.18	4.02	40.52
Fast	0.7	5	23.00	15.75	6.57	4.60	4.98	1.91	33.81
Polyurethane (material #22)									
Brush 2 coats									
Slow	1.7	5.0	34.30	17.85	5.07	6.86	9.23	5.85	44.86
Medium	1.5	4.5	30.00	25.13	8.76	6.67	9.94	5.56	56.06
Fast	1.2	4.0	25.70	27.00	11.27	6.43	8.27	3.18	56.15
Marine spar varnish, flat or gloss (material #23)									
Brush 2 coats									
Slow	1.4	7	39.40	14.70	4.17	5.63	7.60	4.82	36.92
Medium	1.1	6	34.50	18.43	6.42	5.75	7.50	4.19	42.29
Fast	0.7	5	29.50	15.75	6.57	5.90	5.22	2.01	35.45
ADD - Preparation for spar varnish									
Steel wool buff	0.2	--	--	3.35	1.17	--	1.11	.62	6.25
Wax application	0.2	--	--	3.35	1.17	--	1.11	.62	6.25

Use these figures for painting exterior louvered doors. These doors require a two coat system, and these costs are included with the exterior door take-off. Add minimum preparation time for varnishing as indicated above. These costs are in addition to the costs that are included in the interior take-off as explained by the example in the previous section. "Slow" work is based on a $10.50 hourly wage, "Medium" work on a $16.75 hourly wage, and "Fast" work on a $22.50 hourly wage. Other qualifications that apply to this table are on page 9.

	Manhours per door	Doors per gallon	Material cost per gallon	Labor cost per door	Labor burden door	Material cost per door	Overhead per door	Profit per door	Total price per door
Doors, exterior, panel (entry), two coat system									
Exterior enamel, 2 coats, water base (material #24)									
Roll & brush each coat									
Slow	1.4	4	27.30	14.70	4.17	6.83	7.97	5.05	38.72
Medium	1.1	3	23.90	18.43	6.42	7.97	8.04	4.49	45.35
Fast	0.8	2	20.50	18.00	7.51	10.25	6.62	2.54	44.92
Exterior enamel, 2 coats, oil base (material #25)									
Roll & brush each coat									
Slow	1.4	4	30.70	14.70	4.17	7.68	8.23	5.22	40.00
Medium	1.1	3	26.80	18.43	6.42	8.93	8.28	4.63	46.69
Fast	0.8	2	23.00	18.00	7.51	11.50	6.85	2.63	46.49
Polyurethane (material #22)									
Brush 2 coats									
Slow	1.7	4	34.30	17.85	5.07	8.58	9.77	6.19	47.46
Medium	1.5	3	30.00	25.13	8.76	10.00	10.75	6.01	60.65
Fast	1.2	2	25.70	27.00	11.27	12.85	9.46	3.63	64.21
Marine spar varnish, flat or gloss (material #23)									
Brush 2 coats									
Slow	1.4	4	39.40	14.70	4.17	9.85	8.91	5.65	43.28
Medium	1.1	3	34.50	18.43	6.42	11.50	8.91	4.98	50.24
Fast	0.8	2	16.00	18.00	7.51	8.00	6.20	2.38	42.09
ADD - Preparation for spar varnish									
Steel wool buff	0.2	--	--	3.35	1.17	--	1.11	.62	6.25
Wax application	0.2	--	--	3.35	1.17	--	1.11	.62	6.25

Use these figures for painting typical exterior paneled doors. These doors require a two coat system, and these costs are included with the exterior door take-off. Add minimum preparation time for varnishing as indicated above. These costs are in addition to the costs that are included in the interior take-off as explained by the example in the previous section. "Slow" work is based on a $10.50 hourly wage, "Medium" work on a $16.75 hourly wage, and "Fast" work on a $22.50 hourly wage. Other qualifications that apply to this table are on page 9.

Doors, Opening count method

Many painting companies estimate paint grade doors, jambs, frames, wood windows, pullmans, linens, bookcases, wine racks and other interior surfaces that take an undercoat and enamel finish by the "opening." Each opening is considered to take the same time regardless of whether it's a door, window, pullman, etc. These figures are based on the number of openings finished per 8-hour day and the material required per opening. The opening count method of estimating involves counting the quantity of all openings (including exterior doors) based on the opening allowance table at Figure 20 below. After you determine the number of openings use the following table in accumulated multiples of 10 for applying undercoat and enamel. The undercoat process is based on 11 to 13 openings per gallon and enamel is based on 10 to 12 openings per gallon. As an example, using the medium rate for water based material on 12 openings with 1 coat of undercoat and 1 coat of enamel, add the 10 opening figures for each coat to the 2 opening figures for each coat as follows:

	Undercoat	Enamel
10 openings	$96.40	$102.14
2 openings	19.97	20.93
12 openings total	$116.37	$123.07

Item		Opening Count
Closets	Molding at closet perimeter	Count 1 opening per 25'0" length
	Poles, stain	Count 1 opening per 10'0" length
	Shelf & pole (undercoat or enamel)	Count 1 opening per 6'0" length
	Shelves (undercoat or enamel)	Count 1 opening per 10'0" length
Doors	Bifold doors & frames	Count 1 opening per door
	Dutch doors & frames	Count 2 openings per door
	Entry doors & frames	Count 1 opening per door
	Forced air unit doors & frames	Count 1 opening per door
	French doors & frames	Count 1.5 openings per door
	Linen doors with face frame	Count 1 opening per 2'0" width
	Louvered bifold doors & frames	Count 1 opening per door panel
	false	Count 1 opening per door panel
	real	Count 1 opening per door or per 1'6" width
	Passage doors & frames	
	flush	Count 1 opening per door
	paneled	Count 1.25 openings per door
	Wardrobe doors	Count 1 opening per door
	Split coat operation, doors & frames	Count 1 opening per door
	Tipoff operation (doors only)	Count .5 opening per door
Pullman cabinets		Count 1 opening per lavatory or per 4'0" width
Windows, wood		Count 1 opening per 6 SF of window

Figure 20
Interior opening count allowance table

	Manhours per opening(s)	Gallons per opening(s)	Material cost per gallon	Labor cost per opening(s)	Labor burden opening(s)	Material cost per opening(s)	Overhead per opening(s)	Profit per opening(s)	Total price per opening(s)
Doors, interior openings, based on opening count method									
1 opening total									
Undercoat, water base (material #3)									
Roll & brush 1 coat									
Slow	0.4	0.080	21.10	4.20	1.19	1.69	2.19	1.39	10.66
Medium	0.3	0.085	18.50	5.03	1.75	1.57	2.05	1.14	11.54
Fast	0.2	0.090	15.80	4.50	1.88	1.42	1.44	.55	9.79
Undercoat, oil base (material #4)									
Roll & brush 1 coat									
Slow	0.4	0.080	21.60	4.20	1.19	1.73	2.21	1.40	10.73
Medium	0.3	0.085	18.90	5.03	1.75	1.61	2.06	1.15	11.60
Fast	0.2	0.090	16.20	4.50	1.88	1.46	1.45	.56	9.85
Enamel, water base (material #9)									
Roll & brush 1 coat									
Slow	0.4	0.08	24.50	4.20	1.19	1.96	2.28	1.44	11.07
Medium	0.3	0.09	21.40	5.03	1.75	1.93	2.13	1.19	12.03
Fast	0.2	0.10	18.40	4.50	1.88	1.84	1.52	.58	10.32
Enamel, oil base (material #10)									
Roll & brush 1 coat									
Slow	0.4	0.08	26.60	4.20	1.19	2.13	2.33	1.48	11.33
Medium	0.3	0.09	23.30	5.03	1.75	2.10	2.18	1.22	12.28
Fast	0.2	0.10	20.00	4.50	1.88	2.00	1.55	.60	10.53
2 openings total									
Undercoat, water base (material #3)									
Roll & brush 1 coat									
Slow	0.7	0.16	21.10	7.35	2.09	3.38	3.97	2.52	19.31
Medium	0.5	0.17	18.50	8.38	2.92	3.15	3.54	1.98	19.97
Fast	0.3	0.18	15.80	6.75	2.82	2.84	2.30	.88	15.59
Undercoat, oil base (material #4)									
Roll & brush 1 coat									
Slow	0.7	0.16	21.60	7.35	2.09	3.46	4.00	2.54	19.44
Medium	0.5	0.17	18.90	8.38	2.92	3.21	3.55	1.99	20.05
Fast	0.3	0.18	16.20	6.75	2.82	2.92	2.31	.89	15.69
Enamel, water base (material #9)									
Roll & brush 1 coat									
Slow	0.7	0.17	24.50	7.35	2.09	4.17	4.22	2.67	20.50
Medium	0.5	0.18	21.40	8.38	2.92	3.85	3.71	2.07	20.93
Fast	0.3	0.20	18.40	6.75	2.82	3.68	2.45	.94	16.64

	Manhours per opening(s)	Gallons per opening(s)	Material cost per gallon	Labor cost per opening(s)	Labor burden opening(s)	Material cost per opening(s)	Overhead per opening(s)	Profit per opening(s)	Total price per opening(s)
Enamel, oil base (material #10)									
Roll & brush 1 coat									
Slow	0.7	0.17	26.60	7.35	2.09	4.52	4.33	2.74	21.03
Medium	0.5	0.18	23.30	8.38	2.92	4.19	3.80	2.12	21.41
Fast	0.3	0.20	20.00	6.75	2.82	4.00	2.51	.96	17.04
3 openings total									
Undercoat, water base (material #3)									
Roll & brush 1 coat									
Slow	1.00	0.23	21.10	10.50	2.98	4.85	5.69	3.60	27.62
Medium	0.75	0.25	18.50	12.56	4.38	4.63	5.28	2.95	29.80
Fast	0.50	0.27	15.80	11.25	4.70	4.27	3.74	1.44	25.40
Undercoat, oil base (material #4)									
Roll & brush 1 coat									
Slow	1.00	0.23	21.60	10.50	2.98	4.97	5.72	3.63	27.80
Medium	0.75	0.25	18.90	12.56	4.38	4.73	5.31	2.97	29.95
Fast	0.50	0.27	16.20	11.25	4.70	4.37	3.76	1.44	25.52
Enamel, water base (material #9)									
Roll & brush 1 coat									
Slow	1.00	0.25	24.50	10.50	2.98	6.13	6.08	3.86	29.55
Medium	0.75	0.27	21.40	12.56	4.38	5.78	5.57	3.11	31.40
Fast	0.50	0.30	18.40	11.25	4.70	5.52	3.97	1.53	26.97
Enamel, oil base (material #10)									
Roll & brush 1 coat									
Slow	1.00	0.25	26.60	10.50	2.98	6.65	6.24	3.96	30.33
Medium	0.75	0.27	23.30	12.56	4.38	6.29	5.69	3.18	32.10
Fast	0.50	0.30	20.00	11.25	4.70	6.00	4.06	1.56	27.57
4 openings total									
Undercoat, water base (material #3)									
Roll & brush 1 coat									
Slow	1.3	0.31	21.10	13.65	3.87	6.54	7.46	4.73	36.25
Medium	1.0	0.33	18.50	16.75	5.84	6.11	7.03	3.93	39.66
Fast	0.7	0.36	15.80	15.75	6.57	5.69	5.18	1.99	35.18
Undercoat, oil base (material #4)									
Roll & brush 1 coat									
Slow	1.3	0.31	21.60	13.65	3.87	6.70	7.51	4.76	36.49
Medium	1.0	0.33	18.90	16.75	5.84	6.24	7.06	3.95	39.84
Fast	0.7	0.36	16.20	15.75	6.57	5.83	5.21	2.00	35.36

	Manhours per opening(s)	Gallons per opening(s)	Material cost per gallon	Labor cost per opening(s)	Labor burden opening(s)	Material cost per opening(s)	Overhead per opening(s)	Profit per opening(s)	Total price per opening(s)
Enamel, water base (material #9)									
Roll & brush 1 coat									
Slow	1.3	0.33	24.50	13.65	3.87	8.09	7.94	5.03	38.58
Medium	1.0	0.36	21.40	16.75	5.84	7.70	7.42	4.15	41.86
Fast	0.7	0.40	18.40	15.75	6.57	7.36	5.49	2.11	37.28
Enamel, oil base (material #10)									
Roll & brush 1 coat									
Slow	1.3	0.33	26.60	13.65	3.87	8.78	8.16	5.17	39.63
Medium	1.0	0.36	23.30	16.75	5.84	8.39	7.59	4.24	42.81
Fast	0.7	0.40	20.00	15.75	6.57	8.00	5.61	2.16	38.09
5 openings total									
Undercoat, water base (material #3)									
Roll & brush 1 coat									
Slow	1.6	0.38	21.10	16.80	4.77	8.02	9.18	5.82	44.59
Medium	1.3	0.41	18.50	21.78	7.59	7.59	9.06	5.06	51.08
Fast	0.9	0.45	15.80	20.25	8.45	7.11	6.62	2.55	44.98
Undercoat, oil base (material #4)									
Roll & brush 1 coat									
Slow	1.6	0.38	21.60	16.80	4.77	8.21	9.23	5.85	44.86
Medium	1.3	0.41	18.90	21.78	7.59	7.75	9.09	5.08	51.29
Fast	0.9	0.45	16.20	20.25	8.45	7.29	6.66	2.56	45.21
Enamel, water base (material #9)									
Roll & brush 1 coat									
Slow	1.6	0.42	24.50	16.80	4.77	10.29	9.88	6.26	48.00
Medium	1.3	0.46	21.40	21.78	7.59	9.84	9.61	5.37	54.19
Fast	0.9	0.50	18.40	20.25	8.45	9.20	7.01	2.69	47.60
Enamel, oil base (material #10)									
Roll & brush 1 coat									
Slow	1.6	0.42	26.60	16.80	4.77	11.17	10.15	6.44	49.33
Medium	1.3	0.46	23.30	21.78	7.59	10.72	9.82	5.49	55.40
Fast	0.9	0.50	20.00	20.25	8.45	10.00	7.16	2.75	48.61
6 openings total									
Undercoat, water base (material #3)									
Roll & brush 1 coat									
Slow	1.90	0.46	21.10	19.95	5.66	9.71	10.96	6.95	53.23
Medium	1.45	0.50	18.50	24.29	8.47	9.25	10.29	5.75	58.05
Fast	1.00	0.54	15.80	22.50	9.39	8.53	7.48	2.87	50.77

	Manhours per opening(s)	Gallons per opening(s)	Material cost per gallon	Labor cost per opening(s)	Labor burden opening(s)	Material cost per opening(s)	Overhead per opening(s)	Profit per opening(s)	Total price per opening(s)
Undercoat, oil base (material #4)									
Roll & brush 1 coat									
Slow	1.90	0.46	21.60	19.95	5.66	9.94	11.03	6.99	53.57
Medium	1.45	0.50	18.90	24.29	8.47	9.45	10.34	5.78	58.33
Fast	1.00	0.54	16.20	22.50	9.39	8.75	7.52	2.89	51.05
Enamel, water base (material #9)									
Roll & brush 1 coat									
Slow	1.90	0.50	24.50	19.95	5.66	12.25	11.74	7.44	57.04
Medium	1.45	0.55	21.40	24.29	8.47	11.77	10.91	6.10	61.54
Fast	1.00	0.60	18.40	22.50	9.39	11.04	7.94	3.05	53.92
Enamel, oil base (material #10)									
Roll & brush 1 coat									
Slow	1.90	0.50	26.60	19.95	5.66	13.30	12.07	7.65	58.63
Medium	1.45	0.55	23.30	24.29	8.47	12.82	11.17	6.24	62.99
Fast	1.00	0.60	20.00	22.50	9.39	12.00	8.12	3.12	55.13
7 openings total									
Undercoat, water base (material #3)									
Roll & brush 1 coat									
Slow	2.2	0.54	21.10	23.10	6.56	11.39	12.73	8.07	61.85
Medium	1.7	0.59	18.50	28.48	9.92	10.92	12.09	6.76	68.17
Fast	1.2	0.64	15.80	27.00	11.27	10.11	8.95	3.44	60.77
Undercoat, oil base (material #4)									
Roll & brush 1 coat									
Slow	2.2	0.54	21.60	23.10	6.56	11.66	12.81	8.12	62.25
Medium	1.7	0.59	18.90	28.48	9.92	11.15	12.14	6.79	68.48
Fast	1.2	0.64	16.20	27.00	11.27	10.37	9.00	3.46	61.10
Enamel, water base (material #9)									
Roll & brush 1 coat									
Slow	2.2	0.58	24.50	23.10	6.56	14.21	13.60	8.62	66.09
Medium	1.7	0.64	21.40	28.48	9.92	13.70	12.77	7.14	72.01
Fast	1.2	0.70	18.40	27.00	11.27	12.88	9.46	3.64	64.25
Enamel, oil base (material #10)									
Roll & brush 1 coat									
Slow	2.2	0.58	26.60	23.10	6.56	15.43	13.98	8.86	67.93
Medium	1.7	0.64	23.30	28.48	9.92	14.91	13.06	7.30	73.67
Fast	1.2	0.70	20.00	27.00	11.27	14.00	9.67	3.72	65.66

	Manhours per opening(s)	Gallons per opening(s)	Material cost per gallon	Labor cost per opening(s)	Labor burden opening(s)	Material cost per opening(s)	Overhead per opening(s)	Profit per opening(s)	Total price per opening(s)
8 openings total									
Undercoat, water base (material #3)									
Roll & brush 1 coat									
Slow	2.50	0.62	21.10	26.25	7.45	13.08	14.51	9.20	70.49
Medium	1.95	0.67	18.50	32.66	11.39	12.40	13.83	7.73	78.01
Fast	1.40	0.73	15.80	31.50	13.15	11.53	10.39	3.99	70.56
Undercoat, oil base (material #4)									
Roll & brush 1 coat									
Slow	2.50	0.62	21.60	26.25	7.45	13.39	14.60	9.26	70.95
Medium	1.95	0.67	18.90	32.66	11.39	12.66	13.89	7.76	78.36
Fast	1.40	0.73	16.20	31.50	13.15	11.83	10.45	4.02	70.95
Enamel, water base (material #9)									
Roll & brush 1 coat									
Slow	2.50	0.67	24.50	26.25	7.45	16.42	15.54	9.85	75.51
Medium	1.95	0.74	21.40	32.66	11.39	15.84	14.67	8.20	82.76
Fast	1.40	0.80	18.40	31.50	13.15	14.72	10.98	4.22	74.57
Enamel, oil base (material #10)									
Roll & brush 1 coat									
Slow	2.50	0.67	26.60	26.25	7.45	17.82	15.98	10.13	77.63
Medium	1.95	0.74	23.30	32.66	11.39	17.24	15.01	8.39	84.69
Fast	1.40	0.80	20.00	31.50	13.15	16.00	11.22	4.31	76.18
9 openings total									
Undercoat, water base (material #3)									
Roll & brush 1 coat									
Slow	2.80	0.69	21.10	29.40	8.34	14.56	16.22	10.28	78.80
Medium	2.15	0.75	18.50	36.01	12.56	13.88	15.30	8.55	86.30
Fast	1.50	0.81	15.80	33.75	14.09	12.80	11.22	4.31	76.17
Undercoat, oil base (material #4)									
Roll & brush 1 coat									
Slow	2.80	0.69	21.60	29.40	8.34	14.90	16.32	10.35	79.31
Medium	2.15	0.75	18.90	36.01	12.56	14.18	15.37	8.59	86.71
Fast	1.50	0.81	16.20	33.75	14.09	13.12	11.28	4.33	76.57
Enamel, water base (material #9)									
Roll & brush 1 coat									
Slow	2.80	0.75	24.50	29.40	8.34	18.38	17.40	11.03	84.55
Medium	2.15	0.82	21.40	36.01	12.56	17.55	16.20	9.05	91.37
Fast	1.50	0.90	18.40	33.75	14.09	16.56	11.91	4.58	80.89

	Manhours per opening(s)	Gallons per opening(s)	Material cost per gallon	Labor cost per opening(s)	Labor burden opening(s)	Material cost per opening(s)	Overhead per opening(s)	Profit per opening(s)	Total price per opening(s)
Enamel, oil base (material #10)									
Roll & brush 1 coat									
Slow	2.80	0.75	26.60	29.40	8.34	19.95	17.89	11.34	86.92
Medium	2.15	0.82	23.30	36.01	12.56	19.11	16.58	9.27	93.53
Fast	1.50	0.90	20.00	33.75	14.09	18.00	12.18	4.68	82.70
10 openings total									
Undercoat, water base (material #3)									
Roll & brush 1 coat									
Slow	3.1	0.77	21.10	32.55	9.24	16.25	18.00	11.41	87.45
Medium	2.4	0.84	18.50	40.20	14.02	15.54	17.09	9.55	96.40
Fast	1.7	0.90	15.80	38.25	15.96	14.22	12.66	4.87	85.96
Undercoat, oil base (material #4)									
Roll & brush 1 coat									
Slow	3.1	0.77	21.60	32.55	9.24	16.63	18.12	11.48	88.02
Medium	2.4	0.84	18.90	40.20	14.02	15.88	17.17	9.60	96.87
Fast	1.7	0.90	16.20	38.25	15.96	14.58	12.73	4.89	86.41
Enamel, water base (material #9)									
Slow	3.1	0.83	24.50	32.55	9.24	20.34	19.27	12.21	93.61
Medium	2.4	0.92	21.40	40.20	14.02	19.69	18.11	10.12	102.14
Fast	1.7	1.00	18.40	38.25	15.96	18.40	13.43	5.16	91.20
Enamel, oil base (material #10)									
Roll & brush 1 coat									
Slow	3.1	0.83	26.60	32.55	9.24	22.08	19.81	12.56	96.24
Medium	2.4	0.92	23.30	40.20	14.02	21.44	18.53	10.36	104.55
Fast	1.7	1.00	20.00	38.25	15.96	20.00	13.73	5.28	93.22

Use these figures for painting interior doors, pullmans, linens and other surfaces described in Figure 20 on page 100. "Slow" work is based on a $10.50 hourly wage, "Medium" work on a $16.75 hourly wage, and "Fast" work on a $22.50 hourly wage. Other qualifications that apply to this table are on page 9.

	Manhours per door	Doors per gallon	Material cost per gallon	Labor cost per door	Labor burden door	Material cost per door	Overhead per door	Profit per door	Total price per door
Doors, interior, flush, paint grade, roll & brush, per door									
Undercoat, water base (material #3)									
Roll & brush 1 coat									
Slow	0.40	13.0	21.10	4.20	1.19	1.62	2.17	1.38	10.56
Medium	0.30	11.5	18.50	5.03	1.75	1.61	2.06	1.15	11.60
Fast	0.20	10.0	15.80	4.50	1.88	1.58	1.47	.57	10.00
Undercoat, oil base (material #4)									
Roll & brush 1 coat									
Slow	0.40	13.0	21.60	4.20	1.19	1.66	2.19	1.39	10.63
Medium	0.30	11.5	18.90	5.03	1.75	1.64	2.06	1.15	11.63
Fast	0.20	10.0	16.20	4.50	1.88	1.62	1.48	.57	10.05
Enamel, water base (material #9)									
Roll & brush 1st finish coat									
Slow	0.33	14.0	24.50	3.47	.98	1.75	1.93	1.22	9.35
Medium	0.25	12.5	21.40	4.19	1.46	1.71	1.80	1.01	10.17
Fast	0.17	11.0	18.40	3.83	1.59	1.67	1.31	.50	8.90
Roll & brush additional finish coats									
Slow	0.25	15.0	24.50	2.63	.74	1.63	1.55	.98	7.53
Medium	0.20	13.5	21.40	3.35	1.17	1.59	1.50	.84	8.45
Fast	0.15	12.0	18.40	3.38	1.40	1.53	1.17	.45	7.93
Enamel, oil base (material #10)									
Roll & brush 1st finish coat									
Slow	0.33	14.0	26.60	3.47	.98	1.90	1.97	1.25	9.57
Medium	0.25	12.5	23.30	4.19	1.46	1.86	1.84	1.03	10.38
Fast	0.17	11.0	20.00	3.83	1.59	1.82	1.34	.52	9.10
Roll & brush additional finish coats									
Slow	0.25	15.0	26.60	2.63	.74	1.77	1.60	1.01	7.75
Medium	0.20	13.5	23.30	3.35	1.17	1.73	1.53	.86	8.64
Fast	0.15	12.0	20.00	3.38	1.40	1.67	1.20	.46	8.11

This table is not to be confused with the Opening count method for estimating doors. Use these figures for painting both interior and exterior doors. These costs are in addition to the costs for applying the finish coats to exterior doors as explained on page 95. These figures include coating all six (6) sides of each door along with the frame and the jamb on both sides. These figures include minimal preparation time. Add for masking or extensive preparation time. "Slow" work is based on a $10.50 hourly wage, "Medium" work on a $16.75 hourly wage, and "Fast" work on a $22.50 hourly wage. Other qualifications that apply to this table are on page 9.

	Manhours per door	Doors per gallon	Material cost per gallon	Labor cost per door	Labor burden door	Material cost per door	Overhead per door	Profit per door	Total price per door
Doors, interior, flush, paint grade, spray application, per door									
Undercoat, water base (material #3)									
Spray 1 coat									
Slow	0.10	17	21.10	1.05	.30	1.24	.80	.51	3.90
Medium	0.09	16	18.50	1.51	.52	1.16	.78	.44	4.41
Fast	0.08	15	15.80	1.80	.75	1.05	.67	.26	4.53
Undercoat, oil base (material #4)									
Spray 1 coat									
Slow	0.10	17	21.60	1.05	.30	1.27	.81	.51	3.94
Medium	0.09	16	18.90	1.51	.52	1.18	.79	.44	4.44
Fast	0.08	15	16.20	1.80	.75	1.08	.67	.26	4.56
Enamel, water base (material #9)									
Spray 1st finish coat									
Slow	0.09	18	24.50	.95	.26	1.36	.80	.51	3.88
Medium	0.08	17	21.40	1.34	.47	1.26	.75	.42	4.24
Fast	0.07	16	18.40	1.58	.65	1.15	.63	.24	4.25
Spray 2nd or additional finish coats									
Slow	0.08	19	24.50	.84	.24	1.29	.73	.47	3.57
Medium	0.07	18	21.40	1.17	.41	1.19	.68	.38	3.83
Fast	0.06	17	18.40	1.35	.56	1.08	.55	.21	3.75
Enamel, oil base (material #10)									
Spray 1st finish coat									
Slow	0.09	18	26.60	.95	.26	1.48	.84	.53	4.06
Medium	0.08	17	23.30	1.34	.47	1.37	.78	.44	4.40
Fast	0.07	16	20.00	1.58	.65	1.25	.65	.25	4.38
Spray 2nd or additional finish coats									
Slow	0.08	19	26.60	.84	.24	1.40	.77	.49	3.74
Medium	0.07	18	23.30	1.17	.41	1.29	.70	.39	3.96
Fast	0.06	17	20.00	1.35	.56	1.18	.57	.22	3.88

This table is not to be confused with the Opening count method for estimating doors. Use these figures for painting both interior and exterior doors. These costs are in addition to the costs for applying the finish coats to exterior doors as explained on page 95. These figures include coating all six (6) sides of each door along with the frame and the jamb on both sides. These figures include minimal preparation time. Add for masking or extensive preparation time. "Slow" work is based on a $10.50 hourly wage, "Medium" work on a $16.75 hourly wage, and "Fast" work on a $22.50 hourly wage. Other qualifications that apply to this table are on page 9.

	Manhours per door	Doors per gallon	Material cost per gallon	Labor cost per door	Labor burden door	Material cost per door	Overhead per door	Profit per door	Total price per door

Doors, interior, flush, stain grade, spray application, per door

	Manhours per door	Doors per gallon	Material cost per gallon	Labor cost per door	Labor burden door	Material cost per door	Overhead per door	Profit per door	Total price per door
Complete 7 step stain, seal & 2 coat lacquer system (material #11)									
Spray all coats									
Slow	0.90	6	21.60	9.45	2.68	3.60	4.88	3.09	23.70
Medium	0.80	5	19.60	13.40	4.67	3.92	5.39	3.01	30.39
Fast	0.70	4	16.20	15.75	6.57	4.05	4.88	1.88	33.13

This table is not to be confused with the Opening count method for estimating doors. Use these figures for painting both interior and exterior doors. These costs are in addition to the costs for applying the finish coats to exterior doors as explained on page 95. These figures include coating all six (6) sides of each door along with the frame and the jamb on both sides. These figures include minimal preparation time. Add for masking or extensive preparation time. "Slow" work is based on a $10.50 hourly wage, "Medium" work on a $16.75 hourly wage, and "Fast" work on a $22.50 hourly wage. Other qualifications that apply to this table are on page 9.

Doors, interior, French, paint grade, roll & brush, per door

	Manhours per door	Doors per gallon	Material cost per gallon	Labor cost per door	Labor burden door	Material cost per door	Overhead per door	Profit per door	Total price per door
Undercoat, water base (material #3)									
Roll & brush 1 coat									
Slow	0.45	14	21.10	4.73	1.34	1.51	2.35	1.49	11.42
Medium	0.38	13	18.50	6.37	2.21	1.42	2.45	1.37	13.82
Fast	0.30	12	15.80	6.75	2.82	1.32	2.01	.77	13.67
Undercoat, oil base (material #4)									
Roll & brush 1 coat									
Slow	0.45	14	21.60	4.73	1.34	1.54	2.36	1.50	11.47
Medium	0.38	13	18.90	6.37	2.21	1.45	2.46	1.38	13.87
Fast	0.30	12	16.20	6.75	2.82	1.35	2.02	.78	13.72
Enamel, water base (material #9)									
Roll & brush 1st finish coat									
Slow	0.43	15	24.50	4.52	1.28	1.63	2.31	1.46	11.20
Medium	0.35	14	21.40	5.86	2.05	1.53	2.31	1.29	13.04
Fast	0.28	13	18.40	6.30	2.63	1.42	1.91	.74	13.00
Roll & brush 2nd or additional finish coats									
Slow	0.40	16	24.50	4.20	1.19	1.53	2.15	1.36	10.43
Medium	0.33	15	21.40	5.53	1.92	1.43	2.18	1.22	12.28
Fast	0.25	14	18.40	5.63	2.34	1.31	1.72	.66	11.66

	Manhours per door	Doors per gallon	Material cost per gallon	Labor cost per door	Labor burden door	Material cost per door	Overhead per door	Profit per door	Total price per door
Enamel, oil base (material #10)									
Roll & brush 1st finish coat									
Slow	0.43	15	26.60	4.52	1.28	1.77	2.35	1.49	11.41
Medium	0.35	14	23.30	5.86	2.05	1.66	2.34	1.31	13.22
Fast	0.28	13	20.00	6.30	2.63	1.54	1.94	.74	13.15
Roll & brush 2nd or additional finish coats									
Slow	0.40	16	26.60	4.20	1.19	1.66	2.19	1.39	10.63
Medium	0.33	15	23.30	5.53	1.92	1.55	2.21	1.23	12.44
Fast	0.25	14	20.00	5.63	2.34	1.43	1.74	.67	11.81

This table is not to be confused with the Opening count method for estimating doors. Use these figures for painting both interior and exterior doors. These costs are in addition to the costs for applying the finish coats to exterior doors as explained on page 95. These figures include coating all six (6) sides of each door along with the frame and the jamb on both sides. These figures include minimal preparation time. Add for masking or extensive preparation time. "Slow" work is based on a $10.50 hourly wage, "Medium" work on a $16.75 hourly wage, and "Fast" work on a $22.50 hourly wage. Other qualifications that apply to this table are on page 9.

	Manhours per door	Doors per gallon	Material cost per gallon	Labor cost per door	Labor burden door	Material cost per door	Overhead per door	Profit per door	Total price per door
Doors, interior, French, stain grade, spray application, per door									
Complete 7 step stain, seal & 2 coat lacquer system (material #11)									
Spray all coats									
Slow	1.50	12	21.60	15.75	4.47	1.80	6.83	4.33	33.18
Medium	1.25	11	19.60	20.94	7.30	1.78	7.35	4.11	41.48
Fast	1.00	10	16.20	22.50	9.39	1.62	6.20	2.38	42.09

This table is not to be confused with the Opening count method for estimating doors. Use these figures for painting both interior and exterior doors. These costs are in addition to the costs for applying the finish coats to exterior doors as explained on page 95. These figures include coating all six (6) sides of each door along with the frame and the jamb on both sides. These figures include minimal preparation time. Add for masking or extensive preparation time. "Slow" work is based on a $10.50 hourly wage, "Medium" work on a $16.75 hourly wage, and "Fast" work on a $22.50 hourly wage. Other qualifications that apply to this table are on page 9.

Doors, interior, louvered, paint grade, roll & brush, per door

	Manhours per door	Doors per gallon	Material cost per gallon	Labor cost per door	Labor burden door	Material cost per door	Overhead per door	Profit per door	Total price per door
Undercoat, water base (material #3)									
Roll & brush 1 coat									
Slow	0.67	8	21.10	7.04	1.99	2.64	3.62	2.30	17.59
Medium	0.54	7	18.50	9.05	3.15	2.64	3.64	2.03	20.51
Fast	0.40	6	15.80	9.00	3.76	2.63	2.85	1.09	19.33
Undercoat, oil base (material #4)									
Roll & brush 1 coat									
Slow	0.67	8	21.60	7.04	1.99	2.70	3.64	2.31	17.68
Medium	0.54	7	18.90	9.05	3.15	2.70	3.65	2.04	20.59
Fast	0.40	6	16.20	9.00	3.76	2.70	2.86	1.10	19.42
Enamel, water base (material #9)									
Roll & brush 1st finish coat									
Slow	0.50	9	24.50	5.25	1.49	2.72	2.93	1.86	14.25
Medium	0.42	8	21.40	7.04	2.45	2.68	2.98	1.67	16.82
Fast	0.33	7	18.40	7.43	3.09	2.63	2.43	.94	16.52
Roll & brush 2nd or additional finish coats									
Slow	0.40	10	24.50	4.20	1.19	2.45	2.43	1.54	11.81
Medium	0.30	9	21.40	5.03	1.75	2.38	2.24	1.25	12.65
Fast	0.20	8	18.40	4.50	1.88	2.30	1.61	.62	10.91
Enamel, oil base (material #10)									
Roll & brush 1st finish coat									
Slow	0.50	9	26.60	5.25	1.49	2.96	3.01	1.91	14.62
Medium	0.42	8	23.30	7.04	2.45	2.91	3.04	1.70	17.14
Fast	0.33	7	20.00	7.43	3.09	2.86	2.48	.95	16.81
Roll & brush 2nd or additional finish coats									
Slow	0.40	10	26.60	4.20	1.19	2.66	2.50	1.58	12.13
Medium	0.30	9	23.30	5.03	1.75	2.59	2.30	1.28	12.95
Fast	0.20	8	20.00	4.50	1.88	2.50	1.64	.63	11.15

This table is not to be confused with the Opening count method for estimating doors. Use these figures for painting both interior and exterior doors. These costs are in addition to the costs for applying the finish coats to exterior doors as explained on page 95. These figures include coating all six (6) sides of each door along with the frame and the jamb on both sides. These figures include minimal preparation time. Add for masking or extensive preparation time. "Slow" work is based on a $10.50 hourly wage, "Medium" work on a $16.75 hourly wage, and "Fast" work on a $22.50 hourly wage. Other qualifications that apply to this table are on page 9.

	Manhours per door	Doors per gallon	Material cost per gallon	Labor cost per door	Labor burden door	Material cost per door	Overhead per door	Profit per door	Total price per door

Doors, interior, louvered, paint grade, spray application, per door

	Manhours per door	Doors per gallon	Material cost per gallon	Labor cost per door	Labor burden door	Material cost per door	Overhead per door	Profit per door	Total price per door
Undercoat, water base (material #3)									
Spray 1 coat									
Slow	0.17	12	21.10	1.79	.50	1.76	1.26	.80	6.11
Medium	0.14	11	18.50	2.35	.81	1.68	1.19	.66	6.69
Fast	0.12	10	15.80	2.70	1.13	1.58	1.00	.38	6.79
Undercoat, oil base (material #4)									
Spray 1 coat									
Slow	0.17	12	21.60	1.79	.50	1.80	1.27	.81	6.17
Medium	0.14	11	18.90	2.35	.81	1.72	1.20	.67	6.75
Fast	0.12	10	16.20	2.70	1.13	1.62	1.01	.39	6.85
Enamel, water base (material #9)									
Spray 1st finish coat									
Slow	0.13	13	24.50	1.37	.38	1.88	1.13	.72	5.48
Medium	0.11	12	21.40	1.84	.64	1.78	1.04	.58	5.88
Fast	0.09	11	18.40	2.03	.84	1.67	.84	.32	5.70
Spray 2nd or additional finish coats									
Slow	0.10	14	24.50	1.05	.30	1.75	.96	.61	4.67
Medium	0.09	13	21.40	1.51	.52	1.65	.90	.50	5.08
Fast	0.08	12	18.40	1.80	.75	1.53	.75	.29	5.12
Enamel, oil base (material #10)									
Spray 1st finish coat									
Slow	0.13	13	26.60	1.37	.38	2.05	1.18	.75	5.73
Medium	0.11	12	23.30	1.84	.64	1.94	1.08	.61	6.11
Fast	0.09	11	20.00	2.03	.84	1.82	.87	.33	5.89
Spray 2nd or additional finish coats									
Slow	0.10	14	26.60	1.05	.30	1.90	1.01	.64	4.90
Medium	0.09	13	23.30	1.51	.52	1.79	.94	.52	5.28
Fast	0.08	12	20.00	1.80	.75	1.67	.78	.30	5.30

This table is not to be confused with the Opening count method for estimating doors. Use these figures for painting both interior and exterior doors. These costs are in addition to the costs for applying the finish coats to exterior doors as explained on page 95. These figures include coating all six (6) sides of each door along with the frame and the jamb on both sides. These figures include minimal preparation time. Add for masking or extensive preparation time. "Slow" work is based on a $10.50 hourly wage, "Medium" work on a $16.75 hourly wage, and "Fast" work on a $22.50 hourly wage. Other qualifications that apply to this table are on page 9.

	Manhours per door	Doors per gallon	Material cost per gallon	Labor cost per door	Labor burden door	Material cost per door	Overhead per door	Profit per door	Total price per door

Doors, interior, louvered, stain grade, spray application, per door

Complete 7 step stain, seal & 2 coat lacquer system (material #11)

Spray all coats									
Slow	1.70	5	21.60	17.85	5.07	4.32	8.45	5.36	41.05
Medium	1.45	4	19.60	24.29	8.47	4.90	9.23	5.16	52.05
Fast	1.20	3	16.20	27.00	11.27	5.40	8.08	3.11	54.86

This table is not to be confused with the Opening count method for estimating doors. Use these figures for painting both interior and exterior doors. These costs are in addition to the costs for applying the finish coats to exterior doors as explained on page 95. These figures include coating all six (6) sides of each door along with the frame and the jamb on both sides. These figures include minimal preparation time. Add for masking or extensive preparation time. "Slow" work is based on a $10.50 hourly wage, "Medium" work on a $16.75 hourly wage, and "Fast" work on a $22.50 hourly wage. Other qualifications that apply to this table are on page 9.

	Manhours per door	Doors per gallon	Material cost per gallon	Labor cost per door	Labor burden door	Material cost per door	Overhead per door	Profit per door	Total price per door
Doors, interior, panel, paint grade, roll & brush, per door									
Undercoat, water base (material #3)									
Roll & brush 1 coat									
Slow	0.50	8	21.10	5.25	1.49	2.64	2.91	1.84	14.13
Medium	0.33	7	18.50	5.53	1.92	2.64	2.47	1.38	13.94
Fast	0.25	6	15.80	5.63	2.34	2.63	1.96	.75	13.31
Undercoat, oil base (material #4)									
Roll & brush 1 coat									
Slow	0.50	8	21.60	5.25	1.49	2.70	2.93	1.86	14.23
Medium	0.33	7	18.90	5.53	1.92	2.70	2.49	1.39	14.03
Fast	0.25	6	16.20	5.63	2.34	2.70	1.98	.76	13.41
Enamel, water base (material #9)									
Roll & brush 1st finish coat									
Slow	0.40	11	24.50	4.20	1.19	2.23	2.36	1.50	11.48
Medium	0.30	10	21.40	5.03	1.75	2.14	2.19	1.22	12.33
Fast	0.20	9	18.40	4.50	1.88	2.04	1.56	.60	10.58
Roll & brush 2nd or additional finish coats									
Slow	0.33	12	24.50	3.47	.98	2.04	2.02	1.28	9.79
Medium	0.25	11	21.40	4.19	1.46	1.95	1.86	1.04	10.50
Fast	0.17	10	18.40	3.83	1.59	1.84	1.34	.52	9.12
Enamel, oil base (material #10)									
Roll & brush 1st finish coat									
Slow	0.40	11	26.60	4.20	1.19	2.42	2.42	1.53	11.76
Medium	0.30	10	23.30	5.03	1.75	2.33	2.23	1.25	12.59
Fast	0.20	9	20.00	4.50	1.88	2.22	1.59	.61	10.80
Roll & brush 2nd or additional finish coats									
Slow	0.33	12	26.60	3.47	.98	2.22	2.07	1.31	10.05
Medium	0.25	11	23.30	4.19	1.46	2.12	1.90	1.06	10.73
Fast	0.17	10	20.00	3.83	1.59	2.00	1.37	.53	9.32

This table is not to be confused with the Opening count method for estimating doors. Use these figures for painting both interior and exterior doors. These costs are in addition to the costs for applying the finish coats to exterior doors as explained on page 95. These figures include coating all six (6) sides of each door along with the frame and the jamb on both sides. These figures include minimal preparation time. Add for masking or extensive preparation time. "Slow" work is based on a $10.50 hourly wage, "Medium" work on a $16.75 hourly wage, and "Fast" work on a $22.50 hourly wage. Other qualifications that apply to this table are on page 9.

Doors, interior, panel, paint grade, spray application, per door

	Manhours per door	Doors per gallon	Material cost per gallon	Labor cost per door	Labor burden door	Material cost per door	Overhead per door	Profit per door	Total price per door
Undercoat, water base (material #3)									
Spray 1 coat									
Slow	0.13	15	21.10	1.37	.38	1.41	.98	.62	4.76
Medium	0.11	14	18.50	1.84	.64	1.32	.93	.52	5.25
Fast	0.10	13	15.80	2.25	.94	1.22	.82	.31	5.54
Undercoat, oil base (material #4)									
Spray 1 coat									
Slow	0.13	15	21.60	1.37	.38	1.44	.99	.63	4.81
Medium	0.11	14	18.90	1.84	.64	1.35	.94	.52	5.29
Fast	0.10	13	16.20	2.25	.94	1.25	.82	.32	5.58
Enamel, water base (material #9)									
Spray 1st finish coat									
Slow	0.09	16	24.50	.95	.26	1.53	.85	.54	4.13
Medium	0.08	15	21.40	1.34	.47	1.43	.79	.44	4.47
Fast	0.08	14	18.40	1.80	.75	1.31	.71	.27	4.84
Spray 2nd or additional finish coats									
Slow	0.08	17	24.50	.84	.24	1.44	.78	.50	3.80
Medium	0.08	16	21.40	1.34	.47	1.34	.77	.43	4.35
Fast	0.07	15	18.40	1.58	.65	1.23	.64	.25	4.35
Enamel, oil base (material #10)									
Spray 1st finish coat									
Slow	0.09	16	26.60	.95	.26	1.66	.89	.57	4.33
Medium	0.08	15	23.30	1.34	.47	1.55	.82	.46	4.64
Fast	0.08	14	20.00	1.80	.75	1.43	.74	.28	5.00
Spray 2nd or additional finish coats									
Slow	0.08	17	26.60	.84	.24	1.56	.82	.52	3.98
Medium	0.08	16	23.30	1.34	.47	1.46	.80	.45	4.52
Fast	0.07	15	20.00	1.58	.65	1.33	.66	.25	4.47

This table is not to be confused with the Opening count method for estimating doors. Use these figures for painting both interior and exterior doors. These costs are in addition to the costs for applying the finish coats to exterior doors as explained on page 95. These figures include coating all six (6) sides of each door along with the frame and the jamb on both sides. These figures include minimal preparation time. Add for masking or extensive preparation time. "Slow" work is based on a $10.50 hourly wage, "Medium" work on a $16.75 hourly wage, and "Fast" work on a $22.50 hourly wage. Other qualifications that apply to this table are on page 9.

	Manhours per door	Doors per gallon	Material cost per gallon	Labor cost per door	Labor burden door	Material cost per door	Overhead per door	Profit per door	Total price per door

Doors, interior, panel, stain grade, spray application, per door

Complete 7 step stain, seal & 2 coat lacquer system (material #11)									
Spray all coats									
Slow	1.25	6	21.60	13.13	3.72	3.60	6.35	4.02	30.82
Medium	1.08	5	19.60	18.09	6.31	3.92	6.94	3.88	39.14
Fast	0.90	4	16.20	20.25	8.45	4.05	6.06	2.33	41.14

This table is not to be confused with the Opening count method for estimating doors. Use these figures for painting both interior and exterior doors. These costs are in addition to the costs for applying the finish coats to exterior doors as explained on page 95. These figures include coating all six (6) sides of each door along with the frame and the jamb on both sides. These figures include minimal preparation time. Add for masking or extensive preparation time. "Slow" work is based on a $10.50 hourly wage, "Medium" work on a $16.75 hourly wage, and "Fast" work on a $22.50 hourly wage. Other qualifications that apply to this table are on page 9.

	Labor LF per manhour	Material coverage LF/gallon	Material cost per gallon	Labor cost per 100 LF	Labor burden 100 LF	Material cost per 100 LF	Overhead per 100 LF	Profit per 100 LF	Total price per 100 LF

Fascia, 2" x 4", brush one coat stain "to cover"

Solid body stain, water or oil base (material #18 or #19)									
Brush each coat									
Slow	80	170	22.70	13.13	3.72	13.35	9.37	5.94	45.51
Medium	105	160	19.85	15.95	5.56	12.41	8.31	4.65	46.88
Fast	130	150	17.00	17.31	7.21	11.33	6.64	2.55	45.04
Semi-transparent stain, water or oil base (material #20 or #21)									
Brush each coat									
Slow	95	195	23.75	11.05	3.14	12.18	8.17	5.18	39.72
Medium	120	185	20.80	13.96	4.86	11.24	7.37	4.12	41.55
Fast	145	175	17.80	15.52	6.48	10.17	5.95	2.29	40.41

Use these figures for brushing stain on 2" x 4" fascia board on two sides - the face and the lower edge. The back side is calculated with the overhang (eaves) operation. These figures include one full coat and any touchup required to meet the "to cover" specification. They also include minimum preparation time. Add time for extensive preparation. Measurements are based on continuous linear feet. "Slow" work is based on a $10.50 hourly wage, "Medium" work on a $16.75 hourly wage, and "Fast" work on a $22.50 hourly wage. Other qualifications that apply to this table are on page 9.

	Labor LF per manhour	Material coverage LF/gallon	Material cost per gallon	Labor cost per 100 LF	Labor burden 100 LF	Material cost per 100 LF	Overhead per 100 LF	Profit per 100 LF	Total price per 100 LF

Fascia, 2" x 4", roll one coat stain "to cover"

	Labor LF per manhour	Material coverage LF/gallon	Material cost per gallon	Labor cost per 100 LF	Labor burden 100 LF	Material cost per 100 LF	Overhead per 100 LF	Profit per 100 LF	Total price per 100 LF
Solid body stain, water or oil base (material #18 or #19)									
Roll each coat									
Slow	180	140	22.70	5.83	1.66	16.21	7.35	4.66	35.71
Medium	205	130	19.85	8.17	2.85	15.27	6.44	3.60	36.33
Fast	230	120	17.00	9.78	4.09	14.17	5.19	1.99	35.22
Semi-transparent stain, water or oil base (material #20 or #21)									
Roll each coat									
Slow	200	160	23.75	5.25	1.49	14.84	6.69	4.24	32.51
Medium	225	150	20.80	7.44	2.59	13.87	5.86	3.27	33.03
Fast	250	140	17.80	9.00	3.76	12.71	4.71	1.81	31.99

Use these figures for rolling stain on 2" x 4" fascia board on two sides - the face and the lower edge. The back side is calculated with the overhang (eaves) operation. These figures include one full coat and any touchup required to meet the "to cover" specification. They also include minimum preparation time. Add time for extensive preparation. Measurements are based on continuous linear feet. "Slow" work is based on a $10.50 hourly wage, "Medium" work on a $16.75 hourly wage, and "Fast" work on a $22.50 hourly wage. Other qualifications that apply to this table are on page 9.

Fascia, 2" x 4", spray one coat stain "to cover"

	Labor LF per manhour	Material coverage LF/gallon	Material cost per gallon	Labor cost per 100 LF	Labor burden 100 LF	Material cost per 100 LF	Overhead per 100 LF	Profit per 100 LF	Total price per 100 LF
Solid body stain, water or oil base (material #18 or #19)									
Spray each coat									
Slow	275	110	22.70	3.82	1.09	20.64	7.92	5.02	38.49
Medium	325	100	19.85	5.15	1.81	19.85	6.56	3.67	37.04
Fast	375	90	17.00	6.00	2.51	18.89	5.07	1.95	34.42
Semi-transparent stain, water or oil base (material #20 or #21)									
Spray each coat									
Slow	300	125	23.75	3.50	.99	19.00	7.29	4.62	35.40
Medium	350	115	20.80	4.79	1.67	18.09	6.01	3.36	33.92
Fast	400	105	17.80	5.63	2.34	16.95	4.61	1.77	31.30

Use these figures for spraying stain on 2" x 4" fascia board on two sides - the face and the lower edge. The back side is calculated with the overhang (eaves) operation. These figures include one full coat and any touchup required to meet the "to cover" specification. They also include minimum preparation time. Add time for extensive preparation. Measurements are based on continuous linear feet. "Slow" work is based on a $10.50 hourly wage, "Medium" work on a $16.75 hourly wage, and "Fast" work on a $22.50 hourly wage. Other qualifications that apply to this table are on page 9.

	Labor LF per manhour	Material coverage LF/gallon	Material cost per gallon	Labor cost per 100 LF	Labor burden 100 LF	Material cost per 100 LF	Overhead per 100 LF	Profit per 100 LF	Total price per 100 LF

Fascia, 2" x 6" to 2" x 10", brush one coat stain "to cover"

	Labor LF per manhour	Material coverage LF/gallon	Material cost per gallon	Labor cost per 100 LF	Labor burden 100 LF	Material cost per 100 LF	Overhead per 100 LF	Profit per 100 LF	Total price per 100 LF
Solid body stain, water or oil base (material #18 or #19)									
Brush each coat									
Slow	70	140	22.70	15.00	4.26	16.21	11.00	6.97	53.44
Medium	90	130	19.85	18.61	6.49	15.27	9.89	5.53	55.79
Fast	110	120	17.00	20.45	8.54	14.17	7.98	3.07	54.21
Semi-transparent stain, water or oil base (material #20 or #21)									
Brush each coat									
Slow	85	165	23.75	12.35	3.50	14.39	9.38	5.94	45.56
Medium	105	155	20.80	15.95	5.56	13.42	8.56	4.78	48.27
Fast	125	145	17.80	18.00	7.51	12.28	6.99	2.69	47.47

Use these figures for brushing stain on 2" x 6" to 2" x 10" fascia board on two sides - the face and the lower edge. The back side is calculated with the overhang (eaves) operation. These figures include one full coat and any touchup required to meet the "to cover" specification. They also include minimum preparation time. Add time for extensive preparation. Measurements are based on continuous linear feet. "Slow" work is based on a $10.50 hourly wage, "Medium" work on a $16.75 hourly wage, and "Fast" work on a $22.50 hourly wage. Other qualifications that apply to this table are on page 9.

Fascia, 2" x 6" to 2" x 10", roll one coat stain "to cover"

	Labor LF per manhour	Material coverage LF/gallon	Material cost per gallon	Labor cost per 100 LF	Labor burden 100 LF	Material cost per 100 LF	Overhead per 100 LF	Profit per 100 LF	Total price per 100 LF
Solid body stain, water or oil base (material #18 or #19)									
Roll each coat									
Slow	150	120	22.70	7.00	1.99	18.92	8.65	5.48	42.04
Medium	175	110	19.85	9.57	3.33	18.05	7.59	4.24	42.78
Fast	200	100	17.00	11.25	4.70	17.00	6.10	2.34	41.39
Semi-transparent stain, water or oil base (material #20 or #21)									
Roll each coat									
Slow	170	140	23.75	6.18	1.75	16.96	7.72	4.89	37.50
Medium	195	130	20.80	8.59	3.00	16.00	6.76	3.78	38.13
Fast	220	120	17.80	10.23	4.28	14.83	5.43	2.09	36.86

Use these figures for rolling stain on 2" x 6" to 2" x 10" fascia board on two sides - the face and the lower edge. The back side is calculated with the overhang (eaves) operation. These figures include one full coat and any touchup required to meet the "to cover" specification. They also include minimum preparation time. Add time for extensive preparation. Measurements are based on continuous linear feet. "Slow" work is based on a $10.50 hourly wage, "Medium" work on a $16.75 hourly wage, and "Fast" work on a $22.50 hourly wage. Other qualifications that apply to this table are on page 9.

	Labor LF per manhour	Material coverage LF/gallon	Material cost per gallon	Labor cost per 100 LF	Labor burden 100 LF	Material cost per 100 LF	Overhead per 100 LF	Profit per 100 LF	Total price per 100 LF

Fascia, 2" x 6" to 2" x 10", spray one coat stain "to cover"

	Labor LF per manhour	Material coverage LF/gallon	Material cost per gallon	Labor cost per 100 LF	Labor burden 100 LF	Material cost per 100 LF	Overhead per 100 LF	Profit per 100 LF	Total price per 100 LF
Solid body stain, water or oil base (material #18 or #19)									
Spray each coat									
Slow	225	90	22.70	4.67	1.32	25.22	9.68	6.14	47.03
Medium	300	80	19.85	5.58	1.94	24.81	7.92	4.43	44.68
Fast	350	70	17.00	6.43	2.69	24.29	6.18	2.37	41.96
Semi-transparent stain, water or oil base (material #20 or #21)									
Spray each coat									
Slow	250	105	23.75	4.20	1.19	22.62	8.68	5.50	42.19
Medium	313	95	20.80	5.35	1.86	21.89	7.13	3.99	40.22
Fast	375	85	17.80	6.00	2.51	20.94	5.45	2.09	36.99

Use these figures for spraying stain on 2" x 6" to 2" x 10" fascia board on two sides - the face and the lower edge. The back side is calculated with the overhang (eaves) operation. These figures include one full coat and any touchup required to meet the "to cover" specification. They also include minimum preparation. Add time for extensive preparation. Measurements are based on continuous linear feet. "Slow" work is based on a $10.50 hourly wage, "Medium" work on a $16.75 hourly wage, and "Fast" work on a $22.50 hourly wage. Other qualifications that apply to this table are on page 9.

Fascia, 2" x 12", brush one coat stain "to cover"

	Labor LF per manhour	Material coverage LF/gallon	Material cost per gallon	Labor cost per 100 LF	Labor burden 100 LF	Material cost per 100 LF	Overhead per 100 LF	Profit per 100 LF	Total price per 100 LF
Solid body stain, water or oil base (material #18 or #19)									
Brush each coat									
Slow	60	100	22.70	17.50	4.97	22.70	14.01	8.88	68.06
Medium	80	90	19.85	20.94	7.30	22.06	12.32	6.89	69.51
Fast	100	80	17.00	22.50	9.39	21.25	9.83	3.78	66.75
Semi-transparent stain, water or oil base (material #20 or #21)									
Brush each coat									
Slow	75	125	23.75	14.00	3.97	19.00	11.46	7.27	55.70
Medium	95	115	20.80	17.63	6.16	18.09	10.26	5.73	57.87
Fast	105	105	17.80	21.43	8.93	16.95	8.76	3.37	59.44

Use these figures for brushing stain on 2" x 12" fascia board on two sides - the face and the lower edge. The back side is calculated with the overhang (eaves) operation. These figures include one full coat and any touchup required to meet the "to cover" specification. They also include minimum preparation. Add time for extensive preparation. Measurements are based on continuous linear feet. "Slow" work is based on a $10.50 hourly wage, "Medium" work on a $16.75 hourly wage, and "Fast" work on a $22.50 hourly wage. Other qualifications that apply to this table are on page 9.

	Labor LF per manhour	Material coverage LF/gallon	Material cost per gallon	Labor cost per 100 LF	Labor burden 100 LF	Material cost per 100 LF	Overhead per 100 LF	Profit per 100 LF	Total price per 100 LF
Fascia, 2" x 12", roll one coat stain "to cover"									
Solid body stain, water or oil base (material #18 or #19)									
Roll each coat									
Slow	110	90	22.70	9.55	2.70	25.22	11.62	7.37	56.46
Medium	130	75	19.85	12.88	4.49	26.47	10.74	6.00	60.58
Fast	150	60	17.00	15.00	6.27	28.33	9.17	3.53	62.30
Semi-transparent stain, water or oil base (material #20 or #21)									
Roll each coat									
Slow	130	110	23.75	8.08	2.29	21.59	9.91	6.28	48.15
Medium	150	95	20.80	11.17	3.90	21.89	9.05	5.06	51.07
Fast	170	80	17.80	13.24	5.51	22.25	7.59	2.92	51.51

Use these figures for rolling stain on 2" x 12" fascia board on two sides - the face and the lower edge. The back side is calculated with the overhang (eaves) operation. These figures include one full coat and any touchup required to meet the "to cover" specification. They also include minimum preparation. Add time for extensive preparation. Measurements are based on continuous linear feet. "Slow" work is based on a $10.50 hourly wage, "Medium" work on a $16.75 hourly wage, and "Fast" work on a $22.50 hourly wage. Other qualifications that apply to this table are on page 9.

	Labor LF per manhour	Material coverage LF/gallon	Material cost per gallon	Labor cost per 100 LF	Labor burden 100 LF	Material cost per 100 LF	Overhead per 100 LF	Profit per 100 LF	Total price per 100 LF
Fascia, 2" x 12", spray one coat stain "to cover"									
Solid body stain, water or oil base (material #18 or #19)									
Spray each coat									
Slow	200	60	22.70	5.25	1.49	37.83	13.82	8.76	67.15
Medium	263	50	19.85	6.37	2.21	39.70	11.83	6.61	66.72
Fast	325	40	17.00	6.92	2.90	42.50	9.68	3.72	65.72
Semi-transparent stain, water or oil base (material #20 or #21)									
Spray each coat									
Slow	225	75	23.75	4.67	1.32	31.67	11.68	7.40	56.74
Medium	288	65	20.80	5.82	2.02	32.00	9.76	5.46	55.06
Fast	350	55	17.80	6.43	2.69	32.36	7.67	2.95	52.10

Use these figures for spraying stain on 2" x 12" fascia board on two sides - the face and the lower edge. The back side is calculated with the overhang (eaves) operation. These figures include one full coat and any touchup required to meet the "to cover" specification. They also include minimum preparation. Add time for extensive preparation. Measurements are based on continuous linear feet. "Slow" work is based on a $10.50 hourly wage, "Medium" work on a $16.75 hourly wage, and "Fast" work on a $22.50 hourly wage. Other qualifications that apply to this table are on page 9.

	Labor SF per manhour	Material coverage SF/gallon	Material cost per gallon	Labor cost per 100 SF	Labor burden 100 SF	Material cost per 100 SF	Overhead per 100 SF	Profit per 100 SF	Total price per 100 SF
Fence, chain link or wire mesh									
Solid body stain, water or oil base (material #18 or #19)									
Brush 1st coat									
Slow	90	600	22.70	11.67	3.31	3.78	5.82	3.69	28.27
Medium	110	550	19.85	15.23	5.30	3.61	5.92	3.31	33.37
Fast	125	500	17.00	18.00	7.51	3.40	5.35	2.06	36.32
Brush 2nd coat									
Slow	130	650	22.70	8.08	2.29	3.49	4.30	2.73	20.89
Medium	145	600	19.85	11.55	4.04	3.31	4.63	2.59	26.12
Fast	160	550	17.00	14.06	5.87	3.09	4.26	1.64	28.92
Roll 1st coat									
Slow	260	575	22.70	4.04	1.15	3.95	2.83	1.80	13.77
Medium	275	525	19.85	6.09	2.13	3.78	2.94	1.64	16.58
Fast	290	475	17.00	7.76	3.24	3.58	2.70	1.04	18.32
Roll 2nd coat									
Slow	280	625	22.70	3.75	1.06	3.63	2.62	1.66	12.72
Medium	300	575	19.85	5.58	1.94	3.45	2.69	1.50	15.16
Fast	320	525	17.00	7.03	2.95	3.24	2.44	.94	16.60

Use these figures for chain link or wire mesh fencing. The figures are based on painting both sides to meet the "to cover" specification. These figures include minimum preparation time. Add time for extensive preparation. To calculate the area, base measurements on the square feet (length times width) of one side of the fence then multiply by a difficulty factor of 3 and use the figures in the above table. For example, if the fence is 100' long x 3' high, the area is 300 SF. Multiply 300 SF x 3 to arrive at 900 SF which is the total to be used with the table. "Slow" work is based on a $10.50 hourly wage, "Medium" work on a $16.75 hourly wage, and "Fast" work on a $22.50 hourly wage. Other qualifications that apply to this table are on page 9.

Fence, wood

For solid plank fence, measure the surface area of one side and multiply by 2 to find the area for both sides. Then use the cost table for Siding, exterior. For good neighbor fence (planks on alternate sides of the rail), measure the surface area of one side and multiply by 2 to find the area for both sides. Then multiply by a difficulty factor of 1.5 and use the pricing table for Siding, exterior.

	Labor SF per manhour	Material coverage SF/gallon	Material cost per gallon	Labor cost per 100 SF	Labor burden 100 SF	Material cost per 100 SF	Overhead per 100 SF	Profit per 100 SF	Total price per 100 SF

Fence, picket, brush application

Solid body or semi-transparent stain, water base (material #18 or #20)

	Labor SF per manhour	Material coverage SF/gallon	Material cost per gallon	Labor cost per 100 SF	Labor burden 100 SF	Material cost per 100 SF	Overhead per 100 SF	Profit per 100 SF	Total price per 100 SF
Brush 1st coat									
Slow	75	400	21.45	14.00	3.97	5.36	7.24	4.59	35.16
Medium	113	388	18.75	14.82	5.17	4.83	6.08	3.40	34.30
Fast	150	375	16.05	15.00	6.27	4.28	4.72	1.82	32.09
Brush 2nd or additional coats									
Slow	120	450	21.45	8.75	2.48	4.77	4.96	3.15	24.11
Medium	145	438	18.75	11.55	4.04	4.28	4.87	2.72	27.46
Fast	170	425	16.05	13.24	5.51	3.78	4.17	1.60	28.30

Solid body or semi-transparent stain, oil base (material #19 or #21)

	Labor SF per manhour	Material coverage SF/gallon	Material cost per gallon	Labor cost per 100 SF	Labor burden 100 SF	Material cost per 100 SF	Overhead per 100 SF	Profit per 100 SF	Total price per 100 SF
Brush 1st coat									
Slow	75	450	25.00	14.00	3.97	5.56	7.30	4.63	35.46
Medium	113	438	21.90	14.82	5.17	5.00	6.12	3.42	34.53
Fast	150	425	18.75	15.00	6.27	4.41	4.75	1.83	32.26
Brush 2nd or additional coats									
Slow	120	500	25.00	8.75	2.48	5.00	5.03	3.19	24.45
Medium	145	488	21.90	11.55	4.04	4.49	4.92	2.75	27.75
Fast	170	475	18.75	13.24	5.51	3.95	4.20	1.62	28.52

For picket fence, measure the overall area of one side and multiply by 4 for painting both sides. Then apply these cost figures. "Slow" work is based on a $10.50 hourly wage, "Medium" work on a $16.75 hourly wage, and "Fast" work on a $22.50 hourly wage. Other qualifications that apply to this table are on page 9.

	Labor SF per manhour	Material coverage SF/gallon	Material cost per gallon	Labor cost per 100 SF	Labor burden 100 SF	Material cost per 100 SF	Overhead per 100 SF	Profit per 100 SF	Total price per 100 SF

Fence, picket, roll application

Solid body or semi-transparent stain, water base (material #18 or #20)

	Labor SF per manhour	Material coverage SF/gallon	Material cost per gallon	Labor cost per 100 SF	Labor burden 100 SF	Material cost per 100 SF	Overhead per 100 SF	Profit per 100 SF	Total price per 100 SF
Roll 1st coat									
Slow	120	360	21.45	8.75	2.48	5.96	5.33	3.38	25.90
Medium	145	343	18.75	11.55	4.04	5.47	5.16	2.88	29.10
Fast	170	325	16.05	13.24	5.51	4.94	4.39	1.69	29.77
Roll 2nd or additional coats									
Slow	200	400	21.45	5.25	1.49	5.36	3.75	2.38	18.23
Medium	225	388	18.75	7.44	2.59	4.83	3.64	2.04	20.54
Fast	250	375	16.05	9.00	3.76	4.28	3.15	1.21	21.40

Solid body or semi-transparent stain, oil base (material #19 or #21)

	Labor SF per manhour	Material coverage SF/gallon	Material cost per gallon	Labor cost per 100 SF	Labor burden 100 SF	Material cost per 100 SF	Overhead per 100 SF	Profit per 100 SF	Total price per 100 SF
Roll 1st coat									
Slow	120	400	25.00	8.75	2.48	6.25	5.42	3.44	26.34
Medium	145	388	21.90	11.55	4.04	5.64	5.20	2.91	29.34
Fast	170	375	18.75	13.24	5.51	5.00	4.40	1.69	29.84
Roll 2nd or additional coats									
Slow	200	450	25.00	5.25	1.49	5.56	3.81	2.42	18.53
Medium	225	438	21.90	7.44	2.59	5.00	3.68	2.06	20.77
Fast	250	425	18.75	9.00	3.76	4.41	3.18	1.22	21.57

For picket fence, measure the overall area of one side and multiply by 4 for painting both sides. Then apply these cost figures. "Slow" work is based on a $10.50 hourly wage, "Medium" work on a $16.75 hourly wage, and "Fast" work on a $22.50 hourly wage. Other qualifications that apply to this table are on page 9.

	Labor SF per manhour	Material coverage SF/gallon	Material cost per gallon	Labor cost per 100 SF	Labor burden 100 SF	Material cost per 100 SF	Overhead per 100 SF	Profit per 100 SF	Total price per 100 SF

Fence, picket, spray application

Solid body or semi-transparent stain, water base (material #18 or #20)

	Labor SF per manhour	Material coverage SF/gallon	Material cost per gallon	Labor cost per 100 SF	Labor burden 100 SF	Material cost per 100 SF	Overhead per 100 SF	Profit per 100 SF	Total price per 100 SF
Spray 1st coat									
Slow	400	300	21.45	2.63	.74	7.15	3.26	2.07	15.85
Medium	500	275	18.75	3.35	1.17	6.82	2.78	1.55	15.67
Fast	600	250	16.05	3.75	1.58	6.42	2.17	.83	14.75
Spray 2nd or additional coats									
Slow	500	350	21.45	2.10	.60	6.13	2.74	1.74	13.31
Medium	600	325	18.75	2.79	.98	5.77	2.33	1.30	13.17
Fast	700	300	16.05	3.21	1.35	5.35	1.83	.70	12.44

Solid body or semi-transparent stain, oil base (material #19 or #21)

	Labor SF per manhour	Material coverage SF/gallon	Material cost per gallon	Labor cost per 100 SF	Labor burden 100 SF	Material cost per 100 SF	Overhead per 100 SF	Profit per 100 SF	Total price per 100 SF
Spray 1st coat									
Slow	400	350	25.00	2.63	.74	7.14	3.26	2.07	15.84
Medium	500	325	21.90	3.35	1.17	6.74	2.76	1.54	15.56
Fast	600	300	18.75	3.75	1.58	6.25	2.14	.82	14.54
Spray 2nd or additional coats									
Slow	500	425	25.00	2.10	.60	5.88	2.66	1.69	12.93
Medium	600	400	21.90	2.79	.98	5.48	2.26	1.27	12.78
Fast	700	375	18.75	3.21	1.35	5.00	1.77	.68	12.01

For picket fence, measure the overall area of one side and multiply by 4 for painting both sides. Then apply these cost figures. "Slow" work is based on a $10.50 hourly wage, "Medium" work on a $16.75 hourly wage, and "Fast" work on a $22.50 hourly wage. Other qualifications that apply to this table are on page 9.

	Labor SF per manhour	Material coverage SF/gallon	Material cost per gallon	Labor cost per 100 SF	Labor burden 100 SF	Material cost per 100 SF	Overhead per 100 SF	Profit per 100 SF	Total price per 100 SF
Fireplace masonry, interior, smooth surface masonry									
Masonry paint, water base (material #31)									
Brush each coat									
Slow	70	140	20.90	15.00	4.26	14.93	10.60	6.72	51.51
Medium	75	130	18.30	22.33	7.78	14.08	10.83	6.05	61.07
Fast	80	120	15.70	28.13	11.73	13.08	9.80	3.77	66.51
Masonry paint, oil base (material #32)									
Brush each coat									
Slow	70	165	28.30	15.00	4.26	17.15	11.29	7.16	54.86
Medium	75	155	24.80	22.33	7.78	16.00	11.30	6.32	63.73
Fast	80	145	21.20	28.13	11.73	14.62	10.08	3.87	68.43
Masonry paint, water base (material #31)									
Roll each coat									
Slow	140	120	20.90	7.50	2.12	17.42	8.39	5.32	40.75
Medium	150	110	18.30	11.17	3.90	16.64	7.77	4.34	43.82
Fast	160	100	15.70	14.06	5.87	15.70	6.59	2.53	44.75
Masonry paint, oil base (material #32)									
Roll each coat									
Slow	140	140	28.30	7.50	2.12	20.21	9.25	5.86	44.94
Medium	150	130	24.80	11.17	3.90	19.08	8.36	4.68	47.19
Fast	160	120	21.20	14.06	5.87	17.67	6.96	2.67	47.23
Masonry paint, water base (material #31)									
Spray each coat									
Slow	400	105	20.90	2.63	.74	19.90	7.22	4.58	35.07
Medium	450	100	18.30	3.72	1.29	18.30	5.71	3.19	32.21
Fast	500	90	15.70	4.50	1.88	17.44	4.41	1.69	29.92
Masonry paint, oil base (material #32)									
Spray each coat									
Slow	400	125	28.30	2.63	.74	22.64	8.07	5.11	39.19
Medium	450	120	24.80	3.72	1.29	20.67	6.29	3.52	35.49
Fast	500	110	21.20	4.50	1.88	19.27	4.75	1.82	32.22

Measurements are based on square feet of the surface area (length times width) to be painted. For fireplace exteriors, use the Masonry cost table which applies. "Slow" work is based on a $10.50 hourly wage, "Medium" work on a $16.75 hourly wage, and "Fast" work on a $22.50 hourly wage. Other qualifications that apply to this table are on page 9.

	Labor LF per manhour	Material coverage LF/gallon	Material cost per gallon	Labor cost per 100 LF	Labor burden 100 LF	Material cost per 100 LF	Overhead per 100 LF	Profit per 100 LF	Total price per 100 LF
Fireplace trim, wood, roll & brush each coat									
Mantle, rough sawn 4' x 12"									
Solid body or semi-transparent stain, water or oil base (Material #18 or #19 or #20 or #21)									
Roll & brush each coat									
Slow	15	50	23.23	70.00	19.87	46.46	42.28	26.80	205.41
Medium	18	45	20.33	93.06	32.45	45.18	41.81	23.37	235.87
Fast	20	40	17.40	112.50	46.95	43.50	37.55	14.43	254.93
Plant-on trim, interior									
Solid body or semi-transparent stain, water or oil base (Material #18 or #19 or #20 or #21)									
Roll & brush each coat									
Slow	75	135	23.23	14.00	3.97	17.21	10.91	6.92	53.01
Medium	80	130	20.33	20.94	7.30	15.64	10.75	6.01	60.64
Fast	85	125	17.40	26.47	11.03	13.92	9.52	3.66	64.60
Siding, interior, tongue & groove									
Solid body or semi-transparent stain, water or oil base (Material #18 or #19 or #20 or #21)									
Roll & brush each coat									
Slow	50	100	23.23	21.00	5.96	23.23	15.56	9.86	75.61
Medium	75	95	20.33	22.33	7.78	21.40	12.62	7.05	71.18
Fast	100	90	17.40	22.50	9.39	19.33	9.48	3.64	64.34

"Slow" work is based on a $10.50 hourly wage, "Medium" work on a $16.75 hourly wage, and "Fast" work on a $22.50 hourly wage. Other qualifications that apply to this table are on page 9.

	Manhours per box	Material coverage gallons/box	Material cost per gallon	Labor cost per box	Labor burden box	Material cost per box	Overhead per box	Profit per box	Total price per box
Firewood boxes, wood, brush each coat									
Boxes of rough sawn wood, 3'0" x 3'0" x 3'0" deep									
Solid body or semi-transparent stain, water or oil base (Material #18 or #19 or #20 or #21)									
Roll & brush each coat									
Slow	0.40	0.20	23.23	4.20	1.19	4.65	3.11	1.97	15.12
Medium	0.35	0.23	20.33	5.86	2.05	4.68	3.08	1.72	17.39
Fast	0.30	0.25	17.40	6.75	2.82	4.35	2.58	.99	17.49

"Slow" work is based on a $10.50 hourly wage, "Medium" work on a $16.75 hourly wage, and "Fast" work on a $22.50 hourly wage. Other qualifications that apply to this table are on page 9.

	Labor SF per manhour	Material coverage SF/gallon	Material cost per gallon	Labor cost per 100 SF	Labor burden 100 SF	Material cost per 100 SF	Overhead per 100 SF	Profit per 100 SF	Total price per 100 SF
Floors, concrete, brush, interior or exterior									
Masonry (concrete) paint, water base (material #31)									
Brush 1st coat									
Slow	90	250	20.90	11.67	3.31	8.36	7.24	4.59	35.17
Medium	145	238	18.30	11.55	4.04	7.69	5.70	3.19	32.17
Fast	200	225	15.70	11.25	4.70	6.98	4.24	1.63	28.80
Brush 2nd coat									
Slow	125	375	20.90	8.40	2.38	5.57	5.07	3.21	24.63
Medium	200	325	18.30	8.38	2.92	5.63	4.15	2.32	23.40
Fast	275	275	15.70	8.18	3.43	5.71	3.20	1.23	21.75
Brush 3rd or additional coats									
Slow	150	335	20.90	7.00	1.99	6.24	4.72	2.99	22.94
Medium	225	310	18.30	7.44	2.59	5.90	3.90	2.18	22.01
Fast	300	285	15.70	7.50	3.12	5.51	2.99	1.15	20.27
Masonry (concrete) paint, oil base (material #32)									
Brush 1st coat									
Slow	90	300	28.30	11.67	3.31	9.43	7.57	4.80	36.78
Medium	145	288	24.80	11.55	4.04	8.61	5.93	3.31	33.44
Fast	200	275	21.20	11.25	4.70	7.71	4.38	1.68	29.72
Brush 2nd coat									
Slow	125	400	28.30	8.40	2.38	7.08	5.54	3.51	26.91
Medium	200	388	24.80	8.38	2.92	6.39	4.33	2.42	24.44
Fast	275	375	21.20	8.18	3.43	5.65	3.19	1.23	21.68
Brush 3rd or additional coats									
Slow	150	550	28.30	7.00	1.99	5.15	4.38	2.78	21.30
Medium	225	525	24.80	7.44	2.59	4.72	3.61	2.02	20.38
Fast	300	500	21.20	7.50	3.12	4.24	2.75	1.06	18.67
Epoxy, 1 part, water base (material #28)									
Brush each coat									
Slow	125	400	37.30	8.40	2.38	9.33	6.24	3.95	30.30
Medium	163	388	32.60	10.28	3.57	8.40	5.45	3.05	30.75
Fast	200	375	28.00	11.25	4.70	7.47	4.33	1.67	29.42
Epoxy, 2 part system (material #29)									
Brush each coat									
Slow	100	400	73.20	10.50	2.98	18.30	9.85	6.25	47.88
Medium	138	388	64.10	12.14	4.24	16.52	8.06	4.50	45.46
Fast	175	375	54.90	12.86	5.35	14.64	6.08	2.34	41.27

Also use these formulas for concrete steps, stair treads, porches and patios. "Slow" work is based on a $10.50 hourly wage, "Medium" work on a $16.75 hourly wage, and "Fast" work on a $22.50 hourly wage. Other qualifications that apply to this table are on page 9.

	Labor SF per manhour	Material coverage SF/gallon	Material cost per gallon	Labor cost per 100 SF	Labor burden 100 SF	Material cost per 100 SF	Overhead per 100 SF	Profit per 100 SF	Total price per 100 SF
Floors, concrete, roll, interior or exterior									
Masonry (concrete) paint, water base (material #31)									
Roll 1st coat									
Slow	135	275	20.90	7.78	2.21	7.60	5.45	3.46	26.50
Medium	218	263	18.30	7.68	2.69	6.96	4.24	2.37	23.94
Fast	300	250	15.70	7.50	3.12	6.28	3.13	1.20	21.23
Roll 2nd coat									
Slow	195	350	20.90	5.38	1.54	5.97	3.99	2.53	19.41
Medium	268	325	18.30	6.25	2.18	5.63	3.44	1.93	19.43
Fast	340	300	15.70	6.62	2.76	5.23	2.70	1.04	18.35
Roll 3rd or additional coats									
Slow	210	375	20.90	5.00	1.42	5.57	3.72	2.36	18.07
Medium	300	350	18.30	5.58	1.94	5.23	3.12	1.75	17.62
Fast	390	325	15.70	5.77	2.39	4.83	2.41	.93	16.33
Masonry (concrete) paint, oil base (material #32)									
Roll 1st coat									
Slow	135	370	28.30	7.78	2.21	7.65	5.47	3.47	26.58
Medium	218	345	24.80	7.68	2.69	7.19	4.30	2.40	24.26
Fast	300	320	21.20	7.50	3.12	6.63	3.19	1.23	21.67
Roll 2nd coat									
Slow	195	500	28.30	5.38	1.54	5.66	3.90	2.47	18.95
Medium	268	475	24.80	6.25	2.18	5.22	3.34	1.87	18.86
Fast	340	450	21.20	6.62	2.76	4.71	2.61	1.00	17.70
Roll 3rd or additional coats									
Slow	210	550	28.30	5.00	1.42	5.15	3.59	2.27	17.43
Medium	300	525	24.80	5.58	1.94	4.72	3.00	1.68	16.92
Fast	390	500	21.20	5.77	2.39	4.24	2.30	.88	15.58
Epoxy, 1 part, water base (material #28)									
Roll each coat									
Slow	150	500	37.30	7.00	1.99	7.46	5.10	3.23	24.78
Medium	225	488	32.60	7.44	2.59	6.68	4.09	2.29	23.09
Fast	300	475	28.00	7.50	3.12	5.89	3.06	1.17	20.74
Epoxy, 2 part system (material #29)									
Roll each coat									
Slow	135	500	73.20	7.78	2.21	14.64	7.64	4.84	37.11
Medium	208	488	64.10	8.05	2.82	13.14	5.88	3.29	33.18
Fast	250	475	54.90	9.00	3.76	11.56	4.50	1.73	30.55

Also use these formulas for concrete steps, stair treads, porches and patios. "Slow" work is based on a $10.50 hourly wage, "Medium" work on a $16.75 hourly wage, and "Fast" work on a $22.50 hourly wage. Other qualifications that apply to this table are on page 9.

	Labor SF per manhour	Material coverage SF/gallon	Material cost per gallon	Labor cost per 100 SF	Labor burden 100 SF	Material cost per 100 SF	Overhead per 100 SF	Profit per 100 SF	Total price per 100 SF
Floors, concrete, spray, interior or exterior									
Masonry (concrete) paint, water base (material #31)									
Spray 1st coat									
Slow	800	175	20.90	1.31	.38	11.94	4.22	2.68	20.53
Medium	900	163	18.30	1.86	.65	11.23	3.37	1.88	18.99
Fast	1000	150	15.70	2.25	.94	10.47	2.53	.97	17.16
Spray 2nd coat									
Slow	900	275	20.90	1.17	.33	7.60	2.82	1.79	13.71
Medium	1000	263	18.30	1.68	.58	6.96	2.26	1.26	12.74
Fast	1100	250	15.70	2.05	.85	6.28	1.70	.65	11.53
Spray 3rd or additional coats									
Slow	1000	325	20.90	1.05	.30	6.43	2.41	1.53	11.72
Medium	1100	313	18.30	1.52	.54	5.85	1.94	1.08	10.93
Fast	1200	300	15.70	1.88	.77	5.23	1.46	.56	9.90
Masonry (concrete) paint, oil base (material #32)									
Spray 1st coat									
Slow	800	200	28.30	1.31	.38	14.15	4.91	3.11	23.86
Medium	900	188	24.80	1.86	.65	13.19	3.85	2.15	21.70
Fast	1000	175	21.20	2.25	.94	12.11	2.83	1.09	19.22
Spray 2nd coat									
Slow	900	300	28.30	1.17	.33	9.43	3.39	2.15	16.47
Medium	1000	288	24.80	1.68	.58	8.61	2.67	1.49	15.03
Fast	1100	275	21.20	2.05	.85	7.71	1.96	.75	13.32
Spray 3rd or additional coats									
Slow	1000	350	28.30	1.05	.30	8.09	2.93	1.86	14.23
Medium	1100	338	24.80	1.52	.54	7.34	2.30	1.29	12.99
Fast	1200	325	21.20	1.88	.77	6.52	1.70	.65	11.52

Also use these formulas for concrete steps, stair treads, porches and patios. "Slow" work is based on a $10.50 hourly wage, "Medium" work on a $16.75 hourly wage, and "Fast" work on a $22.50 hourly wage. Other qualifications that apply to this table are on page 9.

	Labor SF per manhour	Material coverage SF/gallon	Material cost per gallon	Labor cost per 100 SF	Labor burden 100 SF	Material cost per 100 SF	Overhead per 100 SF	Profit per 100 SF	Total price per 100 SF
Floors, concrete, penetrating stain, interior or exterior									
Penetrating oil stain (material #13)									
Roll 1st coat									
Slow	225	450	29.00	4.67	1.32	6.44	3.86	2.45	18.74
Medium	250	425	25.40	6.70	2.34	5.98	3.68	2.06	20.76
Fast	275	400	21.80	8.18	3.43	5.45	3.15	1.21	21.42
Roll 2nd coat									
Slow	325	500	29.00	3.23	.92	5.80	3.08	1.95	14.98
Medium	345	475	25.40	4.86	1.69	5.35	2.92	1.63	16.45
Fast	365	450	21.80	6.16	2.58	4.84	2.51	.96	17.05
Roll 3rd and additional coats									
Slow	365	525	29.00	2.88	.81	5.52	2.86	1.81	13.88
Medium	383	500	25.40	4.37	1.53	5.08	2.69	1.50	15.17
Fast	400	475	21.80	5.63	2.34	4.59	2.33	.89	15.78

Also use these formulas for concrete steps, stair treads, porches and patios. "Slow" work is based on a $10.50 hourly wage, "Medium" work on a $16.75 hourly wage, and "Fast" work on a $22.50 hourly wage. Other qualifications that apply to this table are on page 9.

	Labor SF per manhour	Material coverage SF/gallon	Material cost per gallon	Labor cost per 100 SF	Labor burden 100 SF	Material cost per 100 SF	Overhead per 100 SF	Profit per 100 SF	Total price per 100 SF
Floors, wood, interior or exterior, paint grade, brush application									
Undercoat, water base (material #3)									
Brush prime coat									
Slow	275	450	21.10	3.82	1.09	4.69	2.98	1.89	14.47
Medium	300	425	18.50	5.58	1.94	4.35	2.91	1.63	16.41
Fast	325	400	15.80	6.92	2.90	3.95	2.55	.98	17.30
Undercoat, oil base (material #4)									
Brush prime coat									
Slow	275	500	21.60	3.82	1.09	4.32	2.86	1.81	13.90
Medium	300	475	18.90	5.58	1.94	3.98	2.82	1.58	15.90
Fast	325	450	16.20	6.92	2.90	3.60	2.48	.95	16.85
Porch & deck enamel, water base (material #26)									
Brush 1st and additional finish coats									
Slow	300	475	27.60	3.50	.99	5.81	3.20	2.03	15.53
Medium	325	450	24.10	5.15	1.81	5.36	3.01	1.68	17.01
Fast	350	425	20.70	6.43	2.69	4.87	2.59	.99	17.57
Porch & deck enamel, oil base (material #27)									
Brush 1st and additional finish coats									
Slow	300	550	30.10	3.50	.99	5.47	3.09	1.96	15.01
Medium	325	525	26.40	5.15	1.81	5.03	2.93	1.64	16.56
Fast	350	500	22.60	6.43	2.69	4.52	2.52	.97	17.13
Epoxy, 1 part, water base (material #28)									
Brush each coat									
Slow	125	450	37.30	8.40	2.38	8.29	5.91	3.75	28.73
Medium	163	425	32.60	10.28	3.57	7.67	5.27	2.95	29.74
Fast	200	400	28.00	11.25	4.70	7.00	4.25	1.63	28.83
Epoxy, 2 part system (material #29)									
Brush each coat									
Slow	100	425	73.20	10.50	2.98	17.22	9.52	6.03	46.25
Medium	138	400	64.10	12.14	4.24	16.03	7.94	4.44	44.79
Fast	175	375	54.90	12.86	5.35	14.64	6.08	2.34	41.27

"Slow" work is based on a $10.50 hourly wage, "Medium" work on a $16.75 hourly wage, and "Fast" work on a $22.50 hourly wage. Other qualifications that apply to this table are on page 9.

	Labor SF per manhour	Material coverage SF/gallon	Material cost per gallon	Labor cost per 100 SF	Labor burden 100 SF	Material cost per 100 SF	Overhead per 100 SF	Profit per 100 SF	Total price per 100 SF
Floors, wood, interior or exterior, paint grade, roll application									
Undercoat, water base (material #3)									
Roll prime coat									
Slow	400	425	21.10	2.63	.74	4.96	2.59	1.64	12.56
Medium	438	400	18.50	3.82	1.33	4.63	2.40	1.34	13.52
Fast	475	375	15.80	4.74	1.99	4.21	2.02	.78	13.74
Undercoat, oil base (material #4)									
Roll prime coat									
Slow	400	475	21.60	2.63	.74	4.55	2.46	1.56	11.94
Medium	438	450	18.90	3.82	1.33	4.20	2.29	1.28	12.92
Fast	475	425	16.20	4.74	1.99	3.81	1.95	.75	13.24
Porch & deck enamel, water base (material #26)									
Roll 1st or additional finish coats									
Slow	425	475	27.60	2.47	.70	5.81	2.78	1.76	13.52
Medium	463	450	24.10	3.62	1.26	5.36	2.51	1.40	14.15
Fast	500	425	20.70	4.50	1.88	4.87	2.08	.80	14.13
Porch & deck enamel, oil base (material #27)									
Roll 1st or additional finish coats									
Slow	425	525	30.10	2.47	.70	5.73	2.76	1.75	13.41
Medium	463	500	26.40	3.62	1.26	5.28	2.49	1.39	14.04
Fast	500	475	22.60	4.50	1.88	4.76	2.06	.79	13.99
Epoxy, 1 part, water base (material #28)									
Brush each coat									
Slow	200	425	37.30	5.25	1.49	8.78	4.81	3.05	23.38
Medium	250	400	32.60	6.70	2.34	8.15	4.21	2.35	23.75
Fast	300	375	28.00	7.50	3.12	7.47	3.35	1.29	22.73
Epoxy, 2 part system (material #29)									
Brush each coat									
Slow	175	400	73.20	6.00	1.70	18.30	8.06	5.11	39.17
Medium	225	375	64.10	7.44	2.59	17.09	6.64	3.71	37.47
Fast	275	350	54.90	8.18	3.43	15.69	5.05	1.94	34.29

"Slow" work is based on a $10.50 hourly wage, "Medium" work on a $16.75 hourly wage, and "Fast" work on a $22.50 hourly wage. Other qualifications that apply to this table are on page 9.

	Labor SF per manhour	Material coverage SF/gallon	Material cost per gallon	Labor cost per 100 SF	Labor burden 100 SF	Material cost per 100 SF	Overhead per 100 SF	Profit per 100 SF	Total price per 100 SF
Floors, wood, interior or exterior, stain grade									
Wiping stain, varnish, oil base (material #30a)									
Stain, brush 1st coat, wipe & fill									
Slow	225	500	30.60	4.67	1.32	6.12	3.76	2.38	18.25
Medium	250	475	26.80	6.70	2.34	5.64	3.59	2.01	20.28
Fast	275	450	22.90	8.18	3.43	5.09	3.09	1.19	20.98
Stain, brush 2nd coat, wipe & fill									
Slow	400	525	30.60	2.63	.74	5.83	2.86	1.81	13.87
Medium	425	500	26.80	3.94	1.37	5.36	2.61	1.46	14.74
Fast	450	475	22.90	5.00	2.08	4.82	2.20	.85	14.95
Stain, brush 3rd or additional coats, wipe & fill									
Slow	425	550	30.60	2.47	.70	5.56	2.71	1.72	13.16
Medium	450	525	26.80	3.72	1.29	5.10	2.48	1.39	13.98
Fast	475	500	22.90	4.74	1.99	4.58	2.09	.80	14.20
Sanding sealer, varnish (material #30b)									
Maple or pine, brush 1 coat									
Slow	375	475	33.40	2.80	.80	7.03	3.30	2.09	16.02
Medium	400	450	29.30	4.19	1.46	6.51	2.98	1.67	16.81
Fast	425	425	25.10	5.29	2.20	5.91	2.48	.95	16.83
Maple or pine, brush 2nd or additional coats									
Slow	425	550	33.40	2.47	.70	6.07	2.86	1.82	13.92
Medium	450	525	29.30	3.72	1.29	5.58	2.60	1.45	14.64
Fast	475	500	25.10	4.74	1.99	5.02	2.17	.83	14.75
Oak, brush 1 coat									
Slow	400	525	33.40	2.63	.74	6.36	3.02	1.91	14.66
Medium	425	500	29.30	3.94	1.37	5.86	2.74	1.53	15.44
Fast	450	475	25.10	5.00	2.08	5.28	2.29	.88	15.53
Oak, brush 2nd or additional coats									
Slow	500	625	33.40	2.10	.60	5.34	2.49	1.58	12.11
Medium	525	600	29.30	3.19	1.10	4.88	2.25	1.26	12.68
Fast	550	575	25.10	4.09	1.71	4.37	1.88	.72	12.77
Shellac, clear (material #12)									
Brush 1st coat									
Slow	275	475	25.30	3.82	1.09	5.33	3.17	2.01	15.42
Medium	300	450	22.10	5.58	1.94	4.91	3.05	1.70	17.18
Fast	325	425	18.90	6.92	2.90	4.45	2.64	1.01	17.92
Brush 2nd or additional coats									
Slow	400	500	25.30	2.63	.74	5.06	2.62	1.66	12.71
Medium	425	475	22.10	3.94	1.37	4.65	2.44	1.36	13.76
Fast	450	450	18.90	5.00	2.08	4.20	2.09	.80	14.17

	Labor SF per manhour	Material coverage SF/gallon	Material cost per gallon	Labor cost per 100 SF	Labor burden 100 SF	Material cost per 100 SF	Overhead per 100 SF	Profit per 100 SF	Total price per 100 SF
Varnish, gloss or flat (material #30c)									
Brush 1st coat									
Slow	275	475	35.40	3.82	1.09	7.45	3.83	2.43	18.62
Medium	300	450	31.00	5.58	1.94	6.89	3.53	1.97	19.91
Fast	325	425	26.60	6.92	2.90	6.26	2.97	1.14	20.19
Brush 2nd or additional coats									
Slow	350	600	35.40	3.00	.86	5.90	3.02	1.92	14.70
Medium	375	575	31.00	4.47	1.56	5.39	2.80	1.56	15.78
Fast	400	550	26.60	5.63	2.34	4.84	2.37	.91	16.09
Penetrating stain wax & wipe (material #14)									
Stain, brush 1st coat & wipe									
Slow	200	550	27.20	5.25	1.49	4.95	3.62	2.30	17.61
Medium	250	525	23.80	6.70	2.34	4.53	3.32	1.86	18.75
Fast	300	500	20.40	7.50	3.12	4.08	2.72	1.05	18.47
Stain, brush 2nd or additional coats & wipe									
Slow	250	600	27.20	4.20	1.19	4.53	3.08	1.95	14.95
Medium	300	575	23.80	5.58	1.94	4.14	2.86	1.60	16.12
Fast	350	550	20.40	6.43	2.69	3.71	2.37	.91	16.11
Wax & polish (material #15)									
Hand apply 1 coat									
Slow	175	1000	8.00	6.00	1.70	.80	2.64	1.67	12.81
Medium	200	950	7.00	8.38	2.92	.74	2.95	1.65	16.64
Fast	225	900	6.00	10.00	4.16	.67	2.75	1.06	18.64
Buffing with machine									
Slow	400	--	--	2.63	.74	--	1.05	.66	5.08
Medium	450	--	--	3.72	1.29	--	1.23	.69	6.93
Fast	500	--	--	4.50	1.88	--	1.18	.45	8.01

"Slow" work is based on a $10.50 hourly wage, "Medium" work on a $16.75 hourly wage, and "Fast" work on a $22.50 hourly wage. Other qualifications that apply to this table are on page 9.

	Manhours per door	Gallons per Door	Material cost per gallon	Labor cost per door	Labor burden door	Material cost per door	Overhead per door	Profit per door	Total price per door

Garage door backs, seal coat, spray one coat

	Manhours per door	Gallons per Door	Material cost per gallon	Labor cost per door	Labor burden door	Material cost per door	Overhead per door	Profit per door	Total price per door
Sanding sealer, lacquer (material #11b)									
1 car garage, 8' x 7'									
Slow	0.30	0.40	17.20	3.15	.89	6.88	3.39	2.15	16.46
Medium	0.25	0.50	15.00	4.19	1.46	7.50	3.22	1.80	18.17
Fast	0.20	0.60	12.90	4.50	1.88	7.74	2.61	1.00	17.73
2 car garage, 16' x 7'									
Slow	0.40	0.80	17.20	4.20	1.19	13.76	5.94	3.76	28.85
Medium	0.35	0.90	15.00	5.86	2.05	13.50	5.24	2.93	29.58
Fast	0.30	1.00	12.90	6.75	2.82	12.90	4.16	1.60	28.23
3 car garage, 16' x 7' + 8' x 7'									
Slow	0.60	1.00	17.20	6.30	1.79	17.20	7.84	4.97	38.10
Medium	0.55	1.10	15.00	9.21	3.21	16.50	7.09	3.96	39.97
Fast	0.50	1.20	12.90	11.25	4.70	15.48	5.81	2.23	39.47

Use the figures for Siding when estimating the cost of painting garage door fronts. These figures assume a one-car garage door measures 7' x 8' and a two-car garage door measures 7' x 16'. A three-car garage has one single and one double door. Government funded projects (FHA, VA, HUD) usually require sealing the garage door back on new construction projects. The doors are usually sprayed along with the cabinet sealer coat (as used in this table) or stained along with the exterior trim. "Slow" work is based on a $10.50 hourly wage, "Medium" work on a $16.75 hourly wage, and "Fast" work on a $22.50 hourly wage. Other qualifications that apply to this table are on page 9.

	Labor LF per manhour	Material coverage LF/gallon	Material cost per gallon	Labor cost per 100 LF	Labor burden 100 LF	Material cost per 100 LF	Overhead per 100 LF	Profit per 100 LF	Total price per 100 LF
Gutters and downspouts (galvanized), brush application									
Gutters									
Metal prime, rust inhibitor, clean metal (material #35)									
Brush prime coat									
Slow	80	400	29.90	13.13	3.72	7.48	7.55	4.79	36.67
Medium	90	375	26.20	18.61	6.49	6.99	7.86	4.39	44.34
Fast	100	350	22.40	22.50	9.39	6.40	7.08	2.72	48.09
Metal prime, rust inhibitor, rusty metal (material #36)									
Brush prime coat									
Slow	80	400	36.60	13.13	3.72	9.15	8.07	5.11	39.18
Medium	90	375	32.10	18.61	6.49	8.56	8.25	4.61	46.52
Fast	100	350	27.50	22.50	9.39	7.86	7.35	2.83	49.93
Metal finish - synthetic enamel (off white), gloss (material #37)									
Brush 1st finish coat									
Slow	100	425	27.90	10.50	2.98	6.56	6.22	3.94	30.20
Medium	110	400	24.40	15.23	5.30	6.10	6.53	3.65	36.81
Fast	120	375	20.90	18.75	7.81	5.57	5.95	2.29	40.37
Brush 2nd or additional finish coats									
Slow	120	450	27.90	8.75	2.48	6.20	5.41	3.43	26.27
Medium	130	425	24.40	12.88	4.49	5.74	5.66	3.16	31.93
Fast	140	400	20.90	16.07	6.70	5.23	5.18	1.99	35.17
Metal finish - synthetic enamel (colors except orange/red), gloss (material #38)									
Brush 1st finish coat									
Slow	100	425	30.00	10.50	2.98	7.06	6.37	4.04	30.95
Medium	110	400	26.30	15.23	5.30	6.58	6.64	3.71	37.46
Fast	120	375	22.50	18.75	7.81	6.00	6.03	2.32	40.91
Brush 2nd or additional finish coats									
Slow	120	450	30.00	8.75	2.48	6.67	5.55	3.52	26.97
Medium	130	425	26.30	12.88	4.49	6.19	5.77	3.23	32.56
Fast	140	400	22.50	16.07	6.70	5.63	5.26	2.02	35.68
Downspouts									
Metal prime, rust inhibitor, clean metal (material #35)									
Brush prime coat									
Slow	30	250	29.90	35.00	9.93	11.96	17.65	11.19	85.73
Medium	35	225	26.20	47.86	16.68	11.64	18.66	10.43	105.27
Fast	40	200	22.40	56.25	23.48	11.20	16.82	6.47	114.22
Metal prime, rust inhibitor, rusty metal (material #36)									
Brush prime coat									
Slow	30	250	36.60	35.00	9.93	14.64	18.48	11.71	89.76
Medium	35	225	32.10	47.86	16.68	14.27	19.31	10.79	108.91
Fast	40	200	27.50	56.25	23.48	13.75	17.29	6.65	117.42

	Labor LF per manhour	Material coverage LF/gallon	Material cost per gallon	Labor cost per 100 LF	Labor burden 100 LF	Material cost per 100 LF	Overhead per 100 LF	Profit per 100 LF	Total price per 100 LF
Metal finish - synthetic enamel (off white), gloss (material #37)									
Brush 1st finish coat									
Slow	50	275	27.90	21.00	5.96	10.15	11.51	7.29	55.91
Medium	60	250	24.40	27.92	9.74	9.76	11.62	6.49	65.53
Fast	70	225	20.90	32.14	13.43	9.29	10.15	3.90	68.91
Brush 2nd or additional finish coats									
Slow	70	300	27.90	15.00	4.26	9.30	8.86	5.61	43.03
Medium	80	275	24.40	20.94	7.30	8.87	9.09	5.08	51.28
Fast	90	250	20.90	25.00	10.43	8.36	8.10	3.11	55.00
Metal finish - synthetic enamel (colors except orange/red), gloss (material #38)									
Brush 1st finish coat									
Slow	50	275	30.00	21.00	5.96	10.91	11.74	7.44	57.05
Medium	60	250	26.30	27.92	9.74	10.52	11.80	6.60	66.58
Fast	70	225	22.50	32.14	13.43	10.00	10.28	3.95	69.80
Brush 2nd or additional finish coats									
Slow	70	300	30.00	15.00	4.26	10.00	9.07	5.75	44.08
Medium	80	275	26.30	20.94	7.30	9.56	9.26	5.18	52.24
Fast	90	250	22.50	25.00	10.43	9.00	8.22	3.16	55.81

NOTE: Oil base material is recommended for any metal surface. Although water base material is often used, it may cause oxidation, corrosion and rust. These figures assume that all exposed surfaces of 5" gutters and 4" downspouts are painted. For ornamental gutters and downspouts, multiply the linear feet by 1.5 before using these figures. "Slow" work is based on a $10.50 hourly wage, "Medium" work on a $16.75 hourly wage, and "Fast" work on a $22.50 hourly wage. Other qualifications that apply to this table are on page 9.

High Time Difficulty Factors

Painting takes longer and may require more material when heights above the floor exceed 8 feet. The additional time and material for working at these heights and using a roller pole or a wand on a spray gun, climbing up and down a ladder or scaffolding is applied by using one of the factors listed below. The wall area above 8 feet is typically referred to as the "Clip." To apply the high time difficulty factor, measure the surface above 8 feet which is to be painted and multiply that figure by the appropriate factor. This measurement can be listed on a separate line of your take-off and the appropriate price can be applied for a total.

For labor calculations only:
- Add 30% to the area for heights between 8 and 13 feet (multiply by 1.3)
- Add 60% to the area for heights from 13 to 17 feet (multiply by 1.6)
- Add 90% to the area for heights from 17 to 19 feet (multiply by 1.9)
- Add 120% to the area for heights from 19 to 21 feet (multiply by 2.2)

EXAMPLE: A 17 x 14 living room has a vaulted ceiling 13 feet high. Your take-off sheet might look like this:

Walls to 8 feet: 136 + 112 + 136 + 112 = 496 SF
Clip: [(5 x 14) / 2] x 2 + (5 x 17) = 70 + 85 = 155 SF
area of two triangles + rectangular area
155 SF x 1.3 (high time difficulty factor) = 202 SF

Then multiply each SF total by the appropriate price per square foot.

Mail box structures, wood, apartment type

Measure the length of each board to be painted and use the manhours and material given for Trellis or Plant-on trim or Siding.

	Labor SF per manhour	Material coverage SF/gallon	Material cost per gallon	Labor cost per 100 SF	Labor burden 100 SF	Material cost per 100 SF	Overhead per 100 SF	Profit per 100 SF	Total price per 100 SF
Masonry, anti-graffiti stain eliminator on smooth or rough surface									
Water base primer and sealer (material #39)									
Roll & brush each coat									
Slow	350	400	25.40	3.00	.86	6.35	3.16	2.00	15.37
Medium	375	375	22.30	4.47	1.56	5.95	2.94	1.64	16.56
Fast	400	350	19.10	5.63	2.34	5.46	2.49	.96	16.88
Oil base primer and sealer (material #40)									
Roll & brush each coat									
Slow	350	375	25.70	3.00	.86	6.85	3.32	2.10	16.13
Medium	375	350	22.50	4.47	1.56	6.43	3.05	1.71	17.22
Fast	400	325	19.30	5.63	2.34	5.94	2.58	.99	17.48
Polyurethane 2 part system (material #41)									
Roll & brush each coat									
Slow	300	375	91.40	3.50	.99	24.37	8.95	5.67	43.48
Medium	325	350	80.00	5.15	1.81	22.86	7.30	4.08	41.20
Fast	350	325	68.60	6.43	2.69	21.11	5.59	2.15	37.97

Use these figures for new brick, used brick, or Concrete Masonry Units (CMU) where the block surfaces are either smooth or rough, porous or unfilled, with joints struck to average depth. The more porous the surface, the rougher the texture, the more time and material will be required. "Slow" work is based on a $10.50 hourly wage, "Medium" work on a $16.75 hourly wage, and "Fast" work on a $22.50 hourly wage. Other qualifications that apply to this table are on page 9.

	Labor SF per manhour	Material coverage SF/gallon	Material cost per gallon	Labor cost per 100 SF	Labor burden 100 SF	Material cost per 100 SF	Overhead per 100 SF	Profit per 100 SF	Total price per 100 SF
Masonry, block filler									
Brush 1 coat (material #33)									
Slow	95	75	15.60	11.05	3.14	20.80	10.85	6.88	52.72
Medium	125	65	13.70	13.40	4.67	21.08	9.59	5.36	54.10
Fast	155	55	11.70	14.52	6.05	21.27	7.74	2.98	52.56
Roll 1 coat (material #33)									
Slow	190	70	15.60	5.53	1.56	22.29	9.11	5.78	44.27
Medium	215	60	13.70	7.79	2.71	22.83	8.17	4.57	46.07
Fast	240	50	11.70	9.38	3.92	23.40	6.79	2.61	46.10
Spray 1 coat (material #33)									
Slow	425	65	15.60	2.47	.70	24.00	8.42	5.34	40.93
Medium	525	55	13.70	3.19	1.10	24.91	7.16	4.00	40.36
Fast	625	45	11.70	3.60	1.50	26.00	5.75	2.21	39.06

Use these figures for using block filler on rough or porous masonry with joints struck to average depth. "Slow" work is based on a $10.50 hourly wage, "Medium" work on a $16.75 hourly wage, and "Fast" work on a $22.50 hourly wage. Other qualifications that apply to this table are on page 9.

	Labor SF per manhour	Material coverage SF/gallon	Material cost per gallon	Labor cost per 100 SF	Labor burden 100 SF	Material cost per 100 SF	Overhead per 100 SF	Profit per 100 SF	Total price per 100 SF
Masonry, brick, new, smooth-surface, brush									
Masonry paint, water base, flat or gloss (material #31)									
Brush 1st coat									
Slow	200	300	20.90	5.25	1.49	6.97	4.25	2.69	20.65
Medium	225	275	18.30	7.44	2.59	6.65	4.09	2.28	23.05
Fast	250	250	15.70	9.00	3.76	6.28	3.52	1.35	23.91
Brush 2nd or additional coats									
Slow	250	325	20.90	4.20	1.19	6.43	3.66	2.32	17.80
Medium	275	300	18.30	6.09	2.13	6.10	3.51	1.96	19.79
Fast	300	275	15.70	7.50	3.12	5.71	3.02	1.16	20.51
Masonry paint, oil base (material #32)									
Brush 1st coat									
Slow	200	350	28.30	5.25	1.49	8.09	4.60	2.91	22.34
Medium	225	325	24.80	7.44	2.59	7.63	4.33	2.42	24.41
Fast	250	300	21.20	9.00	3.76	7.07	3.67	1.41	24.91
Brush 2nd or additional coats									
Slow	250	400	28.30	4.20	1.19	7.08	3.87	2.45	18.79
Medium	275	363	24.80	6.09	2.13	6.83	3.68	2.06	20.79
Fast	300	325	21.20	7.50	3.12	6.52	3.17	1.22	21.53

Use these figures for new smooth-surface brick with joints struck to average depth. "Slow" work is based on a $10.50 hourly wage, "Medium" work on a $16.75 hourly wage, and "Fast" work on a $22.50 hourly wage. Other qualifications that apply to this table are on page 9.

	Labor SF per manhour	Material coverage SF/gallon	Material cost per gallon	Labor cost per 100 SF	Labor burden 100 SF	Material cost per 100 SF	Overhead per 100 SF	Profit per 100 SF	Total price per 100 SF
Masonry, brick, new, smooth-surface, roll									
Masonry paint, water base, flat or gloss (material #31)									
Roll 1st coat									
Slow	325	250	20.90	3.23	.92	8.36	3.88	2.46	18.85
Medium	350	213	18.30	4.79	1.67	8.59	3.69	2.06	20.80
Fast	375	175	15.70	6.00	2.51	8.97	3.23	1.24	21.95
Roll 2nd or additional coats									
Slow	375	275	20.90	2.80	.80	7.60	3.47	2.20	16.87
Medium	400	250	18.30	4.19	1.46	7.32	3.18	1.78	17.93
Fast	425	225	15.70	5.29	2.20	6.98	2.68	1.03	18.18
Masonry paint, oil base (material #32)									
Roll 1st coat									
Slow	325	325	28.30	3.23	.92	8.71	3.99	2.53	19.38
Medium	350	288	24.80	4.79	1.67	8.61	3.69	2.06	20.82
Fast	375	250	21.20	6.00	2.51	8.48	3.14	1.21	21.34
Roll 2nd or additional coats									
Slow	375	350	28.30	2.80	.80	8.09	3.62	2.30	17.61
Medium	400	313	24.80	4.19	1.46	7.92	3.32	1.86	18.75
Fast	425	275	21.20	5.29	2.20	7.71	2.81	1.08	19.09
Waterproofing, clear hydro sealer, oil base (material #34)									
Roll 1st coat									
Slow	200	175	17.80	5.25	1.49	10.17	5.24	3.32	25.47
Medium	225	150	15.60	7.44	2.59	10.40	5.01	2.80	28.24
Fast	250	125	13.30	9.00	3.76	10.64	4.33	1.66	29.39
Roll 2nd or additional coats									
Slow	225	200	17.80	4.67	1.32	8.90	4.62	2.93	22.44
Medium	250	190	15.60	6.70	2.34	8.21	4.22	2.36	23.83
Fast	275	180	13.30	8.18	3.43	7.39	3.51	1.35	23.86

Use these figures for new smooth-surface brick with joints struck to average depth. "Slow" work is based on a $10.50 hourly wage, "Medium" work on a $16.75 hourly wage, and "Fast" work on a $22.50 hourly wage. Other qualifications that apply to this table are on page 9.

	Labor SF per manhour	Material coverage SF/gallon	Material cost per gallon	Labor cost per 100 SF	Labor burden 100 SF	Material cost per 100 SF	Overhead per 100 SF	Profit per 100 SF	Total price per 100 SF
Masonry, brick, new, smooth-surface, spray									
Masonry paint, water base, flat or gloss (material #31)									
Spray 1st coat									
Slow	650	250	20.90	1.62	.46	8.36	3.24	2.05	15.73
Medium	750	225	18.30	2.23	.77	8.13	2.73	1.53	15.39
Fast	850	200	15.70	2.65	1.11	7.85	2.15	.83	14.59
Spray 2nd or additional coats									
Slow	750	275	20.90	1.40	.39	7.60	2.91	1.85	14.15
Medium	825	238	18.30	2.03	.70	7.69	2.56	1.43	14.41
Fast	900	250	15.70	2.50	1.04	6.28	1.82	.70	12.34
Masonry paint, oil base (material #32)									
Spray 1st coat									
Slow	650	275	28.30	1.62	.46	10.29	3.83	2.43	18.63
Medium	750	250	24.80	2.23	.77	9.92	3.17	1.77	17.86
Fast	850	225	21.20	2.65	1.11	9.42	2.44	.94	16.56
Spray 2nd or additional coats									
Slow	750	300	28.30	1.40	.39	9.43	3.48	2.21	16.91
Medium	825	288	24.80	2.03	.70	8.61	2.78	1.55	15.67
Fast	900	275	21.20	2.50	1.04	7.71	2.08	.80	14.13
Waterproofing, clear hydro sealer, oil base (material #34)									
Spray 1st coat									
Slow	700	120	17.80	1.50	.43	14.83	5.20	3.29	25.25
Medium	800	100	15.60	2.09	.73	15.60	4.51	2.52	25.45
Fast	900	80	13.30	2.50	1.04	16.63	3.73	1.43	25.33
Spray 2nd or additional coats									
Slow	800	150	17.80	1.31	.38	11.87	4.20	2.66	20.42
Medium	900	138	15.60	1.86	.65	11.30	3.38	1.89	19.08
Fast	1000	125	13.30	2.25	.94	10.64	2.56	.98	17.37

Use these figures for new smooth-surface brick with joints struck to average depth. "Slow" work is based on a $10.50 hourly wage, "Medium" work on a $16.75 hourly wage, and "Fast" work on a $22.50 hourly wage. Other qualifications that apply to this table are on page 9.

	Labor SF per manhour	Material coverage SF/gallon	Material cost per gallon	Labor cost per 100 SF	Labor burden 100 SF	Material cost per 100 SF	Overhead per 100 SF	Profit per 100 SF	Total price per 100 SF
Masonry, brick, used, rough surface, brush									
Masonry paint, water base, flat or gloss (material #31)									
Brush 1st coat									
Slow	150	300	20.90	7.00	1.99	6.97	4.95	3.14	24.05
Medium	175	275	18.30	9.57	3.33	6.65	4.79	2.68	27.02
Fast	200	250	15.70	11.25	4.70	6.28	4.11	1.58	27.92
Brush 2nd or additional coats									
Slow	200	375	20.90	5.25	1.49	5.57	3.82	2.42	18.55
Medium	225	350	18.30	7.44	2.59	5.23	3.74	2.09	21.09
Fast	250	325	15.70	9.00	3.76	4.83	3.25	1.25	22.09
Masonry paint, oil base (material #32)									
Brush 1st coat									
Slow	150	325	28.30	7.00	1.99	8.71	5.49	3.48	26.67
Medium	175	300	24.80	9.57	3.33	8.27	5.19	2.90	29.26
Fast	200	275	21.20	11.25	4.70	7.71	4.38	1.68	29.72
Brush 2nd or additional coats									
Slow	200	400	28.30	5.25	1.49	7.08	4.28	2.72	20.82
Medium	225	375	24.80	7.44	2.59	6.61	4.08	2.28	23.00
Fast	250	350	21.20	9.00	3.76	6.06	3.48	1.34	23.64

Use these figures for dry pressed used brick, clay brick tile, or adobe block with joints struck to average depth. "Slow" work is based on a $10.50 hourly wage, "Medium" work on a $16.75 hourly wage, and "Fast" work on a $22.50 hourly wage. Other qualifications that apply to this table are on page 9.

	Labor SF per manhour	Material coverage SF/gallon	Material cost per gallon	Labor cost per 100 SF	Labor burden 100 SF	Material cost per 100 SF	Overhead per 100 SF	Profit per 100 SF	Total price per 100 SF
Masonry, brick, used, rough surface, roll									
Masonry paint, water base, flat or gloss (material #31)									
Roll 1st coat									
Slow	300	250	20.90	3.50	.99	8.36	3.99	2.53	19.37
Medium	325	225	18.30	5.15	1.81	8.13	3.69	2.06	20.84
Fast	350	200	15.70	6.43	2.69	7.85	3.14	1.21	21.32
Roll 2nd or additional coats									
Slow	350	325	20.90	3.00	.86	6.43	3.19	2.02	15.50
Medium	375	300	18.30	4.47	1.56	6.10	2.97	1.66	16.76
Fast	400	275	15.70	5.63	2.34	5.71	2.53	.97	17.18
Masonry paint, oil base (material #32)									
Roll 1st coat									
Slow	300	275	28.30	3.50	.99	10.29	4.58	2.91	22.27
Medium	325	250	24.80	5.15	1.81	9.92	4.13	2.31	23.32
Fast	350	225	21.20	6.43	2.69	9.42	3.43	1.32	23.29
Roll 2nd or additional coats									
Slow	350	350	28.30	3.00	.86	8.09	3.70	2.35	18.00
Medium	375	325	24.80	4.47	1.56	7.63	3.35	1.87	18.88
Fast	400	300	21.20	5.63	2.34	7.07	2.78	1.07	18.89
Waterproofing, clear hydro sealer, oil base (material #34)									
Roll 1st coat									
Slow	150	125	17.80	7.00	1.99	14.24	7.20	4.56	34.99
Medium	175	113	15.60	9.57	3.33	13.81	6.55	3.66	36.92
Fast	200	100	13.30	11.25	4.70	13.30	5.41	2.08	36.74
Roll 2nd or additional coats									
Slow	175	150	17.80	6.00	1.70	11.87	6.07	3.85	29.49
Medium	200	138	15.60	8.38	2.92	11.30	5.54	3.10	31.24
Fast	225	125	13.30	10.00	4.16	10.64	4.59	1.76	31.15

Use these figures for dry pressed used brick, clay brick tile, or adobe block with joints struck to average depth. "Slow" work is based on a $10.50 hourly wage, "Medium" work on a $16.75 hourly wage, and "Fast" work on a $22.50 hourly wage. Other qualifications that apply to this table are on page 9.

	Labor SF per manhour	Material coverage SF/gallon	Material cost per gallon	Labor cost per 100 SF	Labor burden 100 SF	Material cost per 100 SF	Overhead per 100 SF	Profit per 100 SF	Total price per 100 SF
Masonry, brick, used, rough surface, spray									
Masonry paint, water base, flat or gloss (material #31)									
Spray 1st coat									
Slow	600	200	20.90	1.75	.50	10.45	3.94	2.50	19.14
Medium	700	175	18.30	2.39	.84	10.46	3.35	1.87	18.91
Fast	800	150	15.70	2.81	1.18	10.47	2.67	1.03	18.16
Spray 2nd or additional coats									
Slow	700	225	20.90	1.50	.43	9.29	3.48	2.21	16.91
Medium	800	213	18.30	2.09	.73	8.59	2.80	1.56	15.77
Fast	900	200	15.70	2.50	1.04	7.85	2.11	.81	14.31
Masonry paint, oil base (material #32)									
Spray 1st coat									
Slow	600	225	28.30	1.75	.50	12.58	4.60	2.91	22.34
Medium	700	200	24.80	2.39	.84	12.40	3.83	2.14	21.60
Fast	800	175	21.20	2.81	1.18	12.11	2.98	1.14	20.22
Spray 2nd or additional coats									
Slow	700	250	28.30	1.50	.43	11.32	4.11	2.60	19.96
Medium	800	238	24.80	2.09	.73	10.42	3.24	1.81	18.29
Fast	900	225	21.20	2.50	1.04	9.42	2.40	.92	16.28
Waterproofing, clear hydro sealer, oil base (material #34)									
Spray 1st coat									
Slow	600	80	17.80	1.75	.50	22.25	7.60	4.82	36.92
Medium	700	75	15.60	2.39	.84	20.80	5.88	3.29	33.20
Fast	800	70	13.30	2.81	1.18	19.00	4.25	1.63	28.87
Spray 2nd coat									
Slow	800	100	17.80	1.31	.38	17.80	6.04	3.83	29.36
Medium	900	90	15.60	1.86	.65	17.33	4.86	2.72	27.42
Fast	1000	80	13.30	2.25	.94	16.63	3.67	1.41	24.90

Use these figures for dry pressed used brick, clay brick tile, or adobe block with joints struck to average depth. "Slow" work is based on a $10.50 hourly wage, "Medium" work on a $16.75 hourly wage, and "Fast" work on a $22.50 hourly wage. Other qualifications that apply to this table are on page 9.

	Labor SF per manhour	Material coverage SF/gallon	Material cost per gallon	Labor cost per 100 SF	Labor burden 100 SF	Material cost per 100 SF	Overhead per 100 SF	Profit per 100 SF	Total price per 100 SF
Masonry, Concrete Masonry Units (CMU), rough, porous surface, brush									
Masonry paint, water base, flat or gloss (material #31)									
Brush 1st coat									
Slow	110	100	20.90	9.55	2.70	20.90	10.28	6.52	49.95
Medium	130	88	18.30	12.88	4.49	20.80	9.35	5.23	52.75
Fast	150	75	15.70	15.00	6.27	20.93	7.81	3.00	53.01
Brush 2nd or additional coats									
Slow	185	180	20.90	5.68	1.61	11.61	5.86	3.72	28.48
Medium	210	168	18.30	7.98	2.77	10.89	5.30	2.96	29.90
Fast	230	155	15.70	9.78	4.09	10.13	4.44	1.71	30.15
Masonry paint, oil base (material #32)									
Brush 1st coat									
Slow	110	130	28.30	9.55	2.70	21.77	10.55	6.69	51.26
Medium	130	120	24.80	12.88	4.49	20.67	9.32	5.21	52.57
Fast	150	110	21.20	15.00	6.27	19.27	7.50	2.88	50.92
Brush 2nd or additional coats									
Slow	185	200	28.30	5.68	1.61	14.15	6.65	4.22	32.31
Medium	208	180	24.80	8.05	2.82	13.78	6.04	3.37	34.06
Fast	230	160	21.20	9.78	4.09	13.25	5.02	1.93	34.07
Epoxy coating, 2 part system clear (material #51)									
Brush 1st coat									
Slow	95	110	58.30	11.05	3.14	53.00	20.83	13.20	101.22
Medium	115	98	51.00	14.57	5.08	52.04	17.56	9.82	99.07
Fast	135	85	43.70	16.67	6.96	51.41	13.88	5.34	94.26
Brush 2nd or additional coats									
Slow	165	200	58.30	6.36	1.81	29.15	11.57	7.33	56.22
Medium	190	188	51.00	8.82	3.06	27.13	9.56	5.34	53.91
Fast	210	175	43.70	10.71	4.47	24.97	7.43	2.85	50.43
Waterproofing, clear hydro sealer, oil base (material #34)									
Brush 1st coat									
Slow	125	90	17.80	8.40	2.38	19.78	9.48	6.01	46.05
Medium	150	80	15.60	11.17	3.90	19.50	8.47	4.73	47.77
Fast	175	70	13.30	12.86	5.35	19.00	6.89	2.65	46.75
Brush 2nd or additional coats									
Slow	230	130	17.80	4.57	1.29	13.69	6.06	3.84	29.45
Medium	275	110	15.60	6.09	2.13	14.18	5.49	3.07	30.96
Fast	295	90	13.30	7.63	3.18	14.78	4.74	1.82	32.15

Use these figures for Concrete Masonry Units (CMU) where the block surfaces are rough, porous or unfilled, with joints struck to average depth. The more porous the surface, the rougher the texture, the more time and material will be required. For heavy waterproofing applications, see Masonry under Industrial Painting Operations. "Slow" work is based on a $10.50 hourly wage, "Medium" work on a $16.75 hourly wage, and "Fast" work on a $22.50 hourly wage. Other qualifications that apply to this table are on page 9.

	Labor SF per manhour	Material coverage SF/gallon	Material cost per gallon	Labor cost per 100 SF	Labor burden 100 SF	Material cost per 100 SF	Overhead per 100 SF	Profit per 100 SF	Total price per 100 SF
Masonry, Concrete Masonry Units (CMU), rough, porous surface, roll									
Masonry paint, water base, flat or gloss (material #31)									
Roll 1st coat									
Slow	245	90	20.90	4.29	1.21	23.22	8.91	5.65	43.28
Medium	300	78	18.30	5.58	1.94	23.46	7.59	4.24	42.81
Fast	350	65	15.70	6.43	2.69	24.15	6.15	2.36	41.78
Roll 2nd or additional coats									
Slow	275	160	20.90	3.82	1.09	13.06	5.57	3.53	27.07
Medium	325	143	18.30	5.15	1.81	12.80	4.84	2.70	27.30
Fast	420	125	15.70	5.36	2.23	12.56	3.73	1.43	25.31
Masonry paint, oil base (material #32)									
Roll 1st coat									
Slow	245	110	28.30	4.29	1.21	25.73	9.68	6.14	47.05
Medium	300	98	24.80	5.58	1.94	25.31	8.04	4.50	45.37
Fast	350	85	21.20	6.43	2.69	24.94	6.30	2.42	42.78
Roll 2nd or additional coats									
Slow	275	185	28.30	3.82	1.09	15.30	6.27	3.97	30.45
Medium	325	170	24.80	5.15	1.81	14.59	5.27	2.95	29.77
Fast	420	155	21.20	5.36	2.23	13.68	3.94	1.51	26.72
Epoxy coating, 2 part system, clear (material #51)									
Roll 1st coat									
Slow	220	100	58.30	4.77	1.36	58.30	19.97	12.66	97.06
Medium	275	88	51.00	6.09	2.13	57.95	16.21	9.06	91.44
Fast	325	75	43.70	6.92	2.90	58.27	12.59	4.84	85.52
Roll 2nd or additional coats									
Slow	250	175	58.30	4.20	1.19	33.31	12.00	7.61	58.31
Medium	300	160	51.00	5.58	1.94	31.88	9.65	5.40	54.45
Fast	395	145	43.70	5.70	2.37	30.14	7.07	2.72	48.00
Waterproofing, clear hydro sealer, oil base (material #34)									
Roll 1st coat									
Slow	170	110	17.80	6.18	1.75	16.18	7.48	4.74	36.33
Medium	200	98	15.60	8.38	2.92	15.92	6.67	3.73	37.62
Fast	245	85	13.30	9.18	3.83	15.65	5.30	2.04	36.00
Roll 2nd or additional coats									
Slow	275	175	17.80	3.82	1.09	10.17	4.67	2.96	22.71
Medium	300	145	15.60	5.58	1.94	10.76	4.48	2.50	25.26
Fast	325	115	13.30	6.92	2.90	11.57	3.96	1.52	26.87

Use these figures for Concrete Masonry Units (CMU) where the block surfaces are rough, porous or unfilled, with joints struck to average depth. The more porous the surface, the rougher the texture, the more time and material will be required. For heavy waterproofing applications, see Masonry under Industrial Painting Operations. "Slow" work is based on a $10.50 hourly wage, "Medium" work on a $16.75 hourly wage, and "Fast" work on a $22.50 hourly wage. Other qualifications that apply to this table are on page 9.

	Labor SF per manhour	Material coverage SF/gallon	Material cost per gallon	Labor cost per 100 SF	Labor burden 100 SF	Material cost per 100 SF	Overhead per 100 SF	Profit per 100 SF	Total price per 100 SF
Masonry, Concrete Masonry Units (CMU), rough, porous surface, spray									
Masonry paint, water base, flat or gloss (material #31)									
Spray 1st coat									
Slow	600	100	20.90	1.75	.50	20.90	7.18	4.55	34.88
Medium	700	78	18.30	2.39	.84	23.46	6.54	3.65	36.88
Fast	800	55	15.70	2.81	1.18	28.55	6.02	2.31	40.87
Spray 2nd or additional coats									
Slow	700	155	20.90	1.50	.43	13.48	4.78	3.03	23.22
Medium	800	133	18.30	2.09	.73	13.76	4.06	2.27	22.91
Fast	900	110	15.70	2.50	1.04	14.27	3.29	1.27	22.37
Masonry paint, oil base (material #32)									
Spray 1st coat									
Slow	600	100	28.30	1.75	.50	28.30	9.47	6.00	46.02
Medium	700	83	24.80	2.39	.84	29.88	8.11	4.53	45.75
Fast	800	65	21.20	2.81	1.18	32.62	6.77	2.60	45.98
Spray 2nd or additional coats									
Slow	700	160	28.30	1.50	.43	17.69	6.08	3.86	29.56
Medium	800	143	24.80	2.09	.73	17.34	4.94	2.76	27.86
Fast	900	125	21.20	2.50	1.04	16.96	3.79	1.46	25.75
Epoxy coating, 2 part system, clear (material #51)									
Spray 1st coat									
Slow	500	85	58.30	2.10	.60	68.59	22.10	14.01	107.40
Medium	600	68	51.00	2.79	.98	75.00	19.30	10.79	108.86
Fast	700	50	43.70	3.21	1.35	87.40	17.01	6.54	115.51
Spray 2nd or additional coats									
Slow	600	145	58.30	1.75	.50	40.21	13.16	8.34	63.96
Medium	700	130	51.00	2.39	.84	39.23	10.40	5.81	58.67
Fast	800	115	43.70	2.81	1.18	38.00	7.77	2.99	52.75
Waterproofing, clear hydro sealer, oil base (material #34)									
Spray 1st coat									
Slow	500	60	17.80	2.10	.60	29.67	10.03	6.36	48.76
Medium	700	50	15.60	2.39	.84	31.20	8.43	4.71	47.57
Fast	900	40	13.30	2.50	1.04	33.25	6.81	2.62	46.22
Spray 2nd or additional coats									
Slow	600	90	17.80	1.75	.50	19.78	6.83	4.33	33.19
Medium	800	75	15.60	2.09	.73	20.80	5.79	3.24	32.65
Fast	1000	60	13.30	2.25	.94	22.17	4.69	1.80	31.85

Use these figures for Concrete Masonry Units (CMU) where the block surfaces are rough, porous or unfilled, with joints struck to average depth. The more porous the surface, the rougher the texture, the more time and material will be required. For heavy waterproofing applications, see Masonry under Industrial Painting Operations. "Slow" work is based on a $10.50 hourly wage, "Medium" work on a $16.75 hourly wage, and "Fast" work on a $22.50 hourly wage. Other qualifications that apply to this table are on page 9.

	Labor SF per manhour	Material coverage SF/gallon	Material cost per gallon	Labor cost per 100 SF	Labor burden 100 SF	Material cost per 100 SF	Overhead per 100 SF	Profit per 100 SF	Total price per 100 SF
Masonry, Concrete Masonry Units (CMU), smooth-surface, brush									
Masonry paint, water base, flat or gloss (material #31)									
Brush 1st coat									
Slow	140	310	20.90	7.50	2.12	6.74	5.07	3.22	24.65
Medium	190	275	18.30	8.82	3.06	6.65	4.54	2.54	25.61
Fast	230	240	15.70	9.78	4.09	6.54	3.77	1.45	25.63
Brush 2nd or additional coats									
Slow	170	410	20.90	6.18	1.75	5.10	4.04	2.56	19.63
Medium	250	370	18.30	6.70	2.34	4.95	3.43	1.92	19.34
Fast	325	340	15.70	6.92	2.90	4.62	2.67	1.03	18.14
Masonry paint, oil base (material #32)									
Brush 1st coat									
Slow	140	350	28.30	7.50	2.12	8.09	5.49	3.48	26.68
Medium	190	320	24.80	8.82	3.06	7.75	4.81	2.69	27.13
Fast	230	290	21.20	9.78	4.09	7.31	3.92	1.51	26.61
Brush 2nd or additional coats									
Slow	170	450	28.30	6.18	1.75	6.29	4.41	2.80	21.43
Medium	250	420	24.80	6.70	2.34	5.90	3.66	2.04	20.64
Fast	325	390	21.20	6.92	2.90	5.44	2.82	1.08	19.16
Epoxy coating, 2 part system clear (material #51)									
Brush 1st coat									
Slow	120	325	58.30	8.75	2.48	17.94	9.05	5.73	43.95
Medium	160	295	51.00	10.47	3.65	17.29	7.70	4.30	43.41
Fast	200	265	43.70	11.25	4.70	16.49	6.00	2.31	40.75
Brush 2nd or additional coats									
Slow	150	425	58.30	7.00	1.99	13.72	7.04	4.46	34.21
Medium	225	395	51.00	7.44	2.59	12.91	5.62	3.14	31.70
Fast	300	365	43.70	7.50	3.12	11.97	4.18	1.61	28.38
Waterproofing, clear hydro sealer, oil base (material #34)									
Brush 1st coat									
Slow	150	100	17.80	7.00	1.99	17.80	8.30	5.26	40.35
Medium	175	88	15.60	9.57	3.33	17.73	7.51	4.20	42.34
Fast	200	75	13.30	11.25	4.70	17.73	6.23	2.39	42.30
Brush 2nd or additional coats									
Slow	230	140	17.80	4.57	1.29	12.71	5.76	3.65	27.98
Medium	275	120	15.60	6.09	2.13	13.00	5.20	2.91	29.33
Fast	295	100	13.30	7.63	3.18	13.30	4.46	1.71	30.28

Use these figures for Concrete Masonry Units (CMU) where the block surfaces are smooth with joints struck to average depth. For heavy waterproofing applications, see Masonry under Industrial Painting Operations. "Slow" work is based on a $10.50 hourly wage, "Medium" work on a $16.75 hourly wage, and "Fast" work on a $22.50 hourly wage. Other qualifications that apply to this table are on page 9.

	Labor SF per manhour	Material coverage SF/gallon	Material cost per gallon	Labor cost per 100 SF	Labor burden 100 SF	Material cost per 100 SF	Overhead per 100 SF	Profit per 100 SF	Total price per 100 SF

Masonry, Concrete Masonry Units (CMU), smooth-surface, roll

	Labor SF per manhour	Material coverage SF/gallon	Material cost per gallon	Labor cost per 100 SF	Labor burden 100 SF	Material cost per 100 SF	Overhead per 100 SF	Profit per 100 SF	Total price per 100 SF
Masonry paint, water base, flat or gloss (material #31)									
Roll 1st coat									
Slow	300	240	20.90	3.50	.99	8.71	4.10	2.60	19.90
Medium	350	228	18.30	4.79	1.67	8.03	3.55	1.98	20.02
Fast	390	215	15.70	5.77	2.39	7.30	2.86	1.10	19.42
Roll 2nd or additional coats									
Slow	350	325	20.90	3.00	.86	6.43	3.19	2.02	15.50
Medium	400	313	18.30	4.19	1.46	5.85	2.82	1.58	15.90
Fast	450	300	15.70	5.00	2.08	5.23	2.28	.88	15.47
Masonry paint, oil base (material #32)									
Roll 1st coat									
Slow	300	285	28.30	3.50	.99	9.93	4.47	2.84	21.73
Medium	350	273	24.80	4.79	1.67	9.08	3.81	2.13	21.48
Fast	390	260	21.20	5.77	2.39	8.15	3.02	1.16	20.49
Roll 2nd or additional coats									
Slow	350	375	28.30	3.00	.86	7.55	3.53	2.24	17.18
Medium	400	358	24.80	4.19	1.46	6.93	3.08	1.72	17.38
Fast	450	340	21.20	5.00	2.08	6.24	2.47	.95	16.74
Epoxy coating, 2 part system, clear (material #51)									
Roll 1st coat									
Slow	275	265	58.30	3.82	1.09	22.00	8.34	5.29	40.54
Medium	325	253	51.00	5.15	1.81	20.16	6.64	3.71	37.47
Fast	375	240	43.70	6.00	2.51	18.21	4.94	1.90	33.56
Roll 2nd or additional coats									
Slow	325	355	58.30	3.23	.92	16.42	6.38	4.04	30.99
Medium	375	340	51.00	4.47	1.56	15.00	5.15	2.88	29.06
Fast	425	325	43.70	5.29	2.20	13.45	3.88	1.49	26.31
Waterproofing, clear hydro sealer, oil base (material #34)									
Roll 1st coat									
Slow	170	110	17.80	6.18	1.75	16.18	7.48	4.74	36.33
Medium	200	95	15.60	8.38	2.92	16.42	6.79	3.80	38.31
Fast	245	80	13.30	9.18	3.83	16.63	5.48	2.11	37.23
Roll 2nd or additional coats									
Slow	275	150	17.80	3.82	1.09	11.87	5.20	3.30	25.28
Medium	300	125	15.60	5.58	1.94	12.48	4.90	2.74	27.64
Fast	325	100	13.30	6.92	2.90	13.30	4.28	1.64	29.04

Use these figures for Concrete Masonry Units (CMU) where the block surfaces are smooth with joints struck to average depth. For heavy waterproofing applications, see Masonry under Industrial Painting Operations. "Slow" work is based on a $10.50 hourly wage, "Medium" work on a $16.75 hourly wage, and "Fast" work on a $22.50 hourly wage. Other qualifications that apply to this table are on page 9.

	Labor SF per manhour	Material coverage SF/gallon	Material cost per gallon	Labor cost per 100 SF	Labor burden 100 SF	Material cost per 100 SF	Overhead per 100 SF	Profit per 100 SF	Total price per 100 SF
Masonry: Concrete Masonry Units (CMU), smooth-surface, spray									
Masonry paint, water base, flat or gloss (material #31)									
Spray 1st coat									
Slow	725	240	20.90	1.45	.41	8.71	3.28	2.08	15.93
Medium	788	215	18.30	2.13	.74	8.51	2.79	1.56	15.73
Fast	850	190	15.70	2.65	1.11	8.26	2.22	.85	15.09
Spray 2nd or additional coats									
Slow	800	320	20.90	1.31	.38	6.53	2.55	1.61	12.38
Medium	950	295	18.30	1.76	.61	6.20	2.10	1.17	11.84
Fast	1100	270	15.70	2.05	.85	5.81	1.61	.62	10.94
Masonry paint, oil base (material #32)									
Spray 1st coat									
Slow	725	280	28.30	1.45	.41	10.11	3.71	2.35	18.03
Medium	788	255	24.80	2.13	.74	9.73	3.09	1.73	17.42
Fast	850	230	21.20	2.65	1.11	9.22	2.40	.92	16.30
Spray 2nd or additional coats									
Slow	800	345	28.30	1.31	.38	8.20	3.06	1.94	14.89
Medium	950	328	24.80	1.76	.61	7.56	2.43	1.36	13.72
Fast	1100	310	21.20	2.05	.85	6.84	1.80	.69	12.23
Epoxy coating, 2 part system clear (material #51)									
Spray 1st coat									
Slow	675	255	58.30	1.56	.44	22.86	7.71	4.89	37.46
Medium	738	238	51.00	2.27	.80	21.43	6.00	3.35	33.85
Fast	800	220	43.70	2.81	1.18	19.86	4.41	1.70	29.96
Spray 2nd or additional coats									
Slow	750	325	58.30	1.40	.39	17.94	6.12	3.88	29.73
Medium	900	308	51.00	1.86	.65	16.56	4.67	2.61	26.35
Fast	1050	290	43.70	2.14	.89	15.07	3.35	1.29	22.74
Waterproofing, clear hydro sealer, oil base (material #34)									
Spray 1st coat									
Slow	500	75	17.80	2.10	.60	23.73	8.19	5.19	39.81
Medium	750	63	15.60	2.23	.77	24.76	6.80	3.80	38.36
Fast	900	50	13.30	2.50	1.04	26.60	5.58	2.14	37.86
Spray 2nd or additional coats									
Slow	600	100	17.80	1.75	.50	17.80	6.22	3.94	30.21
Medium	800	88	15.60	2.09	.73	17.73	5.03	2.81	28.39
Fast	1000	75	13.30	2.25	.94	17.73	3.87	1.49	26.28

Use these figures for Concrete Masonry Units (CMU) where the block surfaces are smooth with joints struck to average depth. For heavy waterproofing applications, see Masonry under Industrial Painting Operations. "Slow" work is based on a $10.50 hourly wage, "Medium" work on a $16.75 hourly wage, and "Fast" work on a $22.50 hourly wage. Other qualifications that apply to this table are on page 9.

	Labor SF per manhour	Material coverage SF/gallon	Material cost per gallon	Labor cost per 100 SF	Labor burden 100 SF	Material cost per 100 SF	Overhead per 100 SF	Profit per 100 SF	Total price per 100 SF

Masonry, stone, marble or granite

Waterproof, clear hydro sealer, oil base (material #34)

	Labor SF per manhour	Material coverage SF/gallon	Material cost per gallon	Labor cost per 100 SF	Labor burden 100 SF	Material cost per 100 SF	Overhead per 100 SF	Profit per 100 SF	Total price per 100 SF
Spray 1st coat									
Slow	600	220	17.80	1.75	.50	8.09	3.21	2.03	15.58
Medium	700	200	15.60	2.39	.84	7.80	2.70	1.51	15.24
Fast	800	180	13.30	2.81	1.18	7.39	2.10	.81	14.29
Spray 2nd coat									
Slow	700	225	17.80	1.50	.43	7.91	3.05	1.93	14.82
Medium	800	213	15.60	2.09	.73	7.32	2.48	1.39	14.01
Fast	900	200	13.30	2.50	1.04	6.65	1.89	.72	12.80

Use these figures for the cost to apply waterproof sealer on stone, marble or granite surfaces. "Slow" work is based on a $10.50 hourly wage, "Medium" work on a $16.75 hourly wage, and "Fast" work on a $22.50 hourly wage. Other qualifications that apply to this table are on page 9.

Molding, interior or exterior, paint grade, smooth-surface

	Labor LF per manhour	Material coverage LF/gallon	Material cost per gallon	Labor cost per 100 LF	Labor burden 100 LF	Material cost per 100 LF	Overhead per 100 LF	Profit per 100 LF	Total price per 100 LF
Undercoat, water or oil base (material #3 or #4)									
Brush prime coat									
Slow	135	700	21.35	7.78	2.21	3.05	4.04	2.56	19.64
Medium	205	600	18.70	8.17	2.85	3.12	3.46	1.94	19.54
Fast	275	500	16.00	8.18	3.43	3.20	2.74	1.05	18.60
Split coat (1/2 undercoat + 1/2 enamel), water or oil base (material #3 and #9 or #4 and #10)									
Brush 1st or additional finish coats									
Slow	125	750	23.45	8.40	2.38	3.13	4.32	2.74	20.97
Medium	135	675	20.53	12.41	4.33	3.04	4.84	2.71	27.33
Fast	145	600	17.60	15.52	6.48	2.93	4.61	1.77	31.31
Enamel, interior, water or oil base (material #9 or #10)									
Brush 1st finish coat									
Slow	125	750	25.55	8.40	2.38	3.41	4.40	2.79	21.38
Medium	160	675	22.35	10.47	3.65	3.31	4.27	2.39	24.09
Fast	200	600	19.20	11.25	4.70	3.20	3.54	1.36	24.05
Brush 2nd or additional finish coats									
Slow	115	750	25.55	9.13	2.60	3.41	4.69	2.97	22.80
Medium	150	675	22.35	11.17	3.90	3.31	4.50	2.52	25.40
Fast	175	600	19.20	12.86	5.35	3.20	3.96	1.52	26.89

	Labor LF per manhour	Material coverage LF/gallon	Material cost per gallon	Labor cost per 100 LF	Labor burden 100 LF	Material cost per 100 LF	Overhead per 100 LF	Profit per 100 LF	Total price per 100 LF
Enamel, exterior, water or oil base (material #24 or #25)									
Brush 1st finish coat									
Slow	125	750	29.00	8.40	2.38	3.87	4.54	2.88	22.07
Medium	160	675	25.35	10.47	3.65	3.76	4.38	2.45	24.71
Fast	200	600	21.75	11.25	4.70	3.63	3.62	1.39	24.59
Brush 2nd or additional finish coats									
Slow	115	750	29.00	9.13	2.60	3.87	4.84	3.07	23.51
Medium	150	675	25.35	11.17	3.90	3.76	4.61	2.58	26.02
Fast	175	600	21.75	12.86	5.35	3.63	4.04	1.55	27.43
Stipple finish									
Slow	80	--	--	13.13	3.72	--	5.23	3.32	25.40
Medium	90	--	--	18.61	6.49	--	6.15	3.44	34.69
Fast	100	--	--	22.50	9.39	--	5.90	2.27	40.06
Glazing or mottling over enamel (material #16)									
Glaze & wipe, brush 1 coat									
Slow	55	950	31.20	19.09	5.42	3.28	8.62	5.46	41.87
Medium	65	900	27.30	25.77	8.97	3.03	9.26	5.17	52.20
Fast	75	850	23.40	30.00	12.51	2.75	8.38	3.22	56.86

Consider all trim to be at least 12" wide (even if it's much less than 12" wide) when calculating the area to be painted. Trim painted the same color as the wall or ceiling behind it may take no more time than painting the wall or ceiling itself. Use the slow rate when cutting-in is required to paint molding that's a different color from the surface behind the molding. "Slow" work is based on a $10.50 hourly wage, "Medium" work on a $16.75 hourly wage, and "Fast" work on a $22.50 hourly wage. Other qualifications that apply to this table are on page 9.

	Labor LF per manhour	Material coverage LF/gallon	Material cost per gallon	Labor cost per 100 LF	Labor burden 100 LF	Material cost per 100 LF	Overhead per 100 LF	Profit per 100 LF	Total price per 100 LF

Molding, interior, stain grade, smooth-surface

Stain, seal and 2 coat lacquer system (7 step process)

	Labor LF per manhour	Material coverage LF/gallon	Material cost per gallon	Labor cost per 100 LF	Labor burden 100 LF	Material cost per 100 LF	Overhead per 100 LF	Profit per 100 LF	Total price per 100 LF
STEP 1: Sand & putty;									
Slow	150	--	--	7.00	1.99	--	2.79	1.77	13.55
Medium	175	--	--	9.57	3.33	--	3.16	1.77	17.83
Fast	200	--	--	11.25	4.70	--	2.95	1.13	20.03
STEP 2 & 3: Wiping stain (material 11a) & wipe									
Brush 1 coat & wipe									
Slow	150	550	30.90	7.00	1.99	5.62	4.53	2.87	22.01
Medium	175	525	27.00	9.57	3.33	5.14	4.42	2.47	24.93
Fast	200	500	23.10	11.25	4.70	4.62	3.81	1.46	25.84
Spray 1 coat & wipe									
Slow	400	250	30.90	2.63	.74	12.36	4.88	3.09	23.70
Medium	425	225	27.00	3.94	1.37	12.00	4.24	2.37	23.92
Fast	450	200	23.10	5.00	2.08	11.55	3.45	1.33	23.41
STEP 4 & 5: Sanding sealer (material 11b) & light sand									
Brush 1 coat									
Slow	260	575	17.20	4.04	1.15	2.99	2.54	1.61	12.33
Medium	280	563	15.00	5.98	2.08	2.66	2.63	1.47	14.82
Fast	300	550	12.90	7.50	3.12	2.35	2.40	.92	16.29
Spray 1 coat									
Slow	450	250	17.20	2.33	.66	6.88	3.06	1.94	14.87
Medium	475	225	15.00	3.53	1.24	6.67	2.80	1.57	15.81
Fast	500	200	12.90	4.50	1.88	6.45	2.37	.91	16.11
STEP 6 & 7: Lacquer (material 11c), 2 coats									
Brush 1st coat									
Slow	200	300	19.30	5.25	1.49	6.43	4.08	2.59	19.84
Medium	275	288	16.80	6.09	2.13	5.83	3.44	1.92	19.41
Fast	350	275	14.40	6.43	2.69	5.24	2.65	1.02	18.03
Brush 2nd coat									
Slow	225	375	19.30	4.67	1.32	5.15	3.46	2.19	16.79
Medium	300	350	16.80	5.58	1.94	4.80	3.02	1.69	17.03
Fast	375	325	14.40	6.00	2.51	4.43	2.39	.92	16.25
Spray 1st coat									
Slow	250	200	19.30	4.20	1.19	9.65	4.66	2.96	22.66
Medium	350	188	16.80	4.79	1.67	8.94	3.77	2.11	21.28
Fast	450	175	14.40	5.00	2.08	8.23	2.83	1.09	19.23
Spray 2nd coat									
Slow	300	250	19.30	3.50	.99	7.72	3.79	2.40	18.40
Medium	388	225	16.80	4.32	1.51	7.47	3.26	1.82	18.38
Fast	475	200	14.40	4.74	1.99	7.20	2.58	.99	17.50

	Labor LF per manhour	Material coverage LF/gallon	Material cost per gallon	Labor cost per 100 LF	Labor burden 100 LF	Material cost per 100 LF	Overhead per 100 LF	Profit per 100 LF	Total price per 100 LF
Complete 7 step process, stain, seal & 2 coat lacquer system (material #11)									
Brush all coats									
Slow	50	60	21.60	21.00	5.96	36.00	19.52	12.37	94.85
Medium	75	50	19.60	22.33	7.78	39.20	16.98	9.49	95.78
Fast	100	40	16.20	22.50	9.39	40.50	13.39	5.15	90.93
Spray all coats									
Slow	100	40	19.30	10.50	2.98	48.25	19.14	12.13	93.00
Medium	150	30	16.80	11.17	3.90	56.00	17.41	9.73	98.21
Fast	200	20	14.40	11.25	4.70	72.00	16.27	6.25	110.47

These figures are based on linear feet for all molding up to 14" wide. The spray figures are based on finishing large quantities of molding in a boneyard before it's installed. Use the brush figures for small quantities. Use the fast brush rate for molding that's finished before it's installed. If the molding is attached and has to be masked off, use the slow brush figures. Masking time is not included. See the Preparation Operations tables for masking rates. Trim stained the same color as the surface behind it will take no more time than staining the wall itself. "Slow" work is based on a $10.50 hourly wage, "Medium" work on a $16.75 hourly wage, and "Fast" work on a $22.50 hourly wage. Other qualifications that apply to this table are on page 9.

	Labor LF per manhour	Material coverage LF/gallon	Material cost per gallon	Labor cost per 100 LF	Labor burden 100 LF	Material cost per 100 LF	Overhead per 100 LF	Profit per 100 LF	Total price per 100 LF
Molding, interior or exterior, stain grade, smooth-surface									
Stain, fill and shellac or varnish									
Wiping stain (material #30a) & fill									
Brush each coat									
Slow	80	550	30.60	13.13	3.72	5.56	6.95	4.41	33.77
Medium	130	525	24.10	12.88	4.49	4.59	5.38	3.01	30.35
Fast	180	500	20.70	12.50	5.23	4.14	4.04	1.55	27.46
Shellac, clear (material #12)									
Brush each coat									
Slow	180	550	25.30	5.83	1.66	4.60	3.75	2.38	18.22
Medium	230	525	22.10	7.28	2.55	4.21	3.44	1.92	19.40
Fast	280	500	18.90	8.04	3.34	3.78	2.81	1.08	19.05
Varnish, gloss (material #30c)									
Brush each coat									
Slow	115	550	35.40	9.13	2.60	6.44	5.63	3.57	27.37
Medium	150	525	31.00	11.17	3.90	5.90	5.14	2.87	28.98
Fast	210	500	26.60	10.71	4.47	5.32	3.79	1.46	25.75
Varnish, flat (material #30c)									
Brush each coat									
Slow	125	550	35.40	8.40	2.38	6.44	5.34	3.39	25.95
Medium	160	525	31.00	10.47	3.65	5.90	4.90	2.74	27.66
Fast	220	500	26.60	10.23	4.28	5.32	3.67	1.41	24.91
Penetrating stain wax (material #14) & wipe									
Stain, brush each coat & wipe									
Slow	225	500	27.20	4.67	1.32	5.44	3.55	2.25	17.23
Medium	275	450	23.80	6.09	2.13	5.29	3.31	1.85	18.67
Fast	325	400	20.40	6.92	2.90	5.10	2.76	1.06	18.74
Polish added coats of wax									
Slow	150	--	--	7.00	1.99	--	2.79	1.77	13.55
Medium	175	--	--	9.57	3.33	--	3.16	1.77	17.83
Fast	200	--	--	11.25	4.70	--	2.95	1.13	20.03

	Labor LF per manhour	Material coverage LF/gallon	Material cost per gallon	Labor cost per 100 LF	Labor burden 100 LF	Material cost per 100 LF	Overhead per 100 LF	Profit per 100 LF	Total price per 100 LF
Steel wool buff by hand									
Slow	100	--	--	10.50	2.98	--	4.18	2.65	20.31
Medium	110	--	--	15.23	5.30	--	5.03	2.81	28.37
Fast	125	--	--	18.00	7.51	--	4.72	1.81	32.04
Wax application & polish, hand apply 1 coat									
Slow	100	--	--	10.50	2.98	--	4.18	2.65	20.31
Medium	110	--	--	15.23	5.30	--	5.03	2.81	28.37
Fast	125	--	--	18.00	7.51	--	4.72	1.81	32.04

These figures are based on linear feet for all molding up to 14" wide. The spray figures are based on finishing large quantities of molding in a boneyard before it's installed. Use the brush figures for small quantities. Use the fast brush rate for molding that's finished before it's installed. If the molding is attached and has to be masked off, use the slow brush figures. Masking time is not included. See the Preparation Operations tables for masking rates. Trim stained the same color as the surface behind it will take no more time than staining the wall itself. "Slow" work is based on a $10.50 hourly wage, "Medium" work on a $16.75 hourly wage, and "Fast" work on a $22.50 hourly wage. Other qualifications that apply to this table are on page 9.

	Boxed eaves		Exposed rafters	
	One color	Two color	One color	Two color

Overhang difficulty factors, eaves, cornice

	Boxed eaves One color	Boxed eaves Two color	Exposed rafters One color	Exposed rafters Two color
One story, repaint				
Standard	1.5	2.0	2.0	2.5
Ornamental	--	--	2.5	3.0
One story, new construction	--	1.5	1.5	2.0
Two story, with scaffolding, repaint				
Standard	1.5	2.0	2.0	2.5
Ornamental	--	--	2.5	3.0
Two story, without scaffolding, repaint				
Standard	3.0	3.5	3.5	4.0
Ornamental	--	--	4.0	4.5
Two story, new construction				
With scaffolding	--	1.5	1.5	2.0
No scaffolding	2.0	2.5	2.5	3.0

Before using the figures in the tables for overhangs, apply these difficulty factors to the surface area (length times width) to be painted. Multiply the factor in this table by the overall surface area of the overhang. This allows for slower work on high eaves and the extra time needed to paint rafter tails. This table adjusts for the kind of eaves, the eave height, the number of colors used, and whether scaffolding is available and erected on site. Boxed eaves have plywood covering the rafter tails.

Overhang, carport, large continuous surface areas

Solid body or semi-transparent stain, water or oil base (material #18 or #19 or #20 or #21)

	Labor SF per manhour	Material coverage SF/gallon	Material cost per gallon	Labor cost per 100 SF	Labor burden 100 SF	Material cost per 100 SF	Overhead per 100 SF	Profit per 100 SF	Total price per 100 SF
Spray 1st coat									
Slow	550	95	23.23	1.91	.54	24.45	8.34	5.29	40.53
Medium	600	90	20.33	2.79	.98	22.59	6.46	3.61	36.43
Fast	650	85	17.40	3.46	1.45	20.47	4.69	1.80	31.87
Spray 2nd coat									
Slow	600	175	23.23	1.75	.50	13.27	4.81	3.05	23.38
Medium	650	163	20.33	2.58	.90	12.47	3.91	2.18	22.04
Fast	700	150	17.40	3.21	1.35	11.60	2.99	1.15	20.30
Spray 3rd or additional coats									
Slow	650	225	23.23	1.62	.46	10.32	3.84	2.44	18.68
Medium	700	213	20.33	2.39	.84	9.54	3.13	1.75	17.65
Fast	750	200	17.40	3.00	1.24	8.70	2.40	.92	16.26

"Slow" work is based on a $10.50 hourly wage, "Medium" work on a $16.75 hourly wage, and "Fast" work on a $22.50 hourly wage. Other qualifications that apply to this table are on page 9.

	Labor SF per manhour	Material coverage SF/gallon	Material cost per gallon	Labor cost per 100 SF	Labor burden 100 SF	Material cost per 100 SF	Overhead per 100 SF	Profit per 100 SF	Total price per 100 SF
Overhang, eaves or rake, widths up to 2.5 feet									
Solid body stain, water or oil base (material #18 or #19)									
Roll & Brush 1st coat									
Slow	95	200	22.70	11.05	3.14	11.35	7.92	5.02	38.48
Medium	125	185	19.85	13.40	4.67	10.73	7.06	3.94	39.80
Fast	150	170	17.00	15.00	6.27	10.00	5.78	2.22	39.27
Roll & Brush 2nd coat									
Slow	140	260	22.70	7.50	2.12	8.73	5.69	3.61	27.65
Medium	185	240	19.85	9.05	3.17	8.27	5.02	2.80	28.31
Fast	220	220	17.00	10.23	4.28	7.73	4.11	1.58	27.93
Roll & Brush 3rd or additional coats									
Slow	170	295	22.70	6.18	1.75	7.69	4.85	3.07	23.54
Medium	225	270	19.85	7.44	2.59	7.35	4.26	2.38	24.02
Fast	275	245	17.00	8.18	3.43	6.94	3.43	1.32	23.30
Solid body stain or semi-transparent stain, water or oil base (material #18 or #19 or #20 or #21)									
Spray 1st coat									
Slow	300	250	23.23	3.50	.99	9.29	4.27	2.71	20.76
Medium	350	213	20.33	4.79	1.67	9.54	3.92	2.19	22.11
Fast	400	175	17.40	5.63	2.34	9.94	3.32	1.27	22.50
Spray 2nd coat									
Slow	375	275	23.23	2.80	.80	8.45	3.74	2.37	18.16
Medium	450	250	20.33	3.72	1.29	8.13	3.22	1.80	18.16
Fast	540	225	17.40	4.17	1.73	7.73	2.52	.97	17.12
Spray 3rd or additional coats									
Slow	450	275	23.23	2.33	.66	8.45	3.55	2.25	17.24
Medium	525	250	20.33	3.19	1.10	8.13	3.05	1.70	17.17
Fast	600	225	17.40	3.75	1.58	7.73	2.41	.93	16.40

Use this table after multiplying the overall area (length time width) by the difficulty factor listed under Overhang, difficulty factor. Remember to add preparation time for masking, caulking, sanding, waterblasting, etc. Slow work is based on a $10.50 hourly wage, "Medium" work on a $16.75 hourly wage, and "Fast" work on a $22.50 hourly wage. Other qualifications that apply to this table are on page 9.

	Labor SF per manhour	Material coverage SF/gallon	Material cost per gallon	Labor cost per 100 SF	Labor burden 100 SF	Material cost per 100 SF	Overhead per 100 SF	Profit per 100 SF	Total price per 100 SF
Overhang, eaves or rake, widths greater than 2.5 feet									
Solid body stain, water or oil base (material #18 or #19)									
Roll & Brush 1st coat									
Slow	125	225	22.70	8.40	2.38	10.09	6.47	4.10	31.44
Medium	180	210	19.85	9.31	3.25	9.45	5.39	3.01	30.41
Fast	225	195	17.00	10.00	4.16	8.72	4.24	1.63	28.75
Roll & Brush 2nd coat									
Slow	170	295	22.70	6.18	1.75	7.69	4.85	3.07	23.54
Medium	225	275	19.85	7.44	2.59	7.22	4.23	2.36	23.84
Fast	275	255	17.00	8.18	3.43	6.67	3.38	1.30	22.96
Roll & Brush 3rd or additional coats									
Slow	210	340	22.70	5.00	1.42	6.68	4.06	2.57	19.73
Medium	280	315	19.85	5.98	2.08	6.30	3.52	1.97	19.85
Fast	345	290	17.00	6.52	2.73	5.86	2.79	1.07	18.97
Solid body stain or semi-transparent stain, water or oil base (material #18 or #19 or #20 or #21)									
Spray 1st coat									
Slow	400	250	23.23	2.63	.74	9.29	3.93	2.49	19.08
Medium	500	213	20.33	3.35	1.17	9.54	3.44	1.93	19.43
Fast	600	175	17.40	3.75	1.58	9.94	2.82	1.08	19.17
Spray 2nd coat									
Slow	450	275	23.23	2.33	.66	8.45	3.55	2.25	17.24
Medium	550	250	20.33	3.05	1.06	8.13	3.00	1.68	16.92
Fast	650	225	17.40	3.46	1.45	7.73	2.34	.90	15.88
Spray 3rd or additional coats									
Slow	550	275	23.23	1.91	.54	8.45	3.38	2.14	16.42
Medium	650	250	20.33	2.58	.90	8.13	2.84	1.59	16.04
Fast	750	225	17.40	3.00	1.24	7.73	2.22	.85	15.04

Use this table after multiplying the overall area (length time width) by the difficulty factor listed under Overhang, difficulty factor. Remember to add preparation time for masking, caulking, sanding, waterblasting, etc. Slow work is based on a $10.50 hourly wage, "Medium" work on a $16.75 hourly wage, and "Fast" work on a $22.50 hourly wage. Other qualifications that apply to this table are on page 9.

	Labor LF per manhour	Material coverage LF/gallon	Material cost per gallon	Labor cost per 100 LF	Labor burden 100 LF	Material cost per 100 LF	Overhead per 100 LF	Profit per 100 LF	Total price per 100 LF

Pass-through shelves, wood top & wrought iron support

Metal primer, rust inhibitor, clean metal (material #35)									
Brush 1st coat									
Slow	30	125	29.90	35.00	9.93	23.92	21.35	13.53	103.73
Medium	35	113	26.20	47.86	16.68	23.19	21.49	12.01	121.23
Fast	40	100	22.40	56.25	23.48	22.40	18.89	7.26	128.28
Metal finish - synthetic enamel (colors except orange/red) (material #38)									
Brush 1st coat									
Slow	30	125	30.00	35.00	9.93	24.00	21.38	13.55	103.86
Medium	35	113	26.30	47.86	16.68	23.27	21.51	12.03	121.35
Fast	40	100	22.50	56.25	23.48	22.50	18.91	7.27	128.41

Use these figures to estimate pass-through shelves which are about 12" wide and 3'0" long. The rule-of-thumb minimum time and material is .2 hours and $1.00 for material per shelf. A two-coat system using oil based material is recommended for any metal surface. Although water based material is often used, it may cause oxidation, corrosion and rust. One coat of oil based, solid body stain is often used on exterior metal, but it may crack, peel or chip without the proper prime coat application. "Slow" work is based on a $10.50 hourly wage, "Medium" work on a $16.75 hourly wage, and "Fast" work on a $22.50 hourly wage. Other qualifications that apply to this table are on page 9.

	Labor LF per manhour	Material coverage LF/gallon	Material cost per gallon	Labor cost per 100 LF	Labor burden 100 LF	Material cost per 100 LF	Overhead per 100 LF	Profit per 100 LF	Total price per 100 LF

Plant-on trim, exterior, 2" x 2" to 2" x 4" wood

2" x 2" to 2" x 4" rough sawn or resawn wood									
Solid body or semi-transparent stain, water base (material #18 or #20)									
Roll & brush 1st coat									
Slow	80	310	21.45	13.13	3.72	6.92	7.37	4.67	35.81
Medium	100	275	18.75	16.75	5.84	6.82	7.21	4.03	40.65
Fast	120	240	16.05	18.75	7.81	6.69	6.15	2.37	41.77
Roll & brush 2nd coat									
Slow	90	350	21.45	11.67	3.31	6.13	6.55	4.15	31.81
Medium	120	325	18.75	13.96	4.86	5.77	6.03	3.37	33.99
Fast	150	300	16.05	15.00	6.27	5.35	4.92	1.89	33.43
Roll & brush 3rd or additional coats									
Slow	100	375	21.45	10.50	2.98	5.72	5.96	3.78	28.94
Medium	140	350	18.75	11.96	4.17	5.36	5.27	2.94	29.70
Fast	180	325	16.05	12.50	5.23	4.94	4.19	1.61	28.47
Solid body or semi-transparent stain, oil base (material #19 or #21)									
Roll & brush 1st coat									
Slow	95	400	25.00	11.05	3.14	6.25	6.34	4.02	30.80
Medium	115	365	21.90	14.57	5.08	6.00	6.28	3.51	35.44
Fast	135	330	18.75	16.67	6.96	5.68	5.42	2.08	36.81

	Labor LF per manhour	Material coverage LF/gallon	Material cost per gallon	Labor cost per 100 LF	Labor burden 100 LF	Material cost per 100 LF	Overhead per 100 LF	Profit per 100 LF	Total price per 100 LF
Roll & brush 2nd coat									
Slow	105	475	25.00	10.00	2.83	5.26	5.61	3.56	27.26
Medium	135	438	21.90	12.41	4.33	5.00	5.32	2.98	30.04
Fast	165	400	18.75	13.64	5.69	4.69	4.44	1.71	30.17
Roll & brush 3rd or additional coats									
Slow	115	500	25.00	9.13	2.60	5.00	5.19	3.29	25.21
Medium	155	463	21.90	10.81	3.76	4.73	4.73	2.64	26.67
Fast	195	425	18.75	11.54	4.82	4.41	3.84	1.48	26.09
Varnish, flat or gloss (material #30c)									
Roll & brush 1st coat									
Slow	70	270	33.40	15.00	4.26	12.37	9.81	6.22	47.66
Medium	90	255	29.30	18.61	6.49	11.49	8.96	5.01	50.56
Fast	110	240	25.10	20.45	8.54	10.46	7.30	2.81	49.56
Roll & brush 2nd coat									
Slow	80	330	33.40	13.13	3.72	10.12	8.37	5.30	40.64
Medium	110	315	29.30	15.23	5.30	9.30	7.31	4.09	41.23
Fast	140	300	25.10	16.07	6.70	8.37	5.76	2.21	39.11
Roll & brush 3rd or additional coats									
Slow	90	350	33.40	11.67	3.31	9.54	7.60	4.82	36.94
Medium	130	335	29.30	12.88	4.49	8.75	6.40	3.58	36.10
Fast	170	320	25.10	13.24	5.51	7.84	4.92	1.89	33.40

Don't add additional time for plant-on trim if it's painted with the same coating as the adjacent siding. For heights above 8 feet, use the High Time Difficulty Factors on page 137. Use slow rates when cutting-in or masking adjacent surfaces. Add preparation time for masking adjacent surfaces or protecting windows. Use fast rates when plant-on trim is finished before it's installed or on new construction projects where a prime coat can be sprayed prior to stucco color coat application. "Slow" work is based on a $10.50 hourly wage, "Medium" work on a $16.75 hourly wage, and "Fast" work on a $22.50 hourly wage. Other qualifications that apply to this table are on page 9.

	Labor LF per manhour	Material coverage LF/gallon	Material cost per gallon	Labor cost per 100 LF	Labor burden 100 LF	Material cost per 100 LF	Overhead per 100 LF	Profit per 100 LF	Total price per 100 LF

Plant-on trim, exterior, 2" x 6" to 2" x 8" wood

2" x 6" to 2" x 8" rough sawn or resawn wood									
Solid body or semi-transparent stain, water base (material #18 or #20)									
Roll & brush 1st coat									
Slow	70	210	21.45	15.00	4.26	10.21	9.14	5.79	44.40
Medium	85	175	18.75	19.71	6.86	10.71	9.14	5.11	51.53
Fast	100	140	16.05	22.50	9.39	11.46	8.02	3.08	54.45
Roll & brush 2nd coat									
Slow	80	250	21.45	13.13	3.72	8.58	7.89	5.00	38.32
Medium	110	225	18.75	15.23	5.30	8.33	7.07	3.95	39.88
Fast	140	200	16.05	16.07	6.70	8.03	5.70	2.19	38.69
Roll & brush 3rd or additional coats									
Slow	90	275	21.45	11.67	3.31	7.80	7.06	4.48	34.32
Medium	130	250	18.75	12.88	4.49	7.50	6.09	3.41	34.37
Fast	170	225	16.05	13.24	5.51	7.13	4.79	1.84	32.51
Solid body or semi-transparent stain, oil base (material #19 or #21)									
Roll & brush 1st coat									
Slow	85	300	25.00	12.35	3.50	8.33	7.50	4.75	36.43
Medium	110	265	21.90	15.23	5.30	8.26	7.06	3.94	39.79
Fast	135	230	18.75	16.67	6.96	8.15	5.88	2.26	39.92
Roll & brush 2nd coat									
Slow	95	375	25.00	11.05	3.14	6.67	6.47	4.10	31.43
Medium	125	338	21.90	13.40	4.67	6.48	6.01	3.36	33.92
Fast	155	300	18.75	14.52	6.05	6.25	4.96	1.91	33.69
Roll & brush 3rd or additional coats									
Slow	105	400	25.00	10.00	2.83	6.25	5.92	3.75	28.75
Medium	145	363	21.90	11.55	4.04	6.03	5.29	2.96	29.87
Fast	185	325	18.75	12.16	5.09	5.77	4.26	1.64	28.92
Varnish, flat or gloss (material #30c)									
Roll & brush 1st coat									
Slow	60	170	33.40	17.50	4.97	19.65	13.06	8.28	63.46
Medium	70	155	29.30	23.93	8.35	18.90	12.54	7.01	70.73
Fast	90	140	25.10	25.00	10.43	17.93	9.87	3.79	67.02

	Labor LF per manhour	Material coverage LF/gallon	Material cost per gallon	Labor cost per 100 LF	Labor burden 100 LF	Material cost per 100 LF	Overhead per 100 LF	Profit per 100 LF	Total price per 100 LF
Roll & brush 2nd coat									
Slow	70	240	33.40	15.00	4.26	13.92	10.29	6.52	49.99
Medium	100	220	29.30	16.75	5.84	13.32	8.80	4.92	49.63
Fast	130	200	25.10	17.31	7.21	12.55	6.86	2.64	46.57
Roll & brush 3rd or additional coats									
Slow	80	260	33.40	13.13	3.72	12.85	9.21	5.84	44.75
Medium	115	240	29.30	14.57	5.08	12.21	7.81	4.36	44.03
Fast	150	220	25.10	15.00	6.27	11.41	6.04	2.32	41.04

Don't add additional time for plant-on trim if it's painted with the same coating as the adjacent siding. For heights above 8 feet, use the High Time Difficulty Factors on page 137. Use slow rates when cutting-in or masking adjacent surfaces. Add preparation time for masking adjacent surfaces or protecting windows. Use fast rates when plant-on trim is finished before it's installed or on new construction projects where a prime coat can be sprayed prior to stucco color coat application. "Slow" work is based on a $10.50 hourly wage, "Medium" work on a $16.75 hourly wage, and "Fast" work on a $22.50 hourly wage. Other qualifications that apply to this table are on page 9.

	Labor LF per manhour	Material coverage LF/gallon	Material cost per gallon	Labor cost per 100 LF	Labor burden 100 LF	Material cost per 100 LF	Overhead per 100 LF	Profit per 100 LF	Total price per 100 LF

Plant-on trim, exterior, 2" x 10" to 2" x 12" wood

2" x 10" to 2" x 12" rough sawn or resawn wood

	Labor LF per manhour	Material coverage LF/gallon	Material cost per gallon	Labor cost per 100 LF	Labor burden 100 LF	Material cost per 100 LF	Overhead per 100 LF	Profit per 100 LF	Total price per 100 LF
Solid body or semi-transparent stain, water base (material #18 or #20)									
Roll & brush 1st coat									
Slow	60	150	21.45	17.50	4.97	14.30	11.40	7.23	55.40
Medium	75	115	18.75	22.33	7.78	16.30	11.37	6.36	64.14
Fast	90	80	16.05	25.00	10.43	20.06	10.27	3.95	69.71
Roll & brush 2nd coat									
Slow	70	190	21.45	15.00	4.26	11.29	9.47	6.00	46.02
Medium	110	165	18.75	15.23	5.30	11.36	7.82	4.37	44.08
Fast	130	140	16.05	17.31	7.21	11.46	6.66	2.56	45.20
Roll & brush 3rd or additional coats									
Slow	80	215	21.45	13.13	3.72	9.98	8.32	5.28	40.43
Medium	120	190	18.75	13.96	4.86	9.87	7.03	3.93	39.65
Fast	160	165	16.05	14.06	5.87	9.73	5.49	2.11	37.26
Solid body or semi-transparent stain, oil base (material #19 or #21)									
Roll & brush 1st coat									
Slow	70	290	25.00	15.00	4.26	8.62	8.65	5.48	42.01
Medium	85	255	21.90	19.71	6.86	8.59	8.62	4.82	48.60
Fast	100	220	18.75	22.50	9.39	8.52	7.48	2.87	50.76
Roll & brush 2nd coat									
Slow	80	315	25.00	13.13	3.72	7.94	7.69	4.88	37.36
Medium	110	278	21.90	15.23	5.30	7.88	6.96	3.89	39.26
Fast	140	240	18.75	16.07	6.70	7.81	5.66	2.18	38.42
Roll & brush 3rd or additional coats									
Slow	90	340	25.00	11.67	3.31	7.35	6.93	4.39	33.65
Medium	130	303	21.90	12.88	4.49	7.23	6.03	3.37	34.00
Fast	170	265	18.75	13.24	5.51	7.08	4.78	1.84	32.45
Varnish, flat or gloss (material #30c)									
Roll & brush 1st coat									
Slow	50	120	33.40	21.00	5.96	27.83	16.99	10.77	82.55
Medium	65	100	29.30	25.77	8.97	29.30	15.69	8.77	88.50
Fast	80	80	25.10	28.13	11.73	31.38	13.18	5.07	89.49

	Labor LF per manhour	Material coverage LF/gallon	Material cost per gallon	Labor cost per 100 LF	Labor burden 100 LF	Material cost per 100 LF	Overhead per 100 LF	Profit per 100 LF	Total price per 100 LF
Roll & brush 2nd coat									
Slow	60	190	33.40	17.50	4.97	17.58	12.42	7.87	60.34
Medium	90	165	29.30	18.61	6.49	17.76	10.50	5.87	59.23
Fast	120	140	25.10	18.75	7.81	17.93	8.23	3.16	55.88
Roll & brush 3rd or additional coats									
Slow	70	200	33.40	15.00	4.26	16.70	11.15	7.07	54.18
Medium	110	175	29.30	15.23	5.30	16.74	9.13	5.11	51.51
Fast	150	150	25.10	15.00	6.27	16.73	7.03	2.70	47.73

Don't add additional time for plant-on trim if it's painted with the same coating as the adjacent siding. For heights above 8 feet, use the High Time Difficulty Factors on page 137. Use slow rates when cutting-in or masking adjacent surfaces. Add preparation time for masking adjacent surfaces or protecting windows. Use fast rates when plant-on trim is finished before it's installed or on new construction projects where a prime coat can be sprayed prior to stucco color coat application. "Slow" work is based on a $10.50 hourly wage, "Medium" work on a $16.75 hourly wage, and "Fast" work on a $22.50 hourly wage. Other qualifications that apply to this table are on page 9.

	Labor SF per manhour	Material coverage SF/gallon	Material cost per gallon	Labor cost per 100 SF	Labor burden 100 SF	Material cost per 100 SF	Overhead per 100 SF	Profit per 100 SF	Total price per 100 SF

Plaster or stucco, exterior, medium texture, brush application

	Labor SF per manhour	Material coverage SF/gallon	Material cost per gallon	Labor cost per 100 SF	Labor burden 100 SF	Material cost per 100 SF	Overhead per 100 SF	Profit per 100 SF	Total price per 100 SF
Masonry paint, water base, flat or gloss (material #31)									
Brush prime coat									
Slow	100	150	20.90	10.50	2.98	13.93	8.50	5.39	41.30
Medium	120	125	18.30	13.96	4.86	14.64	8.20	4.58	46.24
Fast	140	100	15.70	16.07	6.70	15.70	7.12	2.74	48.33
Brush 2nd coat									
Slow	150	250	20.90	7.00	1.99	8.36	5.38	3.41	26.14
Medium	163	225	18.30	10.28	3.57	8.13	5.39	3.01	30.38
Fast	175	200	15.70	12.86	5.35	7.85	4.82	1.85	32.73
Brush 3rd or additional coats									
Slow	160	270	20.90	6.56	1.87	7.74	5.01	3.18	24.36
Medium	173	245	18.30	9.68	3.38	7.47	5.03	2.81	28.37
Fast	185	220	15.70	12.16	5.09	7.14	4.51	1.73	30.63
Masonry paint, oil base (material #32)									
Brush prime coat									
Slow	80	265	28.30	13.13	3.72	10.68	8.54	5.41	41.48
Medium	100	250	24.80	16.75	5.84	9.92	7.96	4.45	44.92
Fast	120	235	21.20	18.75	7.81	9.02	6.59	2.53	44.70
Brush 2nd coat									
Slow	145	300	28.30	7.24	2.06	9.43	5.81	3.68	28.22
Medium	165	275	24.80	10.15	3.54	9.02	5.56	3.11	31.38
Fast	185	250	21.20	12.16	5.09	8.48	4.76	1.83	32.32
Brush 3rd or additional coats									
Slow	155	350	28.30	6.77	1.92	8.09	5.20	3.30	25.28
Medium	175	325	24.80	9.57	3.33	7.63	5.03	2.81	28.37
Fast	185	300	21.20	12.16	5.09	7.07	4.50	1.73	30.55

For heights above 8 feet, use the High Time Difficulty Factors on page 137. For oil base paint and clear hydro sealer, I recommend spraying. For painting interior plaster, see Walls, plaster. "Slow" work is based on a $10.50 hourly wage, "Medium" work on a $16.75 hourly wage, and "Fast" work on a $22.50 hourly wage. Other qualifications that apply to this table are on page 9.

	Labor SF per manhour	Material coverage SF/gallon	Material cost per gallon	Labor cost per 100 SF	Labor burden 100 SF	Material cost per 100 SF	Overhead per 100 SF	Profit per 100 SF	Total price per 100 SF
Plaster or stucco, exterior, medium texture, roll application									
Masonry paint, water base, flat or gloss (material #31)									
Roll prime coat									
Slow	245	225	20.90	4.29	1.21	9.29	4.59	2.91	22.29
Medium	273	188	18.30	6.14	2.13	9.73	4.41	2.47	24.88
Fast	300	150	15.70	7.50	3.12	10.47	3.90	1.50	26.49
Roll 2nd coat									
Slow	300	275	20.90	3.50	.99	7.60	3.75	2.38	18.22
Medium	320	243	18.30	5.23	1.84	7.53	3.57	2.00	20.17
Fast	340	210	15.70	6.62	2.76	7.48	3.12	1.20	21.18
Roll 3rd or additional coats									
Slow	320	300	20.90	3.28	.94	6.97	3.47	2.20	16.86
Medium	340	268	18.30	4.93	1.71	6.83	3.30	1.85	18.62
Fast	360	235	15.70	6.25	2.62	6.68	2.87	1.10	19.52
Masonry paint, oil base (material #32)									
Roll prime coat									
Slow	200	250	28.30	5.25	1.49	11.32	5.60	3.55	27.21
Medium	240	225	24.80	6.98	2.44	11.02	5.01	2.80	28.25
Fast	280	200	21.20	8.04	3.34	10.60	4.07	1.56	27.61
Roll 2nd coat									
Slow	220	275	28.30	4.77	1.36	10.29	5.09	3.23	24.74
Medium	265	250	24.80	6.32	2.20	9.92	4.52	2.53	25.49
Fast	305	225	21.20	7.38	3.08	9.42	3.68	1.41	24.97
Roll 3rd or additional coats									
Slow	235	350	28.30	4.47	1.27	8.09	4.29	2.72	20.84
Medium	285	325	24.80	5.88	2.05	7.63	3.81	2.13	21.50
Fast	335	300	21.20	6.72	2.82	7.07	3.07	1.18	20.86

For heights above 8 feet, use the High Time Difficulty Factors on page 137. For oil base paint and clear hydro sealer, I recommend spraying. For painting interior plaster, see Walls, plaster. "Slow" work is based on a $10.50 hourly wage, "Medium" work on a $16.75 hourly wage, and "Fast" work on a $22.50 hourly wage. Other qualifications that apply to this table are on page 9.

	Labor SF per manhour	Material coverage SF/gallon	Material cost per gallon	Labor cost per 100 SF	Labor burden 100 SF	Material cost per 100 SF	Overhead per 100 SF	Profit per 100 SF	Total price per 100 SF
Plaster or stucco, exterior, medium texture, spray application									
Masonry paint, water base, flat or gloss (material #31)									
Spray prime coat									
Slow	600	200	20.90	1.75	.50	10.45	3.94	2.50	19.14
Medium	675	145	18.30	2.48	.86	12.62	3.91	2.19	22.06
Fast	750	90	15.70	3.00	1.24	17.44	4.01	1.54	27.23
Spray 2nd coat									
Slow	700	225	20.90	1.50	.43	9.29	3.48	2.21	16.91
Medium	800	175	18.30	2.09	.73	10.46	3.25	1.82	18.35
Fast	900	125	15.70	2.50	1.04	12.56	2.98	1.14	20.22
Spray 3rd or additional coats									
Slow	750	235	20.90	1.40	.39	8.89	3.31	2.10	16.09
Medium	850	185	18.30	1.97	.70	9.89	3.07	1.72	17.35
Fast	950	135	15.70	2.37	.98	11.63	2.77	1.07	18.82
Masonry paint, oil base (material #32)									
Spray prime coat									
Slow	550	225	28.30	1.91	.54	12.58	4.66	2.95	22.64
Medium	600	158	24.80	2.79	.98	15.70	4.77	2.67	26.91
Fast	650	90	21.20	3.46	1.45	23.56	5.27	2.02	35.76
Spray 2nd coat									
Slow	650	350	28.30	1.62	.46	8.09	3.15	2.00	15.32
Medium	775	238	24.80	2.16	.75	10.42	3.27	1.83	18.43
Fast	900	125	21.20	2.50	1.04	16.96	3.79	1.46	25.75
Spray 3rd or additional coats									
Slow	700	375	28.30	1.50	.43	7.55	2.94	1.86	14.28
Medium	825	205	24.80	2.03	.70	12.10	3.64	2.03	20.50
Fast	950	135	21.20	2.37	.98	15.70	3.53	1.36	23.94

For heights above 8 feet, use the High Time Difficulty Factors on page 137. For oil base paint and clear hydro sealer, I recommend spraying. For painting interior plaster, see Walls, plaster. "Slow" work is based on a $10.50 hourly wage, "Medium" work on a $16.75 hourly wage, and "Fast" work on a $22.50 hourly wage. Other qualifications that apply to this table are on page 9.

	Labor SF per manhour	Material coverage SF/gallon	Material cost per gallon	Labor cost per 100 SF	Labor burden 100 SF	Material cost per 100 SF	Overhead per 100 SF	Profit per 100 SF	Total price per 100 SF
Plaster or stucco, exterior, medium texture, waterproofing									
Waterproofing, clear hydro sealer (material #34)									
Brush 1st coat									
Slow	125	100	17.80	8.40	2.38	17.80	8.86	5.62	43.06
Medium	150	88	15.60	11.17	3.90	17.73	8.03	4.49	45.32
Fast	175	75	13.30	12.86	5.35	17.73	6.65	2.56	45.15
Brush 2nd or additional coats									
Slow	175	135	17.80	6.00	1.70	13.19	6.48	4.11	31.48
Medium	200	118	15.60	8.38	2.92	13.22	6.01	3.36	33.89
Fast	225	100	13.30	10.00	4.16	13.30	5.08	1.95	34.49
Roll 1st coat									
Slow	325	150	17.80	3.23	.92	11.87	4.97	3.15	24.14
Medium	363	138	15.60	4.61	1.60	11.30	4.29	2.40	24.20
Fast	400	125	13.30	5.63	2.34	10.64	3.44	1.32	23.37
Roll 2nd or additional coats									
Slow	400	200	17.80	2.63	.74	8.90	3.81	2.41	18.49
Medium	425	190	15.60	3.94	1.37	8.21	3.31	1.85	18.68
Fast	450	180	13.30	5.00	2.08	7.39	2.68	1.03	18.18
Spray 1st coat									
Slow	650	225	17.80	1.62	.46	7.91	3.10	1.96	15.05
Medium	725	163	15.60	2.31	.81	9.57	3.11	1.74	17.54
Fast	800	100	13.30	2.81	1.18	13.30	3.20	1.23	21.72
Spray 2nd or additional coats									
Slow	800	250	17.80	1.31	.38	7.12	2.73	1.73	13.27
Medium	900	200	15.60	1.86	.65	7.80	2.53	1.41	14.25
Fast	1000	150	13.30	2.25	.94	8.87	2.23	.86	15.15

For heights above 8 feet, use the High Time Difficulty Factors on page 137. For oil base paint and clear hydro sealer, I recommend spraying. For painting interior plaster, see Walls, plaster. "Slow" work is based on a $10.50 hourly wage, "Medium" work on a $16.75 hourly wage, and "Fast" work on a $22.50 hourly wage. Other qualifications that apply to this table are on page 9.

Plaster, interior: see Walls

	Labor SF per manhour	Material coverage SF/gallon	Material cost per gallon	Labor cost per 100 SF	Labor burden 100 SF	Material cost per 100 SF	Overhead per 100 SF	Profit per 100 SF	Total price per 100 SF
Plaster or stucco, interior/exterior, medium texture, anti-graffiti stain eliminator									
Water base primer and sealer (material #39)									
Roll & brush each coat									
Slow	350	400	25.40	3.00	.86	6.35	3.16	2.00	15.37
Medium	375	375	22.30	4.47	1.56	5.95	2.94	1.64	16.56
Fast	400	350	19.10	5.63	2.34	5.46	2.49	.96	16.88
Oil base primer and sealer (material #40)									
Roll & brush each coat									
Slow	350	375	25.70	3.00	.86	6.85	3.32	2.10	16.13
Medium	375	350	22.50	4.47	1.56	6.43	3.05	1.71	17.22
Fast	400	325	19.30	5.63	2.34	5.94	2.58	.99	17.48
Polyurethane 2 part system (material #41)									
Roll & brush each coat									
Slow	300	375	91.40	3.50	.99	24.37	8.95	5.67	43.48
Medium	325	350	80.00	5.15	1.81	22.86	7.30	4.08	41.20
Fast	350	325	68.60	6.43	2.69	21.11	5.59	2.15	37.97

For heights above 8 feet, use the High Time Difficulty Factors on page 137. For oil base paint and clear hydro sealer, I recommend spraying. For painting interior plaster, see Walls, plaster. "Slow" work is based on a $10.50 hourly wage, "Medium" work on a $16.75 hourly wage, and "Fast" work on a $22.50 hourly wage. Other qualifications that apply to this table are on page 9.

	Labor LF per manhour	Material coverage LF/gallon	Material cost per gallon	Labor cost per 100 LF	Labor burden 100 LF	Material cost per 100 LF	Overhead per 100 LF	Profit per 100 LF	Total price per 100 LF
Pot shelves, 12" to 18" wide									
Solid body or semi-transparent stain, water or oil base (material #18 or #19 or #20 or #21)									
Roll & brush each coat									
Slow	50	50	23.23	21.00	5.96	46.46	22.76	14.43	110.61
Medium	68	45	20.33	24.63	8.60	45.18	19.21	10.74	108.36
Fast	75	40	17.40	30.00	12.51	43.50	15.92	6.12	108.05

These figures are based on painting all sides of exterior or interior pot shelves. "Slow" work is based on a $10.50 hourly wage, "Medium" work on a $16.75 hourly wage, and "Fast" work on a $22.50 hourly wage. Other qualifications that apply to this table are on page 9.

	Labor LF per manhour	Material coverage LF/gallon	Material cost per gallon	Labor cost per 100 LF	Labor burden 100 LF	Material cost per 100 LF	Overhead per 100 LF	Profit per 100 LF	Total price per 100 LF
Railing, exterior, rough sawn or resawn wood									
Solid body or semi-transparent stain, water or oil base (material #18 or #19 or #20 or #21)									
Roll & brush each coat									
Slow	16	20	23.23	65.63	18.62	116.15	62.14	39.39	301.93
Medium	18	18	20.33	93.06	32.45	112.94	58.42	32.65	329.52
Fast	20	15	17.40	112.50	46.95	116.00	50.96	19.59	346.00

Use these costs for finishing railing that's 36" to 42" high and with 2" x 2" verticals spaced 6" on center. These figures include painting the rail cap, baluster, newels and spindles. "Slow" work is based on a $10.50 hourly wage, "Medium" work on a $16.75 hourly wage, and "Fast" work on a $22.50 hourly wage. Other qualifications that apply to this table are on page 9.

	Labor LF per manhour	Material coverage LF/gallon	Material cost per gallon	Labor cost per 100 LF	Labor burden 100 LF	Material cost per 100 LF	Overhead per 100 LF	Profit per 100 LF	Total price per 100 LF
Railing, exterior, stain grade, decorative wood									
STEP 1: Sand & putty									
Slow	50	--	--	21.00	5.96	--	8.36	5.30	40.62
Medium	60	--	--	27.92	9.74	--	9.22	5.16	52.04
Fast	70	--	--	32.14	13.43	--	8.43	3.24	57.24
STEP 2 & 3: Stain (material #30a) & wipe									
Brush & wipe, 1 coat									
Slow	25	60	30.60	42.00	11.92	51.00	32.53	20.62	158.07
Medium	30	55	26.80	55.83	19.46	48.73	30.38	16.98	171.38
Fast	35	50	22.90	64.29	26.82	45.80	25.33	9.74	171.98
Spray & wipe, 1 coat									
Slow	75	35	30.60	14.00	3.97	87.43	32.68	20.71	158.79
Medium	85	30	26.80	19.71	6.86	89.33	28.40	15.87	160.17
Fast	95	25	22.90	23.68	9.90	91.60	23.16	8.90	157.24
STEP 4 & 5: Sanding sealer (material #30b) & light sand									
Brush 1 coat									
Slow	45	65	33.40	23.33	6.62	51.38	25.22	15.99	122.54
Medium	50	60	29.30	33.50	11.68	48.83	23.03	12.87	129.91
Fast	55	55	25.10	40.91	17.07	45.64	19.17	7.37	130.16
Spray 1 coat									
Slow	125	35	33.40	8.40	2.38	95.43	32.93	20.87	160.01
Medium	138	30	29.30	12.14	4.24	97.67	27.94	15.62	157.61
Fast	150	25	25.10	15.00	6.27	100.40	22.51	8.65	152.83

	Labor LF per manhour	Material coverage LF/gallon	Material cost per gallon	Labor cost per 100 LF	Labor burden 100 LF	Material cost per 100 LF	Overhead per 100 LF	Profit per 100 LF	Total price per 100 LF
STEP 6 & 7: Varnish 2 coats (material #30c)									
Varnish, flat or gloss (material #30c)									
Brush 1st coat									
Slow	22	100	35.40	47.73	13.54	35.40	29.98	19.00	145.65
Medium	24	90	31.00	69.79	24.34	34.44	31.49	17.60	177.66
Fast	26	80	26.60	86.54	36.11	33.25	28.85	11.09	195.84
Brush 2nd or additional coats									
Slow	30	120	35.40	35.00	9.93	29.50	23.08	14.63	112.14
Medium	32	110	31.00	52.34	18.25	28.18	24.20	13.53	136.50
Fast	34	100	26.60	66.18	27.61	26.60	22.28	8.56	151.23
Steel wool, hand buff									
Slow	100	--	--	10.50	2.98	--	4.18	2.65	20.31
Medium	113	--	--	14.82	5.17	--	4.90	2.74	27.63
Fast	125	--	--	18.00	7.51	--	4.72	1.81	32.04
Wax & polish by hand									
Slow	100	--	--	10.50	2.98	--	4.18	2.65	20.31
Medium	113	--	--	14.82	5.17	--	4.90	2.74	27.63
Fast	125	--	--	18.00	7.51	--	4.72	1.81	32.04

Use these figures to estimate the cost of applying a natural finish on stain grade railing that's from 36" to 42" high and with spindles spaced at 6" on center. These figures include painting the rail cap, baluster, newels and spindles. "Slow" work is based on a $10.50 hourly wage, "Medium" work on a $16.75 hourly wage, and "Fast" work on a $22.50 hourly wage. Other qualifications that apply to this table are on page 9.

	Labor LF per manhour	Material coverage LF/gallon	Material cost per gallon	Labor cost per 100 LF	Labor burden 100 LF	Material cost per 100 LF	Overhead per 100 LF	Profit per 100 LF	Total price per 100 LF

Railing, interior, handrail, decorative wood

Paint grade

Undercoat or enamel, water or oil base (material #3 or #4 or #9 or #10)

Brush each coat									
Slow	30	120	23.45	35.00	9.93	19.54	20.00	12.68	97.15
Medium	35	110	20.53	47.86	16.68	18.66	20.38	11.39	114.97
Fast	40	100	17.60	56.25	23.48	17.60	18.01	6.92	122.26

Stain grade, 7 step process

Stain, seal and 2 coat lacquer system (material #11)

Spray all coats									
Slow	55	100	21.60	19.09	5.42	21.60	14.30	9.06	69.47
Medium	60	88	19.60	27.92	9.74	22.27	14.68	8.21	82.82
Fast	65	75	16.20	34.62	14.43	21.60	13.07	5.02	88.74

Use these costs for finishing decorative wood, wall mounted handrail. "Slow" work is based on a $10.50 hourly wage, "Medium" work on a $16.75 hourly wage, and "Fast" work on a $22.50 hourly wage. Other qualifications that apply to this table are on page 9.

	Labor LF per manhour	Material coverage LF/gallon	Material cost per gallon	Labor cost per 100 LF	Labor burden 100 LF	Material cost per 100 LF	Overhead per 100 LF	Profit per 100 LF	Total price per 100 LF

Railing, interior, stain grade, decorative wood

Stain, seal and 2 coat lacquer system (7 step process)

STEP 1: Sand & putty									
Slow	50	--	--	21.00	5.96	--	8.36	5.30	40.62
Medium	60	--	--	27.92	9.74	--	9.22	5.16	52.04
Fast	70	--	--	32.14	13.43	--	8.43	3.24	57.24

STEP 2 & 3: Stain (material #11a) & wipe

Brush & wipe, 1 coat									
Slow	25	60	30.90	42.00	11.92	51.50	32.69	20.72	158.83
Medium	30	55	27.00	55.83	19.46	49.09	30.47	17.03	171.88
Fast	35	50	23.10	64.29	26.82	46.20	25.41	9.76	172.48

Spray & wipe, 1 coat									
Slow	75	35	30.90	14.00	3.97	88.29	32.94	20.88	160.08
Medium	85	30	27.00	19.71	6.86	90.00	28.56	15.97	161.10
Fast	95	25	23.10	23.68	9.90	92.40	23.30	8.96	158.24

	Labor LF per manhour	Material coverage LF/gallon	Material cost per gallon	Labor cost per 100 LF	Labor burden 100 LF	Material cost per 100 LF	Overhead per 100 LF	Profit per 100 LF	Total price per 100 LF
STEP 4 & 5: Sanding sealer (material #11b) & light sand									
Brush 1 coat									
Slow	45	65	17.20	23.33	6.62	26.46	17.49	11.09	84.99
Medium	50	60	15.00	33.50	11.68	25.00	17.19	9.61	96.98
Fast	55	55	12.90	40.91	17.07	23.45	15.07	5.79	102.29
Spray 1 coat									
Slow	125	35	17.20	8.40	2.38	49.14	18.58	11.78	90.28
Medium	138	30	15.00	12.14	4.24	50.00	16.26	9.09	91.73
Fast	150	25	12.90	15.00	6.27	51.60	13.48	5.18	91.53
STEP 6 & 7: Lacquer (material #11c), 2 coats									
Brush 1st coat									
Slow	40	65	19.30	26.25	7.45	29.69	19.66	12.46	95.51
Medium	50	60	16.80	33.50	11.68	28.00	17.93	10.02	101.13
Fast	60	55	14.40	37.50	15.66	26.18	14.68	5.64	99.66
Brush 2nd coat									
Slow	45	70	19.30	23.33	6.62	27.57	17.84	11.31	86.67
Medium	55	65	16.80	30.45	10.62	25.85	16.39	9.16	92.47
Fast	65	60	14.40	34.62	14.43	24.00	13.52	5.20	91.77
Spray 1st coat									
Slow	75	55	19.30	14.00	3.97	35.09	16.45	10.43	79.94
Medium	85	50	16.80	19.71	6.86	33.60	14.74	8.24	83.15
Fast	95	45	14.40	23.68	9.90	32.00	12.13	4.66	82.37
Spray 2nd coat									
Slow	85	60	19.30	12.35	3.50	32.17	14.89	9.44	72.35
Medium	95	55	16.80	17.63	6.16	30.55	13.31	7.44	75.09
Fast	105	50	14.40	21.43	8.93	28.80	10.95	4.21	74.32
Complete stain, seal & 2 coat lacquer system (material #11)									
Brush all coats									
Slow	8	30	21.60	131.25	37.25	72.00	74.58	47.28	362.36
Medium	10	25	19.60	167.50	58.40	78.40	74.55	41.67	420.52
Fast	12	20	16.20	187.50	78.24	81.00	64.15	24.66	435.55
Spray all coats									
Slow	16	20	19.30	65.63	18.62	96.50	56.05	35.53	272.33
Medium	20	15	16.80	83.75	29.20	112.00	55.11	30.81	310.87
Fast	24	10	14.40	93.75	39.14	144.00	51.22	19.69	347.80

Use these costs for applying stain, sanding sealer and lacquer to interior railings. Railing is based on a 36" to 42" height and with spindles spaced at 6" on center. These costs include painting the rail cap, baluster, newels and spindles. For rough sawn wood railing with 2" x 2" spindles spaced at 6" on center, see the tables for Exterior wood railing. "Slow" work is based on a $10.50 hourly wage, "Medium" work on a $16.75 hourly wage, and "Fast" work on a $22.50 hourly wage. Other qualifications that apply to this table are on page 9.

	Labor LF per manhour	Material coverage LF/gallon	Material cost per gallon	Labor cost per 100 LF	Labor burden 100 LF	Material cost per 100 LF	Overhead per 100 LF	Profit per 100 LF	Total price per 100 LF
Railing, interior, wood, paint grade, brush application									
Undercoat, water or oil base (material #3 or #4)									
Brush 1 coat									
Slow	11	50	21.35	95.45	27.10	42.70	51.25	32.48	248.98
Medium	14	45	18.70	119.64	41.72	41.56	49.71	27.79	280.42
Fast	17	40	16.00	132.35	55.23	40.00	42.11	16.18	285.87
Split coat (1/2 undercoat + 1/2 enamel), water or oil base (material #3 and #9 or #4 and #10)									
Brush 1 coat									
Slow	15	70	23.45	70.00	19.87	33.50	38.26	24.25	185.88
Medium	18	65	20.53	93.06	32.45	31.58	38.48	21.51	217.08
Fast	21	60	17.60	107.14	44.72	29.33	33.52	12.88	227.59
Enamel, water or oil base (material #9 or #10)									
Brush 1st finish coat									
Slow	13	60	25.55	80.77	22.92	42.58	45.36	28.75	220.38
Medium	16	55	22.35	104.69	36.50	40.64	44.54	24.90	251.27
Fast	18	50	19.20	125.00	52.18	38.40	39.88	15.33	270.79
Brush 2nd or additional finish coats									
Slow	15	70	25.55	70.00	19.87	36.50	39.19	24.84	190.40
Medium	18	65	22.35	93.06	32.45	34.38	39.17	21.89	220.95
Fast	21	60	19.20	107.14	44.72	32.00	34.02	13.07	230.95

Use these costs for applying undercoat or enamel to interior railings. Railing is based on a 36" to 42" height and with spindles spaced at 6" on center. These costs include painting the rail cap, baluster, newels and spindles. For rough sawn wood railing with 2" x 2" spindles spaced at 6" on center, see the tables for exterior wood railing. "Slow" work is based on a $10.50 hourly wage, "Medium" work on a $16.75 hourly wage, and "Fast" work on a $22.50 hourly wage. Other qualifications that apply to this table are on page 9.

	Labor LF per manhour	Material coverage LF/gallon	Material cost per gallon	Labor cost per 100 LF	Labor burden 100 LF	Material cost per 100 LF	Overhead per 100 LF	Profit per 100 LF	Total price per 100 LF
Railing, interior, wood, paint grade, spray application									
Undercoat, water or oil base (material #3 or #4)									
Spray 1 coat									
Slow	50	40	21.35	21.00	5.96	53.38	24.91	15.79	121.04
Medium	60	35	18.70	27.92	9.74	53.43	22.31	12.47	125.87
Fast	70	30	16.00	32.14	13.43	53.33	18.29	7.03	124.22
Split coat (1/2 undercoat + 1/2 enamel), water or oil base (material #3 and #9 or #4 and #10)									
Spray 1 coat									
Slow	65	50	23.45	16.15	4.58	46.90	20.97	13.29	101.89
Medium	75	45	20.53	22.33	7.78	45.62	18.55	10.37	104.65
Fast	85	40	17.60	26.47	11.03	44.00	15.08	5.80	102.38
Enamel, water or oil base (material #9 or #10)									
Spray 1st finish coat									
Slow	55	45	25.55	19.09	5.42	56.78	25.20	15.98	122.47
Medium	65	40	22.35	25.77	8.97	55.88	22.20	12.41	125.23
Fast	75	35	19.20	30.00	12.51	54.86	18.02	6.92	122.31
Spray 2nd or additional finish coats									
Slow	65	50	25.55	16.15	4.58	51.10	22.27	14.12	108.22
Medium	75	45	22.35	22.33	7.78	49.67	19.55	10.93	110.26
Fast	85	40	19.20	26.47	11.03	48.00	15.82	6.08	107.40

Use these costs for applying undercoat or enamel to interior railings. Railing is based on a 36" to 42" height and with spindles spaced at 6" on center. These costs include painting the rail cap, baluster, newels and spindles. For rough sawn wood railing with 2" x 2" spindles spaced at 6" on center, see the tables for exterior wood railing. "Slow" work is based on a $10.50 hourly wage, "Medium" work on a $16.75 hourly wage, and "Fast" work on a $22.50 hourly wage. Other qualifications that apply to this table are on page 9.

	Labor LF per manhour	Material coverage LF/gallon	Material cost per gallon	Labor cost per 100 LF	Labor burden 100 LF	Material cost per 100 LF	Overhead per 100 LF	Profit per 100 LF	Total price per 100 LF
Railing, wrought iron, 42" high bars with wood cap									
Metal primer, rust inhibitor, clean metal (material #35)									
Brush 1 coat									
Slow	15	90	29.90	70.00	19.87	33.22	38.17	24.20	185.46
Medium	18	85	26.20	93.06	32.45	30.82	38.30	21.41	216.04
Fast	20	80	22.40	112.50	46.95	28.00	34.68	13.33	235.46
Metal primer, rust inhibitor, rusty metal (material #36)									
Brush 1 coat									
Slow	15	90	36.60	70.00	19.87	40.67	40.48	25.66	196.68
Medium	18	85	32.10	93.06	32.45	37.76	40.00	22.36	225.63
Fast	20	80	27.50	112.50	46.95	34.38	35.86	13.78	243.47
Metal finish, synthetic enamel, off white, gloss, interior or exterior - (material #37)									
Brush 1st finish coat									
Slow	20	110	27.90	52.50	14.90	25.36	28.77	18.24	139.77
Medium	23	100	24.40	72.83	25.39	24.40	30.04	16.79	169.45
Fast	26	90	20.90	86.54	36.11	23.22	26.99	10.37	183.23
Brush 2nd or additional finish coats									
Slow	30	125	27.90	35.00	9.93	22.32	20.86	13.22	101.33
Medium	35	120	24.40	47.86	16.68	20.33	20.79	11.62	117.28
Fast	40	115	20.90	56.25	23.48	18.17	18.11	6.96	122.97
Metal finish, synthetic enamel, colors (except orange/red), gloss, interior or exterior - (material #38)									
Brush 1st finish coat									
Slow	20	110	30.00	52.50	14.90	27.27	29.36	18.61	142.64
Medium	23	100	26.30	72.83	25.39	26.30	30.50	17.05	172.07
Fast	26	90	22.50	86.54	36.11	25.00	27.32	10.50	185.47
Brush 2nd or additional finish coats									
Slow	30	125	30.00	35.00	9.93	24.00	21.38	13.55	103.86
Medium	35	120	26.30	47.86	16.68	21.92	21.18	11.84	119.48
Fast	40	115	22.50	56.25	23.48	19.57	18.37	7.06	124.73

Use these figures for painting prefabricated wrought iron railing which is 42" high with 1/2" square vertical bars at 6" to 9" on center with a stain grade wood cap supported by a 1/2" by 1-1/2" top rail, and 1" square support posts at 6' to 10' on center and with a 1/2" by 1-1/2" bottom rail, unless otherwise noted. The Metal finish figures include only minor touchup of pre-primed steel or wrought iron prefabricated railings. Note: A two coat system, prime and finish, using oil base material is recommended for any metal surface. Although water base material is often used, it may cause oxidation, corrosion and rust. Using one coat of oil base paint on exterior metal may result in cracking, peeling, or chipping without the proper prime coat application. If off white or other light colored finish paint is specified, make sure the prime coat is a light color also, or more than one finish coat will be necessary. "Slow" work is based on a $10.50 hourly wage,"Medium" work on a $16.75 hourly wage, and "Fast" work on a $22.50 hourly wage. Other qualifications that apply to this table are on page 9.

	Labor LF per manhour	Material coverage LF/gallon	Material cost per gallon	Labor cost per 100 LF	Labor burden 100 LF	Material cost per 100 LF	Overhead per 100 LF	Profit per 100 LF	Total price per 100 LF
Railing, wrought iron, 42" high bars with wrought iron cap									
Metal primer, rust inhibitor, clean metal (material #35)									
Brush 1 coat									
Slow	20	110	29.90	52.50	14.90	27.18	29.33	18.59	142.50
Medium	25	105	26.20	67.00	23.36	24.95	28.25	15.79	159.35
Fast	30	100	22.40	75.00	31.29	22.40	23.81	9.15	161.65
Metal primer, rust inhibitor, rusty metal (material #36)									
Brush 1 coat									
Slow	20	110	36.60	52.50	14.90	33.27	31.22	19.79	151.68
Medium	25	105	32.10	67.00	23.36	30.57	29.63	16.56	167.12
Fast	30	100	27.50	75.00	31.29	27.50	24.75	9.51	168.05
Metal finish, synthetic enamel, off white, gloss, interior or exterior - (material #37)									
Brush 1st finish coat									
Slow	25	120	27.90	42.00	11.92	23.25	23.93	15.17	116.27
Medium	30	115	24.40	55.83	19.46	21.22	23.64	13.22	133.37
Fast	35	110	20.90	64.29	26.82	19.00	20.37	7.83	138.31
Brush 2nd or additional finish coats									
Slow	35	135	27.90	30.00	8.51	20.67	18.36	11.64	89.18
Medium	40	130	24.40	41.88	14.60	18.77	18.44	10.31	104.00
Fast	45	125	20.90	50.00	20.86	16.72	16.21	6.23	110.02
Metal finish, synthetic enamel, colors (except orange/red), gloss, interior or exterior - (material #38)									
Brush 1st finish coat									
Slow	25	120	30.00	42.00	11.92	25.00	24.47	15.51	118.90
Medium	30	115	26.30	55.83	19.46	22.87	24.05	13.44	135.65
Fast	35	110	22.50	64.29	26.82	20.45	20.64	7.93	140.13
Brush 2nd or additional finish coats									
Slow	35	135	30.00	30.00	8.51	22.22	18.84	11.94	91.51
Medium	40	130	26.30	41.88	14.60	20.23	18.79	10.51	106.01
Fast	45	125	22.50	50.00	20.86	18.00	16.44	6.32	111.62

Use these figures for painting prefabricated wrought iron railing which is 42" high with 1/2" square vertical bars at 6" to 9" on center and a 1/2" by 1-1/2" wrought iron cap with 1" square support posts at 6' to 10' on center, and a 1/2" by 1-1/2" bottom rail, unless otherwise noted. The Metal finish figures include only minor touchup of pre-primed steel or wrought iron prefabricated railings. Note: A two coat system, prime and finish, using oil base material is recommended for any metal surface. Although water base material is often used, it may cause oxidation, corrosion and rust. Using one coat of oil base paint on exterior metal may result in cracking, peeling, or chipping without the proper prime coat application. If off white or other light colored finish paint is specified, make sure the prime coat is a light color also, or more than one finish coat will be necessary. "Slow" work is based on a $10.50 hourly wage, "Medium" work on a $16.75 hourly wage, and "Fast" work on a $22.50 hourly wage. Other qualifications that apply to this table are on page 9.

	Labor LF per manhour	Material coverage LF/gallon	Material cost per gallon	Labor cost per 100 LF	Labor burden 100 LF	Material cost per 100 LF	Overhead per 100 LF	Profit per 100 LF	Total price per 100 LF

Railing, wrought iron, 60" to 72" high bars with wrought iron cap

Metal primer, rust inhibitor, clean metal (material #35)									
Brush 1 coat									
Slow	10	90	29.90	105.00	29.80	33.22	52.11	33.03	253.16
Medium	15	85	26.20	111.67	38.94	30.82	44.45	24.84	250.72
Fast	20	80	22.40	112.50	46.95	28.00	34.68	13.33	235.46
Metal primer, rust inhibitor, rusty metal (material #36)									
Brush 1 coat									
Slow	10	90	36.60	105.00	29.80	40.67	54.42	34.49	264.38
Medium	15	85	32.10	111.67	38.94	37.76	46.15	25.80	260.32
Fast	20	80	27.50	112.50	46.95	34.38	35.86	13.78	243.47
Metal finish, synthetic enamel, off white, gloss, interior or exterior - (material #37)									
Brush 1st finish coat									
Slow	15	120	27.90	70.00	19.87	23.25	35.08	22.24	170.44
Medium	20	115	24.40	83.75	29.20	21.22	32.87	18.37	185.41
Fast	25	110	20.90	90.00	37.56	19.00	27.12	10.42	184.10
Brush 2nd or additional finish coats									
Slow	25	135	27.90	42.00	11.92	20.67	23.13	14.66	112.38
Medium	30	130	24.40	55.83	19.46	18.77	23.04	12.88	129.98
Fast	35	125	20.90	64.29	26.82	16.72	19.95	7.67	135.45
Metal finish, synthetic enamel, colors (except orange/red), gloss, interior or exterior - (material #38)									
Brush 1st finish coat									
Slow	15	120	30.00	70.00	19.87	25.00	35.63	22.58	173.08
Medium	20	115	26.30	83.75	29.20	22.87	33.27	18.60	187.69
Fast	25	110	22.50	90.00	37.56	20.45	27.39	10.53	185.93
Brush 2nd or additional finish coats									
Slow	25	135	30.00	42.00	11.92	22.22	23.61	14.97	114.72
Medium	30	130	26.30	55.83	19.46	20.23	23.40	13.08	132.00
Fast	35	125	22.50	64.29	26.82	18.00	20.19	7.76	137.06

Use these figures for painting prefabricated wrought iron railing which is 60" to 72" high with 1/2" square vertical bars at 6" to 9" on center with 1" square support posts at 6' to 10' on center and with a 1/2" by 1-1/2" bottom rail, unless otherwise noted. The Metal finish figures include only minor touchup of pre-primed steel or wrought iron prefabricated railings. Note: A two coat system, prime and finish, using oil base material is recommended for any metal surface. Although water base material is often used, it may cause oxidation, corrosion and rust. Using one coat of oil base paint on exterior metal may result in cracking, peeling, or chipping without the proper prime coat application. If off white or other light colored finish paint is specified, make sure the prime coat is a light color also, or more than one finish coat will be necessary. "Slow" work is based on a $10.50 hourly wage, "Medium" work on a $16.75 hourly wage, and "Fast" work on a $22.50 hourly wage. Other qualifications that apply to this table are on page 9.

	SF floor per manhour	Material SF floor per can	Material cost per can	Labor per 100 SF floor	Labor burden 100 SF	Material per 100 SF floor	Overhead per 100 SF floor	Profit per 100 SF floor	Total per 100 SF floor
Registers, HVAC, per 100 square feet of floor area									
Repaint jobs, spray cans (material #17)									
Spray 1 coat									
Slow	900	700	4.20	1.17	.33	.60	.65	.41	3.16
Medium	950	650	3.70	1.76	.61	.57	.72	.40	4.06
Fast	1000	600	3.10	2.25	.94	.52	.69	.26	4.66
New construction projects, spray cans (material #17)									
Spray 1 coat									
Slow	2500	800	4.20	.42	.12	.53	.33	.21	1.61
Medium	2750	750	3.70	.61	.20	.49	.32	.18	1.80
Fast	3000	700	3.10	.75	.30	.44	.28	.11	1.88

These costs assume HVAC registers are painted with spray cans (bombs) to match the adjacent walls. Costs are based on square footage of the floor area of the building. These rates include time to remove, paint and replace the HVAC registers. Use the square feet of floor area divided by these rates to find manhours and the number of spray bombs needed to paint all the heat registers in a building. Rule of Thumb: 2 minutes per 100 square feet of floor is for new construction projects. "Slow" work is based on a $10.50 hourly wage, "Medium" work on a $16.75 hourly wage, and "Fast" work on a $22.50 hourly wage. Other qualifications that apply to this table are on page 9.

	Manhours per 1500 SF roof	Material gallons 1500 SF	Material cost per gallon	Labor per 1500 SF roof	Labor burden 1500 SF	Material per 1500 SF roof	Overhead per 1500 SF roof	Profit per 1500 SF roof	Total per 1500 SF roof
Roof jacks, per 1500 square feet of roof area									
Metal primer, clean metal (material #35)									
1 story building, brush prime coat									
Slow	0.40	0.30	29.90	4.20	1.19	8.97	4.45	2.82	21.63
Medium	0.35	0.33	26.20	5.86	2.05	8.65	4.05	2.27	22.88
Fast	0.30	0.35	22.40	6.75	2.81	7.84	3.22	1.24	21.86
2 story building, brush prime coat									
Slow	0.50	0.40	29.90	5.25	1.49	11.96	5.80	3.68	28.18
Medium	0.45	0.43	26.20	7.54	2.63	11.27	5.25	2.94	29.63
Fast	0.40	0.45	22.40	9.00	3.76	10.08	4.23	1.62	28.69
Metal primer, rusty metal (material #36)									
1 story building, brush prime coat									
Slow	0.40	0.30	36.60	4.20	1.19	10.98	5.07	3.22	24.66
Medium	0.35	0.33	32.10	5.86	2.05	10.59	4.53	2.53	25.56
Fast	0.30	0.35	27.50	6.75	2.81	9.63	3.55	1.37	24.11
2 story building, brush prime coat									
Slow	0.50	0.40	36.60	5.25	1.49	14.64	6.63	4.20	32.21
Medium	0.45	0.43	32.10	7.54	2.63	13.80	5.87	3.28	33.12
Fast	0.40	0.45	27.50	9.00	3.76	12.38	4.65	1.79	31.58
Metal finish, synthetic enamel - off white, gloss, interior or exterior (material #37)									
1 story building, brush each coat									
Slow	0.30	0.20	27.90	3.15	.90	5.58	2.99	1.89	14.51
Medium	0.25	0.25	24.40	4.19	1.46	6.10	2.88	1.61	16.24
Fast	0.20	0.30	20.90	4.50	1.88	6.27	2.34	.90	15.89
2 story building, brush each coat									
Slow	0.40	0.30	27.90	4.20	1.19	8.37	4.27	2.70	20.73
Medium	0.35	0.33	24.40	5.86	2.05	8.05	3.91	2.18	22.05
Fast	0.30	0.35	20.90	6.75	2.81	7.32	3.12	1.20	21.20

	Manhours per 1500 SF roof	Material gallons 1500 SF	Material cost per gallon	Labor per 1500 SF roof	Labor burden 1500 SF	Material per 1500 SF roof	Overhead per 1500 SF roof	Profit per 1500 SF roof	Total per 1500 SF roof
Metal finish, synthetic enamel, colors (except orange/red), gloss, interior or exterior - (material #38)									
1 story building, brush each coat									
Slow	0.30	0.20	30.00	3.15	.90	6.00	3.12	1.98	15.15
Medium	0.25	0.25	26.30	4.19	1.46	6.58	3.00	1.68	16.91
Fast	0.20	0.30	22.50	4.50	1.88	6.75	2.43	.93	16.49
2 story building, brush each coat									
Slow	0.40	0.30	30.00	4.20	1.19	9.00	4.46	2.83	21.68
Medium	0.35	0.33	26.30	5.86	2.05	8.68	4.06	2.27	22.92
Fast	0.30	0.35	22.50	6.75	2.81	7.88	3.23	1.24	21.91

Production rates and coverage figures are minimum values based on 1 or 2 story roof areas of up to 1500 square feet. For example, to apply metal primer on clean metal roof jacks on a 3000 SF one-story building at a medium rate, use two times the cost of $22.88 or $45.76. This figure includes ladder time. See the paragraphs below on Roof pitch difficulty factors and Roof area conversion factors to adjust for roof slope and type. Note: A two coat system, prime and finish, using oil base material is recommended for any metal surface. Although water base material is often used, it may cause oxidation, corrosion and rust. One coat of oil base solid body stain is often used on exterior metal but it may crack, peel or chip without the proper prime coat application. "Slow" work is based on a $10.50 hourly wage, "Medium" work on a $16.75 hourly wage, and "Fast" work on a $22.50 hourly wage. Other qualifications that apply to this table are on page 9.

Roof area conversion factors

For an arched roof, multiply the building length by the building width, then multiply by 1.5.
For a gambrel roof, multiply the building length by the building width, then multiply by 1.33.

Roof pitch difficulty factors

It's harder to paint on a sloped surface than a flat surface. The steeper the slope, the more difficult the work. Roof slope is usually measured in inches of rise per inch of horizontal run. For example, a 3 in 12 pitch means the roof rises 3 inches for each 12 inches of run, measuring horizontally. Use the difficulty factors that follow when estimating the time needed to paint on a sloping roof.

On a flat roof, or roof with a pitch of less than 3 in 12, calculate the roof area without modification.
If the pitch is 3 in 12, multiply the surface area by 1.1.
If the pitch is 4 in 12, multiply the surface area by 1.2.
If the pitch is 6 in 12, multiply the surface area by 1.3.

	Labor SF per manhour	Material coverage SF/gallon	Material cost per gallon	Labor cost per 100 SF	Labor burden 100 SF	Material cost per 100 SF	Overhead per 100 SF	Profit per 100 SF	Total price per 100 SF
Roofing, composition shingles, brush application									
Solid body stain, water base (material #18)									
Brush 1st coat									
Slow	45	220	21.50	23.33	6.62	9.77	12.32	7.81	59.85
Medium	60	200	18.80	27.92	9.74	9.40	11.53	6.44	65.03
Fast	80	180	16.10	28.13	11.73	8.94	9.03	3.47	61.30
Brush 2nd coat									
Slow	65	330	21.50	16.15	4.58	6.52	8.45	5.36	41.06
Medium	80	210	18.80	20.94	7.30	8.95	9.11	5.09	51.39
Fast	100	290	16.10	22.50	9.39	5.55	6.93	2.66	47.03
Brush 3rd or additional coats									
Slow	85	405	21.50	12.35	3.50	5.31	6.56	4.16	31.88
Medium	100	385	18.80	16.75	5.84	4.88	6.73	3.76	37.96
Fast	120	365	16.10	18.75	7.81	4.41	5.73	2.20	38.90
Solid body stain, oil base (material #19)									
Brush 1st coat									
Slow	45	270	23.90	23.33	6.62	8.85	12.03	7.63	58.46
Medium	60	250	20.90	27.92	9.74	8.36	11.27	6.30	63.59
Fast	80	230	17.90	28.13	11.73	7.78	8.82	3.39	59.85
Brush 2nd coat									
Slow	65	360	23.90	16.15	4.58	6.64	8.49	5.38	41.24
Medium	80	345	20.90	20.94	7.30	6.06	8.40	4.70	47.40
Fast	100	330	17.90	22.50	9.39	5.42	6.90	2.65	46.86
Brush 3rd or additional coats									
Slow	85	425	23.90	12.35	3.50	5.62	6.66	4.22	32.35
Medium	100	405	20.90	16.75	5.84	5.16	6.80	3.80	38.35
Fast	120	385	17.90	18.75	7.81	4.65	5.78	2.22	39.21

Use these figures for repaint jobs only. It has been established that asbestos fibers are a known carcinogen (cancer causing) and it is likely that no new construction projects will specify materials or products which contain asbestos. Roofing and siding products usually contain very little asbestos and are typically non-friable (hand pressure can not crumble, pulverize or reduce to powder when dry). There is danger when asbestos is being removed because of exposure to airborne particulate matter. Apparently, there is little danger when painting asbestos roofing or siding, but it is a good idea to have your painters wear respirators or particle masks for their safety. Coverage figures are based on shingles or shakes with average moisture content. See the paragraphs on Roof pitch difficulty factors and Roof area conversion factors to adjust for roof slope and type. "Slow" work is based on a $10.50 hourly wage, "Medium" work on a $16.75 hourly wage, and "Fast" work on a $22.50 hourly wage. Other qualifications that apply to this table are on page 9.

	Labor SF per manhour	Material coverage SF/gallon	Material cost per gallon	Labor cost per 100 SF	Labor burden 100 SF	Material cost per 100 SF	Overhead per 100 SF	Profit per 100 SF	Total price per 100 SF
Roofing, composition shingles, roll application									
Solid body stain, water base (material #18)									
Roll 1st coat									
Slow	150	190	21.50	7.00	1.99	11.32	6.30	3.99	30.60
Medium	170	180	18.80	9.85	3.43	10.44	5.81	3.25	32.78
Fast	200	170	16.10	11.25	4.70	9.47	4.70	1.81	31.93
Roll 2nd coat									
Slow	250	300	21.50	4.20	1.19	7.17	3.89	2.47	18.92
Medium	305	290	18.80	5.49	1.92	6.48	3.40	1.90	19.19
Fast	360	280	16.10	6.25	2.62	5.75	2.70	1.04	18.36
Roll 3rd or additional coats									
Slow	360	385	21.50	2.92	.83	5.58	2.89	1.83	14.05
Medium	385	375	18.80	4.35	1.52	5.01	2.67	1.49	15.04
Fast	420	365	16.10	5.36	2.23	4.41	2.22	.85	15.07
Solid body stain, oil base (material #19)									
Roll 1st coat									
Slow	150	220	23.90	7.00	1.99	10.86	6.15	3.90	29.90
Medium	170	205	20.90	9.85	3.43	10.20	5.75	3.22	32.45
Fast	200	190	17.90	11.25	4.70	9.42	4.69	1.80	31.86
Roll 2nd coat									
Slow	250	330	23.90	4.20	1.19	7.24	3.92	2.48	19.03
Medium	305	320	20.90	5.49	1.92	6.53	3.41	1.91	19.26
Fast	360	310	17.90	6.25	2.62	5.77	2.71	1.04	18.39
Roll 3rd or additional coats									
Slow	360	405	23.90	2.92	.83	5.90	2.99	1.90	14.54
Medium	385	395	20.90	4.35	1.52	5.29	2.73	1.53	15.42
Fast	420	385	17.90	5.36	2.23	4.65	2.27	.87	15.38

	Labor SF per manhour	Material coverage SF/gallon	Material cost per gallon	Labor cost per 100 SF	Labor burden 100 SF	Material cost per 100 SF	Overhead per 100 SF	Profit per 100 SF	Total price per 100 SF
Waterproofing, clear hydro sealer (material #34)									
Roll 1st coat									
Slow	100	300	17.80	10.50	2.98	5.93	6.02	3.82	29.25
Medium	200	275	15.60	8.38	2.92	5.67	4.16	2.32	23.45
Fast	300	250	13.30	7.50	3.12	5.32	2.95	1.13	20.02
Roll 2nd or additional coats									
Slow	150	350	17.80	7.00	1.99	5.09	4.36	2.77	21.21
Medium	250	325	15.60	6.70	2.34	4.80	3.39	1.89	19.12
Fast	350	300	13.30	6.43	2.69	4.43	2.50	.96	17.01

Use these figures for repaint jobs only. It has been established that asbestos fibers are a known carcinogen (cancer causing) and it is likely that no new construction projects will specify materials or products which contain asbestos. Roofing and siding products usually contain very little asbestos and are typically non-friable (hand pressure can not crumble, pulverize or reduce to powder when dry). There is danger when asbestos is being removed because of exposure to airborne particulate matter. Apparently, there is little danger when painting asbestos roofing or siding, but it is a good idea to have your painters wear respirators or particle masks for their safety. Coverage figures are based on shingles or shakes with average moisture content. See the paragraphs on Roof pitch difficulty factors and Roof area conversion factors to adjust for roof slope and type. "Slow" work is based on a $10.50 hourly wage, "Medium" work on a $16.75 hourly wage, and "Fast" work on a $22.50 hourly wage. Other qualifications that apply to this table are on page 9.

	Labor SF per manhour	Material coverage SF/gallon	Material cost per gallon	Labor cost per 100 SF	Labor burden 100 SF	Material cost per 100 SF	Overhead per 100 SF	Profit per 100 SF	Total price per 100 SF

Roofing, composition shingles, spray application

	Labor SF per manhour	Material coverage SF/gallon	Material cost per gallon	Labor cost per 100 SF	Labor burden 100 SF	Material cost per 100 SF	Overhead per 100 SF	Profit per 100 SF	Total price per 100 SF
Solid body stain, water base (material #18)									
Spray 1st coat									
Slow	325	200	21.50	3.23	.92	10.75	4.62	2.93	22.45
Medium	350	180	18.80	4.79	1.67	10.44	4.14	2.31	23.35
Fast	375	160	16.10	6.00	2.51	10.06	3.44	1.32	23.33
Spray 2nd coat									
Slow	425	290	21.50	2.47	.70	7.41	3.28	2.08	15.94
Medium	450	280	18.80	3.72	1.29	6.71	2.87	1.61	16.20
Fast	475	270	16.10	4.74	1.99	5.96	2.35	.90	15.94
Spray 3rd or additional coats									
Slow	500	365	21.50	2.10	.60	5.89	2.66	1.69	12.94
Medium	538	355	18.80	3.11	1.09	5.30	2.33	1.30	13.13
Fast	575	345	16.10	3.91	1.64	4.67	1.89	.73	12.84
Solid body stain, oil base (material #19)									
Spray 1st coat									
Slow	325	230	23.90	3.23	.92	10.39	4.51	2.86	21.91
Medium	350	205	20.90	4.79	1.67	10.20	4.08	2.28	23.02
Fast	375	180	17.90	6.00	2.51	9.94	3.41	1.31	23.17
Spray 2nd coat									
Slow	425	310	23.90	2.47	.70	7.71	3.37	2.14	16.39
Medium	450	300	20.90	3.72	1.29	6.97	2.94	1.64	16.56
Fast	475	290	17.90	4.74	1.99	6.17	2.38	.92	16.20
Spray 3rd or additional coats									
Slow	500	380	23.90	2.10	.60	6.29	2.79	1.77	13.55
Medium	538	370	20.90	3.11	1.09	5.65	2.41	1.35	13.61
Fast	575	360	17.90	3.91	1.64	4.97	1.94	.75	13.21

	Labor SF per manhour	Material coverage SF/gallon	Material cost per gallon	Labor cost per 100 SF	Labor burden 100 SF	Material cost per 100 SF	Overhead per 100 SF	Profit per 100 SF	Total price per 100 SF
Waterproofing, clear hydro sealer (material #34)									
Spray 1st coat									
Slow	550	100	17.80	1.91	.54	17.80	6.28	3.98	30.51
Medium	600	88	15.60	2.79	.98	17.73	5.27	2.94	29.71
Fast	650	75	13.30	3.46	1.45	17.73	4.19	1.61	28.44
Spray 2nd or additional coats									
Slow	600	150	17.80	1.75	.50	11.87	4.38	2.78	21.28
Medium	650	125	15.60	2.58	.90	12.48	3.91	2.19	22.06
Fast	700	100	13.30	3.21	1.35	13.30	3.30	1.27	22.43

Use these figures for repaint jobs only. It has been established that asbestos fibers are a known carcinogen (cancer causing) and it is likely that no new construction projects will specify materials or products which contain asbestos. Roofing and siding products usually contain very little asbestos and are typically non-friable (hand pressure can not crumble, pulverize or reduce to powder when dry). There is danger when asbestos is being removed because of exposure to airborne particulate matter. Apparently, there is little danger when painting asbestos roofing or siding, but it is a good idea to have your painters wear respirators or particle masks for their safety. Coverage figures are based on shingles or shakes with average moisture content. See the paragraphs on Roof pitch difficulty factors and Roof area conversion factors to adjust for roof slope and type. "Slow" work is based on a $10.50 hourly wage, "Medium" work on a $16.75 hourly wage, and "Fast" work on a $22.50 hourly wage. Other qualifications that apply to this table are on page 9.

	Labor SF per manhour	Material coverage SF/gallon	Material cost per gallon	Labor cost per 100 SF	Labor burden 100 SF	Material cost per 100 SF	Overhead per 100 SF	Profit per 100 SF	Total price per 100 SF
Roofing, wood shingles or shakes, brush application									
Solid body stain, water base (material #18)									
Brush 1st coat									
Slow	100	240	21.50	10.50	2.98	8.96	6.96	4.41	33.81
Medium	155	228	18.80	10.81	3.76	8.25	5.59	3.13	31.54
Fast	210	215	16.10	10.71	4.47	7.49	4.19	1.61	28.47
Brush 2nd or additional coats									
Slow	150	290	21.50	7.00	1.99	7.41	5.08	3.22	24.70
Medium	195	278	18.80	8.59	3.00	6.76	4.49	2.51	25.35
Fast	240	265	16.10	9.38	3.92	6.08	3.59	1.38	24.35
Solid body stain, oil base (material #19)									
Brush 1st coat									
Slow	100	160	23.90	10.50	2.98	14.94	8.81	5.59	42.82
Medium	155	150	20.90	10.81	3.76	13.93	6.98	3.90	39.38
Fast	210	140	17.90	10.71	4.47	12.79	5.17	1.99	35.13
Brush 2nd or additional coats									
Slow	150	260	23.90	7.00	1.99	9.19	5.64	3.57	27.39
Medium	195	250	20.90	8.59	3.00	8.36	4.89	2.73	27.57
Fast	240	240	17.90	9.38	3.92	7.46	3.84	1.48	26.08
Semi-transparent stain, water base (material #20)									
Brush 1st coat									
Slow	120	260	21.40	8.75	2.48	8.23	6.04	3.83	29.33
Medium	175	248	18.70	9.57	3.33	7.54	5.01	2.80	28.25
Fast	230	235	16.00	9.78	4.09	6.81	3.82	1.47	25.97
Brush 2nd or additional coats									
Slow	160	300	21.40	6.56	1.87	7.13	4.82	3.06	23.44
Medium	205	288	18.70	8.17	2.85	6.49	4.29	2.40	24.20
Fast	250	275	16.00	9.00	3.76	5.82	3.44	1.32	23.34
Semi-transparent stain, oil base (material #21)									
Brush 1st coat									
Slow	120	180	26.10	8.75	2.48	14.50	7.98	5.06	38.77
Medium	175	170	22.90	9.57	3.33	13.47	6.46	3.61	36.44
Fast	230	160	19.60	9.78	4.09	12.25	4.83	1.86	32.81
Brush 2nd or additional coats									
Slow	160	280	26.10	6.56	1.87	9.32	5.50	3.49	26.74
Medium	205	270	22.90	8.17	2.85	8.48	4.78	2.67	26.95
Fast	250	260	19.60	9.00	3.76	7.54	3.76	1.44	25.50

	Labor SF per manhour	Material coverage SF/gallon	Material cost per gallon	Labor cost per 100 SF	Labor burden 100 SF	Material cost per 100 SF	Overhead per 100 SF	Profit per 100 SF	Total price per 100 SF
Penetrating oil stain (material #13)									
Brush 1st coat									
Slow	100	160	29.00	10.50	2.98	18.13	9.80	6.21	47.62
Medium	155	150	25.40	10.81	3.76	16.93	7.72	4.32	43.54
Fast	210	140	21.80	10.71	4.47	15.57	5.69	2.19	38.63
Brush 2nd or additional coats									
Slow	150	205	29.00	7.00	1.99	14.15	7.17	4.55	34.86
Medium	195	195	25.40	8.59	3.00	13.03	6.03	3.37	34.02
Fast	240	185	21.80	9.38	3.92	11.78	4.64	1.78	31.50

Coverage figures are based on shingles or shakes with average moisture content. See the paragraphs on Roof pitch difficulty factors and Roof area conversion factors to adjust for roof slope and type. "Slow" work is based on a $10.50 hourly wage, "Medium" work on a $16.75 hourly wage, and "Fast" work on a $22.50 hourly wage. Other qualifications that apply to this table are on page 9.

	Labor SF per manhour	Material coverage SF/gallon	Material cost per gallon	Labor cost per 100 SF	Labor burden 100 SF	Material cost per 100 SF	Overhead per 100 SF	Profit per 100 SF	Total price per 100 SF
Roofing, wood shingles or shakes, roll application									
Solid body stain, water base (material #18)									
Roll 1st coat									
Slow	210	225	21.50	5.00	1.42	9.56	4.95	3.14	24.07
Medium	258	213	18.80	6.49	2.27	8.83	4.31	2.41	24.31
Fast	305	200	16.10	7.38	3.08	8.05	3.42	1.32	23.25
Roll 2nd or additional coats									
Slow	250	275	21.50	4.20	1.19	7.82	4.10	2.60	19.91
Medium	300	263	18.80	5.58	1.94	7.15	3.59	2.01	20.27
Fast	350	250	16.10	6.43	2.69	6.44	2.88	1.11	19.55
Solid body stain, oil base (material #19)									
Roll 1st coat									
Slow	210	150	23.90	5.00	1.42	15.93	6.93	4.39	33.67
Medium	255	140	20.90	6.57	2.29	14.93	5.83	3.26	32.88
Fast	305	130	17.90	7.38	3.08	13.77	4.48	1.72	30.43
Roll 2nd or additional coats									
Slow	250	245	23.90	4.20	1.19	9.76	4.70	2.98	22.83
Medium	300	235	20.90	5.58	1.94	8.89	4.02	2.25	22.68
Fast	350	225	17.90	6.43	2.69	7.96	3.16	1.21	21.45
Semi-transparent stain, water base (material #20)									
Roll 1st coat									
Slow	240	250	21.40	4.38	1.24	8.56	4.40	2.79	21.37
Medium	288	238	18.70	5.82	2.02	7.86	3.85	2.15	21.70
Fast	335	225	16.00	6.72	2.82	7.11	3.08	1.18	20.91
Roll 2nd or additional coats									
Slow	270	290	21.40	3.89	1.10	7.38	3.84	2.43	18.64
Medium	320	278	18.70	5.23	1.84	6.73	3.38	1.89	19.07
Fast	370	255	16.00	6.08	2.53	6.27	2.75	1.06	18.69
Semi-transparent stain, oil base (material #21)									
Roll 1st coat									
Slow	240	175	26.10	4.38	1.24	14.91	6.37	4.04	30.94
Medium	288	165	22.90	5.82	2.02	13.88	5.32	2.98	30.02
Fast	335	155	19.60	6.72	2.82	12.65	4.10	1.58	27.87
Roll 2nd or additional coats									
Slow	270	260	26.10	3.89	1.10	10.04	4.66	2.96	22.65
Medium	320	250	22.90	5.23	1.84	9.16	3.97	2.22	22.42
Fast	370	240	19.60	6.08	2.53	8.17	3.11	1.19	21.08

	Labor SF per manhour	Material coverage SF/gallon	Material cost per gallon	Labor cost per 100 SF	Labor burden 100 SF	Material cost per 100 SF	Overhead per 100 SF	Profit per 100 SF	Total price per 100 SF
Penetrating oil stain (material #13)									
Roll 1st coat									
Slow	210	200	29.00	5.00	1.42	14.50	6.49	4.11	31.52
Medium	258	190	25.40	6.49	2.27	13.37	5.42	3.03	30.58
Fast	305	180	21.80	7.38	3.08	12.11	4.18	1.61	28.36
Roll 2nd or additional coats									
Slow	250	295	29.00	4.20	1.19	9.83	4.72	2.99	22.93
Medium	300	285	25.40	5.58	1.94	8.91	4.03	2.25	22.71
Fast	350	275	21.80	6.43	2.69	7.93	3.15	1.21	21.41

Coverage figures are based on shingles or shakes with average moisture content. See the paragraphs on Roof pitch difficulty factors and Roof area conversion factors to adjust for roof slope and type. "Slow" work is based on a $10.50 hourly wage, "Medium" work on a $16.75 hourly wage, and "Fast" work on a $22.50 hourly wage. Other qualifications that apply to this table are on page 9.

	Labor SF per manhour	Material coverage SF/gallon	Material cost per gallon	Labor cost per 100 SF	Labor burden 100 SF	Material cost per 100 SF	Overhead per 100 SF	Profit per 100 SF	Total price per 100 SF

Roofing, wood shingles or shakes, spray application

	Labor SF per manhour	Material coverage SF/gallon	Material cost per gallon	Labor cost per 100 SF	Labor burden 100 SF	Material cost per 100 SF	Overhead per 100 SF	Profit per 100 SF	Total price per 100 SF
Solid body stain, water base (material #18)									
Spray 1st coat									
Slow	600	230	21.50	1.75	.50	9.35	3.60	2.28	17.48
Medium	700	220	18.80	2.39	.84	8.55	2.88	1.61	16.27
Fast	800	200	16.10	2.81	1.18	8.05	2.23	.86	15.13
Spray 2nd or additional coats									
Slow	700	250	21.50	1.50	.43	8.60	3.26	2.07	15.86
Medium	800	235	18.80	2.09	.73	8.00	2.65	1.48	14.95
Fast	900	220	16.10	2.50	1.04	7.32	2.01	.77	13.64
Solid body stain, oil base (material #19)									
Spray 1st coat									
Slow	600	170	23.90	1.75	.50	14.06	5.06	3.21	24.58
Medium	700	150	20.90	2.39	.84	13.93	4.20	2.35	23.71
Fast	800	130	17.90	2.81	1.18	13.77	3.28	1.26	22.30
Spray 2nd or additional coats									
Slow	700	230	23.90	1.50	.43	10.39	3.82	2.42	18.56
Medium	800	215	20.90	2.09	.73	9.72	3.07	1.72	17.33
Fast	900	200	17.90	2.50	1.04	8.95	2.31	.89	15.69
Semi-transparent stain, water base (material #20)									
Spray 1st coat									
Slow	650	265	21.40	1.62	.46	8.08	3.15	2.00	15.31
Medium	750	253	18.70	2.23	.77	7.39	2.55	1.42	14.36
Fast	850	240	16.00	2.65	1.11	6.67	1.93	.74	13.10
Spray 2nd or additional coats									
Slow	750	305	21.40	1.40	.39	7.02	2.73	1.73	13.27
Medium	850	293	18.70	1.97	.70	6.38	2.21	1.24	12.50
Fast	950	270	16.00	2.37	.98	5.93	1.72	.66	11.66
Semi-transparent stain, oil base (material #21)									
Spray 1st coat									
Slow	650	190	26.10	1.62	.46	13.74	4.90	3.11	23.83
Medium	750	180	22.90	2.23	.77	12.72	3.85	2.15	21.72
Fast	850	170	19.60	2.65	1.11	11.53	2.83	1.09	19.21
Spray 2nd or additional coats									
Slow	750	275	26.10	1.40	.39	9.49	3.50	2.22	17.00
Medium	850	265	22.90	1.97	.70	8.64	2.77	1.55	15.63
Fast	950	255	19.60	2.37	.98	7.69	2.04	.79	13.87

	Labor SF per manhour	Material coverage SF/gallon	Material cost per gallon	Labor cost per 100 SF	Labor burden 100 SF	Material cost per 100 SF	Overhead per 100 SF	Profit per 100 SF	Total price per 100 SF
Penetrating oil stain (material #13)									
Spray 1st coat									
Slow	600	200	29.00	1.75	.50	14.50	5.19	3.29	25.23
Medium	700	170	25.40	2.39	.84	14.94	4.45	2.49	25.11
Fast	800	140	21.80	2.81	1.18	15.57	3.62	1.39	24.57
Spray 2nd or additional coats									
Slow	700	300	29.00	1.50	.43	9.67	3.60	2.28	17.48
Medium	800	260	25.40	2.09	.73	9.77	3.08	1.72	17.39
Fast	900	240	21.80	2.50	1.04	9.08	2.33	.90	15.85
Waterproofing, clear hydro sealer (material #34)									
Spray 1st coat									
Slow	450	75	17.80	2.33	.66	23.73	8.28	5.25	40.25
Medium	475	63	15.60	3.53	1.24	24.76	7.23	4.04	40.80
Fast	500	50	13.30	4.50	1.88	26.60	6.10	2.34	41.42
Spray 2nd or additional coats									
Slow	500	150	17.80	2.10	.60	11.87	4.52	2.86	21.95
Medium	525	138	15.60	3.19	1.10	11.30	3.82	2.14	21.55
Fast	550	125	13.30	4.09	1.71	10.64	3.04	1.17	20.65

Coverage figures are based on shingles or shakes with average moisture content. See the paragraphs on Roof pitch difficulty factors and Roof area conversion factors to adjust for roof slope and type. "Slow" work is based on a $10.50 hourly wage, "Medium" work on a $16.75 hourly wage, and "Fast" work on a $22.50 hourly wage. Other qualifications that apply to this table are on page 9.

	Labor LF per manhour	Material coverage LF/gallon	Material cost per gallon	Labor cost per 100 LF	Labor burden 100 LF	Material cost per 100 LF	Overhead per 100 LF	Profit per 100 LF	Total price per 100 LF
Sheet metal cap or flashing									
3" to 8" wide:									
Metal primer, clean metal (material #35)									
Brush prime coat									
Slow	215	525	29.90	4.88	1.39	5.70	3.71	2.35	18.03
Medium	228	500	26.20	7.35	2.57	5.24	3.71	2.07	20.94
Fast	240	475	22.40	9.38	3.92	4.72	3.33	1.28	22.63
Metal primer, rusty metal (material #36)									
Brush prime coat									
Slow	215	525	36.60	4.88	1.39	6.97	4.10	2.60	19.94
Medium	228	500	32.10	7.35	2.57	6.42	4.00	2.24	22.58
Fast	240	475	27.50	9.38	3.92	5.79	3.53	1.36	23.98
Metal finish, synthetic enamel - off white, gloss, interior or exterior (material #37)									
Brush 1st or additional finish coats									
Slow	275	550	27.90	3.82	1.09	5.07	3.09	1.96	15.03
Medium	288	525	24.40	5.82	2.02	4.65	3.06	1.71	17.26
Fast	300	500	20.90	7.50	3.12	4.18	2.74	1.05	18.59
Metal finish, synthetic enamel, colors (except orange/red), gloss, interior or exterior (material #38)									
Brush 1st or additional finish coats									
Slow	275	550	30.00	3.82	1.09	5.45	3.21	2.04	15.61
Medium	288	525	26.30	5.82	2.02	5.01	3.15	1.76	17.76
Fast	300	500	22.50	7.50	3.12	4.50	2.80	1.08	19.00
8" to 12" wide:									
Metal primer, clean metal (material #35)									
Brush prime coat									
Slow	200	500	29.90	5.25	1.49	5.98	3.94	2.50	19.16
Medium	215	475	26.20	7.79	2.71	5.52	3.92	2.19	22.13
Fast	230	450	22.40	9.78	4.09	4.98	3.49	1.34	23.68
Metal primer, rusty metal (material #36)									
Brush prime coat									
Slow	200	500	36.60	5.25	1.49	7.32	4.36	2.76	21.18
Medium	215	475	32.10	7.79	2.71	6.76	4.23	2.36	23.85
Fast	230	450	27.50	9.78	4.09	6.11	3.69	1.42	25.09

	Labor LF per manhour	Material coverage LF/gallon	Material cost per gallon	Labor cost per 100 LF	Labor burden 100 LF	Material cost per 100 LF	Overhead per 100 LF	Profit per 100 LF	Total price per 100 LF
Metal finish, synthetic enamel - off white, gloss, interior or exterior (material #37)									
Brush 1st or additional finish coats									
Slow	250	525	27.90	4.20	1.19	5.31	3.32	2.10	16.12
Medium	265	500	24.40	6.32	2.20	4.88	3.28	1.83	18.51
Fast	280	475	20.90	8.04	3.34	4.40	2.92	1.12	19.82
Metal finish, synthetic enamel, colors (except orange/red), gloss, interior or exterior (material #38)									
Brush 1st or additional finish coats									
Slow	250	525	30.00	4.20	1.19	5.71	3.44	2.18	16.72
Medium	265	500	26.30	6.32	2.20	5.26	3.38	1.89	19.05
Fast	280	475	22.50	8.04	3.34	4.74	2.99	1.15	20.26

A two coat system, prime and finish, using oil base material is recommended for any metal surface. Although water base material is often used, it may cause oxidation, corrosion and rust. One coat of oil base solid body stain is often used on exterior metal but it may crack, peel or chip without the proper prime coat application. "Slow" work is based on a $10.50 hourly wage, "Medium" work on a $16.75 hourly wage, and "Fast" work on a $22.50 hourly wage. Other qualifications that apply to this table are on page 9.

	Labor LF per manhour	Material coverage LF/gallon	Material cost per gallon	Labor cost per 100 LF	Labor burden 100 LF	Material cost per 100 LF	Overhead per 100 LF	Profit per 100 LF	Total price per 100 LF

Sheet metal, diverters or gravel stop

Up to 3" wide:

	Labor LF per manhour	Material coverage LF/gallon	Material cost per gallon	Labor cost per 100 LF	Labor burden 100 LF	Material cost per 100 LF	Overhead per 100 LF	Profit per 100 LF	Total price per 100 LF
Metal primer, clean metal (material #35)									
Brush prime coat									
Slow	230	600	29.90	4.57	1.29	4.98	3.36	2.13	16.33
Medium	240	550	26.20	6.98	2.44	4.76	3.47	1.94	19.59
Fast	250	500	22.40	9.00	3.76	4.48	3.19	1.23	21.66
Metal primer, rusty metal (material #36)									
Brush prime coat									
Slow	230	600	36.60	4.57	1.29	6.10	3.71	2.35	18.02
Medium	240	550	32.10	6.98	2.44	5.84	3.74	2.09	21.09
Fast	250	500	27.50	9.00	3.76	5.50	3.38	1.30	22.94
Metal finish, synthetic enamel - off white, gloss, interior or exterior (material #37)									
Brush 1st or additional finish coats									
Slow	300	650	27.90	3.50	.99	4.29	2.72	1.73	13.23
Medium	313	600	24.40	5.35	1.86	4.07	2.76	1.54	15.58
Fast	325	550	20.90	6.92	2.90	3.80	2.52	.97	17.11
Metal finish, synthetic enamel, colors (except orange/red), gloss, interior or exterior (material #38)									
Brush 1st or additional finish coats									
Slow	300	650	30.00	3.50	.99	4.62	2.83	1.79	13.73
Medium	313	600	26.30	5.35	1.86	4.38	2.84	1.59	16.02
Fast	325	550	22.50	6.92	2.90	4.09	2.57	.99	17.47

A two coat system, prime and finish, using oil base material is recommended for any metal surface. Although water base material is often used, it may cause oxidation, corrosion and rust. One coat of oil base solid body stain is often used on exterior metal but it may crack, peel or chip without the proper prime coat application. "Slow" work is based on a $10.50 hourly wage, "Medium" work on a $16.75 hourly wage, and "Fast" work on a $22.50 hourly wage. Other qualifications that apply to this table are on page 9.

	Manhours per home	Material gallons per home	Material cost per gallon	Labor cost per home	Labor burden per home	Material cost per home	Overhead per home	Profit per home	Total price per home
Sheet metal, vents & flashing									
Metal primer, clean metal (material #35)									
1 story home, brush prime coat									
Slow	0.7	0.40	29.90	7.35	2.09	11.96	6.63	4.20	32.23
Medium	0.6	0.45	26.20	10.05	3.50	11.79	6.21	3.47	35.02
Fast	0.5	0.50	22.40	11.25	4.70	11.20	5.02	1.93	34.10
2 story home, brush prime coat									
Slow	0.9	0.50	29.90	9.45	2.68	14.95	8.40	5.32	40.80
Medium	0.8	0.55	26.20	13.40	4.67	14.41	7.96	4.45	44.89
Fast	0.7	0.60	22.40	15.75	6.57	13.44	6.62	2.54	44.92
Attached units (per unit), brush 1st or additional finish coats									
Slow	0.6	0.40	29.90	6.30	1.79	11.96	6.22	3.94	30.21
Medium	0.5	0.45	26.20	8.38	2.92	11.79	5.66	3.16	31.91
Fast	0.4	0.50	22.40	9.00	3.76	11.20	4.43	1.70	30.09
Metal primer, rusty metal (material #36)									
1 story home, brush prime coat									
Slow	0.7	0.40	36.60	7.35	2.09	14.64	7.46	4.73	36.27
Medium	0.6	0.45	32.10	10.05	3.50	14.45	6.86	3.83	38.69
Fast	0.5	0.50	27.50	11.25	4.70	13.75	5.49	2.11	37.30
2 story home, brush prime coat									
Slow	0.9	0.50	36.60	9.45	2.68	18.30	9.44	5.98	45.85
Medium	0.8	0.55	32.10	13.40	4.67	17.66	8.75	4.89	49.37
Fast	0.7	0.60	27.50	15.75	6.57	16.50	7.18	2.76	48.76
Attached units (per unit), brush 1st or additional finish coats									
Slow	0.6	0.40	36.60	6.30	1.79	14.64	7.05	4.47	34.25
Medium	0.5	0.45	32.10	8.38	2.92	14.45	6.31	3.53	35.59
Fast	0.4	0.50	27.50	9.00	3.76	13.75	4.90	1.88	33.29
Metal finish, synthetic enamel - off white, gloss, interior or exterior (material #37)									
1 story home, brush 1st or additional finish coats									
Slow	0.7	0.40	27.90	7.35	2.09	11.16	6.39	4.05	31.04
Medium	0.6	0.45	24.40	10.05	3.50	10.98	6.01	3.36	33.90
Fast	0.5	0.50	20.90	11.25	4.70	10.45	4.88	1.88	33.16
2 story home, brush 1st or additional finish coats									
Slow	0.9	0.50	27.90	9.45	2.68	13.95	8.09	5.13	39.30
Medium	0.8	0.55	24.40	13.40	4.67	13.42	7.72	4.31	43.52
Fast	0.7	0.60	20.90	15.75	6.57	12.54	6.45	2.48	43.79

	Manhours per home	Material gallons per home	Material cost per gallon	Labor cost per home	Labor burden per home	Material cost per home	Overhead per home	Profit per home	Total price per home
Attached units (per unit), brush 1st or additional finish coats									
Slow	0.6	0.40	27.90	6.30	1.79	11.16	5.97	3.78	29.00
Medium	0.5	0.45	24.40	8.38	2.92	10.98	5.46	3.05	30.79
Fast	0.4	0.50	20.90	9.00	3.76	10.45	4.29	1.65	29.15
Metal finish, synthetic enamel - colors (except orange/red), gloss, interior or exterior (material #38)									
1 story home, brush 1st or additional finish coats									
Slow	0.7	0.40	30.00	7.35	2.09	12.00	6.65	4.21	32.30
Medium	0.6	0.45	26.30	10.05	3.50	11.84	6.22	3.48	35.09
Fast	0.5	0.50	22.50	11.25	4.70	11.25	5.03	1.93	34.16
2 story home, brush 1st or additional finish coats									
Slow	0.9	0.50	30.00	9.45	2.68	15.00	8.41	5.33	40.87
Medium	0.8	0.55	26.30	13.40	4.67	14.47	7.97	4.46	44.97
Fast	0.7	0.60	22.50	15.75	6.57	13.50	6.63	2.55	45.00
Attached units (per unit), brush 1st or additional finish coats									
Slow	0.6	0.40	30.00	6.30	1.79	12.00	6.23	3.95	30.27
Medium	0.5	0.45	26.30	8.38	2.92	11.84	5.67	3.17	31.98
Fast	0.4	0.50	22.50	9.00	3.76	11.25	4.44	1.71	30.16

Use this table for oil base paint only. This table shows the time needed to paint sheet metal vents on an average home or attached dwelling unit. Use it to estimate the costs for residential units without having to take off each vent or piece of flashing. Use the "attached units" section based on 900 square feet of roof area to estimate commercial buildings. Note: A two coat system, prime and finish, using oil base material is recommended for any metal surface. Although water base material is often used, it may cause oxidation, corrosion and rust. One coat of oil base solid body stain is often used on exterior metal but it may crack, peel or chip without the proper prime coat application. "Slow" work is based on a $10.50 hourly wage, "Medium" work on a $16.75 hourly wage, and "Fast" work on a $22.50 hourly wage. Other qualifications that apply to this table are on page 9.

	Shutters per manhour	Shutters per gallon	Material cost per gallon	Labor cost per shutter	Labor burden per shutter	Material cost per shutter	Overhead per shutter	Profit per shutter	Total price per shutter
Shutters or blinds, 2' x 4' average size									
Brush each coat									
Undercoat, water or oil base (material #3 or #4)									
Slow	2.5	12	21.35	4.20	1.19	1.78	2.22	1.41	10.80
Medium	3.0	11	18.70	5.58	1.94	1.70	2.26	1.26	12.74
Fast	3.5	10	16.00	6.43	2.69	1.60	1.98	.76	13.46
Split coat (1/2 undercoat + 1/2 enamel), water or oil base (material #3 and #24 or #4 and #25)									
Slow	3.0	15	25.18	3.50	.99	1.68	1.92	1.22	9.31
Medium	3.5	14	22.03	4.79	1.67	1.57	1.97	1.10	11.10
Fast	4.0	13	18.88	5.63	2.34	1.45	1.74	.67	11.83
Exterior enamel, water or oil base (material #24 or #25)									
Slow	2.0	15	29.00	5.25	1.49	1.93	2.69	1.70	13.06
Medium	2.5	14	25.35	6.70	2.34	1.81	2.66	1.49	15.00
Fast	3.0	13	21.75	7.50	3.12	1.67	2.28	.87	15.44
Spray each coat									
Undercoat, water or oil base (material #3 or #4)									
Slow	8	10	21.35	1.31	.38	2.14	1.18	.75	5.76
Medium	9	9	18.70	1.86	.65	2.08	1.12	.63	6.34
Fast	10	8	16.00	2.25	.94	2.00	.96	.37	6.52
Split coat (1/2 undercoat + 1/2 enamel), water or oil base (material #3 and #24 or #4 and #25)									
Slow	8	12	25.18	1.31	.38	2.10	1.17	.74	5.70
Medium	10	11	22.03	1.68	.58	2.00	1.05	.59	5.90
Fast	12	10	18.88	1.88	.77	1.89	.84	.32	5.70
Exterior enamel, water or oil base (material #24 or #25)									
Slow	7	12	29.00	1.50	.43	2.42	1.35	.86	6.56
Medium	8	11	25.35	2.09	.73	2.30	1.25	.70	7.07
Fast	9	10	21.75	2.50	1.04	2.18	1.06	.41	7.19

Use these figures to estimate the costs to paint all six sides of solid face, paint grade, interior or exterior shutters or blinds. Costs are based on the number of single-panel 2' x 4' false (solid) shutters or blinds that can be painted in one hour. For real louvered shutters, multiply the quantity of shutters by a difficulty factor of 1.5 and use this table. "Slow" work is based on a $10.50 hourly wage, "Medium" work on a $16.75 hourly wage, and "Fast" work on a $22.50 hourly wage. Other qualifications that apply to this table are on page 9.

	Labor SF per manhour	Material coverage SF/gallon	Material cost per gallon	Labor cost per 100 SF	Labor burden 100 SF	Material cost per 100 SF	Overhead per 100 SF	Profit per 100 SF	Total price per 100 SF
Siding, aluminum									
Metal primer, clean metal (material #35)									
Brush prime coat									
Slow	215	440	29.90	4.88	1.39	6.80	4.05	2.57	19.69
Medium	235	420	26.20	7.13	2.49	6.24	3.88	2.17	21.91
Fast	255	400	22.40	8.82	3.68	5.60	3.35	1.29	22.74
Metal primer, rusty metal (material #36)									
Brush prime coat									
Slow	215	440	36.60	4.88	1.39	8.32	4.52	2.87	21.98
Medium	235	420	32.10	7.13	2.49	7.64	4.23	2.36	23.85
Fast	255	400	27.50	8.82	3.68	6.88	3.59	1.38	24.35
Metal finish, synthetic enamel - off white, gloss, interior or exterior (material #37)									
Brush 1st or additional finish coats									
Slow	265	480	27.90	3.96	1.12	5.81	3.38	2.14	16.41
Medium	285	465	24.40	5.88	2.05	5.25	3.23	1.81	18.22
Fast	305	450	20.90	7.38	3.08	4.64	2.79	1.07	18.96
Metal finish, synthetic enamel - colors (except orange/red), gloss, interior or exterior (material #38)									
Brush 1st or additional finish coats									
Slow	265	480	30.00	3.96	1.12	6.25	3.52	2.23	17.08
Medium	285	465	26.30	5.88	2.05	5.66	3.33	1.86	18.78
Fast	305	450	22.50	7.38	3.08	5.00	2.86	1.10	19.42

Don't deduct for openings under 100 square feet. For heights above 8 feet, use the High Time Difficulty Factors on page 137. Note: A two coat system prime and finish, using oil base material is recommended for any metal surface. Although water base material is often used, it may cause oxidation, corrosion and rust. One coat of oil base solid body stain is often used on exterior metal but it may crack, peel or chip without the proper prime coat application. "Slow" work is based on a $10.50 hourly wage, "Medium" work on a $16.75 hourly wage, and "Fast" work on a $22.50 hourly wage. Other qualifications that apply to this table are on page 9.

	Labor SF per manhour	Material coverage SF/gallon	Material cost per gallon	Labor cost per 100 SF	Labor burden 100 SF	Material cost per 100 SF	Overhead per 100 SF	Profit per 100 SF	Total price per 100 SF
Siding, composition shingle, brush application									
Solid body stain, water base (material #18)									
Brush 1st coat									
Slow	65	240	21.50	16.15	4.58	8.96	9.21	5.84	44.74
Medium	85	220	18.80	19.71	6.86	8.55	8.61	4.81	48.54
Fast	105	200	16.10	21.43	8.93	8.05	7.11	2.73	48.25
Brush 2nd coat									
Slow	90	350	21.50	11.67	3.31	6.14	6.55	4.15	31.82
Medium	110	335	18.80	15.23	5.30	5.61	6.41	3.58	36.13
Fast	130	320	16.10	17.31	7.21	5.03	5.47	2.10	37.12
Brush 3rd or additional coats									
Slow	110	425	21.50	9.55	2.70	5.06	5.37	3.41	26.09
Medium	130	405	18.80	12.88	4.49	4.64	5.39	3.01	30.41
Fast	150	385	16.10	15.00	6.27	4.18	4.71	1.81	31.97
Solid body stain, oil base (material #19)									
Brush 1st coat									
Slow	65	290	23.90	16.15	4.58	8.24	8.98	5.69	43.64
Medium	85	270	20.90	19.71	6.86	7.74	8.41	4.70	47.42
Fast	105	250	17.90	21.43	8.93	7.16	6.94	2.67	47.13
Brush 2nd coat									
Slow	90	390	23.90	11.67	3.31	6.13	6.55	4.15	31.81
Medium	110	375	20.90	15.23	5.30	5.57	6.40	3.58	36.08
Fast	130	360	17.90	17.31	7.21	4.97	5.46	2.10	37.05
Brush 3rd or additional coats									
Slow	110	445	23.90	9.55	2.70	5.37	5.47	3.47	26.56
Medium	130	425	20.90	12.88	4.49	4.92	5.46	3.05	30.80
Fast	150	405	17.90	15.00	6.27	4.42	4.75	1.83	32.27

Use these figures for repaint jobs only. Don't deduct for openings under 100 square feet. For heights above 8 feet, use the High Time Difficulty Factors on page 137. It has been established that asbestos fibers are known carcinogens (cancer causing) and it is likely that materials or products which contain asbestos will not be specified on any new construction projects in the future. Siding and roofing products usually contain very little asbestos and are typically non-friable (hand pressure can not crumble, pulverize or reduce to powder when dry). There is danger when asbestos is being removed because of exposure to airborne particulate matter. Apparently, there is little danger when painting asbestos siding or roofing, but it is a good idea to have your painters wear respirators or particle masks for their safety. "Slow" work is based on on a $10.50 hourly wage, "Medium" work a $16.75 hourly wage, and "Fast" work on a $22.50 hourly wage. Other qualifications that apply to this table are on page 9.

	Labor SF per manhour	Material coverage SF/gallon	Material cost per gallon	Labor cost per 100 SF	Labor burden 100 SF	Material cost per 100 SF	Overhead per 100 SF	Profit per 100 SF	Total price per 100 SF

Siding, composition shingle, roll application

	Labor SF per manhour	Material coverage SF/gallon	Material cost per gallon	Labor cost per 100 SF	Labor burden 100 SF	Material cost per 100 SF	Overhead per 100 SF	Profit per 100 SF	Total price per 100 SF
Solid body stain, water base (material #18)									
Roll 1st coat									
Slow	140	230	21.50	7.50	2.12	9.35	5.88	3.73	28.58
Medium	160	215	18.80	10.47	3.65	8.74	5.60	3.13	31.59
Fast	180	200	16.10	12.50	5.23	8.05	4.77	1.83	32.38
Roll 2nd coat									
Slow	190	330	21.50	5.53	1.56	6.52	4.22	2.68	20.51
Medium	210	315	18.80	7.98	2.77	5.97	4.10	2.29	23.11
Fast	230	300	16.10	9.78	4.09	5.37	3.56	1.37	24.17
Roll 3rd or additional coats									
Slow	250	395	21.50	4.20	1.19	5.44	3.36	2.13	16.32
Medium	280	385	18.80	5.98	2.08	4.88	3.17	1.77	17.88
Fast	300	375	16.10	7.50	3.12	4.29	2.76	1.06	18.73
Solid body stain, oil base (material #19)									
Roll 1st coat									
Slow	140	260	23.90	7.50	2.12	9.19	5.83	3.70	28.34
Medium	160	240	20.90	10.47	3.65	8.71	5.59	3.13	31.55
Fast	180	220	17.90	12.50	5.23	8.14	4.78	1.84	32.49
Roll 2nd coat									
Slow	190	360	23.90	5.53	1.56	6.64	4.26	2.70	20.69
Medium	210	345	20.90	7.98	2.77	6.06	4.12	2.30	23.23
Fast	230	330	17.90	9.78	4.09	5.42	3.57	1.37	24.23
Roll 3rd or additional coats									
Slow	250	420	23.90	4.20	1.19	5.69	3.43	2.18	16.69
Medium	280	408	20.90	5.98	2.08	5.12	3.23	1.81	18.22
Fast	300	395	17.90	7.50	3.12	4.53	2.80	1.08	19.03

	Labor SF per manhour	Material coverage SF/gallon	Material cost per gallon	Labor cost per 100 SF	Labor burden 100 SF	Material cost per 100 SF	Overhead per 100 SF	Profit per 100 SF	Total price per 100 SF
Waterproofing, clear hydro seal, oil base (material #34)									
Roll 1st coat									
Slow	75	275	17.80	14.00	3.97	6.47	7.58	4.80	36.82
Medium	150	250	15.60	11.17	3.90	6.24	5.22	2.92	29.45
Fast	225	225	13.30	10.00	4.16	5.91	3.72	1.43	25.22
Roll 2nd or additional coats									
Slow	125	325	17.80	8.40	2.38	5.48	5.04	3.20	24.50
Medium	200	300	15.60	8.38	2.92	5.20	4.04	2.26	22.80
Fast	275	275	13.30	8.18	3.43	4.84	3.04	1.17	20.66

Use these figures for repaint jobs only. Don't deduct for openings under 100 square feet. For heights above 8 feet, use the High Time Difficulty Factors on page 137. It has been established that asbestos fibers are known carcinogens (cancer causing) and it is likely that materials or products which contain asbestos will not be specified on any new construction projects in the future. Siding and roofing products usually contain very little asbestos and are typically non-friable (hand pressure can not crumble, pulverize or reduce to powder when dry). There is danger when asbestos is being removed because of exposure to airborne particulate matter. Apparently, there is little danger when painting asbestos siding or roofing, but it is a good idea to have your painters wear respirators or particle masks for their safety. "Slow" work is basedon on a $10.50 hourly wage, "Medium" work a $16.75 hourly wage, and "Fast" work on a $22.50 hourly wage. Other qualifications that apply to this table are on page 9.

	Labor SF per manhour	Material coverage SF/gallon	Material cost per gallon	Labor cost per 100 SF	Labor burden 100 SF	Material cost per 100 SF	Overhead per 100 SF	Profit per 100 SF	Total price per 100 SF

Siding, composition shingle, spray application

Solid body stain, water base (material #18)									
Spray 1st coat									
Slow	325	160	21.50	3.23	.92	13.44	5.45	3.46	26.50
Medium	350	150	18.80	4.79	1.67	12.53	4.65	2.60	26.24
Fast	375	140	16.10	6.00	2.51	11.50	3.70	1.42	25.13
Spray 2nd coat									
Slow	350	240	21.50	3.00	.86	8.96	3.97	2.52	19.31
Medium	375	228	18.80	4.47	1.56	8.25	3.50	1.96	19.74
Fast	400	215	16.10	5.63	2.34	7.49	2.86	1.10	19.42
Spray 3rd or additional coats									
Slow	425	300	21.50	2.47	.70	7.17	3.21	2.03	15.58
Medium	450	288	18.80	3.72	1.29	6.53	2.83	1.58	15.95
Fast	475	275	16.10	4.74	1.99	5.85	2.33	.89	15.80
Solid body stain, oil base (material #19)									
Spray 1st coat									
Slow	325	190	23.90	3.23	.92	12.58	5.19	3.29	25.21
Medium	350	180	20.90	4.79	1.67	11.61	4.43	2.48	24.98
Fast	375	170	17.90	6.00	2.51	10.53	3.52	1.35	23.91
Spray 2nd coat									
Slow	350	270	23.90	3.00	.86	8.85	3.94	2.50	19.15
Medium	375	258	20.90	4.47	1.56	8.10	3.46	1.93	19.52
Fast	400	245	17.90	5.63	2.34	7.31	2.83	1.09	19.20
Spray 3rd or additional coats									
Slow	425	320	23.90	2.47	.70	7.47	3.30	2.09	16.03
Medium	450	308	20.90	3.72	1.29	6.79	2.89	1.62	16.31
Fast	475	295	17.90	4.74	1.99	6.07	2.37	.91	16.08

	Labor SF per manhour	Material coverage SF/gallon	Material cost per gallon	Labor cost per 100 SF	Labor burden 100 SF	Material cost per 100 SF	Overhead per 100 SF	Profit per 100 SF	Total price per 100 SF
Waterproofing, clear hydro seal, oil base (material #34)									
Spray 1st coat									
Slow	475	100	17.80	2.21	.63	17.80	6.40	4.06	31.10
Medium	525	88	15.60	3.19	1.10	17.73	5.40	3.02	30.44
Fast	575	75	13.30	3.91	1.64	17.73	4.30	1.65	29.23
Spray 2nd or additional coats									
Slow	500	150	17.80	2.10	.60	11.87	4.52	2.86	21.95
Medium	550	125	15.60	3.05	1.06	12.48	4.06	2.27	22.92
Fast	600	100	13.30	3.75	1.58	13.30	3.44	1.32	23.39

Use these figures for repaint jobs only. Don't deduct for openings under 100 square feet. For heights above 8 feet, use the High Time Difficulty Factors on page 137. It has been established that asbestos fibers are known carcinogens (cancer causing) and it is likely that materials or products which contain asbestos will not be specified on any new construction projects in the future. Siding and roofing products usually contain very little asbestos and are typically non-friable (hand pressure can not crumble, pulverize or reduce to powder when dry). There is danger when asbestos is being removed because of exposure to airborne particulate matter. Apparently, there is little danger when painting asbestos siding or roofing, but it is a good idea to have your painters wear respirators or particle masks for their safety. "Slow" work is basedon on a $10.50 hourly wage, "Medium" work a $16.75 hourly wage, and "Fast" work on a $22.50 hourly wage. Other qualifications that apply to this table are on page 9.

	Labor SF per manhour	Material coverage SF/gallon	Material cost per gallon	Labor cost per 100 SF	Labor burden 100 SF	Material cost per 100 SF	Overhead per 100 SF	Profit per 100 SF	Total price per 100 SF
Siding, rough sawn or resawn wood, brush									
Solid body or semi-transparent stain, water base (material #18 or #20)									
Brush 1st coat									
Slow	100	250	21.45	10.50	2.98	8.58	6.84	4.34	33.24
Medium	135	238	18.75	12.41	4.33	7.88	6.03	3.37	34.02
Fast	170	225	16.05	13.24	5.51	7.13	4.79	1.84	32.51
Brush 2nd coat									
Slow	135	300	21.45	7.78	2.21	7.15	5.31	3.37	25.82
Medium	168	288	18.75	9.97	3.47	6.51	4.89	2.73	27.57
Fast	200	275	16.05	11.25	4.70	5.84	4.03	1.55	27.37
Brush 3rd or additional coats									
Slow	150	335	21.45	7.00	1.99	6.40	4.77	3.02	23.18
Medium	183	323	18.75	9.15	3.18	5.80	4.44	2.48	25.05
Fast	215	310	16.05	10.47	4.36	5.18	3.70	1.42	25.13
Solid body or semi-transparent stain, oil base (material #19 or #21)									
Brush 1st coat									
Slow	100	275	25.00	10.50	2.98	9.09	7.00	4.44	34.01
Medium	135	250	21.90	12.41	4.33	8.76	6.25	3.49	35.24
Fast	170	225	18.75	13.24	5.51	8.33	5.01	1.93	34.02
Brush 2nd coat									
Slow	135	350	25.00	7.78	2.21	7.14	5.31	3.37	25.81
Medium	168	325	21.90	9.97	3.47	6.74	4.94	2.76	27.88
Fast	200	300	18.75	11.25	4.70	6.25	4.11	1.58	27.89
Brush 3rd or additional coats									
Slow	150	400	25.00	7.00	1.99	6.25	4.72	2.99	22.95
Medium	183	375	21.90	9.15	3.18	5.84	4.45	2.49	25.11
Fast	215	350	18.75	10.47	4.36	5.36	3.74	1.44	25.37
Penetrating oil stain (material #13)									
Brush 1st coat									
Slow	100	230	29.00	10.50	2.98	12.61	8.09	5.13	39.31
Medium	135	210	25.40	12.41	4.33	12.10	7.06	3.95	39.85
Fast	170	190	21.80	13.24	5.51	11.47	5.59	2.15	37.96

	Labor SF per manhour	Material coverage SF/gallon	Material cost per gallon	Labor cost per 100 SF	Labor burden 100 SF	Material cost per 100 SF	Overhead per 100 SF	Profit per 100 SF	Total price per 100 SF
Brush 2nd coat									
Slow	135	275	29.00	7.78	2.21	10.55	6.37	4.04	30.95
Medium	168	255	25.40	9.97	3.47	9.96	5.73	3.20	32.33
Fast	200	235	21.80	11.25	4.70	9.28	4.67	1.79	31.69
Brush 3rd or additional coats									
Slow	150	350	29.00	7.00	1.99	8.29	5.36	3.40	26.04
Medium	183	330	25.40	9.15	3.18	7.70	4.91	2.74	27.68
Fast	215	310	21.80	10.47	4.36	7.03	4.05	1.56	27.47

Use this table to estimate the cost of painting shingle, shake, resawn or rough sawn wood or plywood siding with average moisture content. Don't deduct for openings under 100 square feet. For heights above 8 feet, use the High Time Difficulty Factors on page 137. For wood or composition drop siding with exposed bevel edges, multiply the surface area by 1.12 to allow for the extra time and material needed to paint the underside of each board. "Slow" work is based on a $10.50 hourly wage, "Medium" work on a $16.75 hourly wage, and "Fast" work on a $22.50 hourly wage. Other qualifications that apply to this table are on page 9.

	Labor SF per manhour	Material coverage SF/gallon	Material cost per gallon	Labor cost per 100 SF	Labor burden 100 SF	Material cost per 100 SF	Overhead per 100 SF	Profit per 100 SF	Total price per 100 SF
Siding, rough sawn or resawn wood, roll									
Solid body or semi-transparent stain, water base (material #18 or #20)									
Roll 1st coat									
Slow	150	235	21.45	7.00	1.99	9.13	5.62	3.56	27.30
Medium	225	213	18.75	7.44	2.59	8.80	4.61	2.58	26.02
Fast	275	210	16.05	8.18	3.43	7.64	3.56	1.37	24.18
Roll 2nd coat									
Slow	200	285	21.45	5.25	1.49	7.53	4.42	2.80	21.49
Medium	275	273	18.75	6.09	2.13	6.87	3.69	2.06	20.84
Fast	350	260	16.05	6.43	2.69	6.17	2.83	1.09	19.21
Roll 3rd or additional coats									
Slow	260	335	21.45	4.04	1.15	6.40	3.59	2.28	17.46
Medium	335	323	18.75	5.00	1.75	5.80	3.07	1.72	17.34
Fast	410	310	16.05	5.49	2.29	5.18	2.40	.92	16.28
Solid body or semi-transparent stain, oil base (material #19 or #21)									
Roll 1st coat									
Slow	150	250	25.00	7.00	1.99	10.00	5.89	3.73	28.61
Medium	225	225	21.90	7.44	2.59	9.73	4.84	2.71	27.31
Fast	275	200	18.75	8.18	3.43	9.38	3.88	1.49	26.36
Roll 2nd coat									
Slow	200	330	25.00	5.25	1.49	7.58	4.44	2.81	21.57
Medium	275	305	21.90	6.09	2.13	7.18	3.77	2.11	21.28
Fast	350	280	18.75	6.43	2.69	6.70	2.92	1.12	19.86
Roll 3rd or additional coats									
Slow	260	415	25.00	4.04	1.15	6.02	3.48	2.20	16.89
Medium	335	390	21.90	5.00	1.75	5.62	3.03	1.69	17.09
Fast	410	365	18.75	5.49	2.29	5.14	2.39	.92	16.23
Penetrating oil stain (material #13)									
Roll 1st coat									
Slow	150	150	29.00	7.00	1.99	19.33	8.78	5.57	42.67
Medium	225	125	25.40	7.44	2.59	20.32	7.44	4.16	41.95
Fast	275	100	21.80	8.18	3.43	21.80	6.18	2.37	41.96

	Labor SF per manhour	Material coverage SF/gallon	Material cost per gallon	Labor cost per 100 SF	Labor burden 100 SF	Material cost per 100 SF	Overhead per 100 SF	Profit per 100 SF	Total price per 100 SF
Roll 2nd coat									
Slow	200	280	29.00	5.25	1.49	10.36	5.30	3.36	25.76
Medium	275	240	25.40	6.09	2.13	10.58	4.60	2.57	25.97
Fast	350	200	21.80	6.43	2.69	10.90	3.70	1.42	25.14
Roll 3rd or additional coats									
Slow	260	365	29.00	4.04	1.15	7.95	4.07	2.58	19.79
Medium	335	330	25.40	5.00	1.75	7.70	3.54	1.98	19.97
Fast	410	295	21.80	5.49	2.29	7.39	2.81	1.08	19.06

Use this table to estimate the cost of painting shingle, shake, resawn or rough sawn wood or plywood siding with average moisture content. Don't deduct for openings under 100 square feet. For heights above 8 feet, use the High Time Difficulty Factors on page 137. For wood or composition drop siding with exposed bevel edges, multiply the surface area by 1.12 to allow for the extra time and material needed to paint the underside of each board. "Slow" work is based on a $10.50 hourly wage, "Medium" work on a $16.75 hourly wage, and "Fast" work on a $22.50 hourly wage. Other qualifications that apply to this table are on page 9.

	Labor SF per manhour	Material coverage SF/gallon	Material cost per gallon	Labor cost per 100 SF	Labor burden 100 SF	Material cost per 100 SF	Overhead per 100 SF	Profit per 100 SF	Total price per 100 SF

Siding, rough sawn or resawn wood, spray

Solid body or semi-transparent stain, water base (material #18 or #20)

	Labor SF per manhour	Material coverage SF/gallon	Material cost per gallon	Labor cost per 100 SF	Labor burden 100 SF	Material cost per 100 SF	Overhead per 100 SF	Profit per 100 SF	Total price per 100 SF
Spray 1st coat									
Slow	400	280	21.45	2.63	.74	7.66	3.42	2.17	16.62
Medium	500	260	18.75	3.35	1.17	7.21	2.87	1.61	16.21
Fast	600	240	16.05	3.75	1.58	6.69	2.22	.85	15.09
Spray 2nd coat									
Slow	450	330	21.45	2.33	.66	6.50	2.94	1.86	14.29
Medium	550	310	18.75	3.05	1.06	6.05	2.49	1.39	14.04
Fast	650	290	16.05	3.46	1.45	5.53	1.93	.74	13.11
Spray 3rd or additional coats									
Slow	550	380	21.45	1.91	.54	5.64	2.51	1.59	12.19
Medium	650	360	18.75	2.58	.90	5.21	2.13	1.19	12.01
Fast	750	340	16.05	3.00	1.24	4.72	1.66	.64	11.26

Solid body or semi-transparent stain, oil base (material #19 or #21)

	Labor SF per manhour	Material coverage SF/gallon	Material cost per gallon	Labor cost per 100 SF	Labor burden 100 SF	Material cost per 100 SF	Overhead per 100 SF	Profit per 100 SF	Total price per 100 SF
Spray 1st coat									
Slow	400	170	25.00	2.63	.74	14.71	5.61	3.56	27.25
Medium	500	150	21.90	3.35	1.17	14.60	4.68	2.62	26.42
Fast	600	130	18.75	3.75	1.58	14.42	3.65	1.40	24.80
Spray 2nd coat									
Slow	450	255	25.00	2.33	.66	9.80	3.96	2.51	19.26
Medium	550	235	21.90	3.05	1.06	9.32	3.29	1.84	18.56
Fast	650	215	18.75	3.46	1.45	8.72	2.52	.97	17.12
Spray 3rd or additional coats									
Slow	550	305	25.00	1.91	.54	8.20	3.30	2.09	16.04
Medium	650	285	21.90	2.58	.90	7.68	2.73	1.53	15.42
Fast	750	245	18.75	3.00	1.24	7.65	2.20	.85	14.94

Penetrating oil stain (material #13)

	Labor SF per manhour	Material coverage SF/gallon	Material cost per gallon	Labor cost per 100 SF	Labor burden 100 SF	Material cost per 100 SF	Overhead per 100 SF	Profit per 100 SF	Total price per 100 SF
Spray 1st coat									
Slow	400	200	29.00	2.63	.74	14.50	5.54	3.51	26.92
Medium	500	180	25.40	3.35	1.17	14.11	4.56	2.55	25.74
Fast	600	160	21.80	3.75	1.58	13.63	3.51	1.35	23.82
Spray 2nd coat									
Slow	450	290	29.00	2.33	.66	10.00	4.03	2.55	19.57
Medium	550	245	25.40	3.05	1.06	10.37	3.55	1.98	20.01
Fast	650	200	21.80	3.46	1.45	10.90	2.92	1.12	19.85
Spray 3rd or additional coats									
Slow	550	340	29.00	1.91	.54	8.53	3.40	2.16	16.54
Medium	650	310	25.40	2.58	.90	8.19	2.86	1.60	16.13
Fast	750	280	21.80	3.00	1.24	7.79	2.23	.86	15.12

	Labor SF per manhour	Material coverage SF/gallon	Material cost per gallon	Labor cost per 100 SF	Labor burden 100 SF	Material cost per 100 SF	Overhead per 100 SF	Profit per 100 SF	Total price per 100 SF
Waterproofing, clear hydro seal, oil base (material #34)									
Spray 1st coat									
Slow	400	150	17.80	2.63	.74	11.87	4.73	2.20	22.17
Medium	500	113	15.60	3.35	1.17	13.81	4.49	2.51	25.33
Fast	600	75	13.30	3.75	1.58	17.73	4.26	1.64	28.96
Spray 2nd coat									
Slow	650	175	17.80	1.62	.46	10.17	3.80	2.41	18.46
Medium	675	150	15.60	2.48	.86	10.40	3.37	1.88	18.99
Fast	700	125	13.30	3.21	1.35	10.64	2.81	1.08	19.09
Spray 3rd or additional coats									
Slow	700	200	17.80	1.50	.43	8.90	3.36	2.13	16.32
Medium	725	175	15.60	2.31	.81	8.91	2.95	1.65	16.63
Fast	750	150	13.30	3.00	1.24	8.87	2.43	.93	16.47

Use this table to estimate the cost of painting shingle, shake, resawn or rough sawn wood or plywood siding with average moisture content. Don't deduct for openings under 100 square feet. For heights above 8 feet, use the High Time Difficulty Factors on page 137. For wood or composition drop siding with exposed bevel edges, multiply the surface area by 1.12 to allow for the extra time and material needed to paint the underside of each board. "Slow" work is based on a $10.50 hourly wage, "Medium" work on a $16.75 hourly wage, and "Fast" work on a $22.50 hourly wage. Other qualifications that apply to this table are on page 9.

	Labor SF per manhour	Material coverage SF/gallon	Material cost per gallon	Labor cost per 100 SF	Labor burden 100 SF	Material cost per 100 SF	Overhead per 100 SF	Profit per 100 SF	Total price per 100 SF
Siding, smooth wood, brush									
Solid body or semi-transparent stain, water base (material #18 or #20)									
Brush 1st coat									
Slow	100	275	21.45	10.50	2.98	7.80	6.60	4.18	32.06
Medium	125	363	18.75	13.40	4.67	5.17	5.69	3.18	32.11
Fast	150	250	16.05	15.00	6.27	6.42	5.12	1.97	34.78
Brush 2nd coat									
Slow	135	350	21.45	7.78	2.21	6.13	5.00	3.17	24.29
Medium	168	325	18.75	9.97	3.47	5.77	4.71	2.63	26.55
Fast	200	300	16.05	11.25	4.70	5.35	3.94	1.51	26.75
Brush 3rd or additional coats									
Slow	150	425	21.45	7.00	1.99	5.05	4.35	2.76	21.15
Medium	188	400	18.75	8.91	3.11	4.69	4.09	2.29	23.09
Fast	215	375	16.05	10.47	4.36	4.28	3.54	1.36	24.01
Solid body or semi-transparent stain, oil base (material #19 or #21)									
Brush 1st coat									
Slow	100	400	25.00	10.50	2.98	6.25	6.12	3.88	29.73
Medium	125	363	21.90	13.40	4.67	6.03	5.90	3.30	33.30
Fast	150	325	18.75	15.00	6.27	5.77	5.00	1.92	33.96
Brush 2nd coat									
Slow	135	450	25.00	7.78	2.21	5.56	4.82	3.06	23.43
Medium	168	408	21.90	9.97	3.47	5.37	4.61	2.58	26.00
Fast	200	375	18.75	11.25	4.70	5.00	3.88	1.49	26.32
Brush 3rd or additional coats									
Slow	150	525	25.00	7.00	1.99	4.76	4.26	2.70	20.71
Medium	188	438	21.90	8.91	3.11	5.00	4.17	2.33	23.52
Fast	215	450	18.75	10.47	4.36	4.17	3.52	1.35	23.87
Penetrating oil stain (material #13)									
Brush 1st coat									
Slow	100	315	29.00	10.50	2.98	9.21	7.04	4.46	34.19
Medium	125	303	25.40	13.40	4.67	8.38	6.48	3.62	36.55
Fast	150	290	21.80	15.00	6.27	7.52	5.32	2.05	36.16

	Labor SF per manhour	Material coverage SF/gallon	Material cost per gallon	Labor cost per 100 SF	Labor burden 100 SF	Material cost per 100 SF	Overhead per 100 SF	Profit per 100 SF	Total price per 100 SF
Brush 2nd coat									
Slow	135	355	29.00	7.78	2.21	8.17	5.63	3.57	27.36
Medium	168	343	25.40	9.97	3.47	7.41	5.11	2.86	28.82
Fast	200	330	21.80	11.25	4.70	6.61	4.17	1.60	28.33
Brush 3rd or additional coats									
Slow	150	430	29.00	7.00	1.99	6.74	4.88	3.09	23.70
Medium	188	418	25.40	8.91	3.11	6.08	4.43	2.48	25.01
Fast	215	405	21.80	10.47	4.36	5.38	3.74	1.44	25.39

Use this table for butt or tongue and groove siding, joint lap, drop, beveled or board and batten siding in redwood, plywood, fir, hemlock or pine. Don't deduct for openings under 100 square feet. For heights above 8 feet, use the High Time Difficulty Factors on page 137. For wood or composition drop siding with exposed bevel edges, multiply the surface area by 1.12 to allow for the extra time and material needed to paint the underside of each board. "Slow" work is based on a $10.50 hourly wage, "Medium" work on a $16.75 hourly wage, and "Fast" work on a $22.50 hourly wage. Other qualifications that apply to this table are on page 9.

	Labor SF per manhour	Material coverage SF/gallon	Material cost per gallon	Labor cost per 100 SF	Labor burden 100 SF	Material cost per 100 SF	Overhead per 100 SF	Profit per 100 SF	Total price per 100 SF
Siding, smooth wood, roll									
Solid body or semi-transparent stain, water base (material #18 or #20)									
Roll 1st coat									
Slow	175	275	21.45	6.00	1.70	7.80	4.81	3.05	23.36
Medium	225	268	18.75	7.44	2.59	7.00	4.17	2.33	23.53
Fast	275	250	16.05	8.18	3.43	6.42	3.33	1.28	22.64
Roll 2nd coat									
Slow	225	350	21.45	4.67	1.32	6.13	3.76	2.38	18.26
Medium	275	325	18.75	6.09	2.13	5.77	3.43	1.92	19.34
Fast	325	300	16.05	6.92	2.90	5.35	2.80	1.08	19.05
Roll 3rd or additional coats									
Slow	260	400	21.45	4.04	1.15	5.36	3.27	2.07	15.89
Medium	335	363	18.75	5.00	1.75	5.17	2.92	1.63	16.47
Fast	410	375	16.05	5.49	2.29	4.28	2.23	.86	15.15
Solid body or semi-transparent stain, oil base (material #19 or #21)									
Roll 1st coat									
Slow	175	350	25.00	6.00	1.70	7.14	4.60	2.92	22.36
Medium	225	325	21.90	7.44	2.59	6.74	4.11	2.30	23.18
Fast	275	300	18.75	8.18	3.43	6.25	3.30	1.27	22.43
Roll 2nd coat									
Slow	225	400	25.00	4.67	1.32	6.25	3.80	2.41	18.45
Medium	275	375	21.90	6.09	2.13	5.84	3.44	1.92	19.42
Fast	325	350	18.75	6.92	2.90	5.36	2.81	1.08	19.07
Roll 3rd or additional coats									
Slow	260	425	25.00	4.04	1.15	5.88	3.43	2.18	16.68
Medium	335	413	21.90	5.00	1.75	5.30	2.95	1.65	16.65
Fast	410	400	18.75	5.49	2.29	4.69	2.31	.89	15.67
Penetrating oil stain (material #13)									
Roll 1st coat									
Slow	175	200	29.00	6.00	1.70	14.50	6.89	4.37	33.46
Medium	225	150	25.40	7.44	2.59	16.93	6.61	3.69	37.26
Fast	275	100	21.80	8.18	3.43	21.80	6.18	2.37	41.96

	Labor SF per manhour	Material coverage SF/gallon	Material cost per gallon	Labor cost per 100 SF	Labor burden 100 SF	Material cost per 100 SF	Overhead per 100 SF	Profit per 100 SF	Total price per 100 SF
Roll 2nd coat									
Slow	225	250	29.00	4.67	1.32	11.60	5.46	3.46	26.51
Medium	275	200	25.40	6.09	2.13	12.70	5.12	2.86	28.90
Fast	325	150	21.80	6.92	2.90	14.53	4.50	1.73	30.58
Roll 3rd or additional coats									
Slow	260	300	29.00	4.04	1.15	9.67	4.61	2.92	22.39
Medium	335	250	25.40	5.00	1.75	10.16	4.14	2.31	23.36
Fast	410	200	21.80	5.49	2.29	10.90	3.46	1.33	23.47

Use this table for butt or tongue and groove siding, joint lap, drop, beveled or board and batten siding in redwood, plywood, fir, hemlock or pine. Don't deduct for openings under 100 square feet. For heights above 8 feet, use the High Time Difficulty Factors on page 137. For wood or composition drop siding with exposed bevel edges, multiply the surface area by 1.12 to allow for the extra time and material needed to paint the underside of each board. "Slow" work is based on a $10.50 hourly wage, "Medium" work on a $16.75 hourly wage, and "Fast" work on a $22.50 hourly wage. Other qualifications that apply to this table are on page 9.

	Labor SF per manhour	Material coverage SF/gallon	Material cost per gallon	Labor cost per 100 SF	Labor burden 100 SF	Material cost per 100 SF	Overhead per 100 SF	Profit per 100 SF	Total price per 100 SF
Siding, smooth wood, spray									
Solid body or semi-transparent stain, water base (material #18 or #20)									
Spray 1st coat									
Slow	400	150	21.45	2.63	.74	14.30	5.48	3.47	26.62
Medium	500	125	18.75	3.35	1.17	15.00	4.78	2.67	26.97
Fast	600	100	16.05	3.75	1.58	16.05	3.95	1.52	26.85
Spray 2nd coat									
Slow	550	250	21.45	1.91	.54	8.58	3.42	2.17	16.62
Medium	725	225	18.75	2.31	.81	8.33	2.81	1.57	15.83
Fast	900	200	16.05	2.50	1.04	8.03	2.14	.82	14.53
Spray 3rd or additional coats									
Slow	650	350	21.45	1.62	.46	6.13	2.55	1.61	12.37
Medium	825	325	18.75	2.03	.70	5.77	2.08	1.16	11.74
Fast	1000	300	16.05	2.25	.94	5.35	1.58	.61	10.73
Solid body or semi-transparent stain, oil base (material #19 or #21)									
Spray 1st coat									
Slow	400	170	25.00	2.63	.74	14.71	5.61	3.56	27.25
Medium	500	150	21.90	3.35	1.17	14.60	4.68	2.62	26.42
Fast	600	130	18.75	3.75	1.58	14.42	3.65	1.40	24.80
Spray 2nd coat									
Slow	550	300	25.00	1.91	.54	8.33	3.34	2.12	16.24
Medium	725	273	21.90	2.31	.81	8.02	2.73	1.53	15.40
Fast	900	245	18.75	2.50	1.04	7.65	2.07	.80	14.06
Spray 3rd or additional coats									
Slow	650	400	25.00	1.62	.46	6.25	2.58	1.64	12.55
Medium	825	373	21.90	2.03	.70	5.87	2.11	1.18	11.89
Fast	1000	345	18.75	2.25	.94	5.43	1.59	.61	10.82
Penetrating oil stain (material #13)									
Spray 1st coat									
Slow	400	150	29.00	2.63	.74	19.33	7.04	4.46	34.20
Medium	500	113	25.40	3.35	1.17	22.48	6.62	3.70	37.32
Fast	600	75	21.80	3.75	1.58	29.07	6.36	2.45	43.21
Spray 2nd coat									
Slow	550	225	29.00	1.91	.54	12.89	4.76	3.02	23.12
Medium	725	188	25.40	2.31	.81	13.51	4.07	2.28	22.98
Fast	900	150	21.80	2.50	1.04	14.53	3.34	1.28	22.69
Spray 3rd or additional coats									
Slow	650	250	29.00	1.62	.46	11.60	4.24	2.69	20.61
Medium	825	225	25.40	2.03	.70	11.29	3.44	1.92	19.38
Fast	1000	200	21.80	2.25	.94	10.90	2.61	1.00	17.70

	Labor SF per manhour	Material coverage SF/gallon	Material cost per gallon	Labor cost per 100 SF	Labor burden 100 SF	Material cost per 100 SF	Overhead per 100 SF	Profit per 100 SF	Total price per 100 SF
Waterproofing, clear hydro seal, oil base (material #34)									
Spray 1st coat									
Slow	550	250	17.80	1.91	.54	7.12	2.97	1.38	13.92
Medium	675	200	15.60	2.48	.86	7.80	2.73	1.53	15.40
Fast	800	150	13.30	2.81	1.18	8.87	2.38	.91	16.15
Spray 2nd coat									
Slow	625	300	17.80	1.68	.48	5.93	2.51	1.59	12.19
Medium	750	250	15.60	2.23	.77	6.24	2.27	1.27	12.78
Fast	875	200	13.30	2.57	1.07	6.65	1.90	.73	12.92
Spray 3rd or additional coats									
Slow	675	325	17.80	1.56	.44	5.48	2.32	1.47	11.27
Medium	800	275	15.60	2.09	.73	5.67	2.08	1.16	11.73
Fast	925	225	13.30	2.43	1.01	5.91	1.73	.66	11.74

Use this table for butt or tongue and groove siding, joint lap, drop, beveled or board and batten siding in redwood, plywood, fir, hemlock or pine. Don't deduct for openings under 100 square feet. For heights above 8 feet, use the High Time Difficulty Factors on page 137. For wood or composition drop siding with exposed bevel edges, multiply the surface area by 1.12 to allow for the extra time and material needed to paint the underside of each board. "Slow" work is based on a $10.50 hourly wage, "Medium" work on a $16.75 hourly wage, and "Fast" work on a $22.50 hourly wage. Other qualifications that apply to this table are on page 9.

Stair steps, interior or exterior, wood

To estimate the cost to paint or stain stairs, find the surface area. Then use the tables for wood siding. To find the surface area of each tread and riser, multiply the length by the width. To find the tread length, add the run, the rise, and the tread nosing. For example, a tread with a 12" run, an 8" rise, and 1" nosing, has a 23" surface area (measured one side). For estimating purposes, figure any length from 14" to 26" as 2 feet. Use the actual width of the tread if the stringers are calculated separately. If the tread in the example is 3 feet wide and you use 2 feet for the length, the surface area is 6 feet. If there are 15 treads, the total top surface area is 90 square feet.

If you're calculating the area to paint the stair treads and stringers in one operation, add 2 feet to the actual tread width to include the stringers. That would make the effective width of the tread in the example 5 feet. Then multiply 5 feet by 2 feet to find the area of each tread. For 15 treads, the total surface area is 150 square feet.

	Labor LF per manhour	Material coverage LF/gallon	Material cost per gallon	Labor cost per 100 LF	Labor burden 100 LF	Material cost per 100 LF	Overhead per 100 LF	Profit per 100 LF	Total price per 100 LF
Stair stringers, exterior, metal, shapes up to 14" wide									
Metal primer - rust inhibitor, clean metal (material #35)									
Roll & brush prime coat									
Slow	50	120	29.90	21.00	5.96	24.92	16.09	10.20	78.17
Medium	55	115	26.20	30.45	10.62	22.78	15.64	8.74	88.23
Fast	60	110	22.40	37.50	15.66	20.36	13.60	5.23	92.35
Metal primer - rust inhibitor, rusty metal (material #36)									
Roll & brush prime coat									
Slow	50	120	36.60	21.00	5.96	30.50	17.82	11.29	86.57
Medium	55	115	32.10	30.45	10.62	27.91	16.90	9.45	95.33
Fast	60	110	27.50	37.50	15.66	25.00	14.46	5.56	98.18
Metal finish - synthetic enamel, off white (material #37)									
Roll & brush 1st or additional finish coats									
Slow	50	135	27.90	21.00	5.96	20.67	14.77	9.36	71.76
Medium	55	130	24.40	30.45	10.62	18.77	14.66	8.19	82.69
Fast	60	125	20.90	37.50	15.66	16.72	12.93	4.97	87.78
Metal finish - synthetic enamel, colors - except orange/red (material #38)									
Roll & brush 1st or additional finish coats									
Slow	50	135	30.00	21.00	5.96	22.22	15.25	9.67	74.10
Medium	55	130	26.30	30.45	10.62	20.23	15.02	8.39	84.71
Fast	60	125	22.50	37.50	15.66	18.00	13.16	5.06	89.38

Use these figures to paint installed stair stringers. Measurements are based on linear feet of each stringer. Note: A two coat system, prime and finish, using oil base material is recommended for any metal surface. Although water base material is often used, it may cause oxidation, corrosion and rust. Using one coat of oil base paint on exterior metal may result in cracking, peeling or chipping without the proper prime coat application. Pre-primed steel or wrought iron generally requires only one coat to cover. The metal finish figures include minor touchup to the prime coat. If off white or other light colored finish paint is specified, make sure the prime coat is a light color also, or more than one finish coat will be necessary. "Slow" work is based on a $10.50 hourly wage, "Medium" work on a $16.75 hourly wage, and "Fast" work on a $22.50 hourly wage. Other qualifications that apply to this table are on page 9.

	Labor LF per manhour	Material coverage LF/gallon	Material cost per gallon	Labor cost per 100 LF	Labor burden 100 LF	Material cost per 100 LF	Overhead per 100 LF	Profit per 100 LF	Total price per 100 LF
Stair stringers, exterior, rough sawn wood up to 4" x 12"									
Solid body or semi-transparent stain, water or oil base (material #18, #19, #20 or #21)									
Roll & brush each coat									
Slow	40	70	23.23	26.25	7.45	33.19	20.74	13.15	100.78
Medium	45	65	20.33	37.22	12.97	31.28	19.96	11.16	112.59
Fast	50	60	17.40	45.00	18.78	29.00	17.17	6.60	116.55

Measurements are based on the linear feet of each stringer. "Slow" work is based on a $10.50 hourly wage, "Medium" work on a $16.75 hourly wage, and "Fast" work on a $22.50 hourly wage. Other qualifications that apply to this table are on page 9.

	Labor LF per manhour	Material coverage LF/gallon	Material cost per gallon	Labor cost per 100 LF	Labor burden 100 LF	Material cost per 100 LF	Overhead per 100 LF	Profit per 100 LF	Total price per 100 LF
Stair stringers, interior, metal, shapes up to 14" wide									
Metal primer - rust inhibitor, clean metal (material #35)									
Roll & brush prime coat									
Slow	45	130	29.90	23.33	6.62	23.00	16.42	10.41	79.78
Medium	50	125	26.20	33.50	11.68	20.96	16.20	9.06	91.40
Fast	55	120	22.40	40.91	17.07	18.67	14.18	5.45	96.28
Metal primer - rust inhibitor, rusty metal (material #36)									
Roll & brush prime coat									
Slow	45	130	36.60	23.33	6.62	28.15	18.02	11.42	87.54
Medium	50	125	32.10	33.50	11.68	25.68	17.36	9.70	97.92
Fast	55	120	27.50	40.91	17.07	22.92	14.97	5.75	101.62
Metal finish - synthetic enamel, off white (material #37)									
Roll & brush 1st or additional finish coats									
Slow	45	145	27.90	23.33	6.62	19.24	15.26	9.67	74.12
Medium	50	140	24.40	33.50	11.68	17.43	15.34	8.57	86.52
Fast	55	135	20.90	40.91	17.07	15.48	13.59	5.22	92.27
Metal finish - synthetic enamel, colors - except orange/red (material #38)									
Roll & brush 1st or additional finish coats									
Slow	45	145	30.00	23.33	6.62	20.69	15.70	9.95	76.29
Medium	50	140	26.30	33.50	11.68	18.79	15.67	8.76	88.40
Fast	55	135	22.50	40.91	17.07	16.67	13.81	5.31	93.77

Use these figures to paint installed stair stringers. Measurements are based on linear feet of each stringer. Note: A two coat system, prime and finish, using oil base material is recommended for any metal surface. Although water base material is often used, it may cause oxidation, corrosion and rust. Using one coat of oil base paint on exterior metal may result in cracking, peeling or chipping without the proper prime coat application. Pre-primed steel or wrought iron generally requires only one coat to cover. The metal finish figures include minor touchup to the prime coat. If off white or other light colored finish paint is specified, make sure the prime coat is a light color also, or more than one finish coat will be necessary. "Slow" work is based on a $10.50 hourly wage, "Medium" work on a $16.75 hourly wage, and "Fast" work on a $22.50 hourly wage. Other qualifications that apply to this table are on page 9.

	Labor LF per manhour	Material coverage LF/gallon	Material cost per gallon	Labor cost per 100 LF	Labor burden 100 LF	Material cost per 100 LF	Overhead per 100 LF	Profit per 100 LF	Total price per 100 LF

Stair stringers, interior, rough sawn wood up to 4" x 12"

Solid body or semi-transparent stain, water or oil base (material #18, #19, #20 or #21)

Roll & brush each coat

	Labor LF per manhour	Material coverage LF/gallon	Material cost per gallon	Labor cost per 100 LF	Labor burden 100 LF	Material cost per 100 LF	Overhead per 100 LF	Profit per 100 LF	Total price per 100 LF
Slow	35	60	23.23	30.00	8.51	38.72	23.95	15.18	116.36
Medium	40	53	20.33	41.88	14.60	38.36	23.24	12.99	131.07
Fast	45	45	17.40	50.00	20.86	38.67	20.27	7.79	137.59

Measurements are based on the linear feet of each stringer. "Slow" work is based on a $10.50 hourly wage, "Medium" work on a $16.75 hourly wage, and "Fast" work on a $22.50 hourly wage. Other qualifications that apply to this table are on page 9.

Stucco: see Plaster and stucco

Touchup, brush as required

	Percentage of total costs
Slow	10%
Medium	9%
Fast	8%

Touchup will be required on all jobs. The skill of your paint crews and the type of job will determine the time needed. The time and materials shown here are average allowances. Figure the touchup time and material needed and then add the appropriate percentage to your estimate. These figures are based on a percentage of the total labor and material needed to touch up all interior and exterior work. If you're painting only the interior or the exterior, divide these figures by 2. "Slow" work is based on a $10.50 hourly wage, "Medium" work on a $16.75 hourly wage, and "Fast" work on a $22.50 hourly wage. Other qualifications that apply to this table are on page 9.

	Labor LF per manhour	Material coverage LF/gallon	Material cost per gallon	Labor cost per 100 LF	Labor burden 100 LF	Material cost per 100 LF	Overhead per 100 LF	Profit per 100 LF	Total price per 100 LF

Trellis or lattice, roll and brush

2" x 2" to 2" x 6", roll & brush all sides, each coat

Solid body or semi-transparent stain, water or oil base (material #18, #19, #20 or #21)

Slow	120	130	23.23	8.75	2.48	17.87	9.02	5.72	43.84
Medium	125	120	20.33	13.40	4.67	16.94	8.58	4.79	48.38
Fast	130	110	17.40	17.31	7.21	15.82	7.47	2.87	50.68

2" x 8" to 4" x 12", roll & brush all sides, each coat

Solid body or semi-transparent stain, water or oil base (material #18, #19, #20 or #21)

Slow	100	100	23.23	10.50	2.98	23.23	11.38	7.22	55.31
Medium	110	90	20.33	15.23	5.30	22.59	10.57	5.91	59.60
Fast	120	80	17.40	18.75	7.81	21.75	8.94	3.44	60.69

Measurements are based on accumulated total linear feet of each trellis or lattice member. These figures are based on roll and brush staining of all four sides and the ends of each member per coat. "Slow" work is based on a $10.50 hourly wage, "Medium" work on a $16.75 hourly wage, and "Fast" work on a $22.50 hourly wage. Other qualifications that apply to this table are on page 9.

	SF surface area per manhour	SF surface area per gallon	Material cost per gallon	Labor cost per 100 SF	Labor burden 100 SF	Material cost per 100 SF	Overhead per 100 SF	Profit per 100 SF	Total price per 100 SF

Trellis or lattice, spray

2" x 2" at 3" on center with 2" x 8" supports, spray all sides, each coat

Solid body or semi-transparent stain, water or oil base (material #18, #19, #20 or #21)

Slow	50	60	23.23	21.00	5.96	38.72	20.36	12.91	98.95
Medium	55	55	20.33	30.45	10.62	36.96	19.11	10.68	107.82
Fast	60	50	17.40	37.50	15.66	34.80	16.27	6.25	110.48

Measurements are based on the square feet of the surface area footprint of the trellis or lattice structure. (The footprint is the surface area seen from the plan or overhead view.) These figures are based on staining all four sides of each member per coat. "Slow" work is based on a $10.50 hourly wage, "Medium" work on a $16.75 hourly wage, and "Fast" work on a $22.50 hourly wage. Other qualifications that apply to this table are on page 9.

	Labor LF per manhour	Material coverage LF/gallon	Material cost per gallon	Labor cost per 100 LF	Labor burden 100 LF	Material cost per 100 LF	Overhead per 100 LF	Profit per 100 LF	Total price per 100 LF

Valances for light fixtures

Solid body or semi-transparent stain, water or oil base (material #18, #19, #20 or #21)

Brush each coat									
Slow	30	100	23.23	35.00	9.93	23.23	21.14	13.40	102.70
Medium	35	95	20.33	47.86	16.68	21.40	21.06	11.77	118.77
Fast	40	90	17.40	56.25	23.48	19.33	18.33	7.04	124.43

Rough sawn or resawn wood valances are commonly found in baths and kitchens surrounding light fixtures or supporting plastic cracked ice diffusers. Measurements are based on the linear feet of the valance. "Slow" work is based on a $10.50 hourly wage, "Medium" work on a $16.75 hourly wage, and "Fast" work on a $22.50 hourly wage. Other qualifications that apply to this table are on page 9.

	Labor SF per manhour	Material coverage SF/gallon	Material cost per gallon	Labor cost per 100 SF	Labor burden 100 SF	Material cost per 100 SF	Overhead per 100 SF	Profit per 100 SF	Total price per 100 SF
Walls, gypsum drywall, anti-graffiti stain eliminator, per 100 square feet of wall									
Water base primer and sealer (material #39)									
Roll & brush each coat									
Slow	375	450	25.40	2.80	.80	5.64	2.86	1.82	13.92
Medium	400	425	22.30	4.19	1.46	5.25	2.67	1.49	15.06
Fast	425	400	19.10	5.29	2.20	4.78	2.27	.87	15.41
Oil base primer and sealer (material #40)									
Roll & brush each coat									
Slow	375	400	25.70	2.80	.80	6.43	3.11	1.97	15.11
Medium	400	388	22.50	4.19	1.46	5.80	2.81	1.57	15.83
Fast	425	375	19.30	5.29	2.20	5.15	2.34	.90	15.88
Polyurethane 2 part system (material #41)									
Roll & brush each coat									
Slow	325	400	91.40	3.23	.92	22.85	8.37	5.31	40.68
Medium	350	375	80.00	4.79	1.67	21.33	6.81	3.81	38.41
Fast	375	350	68.60	6.00	2.51	19.60	5.20	2.00	35.31

Measurements are based on the square feet of wall coated. These figures assume paint products are being applied over a smooth finish. For heights above 8 feet, use the High Time Difficulty Factors on page 137. "Slow" work is based on a $10.50 hourly wage, "Medium" work on a $16.75 hourly wage, and "Fast" work on a $22.50 hourly wage. Other qualifications that apply to this table are on page 9.

	Labor SF per manhour	Material coverage SF/gallon	Material cost per gallon	Labor cost per 100 SF	Labor burden 100 SF	Material cost per 100 SF	Overhead per 100 SF	Profit per 100 SF	Total price per 100 SF
Walls, gypsum drywall, orange peel or knock-down, brush, per 100 SF of wall									
Flat latex, water base (material #5)									
Brush 1st coat									
Slow	150	300	17.10	7.00	1.99	5.70	4.55	2.89	22.13
Medium	175	288	15.00	9.57	3.33	5.21	4.44	2.48	25.03
Fast	200	275	12.90	11.25	4.70	4.69	3.82	1.47	25.93
Brush 2nd coat									
Slow	175	350	17.10	6.00	1.70	4.89	3.91	2.48	18.98
Medium	200	338	15.00	8.38	2.92	4.44	3.86	2.16	21.76
Fast	225	325	12.90	10.00	4.16	3.97	3.36	1.29	22.78
Brush 3rd or additional coats									
Slow	200	400	17.10	5.25	1.49	4.28	3.42	2.17	16.61
Medium	225	375	15.00	7.44	2.59	4.00	3.44	1.92	19.39
Fast	250	350	12.90	9.00	3.76	3.69	3.04	1.17	20.66
Sealer (drywall), water base (material #1)									
Brush prime coat									
Slow	125	300	16.30	8.40	2.38	5.43	5.03	3.19	24.43
Medium	163	288	14.30	10.28	3.57	4.97	4.61	2.58	26.01
Fast	200	275	12.30	11.25	4.70	4.47	3.78	1.45	25.65
Sealer (drywall), oil base (material #2)									
Brush prime coat									
Slow	125	250	22.90	8.40	2.38	9.16	6.18	3.92	30.04
Medium	163	238	20.10	10.28	3.57	8.45	5.47	3.06	30.83
Fast	200	225	17.20	11.25	4.70	7.64	4.36	1.68	29.63
Enamel, water base (material #9)									
Brush 1st finish coat									
Slow	100	300	24.50	10.50	2.98	8.17	6.71	4.26	32.62
Medium	150	288	21.40	11.17	3.90	7.43	5.51	3.08	31.09
Fast	200	275	18.40	11.25	4.70	6.69	4.19	1.61	28.44
Brush 2nd or additional finish coats									
Slow	125	350	24.50	8.40	2.38	7.00	5.51	3.50	26.79
Medium	163	325	21.40	10.28	3.57	6.58	5.01	2.80	28.24
Fast	200	300	18.40	11.25	4.70	6.13	4.08	1.57	27.73
Enamel, oil base (material #10)									
Brush 1st finish coat									
Slow	100	325	26.60	10.50	2.98	8.18	6.72	4.26	32.64
Medium	150	300	23.30	11.17	3.90	7.77	5.59	3.13	31.56
Fast	200	275	20.00	11.25	4.70	7.27	4.30	1.65	29.17

	Labor SF per manhour	Material coverage SF/gallon	Material cost per gallon	Labor cost per 100 SF	Labor burden 100 SF	Material cost per 100 SF	Overhead per 100 SF	Profit per 100 SF	Total price per 100 SF
Brush 2nd or additional finish coats									
Slow	125	350	24.50	8.40	2.38	7.00	5.51	3.50	26.79
Medium	163	325	21.40	10.28	3.57	6.58	5.01	2.80	28.24
Fast	200	300	18.40	11.25	4.70	6.13	4.08	1.57	27.73
Epoxy coating, 2 part system - white (material #52)									
Brush 1st coat									
Slow	175	350	61.70	6.00	1.70	17.63	7.86	4.98	38.17
Medium	200	325	54.00	8.38	2.92	16.62	6.84	3.82	38.58
Fast	225	300	46.30	10.00	4.16	15.43	5.48	2.11	37.18
Brush 2nd or additional coats									
Slow	200	375	61.70	5.25	1.49	16.45	7.19	4.56	34.94
Medium	225	350	54.00	7.44	2.59	15.43	6.24	3.49	35.19
Fast	250	325	46.30	9.00	3.76	14.25	5.00	1.92	33.93

Measurements are based on the square feet of wall coated. These figures assume paint products are being applied over a smooth finish. For heights above 8 feet, use the High Time Difficulty Factors on page 137. "Slow" work is based on a $10.50 hourly wage, "Medium" work on a $16.75 hourly wage, and "Fast" work on a $22.50 hourly wage. Other qualifications that apply to this table are on page 9.

	Labor SF per manhour	Material coverage SF/gallon	Material cost per gallon	Labor cost per 100 SF	Labor burden 100 SF	Material cost per 100 SF	Overhead per 100 SF	Profit per 100 SF	Total price per 100 SF
Walls, gypsum drywall, orange peel or knock down, roll, per 100 SF of wall									
Flat latex, water base (material #5)									
Roll 1st coat									
Slow	400	300	17.10	2.63	.74	5.70	2.81	1.78	13.66
Medium	538	275	15.00	3.11	1.09	5.45	2.36	1.32	13.33
Fast	675	250	12.90	3.33	1.39	5.16	1.83	.70	12.41
Roll 2nd coat									
Slow	500	325	17.10	2.10	.60	5.26	2.47	1.56	11.99
Medium	600	313	15.00	2.79	.98	4.79	2.09	1.17	11.82
Fast	700	300	12.90	3.21	1.35	4.30	1.64	.63	11.13
Roll 3rd or additional coats									
Slow	550	350	17.10	1.91	.54	4.89	2.28	1.44	11.06
Medium	650	338	15.00	2.58	.90	4.44	1.94	1.08	10.94
Fast	750	325	12.90	3.00	1.24	3.97	1.52	.58	10.31
Sealer (drywall), water base (material #1)									
Roll prime coat									
Slow	325	275	16.30	3.23	.92	5.93	3.12	1.98	15.18
Medium	500	263	14.30	3.35	1.17	5.44	2.44	1.36	13.76
Fast	675	250	12.30	3.33	1.39	4.92	1.78	.69	12.11
Sealer (drywall), oil base (material #2)									
Roll prime coat									
Slow	325	275	22.90	3.23	.92	8.33	3.87	2.45	18.80
Medium	500	263	20.10	3.35	1.17	7.64	2.98	1.67	16.81
Fast	675	250	17.20	3.33	1.39	6.88	2.15	.83	14.58
Enamel, water base (material #9)									
Roll 1st finish coat									
Slow	300	285	24.50	3.50	.99	8.60	4.06	2.57	19.72
Medium	450	263	21.40	3.72	1.29	8.14	3.22	1.80	18.17
Fast	600	240	18.40	3.75	1.58	7.67	2.40	.92	16.32
Roll 2nd finish coat									
Slow	325	300	24.50	3.23	.92	8.17	3.82	2.42	18.56
Medium	475	288	21.40	3.53	1.24	7.43	2.99	1.67	16.86
Fast	625	275	18.40	3.60	1.50	6.69	2.18	.84	14.81
Enamel, oil base (material #10)									
Roll 1st finish coat									
Slow	300	250	26.60	3.50	.99	10.64	4.69	2.97	22.79
Medium	450	238	23.30	3.72	1.29	9.79	3.63	2.03	20.46
Fast	600	225	20.00	3.75	1.58	8.89	2.63	1.01	17.86

	Labor SF per manhour	Material coverage SF/gallon	Material cost per gallon	Labor cost per 100 SF	Labor burden 100 SF	Material cost per 100 SF	Overhead per 100 SF	Profit per 100 SF	Total price per 100 SF
Roll 2nd finish coat									
Slow	325	275	26.60	3.23	.92	9.67	4.28	2.72	20.82
Medium	475	263	23.30	3.53	1.24	8.86	3.34	1.87	18.84
Fast	625	250	20.00	3.60	1.50	8.00	2.42	.93	16.45
Epoxy coating, 2 part system - white (material #52)									
Roll 1st coat									
Slow	500	300	61.70	2.10	.60	20.57	7.21	4.57	35.05
Medium	600	288	54.00	2.79	.98	18.75	5.51	3.08	31.11
Fast	700	275	46.30	3.21	1.35	16.84	3.96	1.52	26.88
Roll 2nd or additional coats									
Slow	550	325	61.70	1.91	.54	18.98	6.64	4.21	32.28
Medium	650	313	54.00	2.58	.90	17.25	5.08	2.84	28.65
Fast	750	300	46.30	3.00	1.24	15.43	3.64	1.40	24.71

Measurements are based on the square feet of wall coated. These figures assume paint products are being applied over a smooth finish. For heights above 8 feet, use the High Time Difficulty Factors on page 137. "Slow" work is based on a $10.50 hourly wage, "Medium" work on a $16.75 hourly wage, and "Fast" work on a $22.50 hourly wage. Other qualifications that apply to this table are on page 9.

	SF of floor area per manhour	SF of floor area per gallon	Material cost per gallon	Labor cost per 100 SF	Labor burden 100 SF	Material cost per 100 SF	Overhead per 100 SF	Profit per 100 SF	Total price per 100 SF
Walls, gypsum drywall, orange peel or knock-down, roll, per 100 SF of floor area									
Flat latex, water base (material #5)									
Roll 1st coat on walls only									
Slow	250	175	17.10	4.20	1.19	9.77	4.70	2.98	22.84
Medium	325	158	15.00	5.15	1.81	9.49	4.03	2.25	22.73
Fast	400	140	12.90	5.63	2.34	9.21	3.18	1.22	21.58
Roll 2nd coat on walls only									
Slow	300	200	17.10	3.50	.99	8.55	4.05	2.57	19.66
Medium	400	188	15.00	4.19	1.46	7.98	3.34	1.87	18.84
Fast	500	175	12.90	4.50	1.88	7.37	2.54	.98	17.27
Sealer (drywall), water base (material #1) on walls and ceilings									
Roll prime coat									
Slow	100	100	16.30	10.50	2.98	16.30	9.23	5.85	44.86
Medium	170	88	14.30	9.85	3.43	16.25	7.23	4.04	40.80
Fast	240	75	12.30	9.38	3.92	16.40	5.49	2.11	37.30
Sealer (drywall), oil base (material #2) on walls and ceilings									
Roll prime coat									
Slow	100	100	22.90	10.50	2.98	22.90	11.28	7.15	54.81
Medium	170	88	20.10	9.85	3.43	22.84	8.85	4.95	49.92
Fast	240	75	17.20	9.38	3.92	22.93	6.70	2.58	45.51
Enamel, water base (material #9) on walls and ceilings									
Roll 1st finish coat									
Slow	70	100	24.50	15.00	4.26	24.50	13.57	8.60	65.93
Medium	100	90	21.40	16.75	5.84	23.78	11.36	6.35	64.08
Fast	135	80	18.40	16.67	6.96	23.00	8.63	3.32	58.58
Roll 2nd finish coat									
Slow	125	150	24.50	8.40	2.38	16.33	8.41	5.33	40.85
Medium	175	125	21.40	9.57	3.33	17.12	7.36	4.11	41.49
Fast	225	100	18.40	10.00	4.16	18.40	6.03	2.32	40.91

	SF of floor area per manhour	SF of floor area per gallon	Material cost per gallon	Labor cost per 100 SF	Labor burden 100 SF	Material cost per 100 SF	Overhead per 100 SF	Profit per 100 SF	Total price per 100 SF
Enamel, oil base (material #10) on walls and ceilings									
Roll 1st finish coat									
Slow	70	100	26.60	15.00	4.26	26.60	14.22	9.01	69.09
Medium	100	90	23.30	16.75	5.84	25.89	11.88	6.64	67.00
Fast	135	80	20.00	16.67	6.96	25.00	9.00	3.46	61.09
Roll 2nd finish coat									
Slow	125	150	26.60	8.40	2.38	17.73	8.84	5.60	42.95
Medium	175	125	23.30	9.57	3.33	18.64	7.73	4.32	43.59
Fast	225	100	20.00	10.00	4.16	20.00	6.32	2.43	42.91

Measurements for these costs are based on square feet of floor area. The flatwall figures are for painting walls only but the Sealer and Enamel figures are for painting walls and ceilings in wet areas, i.e. kitchens, baths, utility areas, etc. The floor area measurements are from outside wall to outside wall or from the edge of the concrete slab or edge of an interior wall. This method of figuring the costs to paint the walls and ceilings is not as accurate as measuring the actual surface area of the wall or ceiling area directly, but it is much less time consuming. For heights above 8 feet, use the High Time Difficulty Factors on page 137. "Slow" work is based on a $10.50 hourly wage, "Medium" work on a $16.75 hourly wage, and "Fast" work on a $22.50 hourly wage. Other qualifications that apply to this table are on page 9.

	Labor SF per manhour	Material coverage SF/gallon	Material cost per gallon	Labor cost per 100 SF	Labor burden 100 SF	Material cost per 100 SF	Overhead per 100 SF	Profit per 100 SF	Total price per 100 SF

Walls, gypsum drywall, orange peel or knock-down, spray, per 100 SF of wall

	Labor SF per manhour	Material coverage SF/gallon	Material cost per gallon	Labor cost per 100 SF	Labor burden 100 SF	Material cost per 100 SF	Overhead per 100 SF	Profit per 100 SF	Total price per 100 SF
Flat latex, water base (material #5)									
Spray 1st coat									
Slow	700	250	17.10	1.50	.43	6.84	2.72	1.72	13.21
Medium	800	225	15.00	2.09	.73	6.67	2.33	1.30	13.12
Fast	900	200	12.90	2.50	1.04	6.45	1.85	.71	12.55
Spray 2nd coat									
Slow	800	300	17.10	1.31	.38	5.70	2.29	1.45	11.13
Medium	900	275	15.00	1.86	.65	5.45	1.95	1.09	11.00
Fast	1000	250	12.90	2.25	.94	5.16	1.54	.59	10.48
Spray 3rd or additional coats									
Slow	850	325	17.10	1.24	.35	5.26	2.12	1.35	10.32
Medium	950	300	15.00	1.76	.61	5.00	1.81	1.01	10.19
Fast	1050	275	12.90	2.14	.89	4.69	1.43	.55	9.70
Sealer (drywall), water base (material #1)									
Spray prime coat									
Slow	575	250	16.30	1.83	.52	6.52	2.75	1.74	13.36
Medium	738	225	14.30	2.27	.80	6.36	2.31	1.29	13.03
Fast	900	200	12.30	2.50	1.04	6.15	1.79	.69	12.17
Sealer (drywall), oil base (material #2)									
Spray prime coat									
Slow	575	250	22.90	1.83	.52	9.16	3.57	2.26	17.34
Medium	738	225	20.10	2.27	.80	8.93	2.94	1.64	16.58
Fast	900	200	17.20	2.50	1.04	8.60	2.25	.86	15.25
Enamel, water base (material #9)									
Spray 1st finish coat									
Slow	500	250	24.50	2.10	.60	9.80	3.88	2.46	18.84
Medium	675	238	21.40	2.48	.86	8.99	3.02	1.69	17.04
Fast	850	225	18.40	2.65	1.11	8.18	2.21	.85	15.00
Spray 2nd finish coat									
Slow	525	275	24.50	2.00	.56	8.91	3.56	2.26	17.29
Medium	700	263	21.40	2.39	.84	8.14	2.78	1.56	15.71
Fast	875	250	18.40	2.57	1.07	7.36	2.04	.78	13.82
Spray 3rd or additional finish coats									
Slow	575	425	24.50	1.83	.52	5.76	2.51	1.59	12.21
Medium	775	388	21.40	2.16	.75	5.52	2.07	1.16	11.66
Fast	925	350	18.40	2.43	1.01	5.26	1.61	.62	10.93

	Labor SF per manhour	Material coverage SF/gallon	Material cost per gallon	Labor cost per 100 SF	Labor burden 100 SF	Material cost per 100 SF	Overhead per 100 SF	Profit per 100 SF	Total price per 100 SF
Enamel, oil base (material #10)									
Spray 1st finish coat									
Slow	500	250	26.60	2.10	.60	10.64	4.14	2.62	20.10
Medium	675	225	23.30	2.48	.86	10.36	3.36	1.88	18.94
Fast	850	200	20.00	2.65	1.11	10.00	2.55	.98	17.29
Spray 2nd finish coat									
Slow	525	275	26.60	2.00	.56	9.67	3.79	2.40	18.42
Medium	700	250	23.30	2.39	.84	9.32	3.07	1.72	17.34
Fast	875	225	20.00	2.57	1.07	8.89	2.32	.89	15.74
Spray 3rd or additional finish coat									
Slow	575	425	26.60	1.83	.52	6.26	2.67	1.69	12.97
Medium	775	388	23.30	2.16	.75	6.01	2.19	1.22	12.33
Fast	925	350	20.00	2.43	1.01	5.71	1.69	.65	11.49

Measurements are based on the square feet of wall coated. These figures assume paint products are being applied over a smooth finish. For heights above 8 feet, use the High Time Difficulty Factors on page 137. "Slow" work is based on a $10.50 hourly wage, "Medium" work on a $16.75 hourly wage, and "Fast" work on a $22.50 hourly wage. Other qualifications that apply to this table are on page 9.

	Labor SF per manhour	Material coverage SF/gallon	Material cost per gallon	Labor cost per 100 SF	Labor burden 100 SF	Material cost per 100 SF	Overhead per 100 SF	Profit per 100 SF	Total price per 100 SF

Walls, gypsum drywall, sand finish, brush, per 100 square feet of wall

	Labor SF per manhour	Material coverage SF/gallon	Material cost per gallon	Labor cost per 100 SF	Labor burden 100 SF	Material cost per 100 SF	Overhead per 100 SF	Profit per 100 SF	Total price per 100 SF
Flat latex, water base (material #5)									
Brush 1st coat									
Slow	175	325	17.10	6.00	1.70	5.26	4.02	2.55	19.53
Medium	200	313	15.00	8.38	2.92	4.79	3.94	2.20	22.23
Fast	225	300	12.90	10.00	4.16	4.30	3.42	1.31	23.19
Brush 2nd coat									
Slow	200	400	17.10	5.25	1.49	4.28	3.42	2.17	16.61
Medium	225	375	15.00	7.44	2.59	4.00	3.44	1.92	19.39
Fast	250	350	12.90	9.00	3.76	3.69	3.04	1.17	20.66
Brush 3rd or additional coats									
Slow	225	425	17.10	4.67	1.32	4.02	3.11	1.97	15.09
Medium	250	400	15.00	6.70	2.34	3.75	3.13	1.75	17.67
Fast	275	375	12.90	8.18	3.43	3.44	2.78	1.07	18.90
Sealer (drywall), water base (material #1)									
Brush prime coat									
Slow	140	325	16.30	7.50	2.12	5.02	4.54	2.88	22.06
Medium	183	313	14.30	9.15	3.18	4.57	4.14	2.32	23.36
Fast	225	300	12.30	10.00	4.16	4.10	3.38	1.30	22.94
Sealer (drywall), oil base (material #2)									
Brush prime coat									
Slow	140	350	22.90	7.50	2.12	6.54	5.01	3.18	24.35
Medium	183	338	20.10	9.15	3.18	5.95	4.48	2.50	25.26
Fast	225	325	17.20	10.00	4.16	5.29	3.60	1.38	24.43
Enamel, water base (material #9)									
Brush 1st finish coat									
Slow	125	350	24.50	8.40	2.38	7.00	5.51	3.50	26.79
Medium	175	325	21.40	9.57	3.33	6.58	4.78	2.67	26.93
Fast	225	300	18.40	10.00	4.16	6.13	3.76	1.44	25.49
Brush 2nd or additional finish coats									
Slow	140	350	24.50	7.50	2.12	7.00	5.16	3.27	25.05
Medium	185	325	21.40	9.05	3.17	6.58	4.60	2.57	25.97
Fast	235	300	18.40	9.57	4.02	6.13	3.64	1.40	24.76
Enamel, oil base (material #10)									
Brush 1st finish coat									
Slow	125	350	26.60	8.40	2.38	7.60	5.70	3.61	27.69
Medium	175	325	23.30	9.57	3.33	7.17	4.92	2.75	27.74
Fast	225	300	20.00	10.00	4.16	6.67	3.86	1.48	26.17

	Labor SF per manhour	Material coverage SF/gallon	Material cost per gallon	Labor cost per 100 SF	Labor burden 100 SF	Material cost per 100 SF	Overhead per 100 SF	Profit per 100 SF	Total price per 100 SF
Brush 2nd or additional finish coats									
Slow	140	350	26.60	7.50	2.12	7.60	5.34	3.39	25.95
Medium	185	338	23.30	9.05	3.17	6.89	4.68	2.61	26.40
Fast	235	325	20.00	9.57	4.02	6.15	3.65	1.40	24.79
Epoxy coating, 2 part system - white (material #52)									
Brush 1st coat									
Slow	200	375	61.70	5.25	1.49	16.45	7.19	4.56	34.94
Medium	225	350	54.00	7.44	2.59	15.43	6.24	3.49	35.19
Fast	250	325	46.30	9.00	3.76	14.25	5.00	1.92	33.93
Brush 2nd or additional coats									
Slow	225	400	61.70	4.67	1.32	15.43	6.64	4.21	32.27
Medium	250	375	54.00	6.70	2.34	14.40	5.74	3.21	32.39
Fast	275	350	46.30	8.18	3.43	13.23	4.59	1.77	31.20

Measurements are based on the square feet of wall coated. These figures assume paint products are being applied over a smooth finish. For heights above 8 feet, use the High Time Difficulty Factors on page 137. "Slow" work is based on a $10.50 hourly wage, "Medium" work on a $16.75 hourly wage, and "Fast" work on a $22.50 hourly wage. Other qualifications that apply to this table are on page 9.

	Labor SF per manhour	Material coverage SF/gallon	Material cost per gallon	Labor cost per 100 SF	Labor burden 100 SF	Material cost per 100 SF	Overhead per 100 SF	Profit per 100 SF	Total price per 100 SF

Walls, gypsum drywall, sand finish, roll, per 100 square feet of wall

	Labor SF per manhour	Material coverage SF/gallon	Material cost per gallon	Labor cost per 100 SF	Labor burden 100 SF	Material cost per 100 SF	Overhead per 100 SF	Profit per 100 SF	Total price per 100 SF
Flat latex, water base (material #5)									
Roll 1st coat									
Slow	275	325	17.10	3.82	1.09	5.26	3.15	2.00	15.32
Medium	488	300	15.00	3.43	1.20	5.00	2.36	1.32	13.31
Fast	700	275	12.90	3.21	1.35	4.69	1.71	.66	11.62
Roll 2nd coat									
Slow	350	350	17.10	3.00	.86	4.89	2.71	1.72	13.18
Medium	538	338	15.00	3.11	1.09	4.44	2.11	1.18	11.93
Fast	725	325	12.90	3.10	1.30	3.97	1.55	.59	10.51
Roll 3rd or additional coats									
Slow	425	350	17.10	2.47	.70	4.89	2.50	1.58	12.14
Medium	600	338	15.00	2.79	.98	4.44	2.01	1.12	11.34
Fast	775	325	12.90	2.90	1.21	3.97	1.49	.57	10.14
Sealer (drywall), water base (material #1)									
Roll prime coat									
Slow	225	325	16.30	4.67	1.32	5.02	3.42	2.17	16.60
Medium	463	300	14.30	3.62	1.26	4.77	2.36	1.32	13.33
Fast	700	275	12.30	3.21	1.35	4.47	1.67	.64	11.34
Sealer (drywall), oil base (material #2)									
Roll prime coat									
Slow	225	325	22.90	4.67	1.32	7.05	4.05	2.57	19.66
Medium	463	313	20.10	3.62	1.26	6.42	2.77	1.55	15.62
Fast	700	300	17.20	3.21	1.35	5.73	1.90	.73	12.92
Enamel, water base (material #9)									
Roll 1st finish coat									
Slow	225	300	24.50	4.67	1.32	8.17	4.39	2.78	21.33
Medium	400	288	21.40	4.19	1.46	7.43	3.20	1.79	18.07
Fast	600	275	18.40	3.75	1.58	6.69	2.22	.85	15.09
Roll 2nd or additional finish coats									
Slow	275	300	24.50	3.82	1.09	8.17	4.05	2.57	19.70
Medium	450	288	21.40	3.72	1.29	7.43	3.05	1.71	17.20
Fast	650	275	18.40	3.46	1.45	6.69	2.14	.82	14.56
Enamel, oil base (material #10)									
Roll 1st finish coat									
Slow	225	275	26.60	4.67	1.32	9.67	4.86	3.08	23.60
Medium	400	263	23.30	4.19	1.46	8.86	3.55	1.99	20.05
Fast	600	250	20.00	3.75	1.58	8.00	2.46	.95	16.74

	Labor SF per manhour	Material coverage SF/gallon	Material cost per gallon	Labor cost per 100 SF	Labor burden 100 SF	Material cost per 100 SF	Overhead per 100 SF	Profit per 100 SF	Total price per 100 SF
Roll 2nd or additional finish coats									
Slow	275	300	26.60	3.82	1.09	8.87	4.27	2.71	20.76
Medium	450	288	23.30	3.72	1.29	8.09	3.21	1.80	18.11
Fast	650	275	20.00	3.46	1.45	7.27	2.25	.87	15.30
Epoxy coating, 2 part system - white (material #52)									
Roll 1st coat									
Slow	350	350	61.70	3.00	.86	17.63	6.66	4.22	32.37
Medium	550	325	54.00	3.05	1.06	16.62	5.08	2.84	28.65
Fast	725	300	46.30	3.10	1.30	15.43	3.67	1.41	24.91
Roll 2nd or additional coats									
Slow	425	375	61.70	2.47	.70	16.45	6.08	3.86	29.56
Medium	600	350	54.00	2.79	.98	15.43	4.70	2.63	26.53
Fast	775	325	46.30	2.90	1.21	14.25	3.40	1.31	23.07

Measurements are based on the square feet of wall coated. These figures assume paint products are being applied over a smooth finish. For heights above 8 feet, use the High Time Difficulty Factors on page 137. "Slow" work is based on a $10.50 hourly wage, "Medium" work on a $16.75 hourly wage, and "Fast" work on a $22.50 hourly wage. Other qualifications that apply to this table are on page 9.

	Labor SF per manhour	Material coverage SF/gallon	Material cost per gallon	Labor cost per 100 SF	Labor burden 100 SF	Material cost per 100 SF	Overhead per 100 SF	Profit per 100 SF	Total price per 100 SF

Walls, gypsum drywall, sand finish, spray, per 100 square feet of wall

	Labor SF per manhour	Material coverage SF/gallon	Material cost per gallon	Labor cost per 100 SF	Labor burden 100 SF	Material cost per 100 SF	Overhead per 100 SF	Profit per 100 SF	Total price per 100 SF
Flat latex, water base (material #5)									
Spray 1st coat									
Slow	700	275	17.10	1.50	.43	6.22	2.53	1.60	12.28
Medium	800	250	15.00	2.09	.73	6.00	2.16	1.21	12.19
Fast	900	225	12.90	2.50	1.04	5.73	1.71	.66	11.64
Spray 2nd coat									
Slow	800	325	17.10	1.31	.38	5.26	2.15	1.36	10.46
Medium	900	300	15.00	1.86	.65	5.00	1.84	1.03	10.38
Fast	1000	275	12.90	2.25	.94	4.69	1.46	.56	9.90
Spray 3rd or additional coats									
Slow	850	325	17.10	1.24	.35	5.26	2.12	1.35	10.32
Medium	950	313	15.00	1.76	.61	4.79	1.75	.98	9.89
Fast	1050	300	12.90	2.14	.89	4.30	1.36	.52	9.21
Sealer (drywall), water base (material #1)									
Spray prime coat									
Slow	575	275	16.30	1.83	.52	5.93	2.57	1.63	12.48
Medium	738	250	14.30	2.27	.80	5.72	2.15	1.20	12.14
Fast	900	225	12.30	2.50	1.04	5.47	1.67	.64	11.32
Sealer (drywall), oil base (material #2)									
Spray prime coat									
Slow	575	275	22.90	1.83	.52	8.33	3.31	2.10	16.09
Medium	738	250	20.10	2.27	.80	8.04	2.72	1.52	15.35
Fast	900	225	17.20	2.50	1.04	7.64	2.07	.80	14.05
Enamel, water base (material #9)									
Spray 1st finish coat									
Slow	500	275	24.50	2.10	.60	8.91	3.60	2.28	17.49
Medium	675	250	21.40	2.48	.86	8.56	2.92	1.63	16.45
Fast	850	225	18.40	2.65	1.11	8.18	2.21	.85	15.00
Spray 2nd or additional finish coats									
Slow	525	275	24.50	2.00	.56	8.91	3.56	2.26	17.29
Medium	700	263	21.40	2.39	.84	8.14	2.78	1.56	15.71
Fast	900	250	18.40	2.50	1.04	7.36	2.02	.78	13.70

	Labor SF per manhour	Material coverage SF/gallon	Material cost per gallon	Labor cost per 100 SF	Labor burden 100 SF	Material cost per 100 SF	Overhead per 100 SF	Profit per 100 SF	Total price per 100 SF
Enamel, oil base (material #10)									
Spray 1st finish coat									
Slow	500	250	26.60	2.10	.60	10.64	4.14	2.62	20.10
Medium	675	238	23.30	2.48	.86	9.79	3.22	1.80	18.15
Fast	850	225	20.00	2.65	1.11	8.89	2.34	.90	15.89
Spray 2nd or additional finish coats									
Slow	525	275	26.60	2.00	.56	9.67	3.79	2.40	18.42
Medium	700	263	23.30	2.39	.84	8.86	2.96	1.65	16.70
Fast	900	250	20.00	2.50	1.04	8.00	2.13	.82	14.49

Measurements are based on the square feet of wall coated. These figures assume paint products are being applied over a smooth finish. For heights above 8 feet, use the High Time Difficulty Factors on page 137. "Slow" work is based on a $10.50 hourly wage, "Medium" work on a $16.75 hourly wage, and "Fast" work on a $22.50 hourly wage. Other qualifications that apply to this table are on page 9.

	Labor SF per manhour	Material coverage SF/gallon	Material cost per gallon	Labor cost per 100 SF	Labor burden 100 SF	Material cost per 100 SF	Overhead per 100 SF	Profit per 100 SF	Total price per 100 SF
Walls, gypsum drywall, smooth finish, brush, per 100 square feet of wall									
Flat latex, water base (material #5)									
Brush 1st coat									
Slow	175	325	17.10	6.00	1.70	5.26	4.02	2.55	19.53
Medium	200	313	15.00	8.38	2.92	4.79	3.94	2.20	22.23
Fast	225	300	12.90	10.00	4.16	4.30	3.42	1.31	23.19
Brush 2nd coat									
Slow	225	400	17.10	4.67	1.32	4.28	3.19	2.02	15.48
Medium	250	375	15.00	6.70	2.34	4.00	3.19	1.78	18.01
Fast	275	350	12.90	8.18	3.43	3.69	2.83	1.09	19.22
Brush 3rd or additional coats									
Slow	250	425	17.10	4.20	1.19	4.02	2.92	1.85	14.18
Medium	275	400	15.00	6.09	2.13	3.75	2.93	1.64	16.54
Fast	300	375	12.90	7.50	3.12	3.44	2.60	1.00	17.66
Sealer (drywall), water base (material #1)									
Brush prime coat									
Slow	150	325	16.30	7.00	1.99	5.02	4.34	2.75	21.10
Medium	188	313	14.30	8.91	3.11	4.57	4.06	2.27	22.92
Fast	225	300	12.30	10.00	4.16	4.10	3.38	1.30	22.94
Sealer (drywall), oil base (material #2)									
Brush prime coat									
Slow	150	350	22.90	7.00	1.99	6.54	4.81	3.05	23.39
Medium	188	338	20.10	8.91	3.11	5.95	4.40	2.46	24.83
Fast	225	325	17.20	10.00	4.16	5.29	3.60	1.38	24.43
Enamel, water base (material #9)									
Brush 1st finish coat									
Slow	140	350	24.50	7.50	2.12	7.00	5.16	3.27	25.05
Medium	185	325	21.40	9.05	3.17	6.58	4.60	2.57	25.97
Fast	235	300	18.40	9.57	4.02	6.13	3.64	1.40	24.76
Brush 2nd or additional finish coats									
Slow	150	350	24.50	7.00	1.99	7.00	4.96	3.14	24.09
Medium	200	333	21.40	8.38	2.92	6.43	4.34	2.43	24.50
Fast	250	315	18.40	9.00	3.76	5.84	3.44	1.32	23.36
Enamel, oil base (material #10)									
Brush 1st finish coat									
Slow	140	350	26.60	7.50	2.12	7.60	5.34	3.39	25.95
Medium	185	338	23.30	9.05	3.17	6.89	4.68	2.61	26.40
Fast	235	325	20.00	9.57	4.02	6.15	3.65	1.40	24.79

	Labor SF per manhour	Material coverage SF/gallon	Material cost per gallon	Labor cost per 100 SF	Labor burden 100 SF	Material cost per 100 SF	Overhead per 100 SF	Profit per 100 SF	Total price per 100 SF
Brush 2nd or additional finish coats									
Slow	150	360	26.60	7.00	1.99	7.39	5.08	3.22	24.68
Medium	200	348	23.30	8.38	2.92	6.70	4.41	2.47	24.88
Fast	250	335	20.00	9.00	3.76	5.97	3.47	1.33	23.53
Stipple finish									
Slow	225	--	--	4.67	1.32	--	1.86	1.18	9.03
Medium	250	--	--	6.70	2.34	--	2.21	1.24	12.49
Fast	275	--	--	8.18	3.43	--	2.15	.83	14.59
Epoxy coating, 2 part system - white (material #52)									
Brush 1st coat									
Slow	225	425	61.70	4.67	1.32	14.52	6.36	4.03	30.90
Medium	250	400	54.00	6.70	2.34	13.50	5.52	3.09	31.15
Fast	275	375	46.30	8.18	3.43	12.35	4.43	1.70	30.09
Brush 2nd or additional coats									
Slow	250	450	61.70	4.20	1.19	13.71	5.92	3.75	28.77
Medium	275	425	54.00	6.09	2.13	12.71	5.13	2.87	28.93
Fast	300	400	46.30	7.50	3.12	11.58	4.11	1.58	27.89

Measurements are based on the square feet of wall coated. These figures assume paint products are being applied over a smooth finish. For heights above 8 feet, use the High Time Difficulty Factors on page 137. "Slow" work is based on a $10.50 hourly wage, "Medium" work on a $16.75 hourly wage, and "Fast" work on a $22.50 hourly wage. Other qualifications that apply to this table are on page 9.

	Labor SF per manhour	Material coverage SF/gallon	Material cost per gallon	Labor cost per 100 SF	Labor burden 100 SF	Material cost per 100 SF	Overhead per 100 SF	Profit per 100 SF	Total price per 100 SF

Walls, gypsum drywall, smooth finish, roll, per 100 square feet of wall

	Labor SF per manhour	Material coverage SF/gallon	Material cost per gallon	Labor cost per 100 SF	Labor burden 100 SF	Material cost per 100 SF	Overhead per 100 SF	Profit per 100 SF	Total price per 100 SF
Flat latex, water base (material #5)									
Roll 1st coat									
Slow	300	325	17.10	3.50	.99	5.26	3.03	1.92	14.70
Medium	513	313	15.00	3.27	1.14	4.79	2.25	1.26	12.71
Fast	725	300	12.90	3.10	1.30	4.30	1.61	.62	10.93
Roll 2nd coat									
Slow	375	375	17.10	2.80	.80	4.56	2.53	1.60	12.29
Medium	563	363	15.00	2.98	1.04	4.13	2.00	1.12	11.27
Fast	750	350	12.90	3.00	1.24	3.69	1.47	.56	9.96
Roll 3rd or additional coats									
Slow	450	400	17.10	2.33	.66	4.28	2.25	1.43	10.95
Medium	625	388	15.00	2.68	.93	3.87	1.83	1.02	10.33
Fast	800	375	12.90	2.81	1.18	3.44	1.37	.53	9.33
Sealer (drywall), water base (material #1)									
Roll prime coat									
Slow	245	350	16.30	4.29	1.21	4.66	3.15	2.00	15.31
Medium	485	325	14.30	3.45	1.20	4.40	2.22	1.24	12.51
Fast	725	300	12.30	3.10	1.30	4.10	1.57	.60	10.67
Sealer (drywall), oil base (material #2)									
Roll prime coat									
Slow	245	300	22.90	4.29	1.21	7.63	4.07	2.58	19.78
Medium	485	288	20.10	3.45	1.20	6.98	2.85	1.59	16.07
Fast	725	275	17.20	3.10	1.30	6.25	1.97	.76	13.38
Enamel, water base (material #9)									
Roll 1st finish coat									
Slow	235	325	24.50	4.47	1.27	7.54	4.12	2.61	20.01
Medium	438	313	21.40	3.82	1.33	6.84	2.94	1.64	16.57
Fast	640	300	18.40	3.52	1.45	6.13	2.06	.79	13.95
Roll 2nd or additional finish coats									
Slow	280	350	24.50	3.75	1.06	7.00	3.66	2.32	17.79
Medium	465	338	21.40	3.60	1.26	6.33	2.74	1.53	15.46
Fast	680	325	18.40	3.31	1.38	5.66	1.91	.74	13.00
Enamel, oil base (material #10)									
Roll 1st finish coat									
Slow	235	300	26.60	4.47	1.27	8.87	4.53	2.87	22.01
Medium	438	288	23.30	3.82	1.33	8.09	3.24	1.81	18.29
Fast	640	275	20.00	3.52	1.45	7.27	2.27	.87	15.38

	Labor SF per manhour	Material coverage SF/gallon	Material cost per gallon	Labor cost per 100 SF	Labor burden 100 SF	Material cost per 100 SF	Overhead per 100 SF	Profit per 100 SF	Total price per 100 SF
Roll 2nd or additional finish coats									
Slow	280	325	26.60	3.75	1.06	8.18	4.03	2.55	19.57
Medium	465	313	23.30	3.60	1.26	7.44	3.01	1.68	16.99
Fast	680	300	20.00	3.31	1.38	6.67	2.10	.81	14.27
Epoxy coating, 2 part system - white (material #52)									
Roll 1st coat									
Slow	375	400	61.70	2.80	.80	15.43	5.90	3.74	28.67
Medium	550	375	54.00	3.05	1.06	14.40	4.53	2.53	25.57
Fast	750	350	46.30	3.00	1.24	13.23	3.23	1.24	21.94
Roll 2nd or additional coats									
Slow	450	425	61.70	2.33	.66	14.52	5.43	3.44	26.38
Medium	625	400	54.00	2.68	.93	13.50	4.19	2.34	23.64
Fast	800	375	46.30	2.81	1.18	12.35	3.02	1.16	20.52

Measurements are based on the square feet of wall coated. These figures assume paint products are being applied over a smooth finish. For heights above 8 feet, use the High Time Difficulty Factors on page 137. "Slow" work is based on a $10.50 hourly wage, "Medium" work on a $16.75 hourly wage, and "Fast" work on a $22.50 hourly wage. Other qualifications that apply to this table are on page 9.

	Labor SF per manhour	Material coverage SF/gallon	Material cost per gallon	Labor cost per 100 SF	Labor burden 100 SF	Material cost per 100 SF	Overhead per 100 SF	Profit per 100 SF	Total price per 100 SF
Walls, gypsum drywall, smooth finish, spray, per 100 square feet of wall									
Flat latex, water base (material #5)									
Spray 1st coat									
Slow	750	300	17.10	1.40	.39	5.70	2.33	1.47	11.29
Medium	850	275	15.00	1.97	.70	5.45	1.99	1.11	11.22
Fast	950	250	12.90	2.37	.98	5.16	1.58	.61	10.70
Spray 2nd coat									
Slow	850	350	17.10	1.24	.35	4.89	2.01	1.27	9.76
Medium	950	325	15.00	1.76	.61	4.62	1.71	.96	9.66
Fast	1050	300	12.90	2.14	.89	4.30	1.36	.52	9.21
Spray 3rd or additional coats									
Slow	950	350	17.10	1.11	.31	4.89	1.96	1.24	9.51
Medium	1050	338	15.00	1.60	.55	4.44	1.62	.90	9.11
Fast	1150	325	12.90	1.96	.81	3.97	1.25	.48	8.47
Sealer (drywall), water base (material #1)									
Spray prime coat									
Slow	600	300	16.30	1.75	.50	5.43	2.38	1.51	11.57
Medium	775	275	14.30	2.16	.75	5.20	1.99	1.11	11.21
Fast	950	250	12.90	2.37	.98	5.16	1.58	.61	10.70
Sealer (drywall), oil base (material #2)									
Spray prime coat									
Slow	600	275	22.90	1.75	.50	8.33	3.28	2.08	15.94
Medium	775	250	20.10	2.16	.75	8.04	2.68	1.50	15.13
Fast	950	225	17.20	2.37	.98	7.64	2.04	.78	13.81
Enamel, water base (material #9)									
Spray 1st finish coat									
Slow	525	300	24.50	2.00	.56	8.17	3.33	2.11	16.17
Medium	713	275	21.40	2.35	.81	7.78	2.68	1.50	15.12
Fast	900	250	18.40	2.50	1.04	7.36	2.02	.78	13.70
Spray 2nd or additional finish coats									
Slow	600	300	24.50	1.75	.50	8.17	3.23	2.05	15.70
Medium	788	288	21.40	2.13	.74	7.43	2.52	1.41	14.23
Fast	975	275	18.40	2.31	.97	6.69	1.84	.71	12.52

	Labor SF per manhour	Material coverage SF/gallon	Material cost per gallon	Labor cost per 100 SF	Labor burden 100 SF	Material cost per 100 SF	Overhead per 100 SF	Profit per 100 SF	Total price per 100 SF
Enamel, oil base (material #10)									
Spray 1st finish coat									
Slow	525	275	26.60	2.00	.56	9.67	3.79	2.40	18.42
Medium	713	263	23.30	2.35	.81	8.86	2.95	1.65	16.62
Fast	900	250	20.00	2.50	1.04	8.00	2.13	.82	14.49
Spray 2nd or additional finish coats									
Slow	600	300	26.60	1.75	.50	8.87	3.45	2.19	16.76
Medium	788	288	23.30	2.13	.74	8.09	2.69	1.50	15.15
Fast	975	275	20.00	2.31	.97	7.27	1.95	.75	13.25

Measurements are based on the square feet of wall coated. These figures assume paint products are being applied over a smooth finish. For heights above 8 feet, use the High Time Difficulty Factors on page 137. "Slow" work is based on a $10.50 hourly wage, "Medium" work on a $16.75 hourly wage, and "Fast" work on a $22.50 hourly wage. Other qualifications that apply to this table are on page 9.

Walls, plaster, exterior: see Plaster and stucco

	Labor SF per manhour	Material coverage SF/gallon	Material cost per gallon	Labor cost per 100 SF	Labor burden 100 SF	Material cost per 100 SF	Overhead per 100 SF	Profit per 100 SF	Total price per 100 SF
Walls, plaster, interior, medium texture, per 100 square feet of wall									
Anti-graffiti stain eliminator									
Water base primer and sealer (material #39)									
Roll & brush each coat									
Slow	375	425	25.40	2.80	.80	5.98	2.97	1.88	14.43
Medium	400	400	22.30	4.19	1.46	5.58	2.75	1.54	15.52
Fast	425	375	19.10	5.29	2.20	5.09	2.33	.90	15.81
Oil base primer and sealer (material #40)									
Roll & brush each coat									
Slow	375	400	25.70	2.80	.80	6.43	3.11	1.97	15.11
Medium	400	375	22.50	4.19	1.46	6.00	2.85	1.60	16.10
Fast	425	350	19.30	5.29	2.20	5.51	2.41	.93	16.34
Polyurethane 2 part system (material #41)									
Roll & brush each coat									
Slow	325	375	91.40	3.23	.92	24.37	8.84	5.60	42.96
Medium	350	350	80.00	4.79	1.67	22.86	7.18	4.02	40.52
Fast	375	325	68.60	6.00	2.51	21.11	5.48	2.11	37.21

Measurements are based on the square feet of wall coated. These figures assume paint products are being applied over a smooth finish. For heights above 8 feet, use the High Time Difficulty Factors on page 137. "Slow" work is based on a $10.50 hourly wage, "Medium" work on a $16.75 hourly wage, and "Fast" work on a $22.50 hourly wage. Other qualifications that apply to this table are on page 9.

	Labor SF per manhour	Material coverage SF/gallon	Material cost per gallon	Labor cost per 100 SF	Labor burden 100 SF	Material cost per 100 SF	Overhead per 100 SF	Profit per 100 SF	Total price per 100 SF
Walls, plaster, interior, medium texture, brush, per 100 square feet of wall									
Flat latex, water base (material #5)									
Brush 1st coat									
Slow	125	300	17.10	8.40	2.38	5.70	5.11	3.24	24.83
Medium	150	288	15.00	11.17	3.90	5.21	4.97	2.78	28.03
Fast	175	275	12.90	12.86	5.35	4.69	4.24	1.63	28.77
Brush 2nd coat									
Slow	150	325	17.10	7.00	1.99	5.26	4.42	2.80	21.47
Medium	168	313	15.00	9.97	3.47	4.79	4.47	2.50	25.20
Fast	185	300	12.90	12.16	5.09	4.30	3.98	1.53	27.06
Brush 3rd or additional coats									
Slow	160	350	17.10	6.56	1.87	4.89	4.13	2.62	20.07
Medium	185	338	15.00	9.05	3.17	4.44	4.08	2.28	23.02
Fast	210	325	12.90	10.71	4.47	3.97	3.54	1.36	24.05
Enamel, water base (material #9)									
Brush 1st finish coat									
Slow	100	300	24.50	10.50	2.98	8.17	6.71	4.26	32.62
Medium	125	288	21.40	13.40	4.67	7.43	6.25	3.49	35.24
Fast	150	275	18.40	15.00	6.27	6.69	5.17	1.99	35.12
Brush 2nd finish coat									
Slow	125	325	24.50	8.40	2.38	7.54	5.68	3.60	27.60
Medium	143	313	21.40	11.71	4.08	6.84	5.54	3.10	31.27
Fast	160	300	18.40	14.06	5.87	6.13	4.82	1.85	32.73
Brush 3rd or additional finish coats									
Slow	135	350	24.50	7.78	2.21	7.00	5.27	3.34	25.60
Medium	160	338	21.40	10.47	3.65	6.33	5.01	2.80	28.26
Fast	185	325	18.40	12.16	5.09	5.66	4.24	1.63	28.78
Enamel, oil base (material #10)									
Brush 1st finish coat									
Slow	100	325	26.60	10.50	2.98	8.18	6.72	4.26	32.64
Medium	125	313	23.30	13.40	4.67	7.44	6.25	3.49	35.25
Fast	150	300	20.00	15.00	6.27	6.67	5.17	1.99	35.10
Brush 2nd finish coat									
Slow	125	400	26.60	8.40	2.38	6.65	5.41	3.43	26.27
Medium	143	375	23.30	11.71	4.08	6.21	5.39	3.01	30.40
Fast	160	350	20.00	14.06	5.87	5.71	4.74	1.82	32.20

	Labor SF per manhour	Material coverage SF/gallon	Material cost per gallon	Labor cost per 100 SF	Labor burden 100 SF	Material cost per 100 SF	Overhead per 100 SF	Profit per 100 SF	Total price per 100 SF
Brush 3rd or additional finish coats									
Slow	135	425	26.60	7.78	2.21	6.26	5.04	3.19	24.48
Medium	160	400	23.30	10.47	3.65	5.83	4.89	2.73	27.57
Fast	185	375	20.00	12.16	5.09	5.33	4.18	1.61	28.37
Stipple finish									
Slow	125	--	--	8.40	2.38	--	3.34	2.12	16.24
Medium	143	--	--	11.71	4.08	--	3.87	2.16	21.82
Fast	160	--	--	14.06	5.87	--	3.69	1.42	25.04
Epoxy coating, 2 part system - white (material #52)									
Brush 1st coat									
Slow	150	400	61.70	7.00	1.99	15.43	7.57	4.80	36.79
Medium	165	388	54.00	10.15	3.54	13.92	6.76	3.78	38.15
Fast	185	375	46.30	12.16	5.09	12.35	5.47	2.10	37.17
Brush 2nd or additional coats									
Slow	160	425	61.70	6.56	1.87	14.52	7.11	4.51	34.57
Medium	185	413	54.00	9.05	3.17	13.08	6.19	3.46	34.95
Fast	210	400	46.30	10.71	4.47	11.58	4.95	1.90	33.61
Glazing & mottling over enamel (material #16)									
Brush each coat									
Slow	50	900	31.20	21.00	5.96	3.47	9.44	5.98	45.85
Medium	65	800	27.30	25.77	8.97	3.41	9.35	5.23	52.73
Fast	80	700	23.40	28.13	11.73	3.34	7.99	3.07	54.26
Stipple									
Slow	100	--	--	10.50	2.98	--	4.18	2.65	20.31
Medium	113	--	--	14.82	5.17	--	4.90	2.74	27.63
Fast	125	--	--	18.00	7.51	--	4.72	1.81	32.04

Measurements are based on the square feet of wall coated. These figures assume paint products are being applied over a smooth finish. For heights above 8 feet, use the High Time Difficulty Factors on page 137. "Slow" work is based on a $10.50 hourly wage, "Medium" work on a $16.75 hourly wage, and "Fast" work on a $22.50 hourly wage. Other qualifications that apply to this table are on page 9.

	Labor SF per manhour	Material coverage SF/gallon	Material cost per gallon	Labor cost per 100 SF	Labor burden 100 SF	Material cost per 100 SF	Overhead per 100 SF	Profit per 100 SF	Total price per 100 SF

Walls, plaster, interior, medium texture, roll, per 100 square feet of wall

	Labor SF per manhour	Material coverage SF/gallon	Material cost per gallon	Labor cost per 100 SF	Labor burden 100 SF	Material cost per 100 SF	Overhead per 100 SF	Profit per 100 SF	Total price per 100 SF
Flat latex, water base (material #5)									
Roll 1st coat									
Slow	225	250	17.10	4.67	1.32	6.84	3.98	2.52	19.33
Medium	438	238	15.00	3.82	1.33	6.30	2.81	1.57	15.83
Fast	650	225	12.90	3.46	1.45	5.73	1.97	.76	13.37
Roll 2nd coat									
Slow	250	300	17.10	4.20	1.19	5.70	3.44	2.18	16.71
Medium	463	288	15.00	3.62	1.26	5.21	2.47	1.38	13.94
Fast	675	275	12.90	3.33	1.39	4.69	1.74	.67	11.82
Roll 3rd or additional coats									
Slow	275	325	17.10	3.82	1.09	5.26	3.15	2.00	15.32
Medium	500	313	15.00	3.35	1.17	4.79	2.28	1.27	12.86
Fast	725	300	12.90	3.10	1.30	4.30	1.61	.62	10.93
Enamel, water base (material #9)									
Roll 1st finish coat									
Slow	200	250	24.50	5.25	1.49	9.80	5.13	3.25	24.92
Medium	413	238	21.40	4.06	1.41	8.99	3.54	1.98	19.98
Fast	625	225	18.40	3.60	1.50	8.18	2.46	.94	16.68
Roll 2nd finish coat									
Slow	225	300	24.50	4.67	1.32	8.17	4.39	2.78	21.33
Medium	438	288	21.40	3.82	1.33	7.43	3.08	1.72	17.38
Fast	650	275	18.40	3.46	1.45	6.69	2.14	.82	14.56
Roll 3rd or additional finish coats									
Slow	250	325	24.50	4.20	1.19	7.54	4.01	2.54	19.48
Medium	475	313	21.40	3.53	1.24	6.84	2.84	1.59	16.04
Fast	700	300	18.40	3.21	1.35	6.13	1.98	.76	13.43
Enamel, oil base (material #10)									
Roll 1st finish coat									
Slow	200	275	26.60	5.25	1.49	9.67	5.09	3.23	24.73
Medium	413	263	23.30	4.06	1.41	8.86	3.51	1.96	19.80
Fast	625	250	20.00	3.60	1.50	8.00	2.42	.93	16.45
Roll 2nd finish coat									
Slow	225	350	26.60	4.67	1.32	7.60	4.22	2.67	20.48
Medium	438	325	23.30	3.82	1.33	7.17	3.02	1.69	17.03
Fast	650	300	20.00	3.46	1.45	6.67	2.14	.82	14.54

	Labor SF per manhour	Material coverage SF/gallon	Material cost per gallon	Labor cost per 100 SF	Labor burden 100 SF	Material cost per 100 SF	Overhead per 100 SF	Profit per 100 SF	Total price per 100 SF
Roll 3rd or additional finish coats									
Slow	250	375	26.60	4.20	1.19	7.09	3.87	2.45	18.80
Medium	475	350	23.30	3.53	1.24	6.66	2.80	1.56	15.79
Fast	700	325	20.00	3.21	1.35	6.15	1.98	.76	13.45
Epoxy coating, 2 part system - white (material #52)									
Roll 1st coat									
Slow	250	350	61.70	4.20	1.19	17.63	7.14	4.52	34.68
Medium	463	335	54.00	3.62	1.26	16.12	5.15	2.88	29.03
Fast	675	320	46.30	3.33	1.39	14.47	3.55	1.36	24.10
Roll 2nd or additional coats									
Slow	275	400	61.70	3.82	1.09	15.43	6.31	4.00	30.65
Medium	500	375	54.00	3.35	1.17	14.40	4.64	2.59	26.15
Fast	725	350	46.30	3.10	1.30	13.23	3.26	1.25	22.14

Measurements are based on the square feet of wall coated. These figures assume paint products are being applied over a smooth finish. For heights above 8 feet, use the High Time Difficulty Factors on page 137. "Slow" work is based on a $10.50 hourly wage, "Medium" work on a $16.75 hourly wage, and "Fast" work on a $22.50 hourly wage. Other qualifications that apply to this table are on page 9.

	Labor SF per manhour	Material coverage SF/gallon	Material cost per gallon	Labor cost per 100 SF	Labor burden 100 SF	Material cost per 100 SF	Overhead per 100 SF	Profit per 100 SF	Total price per 100 SF
Walls, plaster, interior, medium texture, spray, per 100 square feet of wall									
Flat latex, water base (material #5)									
Spray 1st coat									
Slow	475	350	17.10	2.21	.63	4.89	2.40	1.52	11.65
Medium	600	313	15.00	2.79	.98	4.79	2.09	1.17	11.82
Fast	725	275	12.90	3.10	1.30	4.69	1.68	.65	11.42
Spray 2nd coat									
Slow	525	400	17.10	2.00	.56	4.28	2.12	1.35	10.31
Medium	675	350	15.00	2.48	.86	4.29	1.87	1.05	10.55
Fast	825	300	12.90	2.73	1.13	4.30	1.51	.58	10.25
Spray 3rd or additional coats									
Slow	575	450	17.10	1.83	.52	3.80	1.91	1.21	9.27
Medium	750	388	15.00	2.23	.77	3.87	1.69	.94	9.50
Fast	925	325	12.90	2.43	1.01	3.97	1.37	.53	9.31
Enamel, water base (material #9)									
Spray 1st finish coat									
Slow	450	350	24.50	2.33	.66	7.00	3.10	1.96	15.05
Medium	575	313	21.40	2.91	1.02	6.84	2.64	1.47	14.88
Fast	700	275	18.40	3.21	1.35	6.69	2.08	.80	14.13
Spray 2nd finish coat									
Slow	500	400	24.50	2.10	.60	6.13	2.74	1.74	13.31
Medium	650	350	21.40	2.58	.90	6.11	2.35	1.31	13.25
Fast	800	300	18.40	2.81	1.18	6.13	1.87	.72	12.71
Spray 3rd or additional finish coats									
Slow	550	450	24.50	1.91	.54	5.44	2.45	1.55	11.89
Medium	750	388	21.40	2.23	.77	5.52	2.09	1.17	11.78
Fast	900	325	18.40	2.50	1.04	5.66	1.70	.65	11.55
Enamel, oil base (material #10)									
Spray 1st finish coat									
Slow	450	400	26.60	2.33	.66	6.65	2.99	1.89	14.52
Medium	575	363	23.30	2.91	1.02	6.42	2.53	1.42	14.30
Fast	700	325	20.00	3.21	1.35	6.15	1.98	.76	13.45
Spray 2nd finish coat									
Slow	500	425	26.60	2.10	.60	6.26	2.78	1.76	13.50
Medium	650	388	23.30	2.58	.90	6.01	2.33	1.30	13.12
Fast	800	350	20.00	2.81	1.18	5.71	1.79	.69	12.18

	Labor SF per manhour	Material coverage SF/gallon	Material cost per gallon	Labor cost per 100 SF	Labor burden 100 SF	Material cost per 100 SF	Overhead per 100 SF	Profit per 100 SF	Total price per 100 SF
Spray 3rd or additional finish coats									
Slow	550	450	26.60	1.91	.54	5.91	2.59	1.64	12.59
Medium	725	400	23.30	2.31	.81	5.83	2.19	1.23	12.37
Fast	900	375	20.00	2.50	1.04	5.33	1.64	.63	11.14
Epoxy coating, 2 part system - white (material #52)									
Spray 1st coat									
Slow	525	325	61.70	2.00	.56	18.98	6.68	4.23	32.45
Medium	675	300	54.00	2.48	.86	18.00	5.23	2.92	29.49
Fast	825	275	46.30	2.73	1.13	16.84	3.83	1.47	26.00
Spray 2nd or additional coats									
Slow	575	350	61.70	1.83	.52	17.63	6.19	3.93	30.10
Medium	725	325	54.00	2.31	.81	16.62	4.84	2.70	27.28
Fast	875	300	46.30	2.57	1.07	15.43	3.53	1.36	23.96

Measurements are based on the square feet of wall coated. These figures assume paint products are being applied over a smooth finish. For heights above 8 feet, use the High Time Difficulty Factors on page 137. "Slow" work is based on a $10.50 hourly wage, "Medium" work on a $16.75 hourly wage, and "Fast" work on a $22.50 hourly wage. Other qualifications that apply to this table are on page 9.

	Labor SF per manhour	Material coverage SF/gallon	Material cost per gallon	Labor cost per 100 SF	Labor burden 100 SF	Material cost per 100 SF	Overhead per 100 SF	Profit per 100 SF	Total price per 100 SF
Walls, plaster, interior, rough texture, per 100 square feet of wall									
Anti-graffiti stain eliminator									
Water base primer and sealer (material #39)									
Roll & brush each coat									
Slow	350	400	25.40	3.00	.86	6.35	3.16	2.00	15.37
Medium	375	375	22.30	4.47	1.56	5.95	2.94	1.64	16.56
Fast	400	350	19.10	5.63	2.34	5.46	2.49	.96	16.88
Oil base primer and sealer (material #40)									
Roll & brush each coat									
Slow	350	375	25.70	3.00	.86	6.85	3.32	2.10	16.13
Medium	375	350	22.50	4.47	1.56	6.43	3.05	1.71	17.22
Fast	400	325	19.30	5.63	2.34	5.94	2.58	.99	17.48
Polyurethane 2 part system (material #41)									
Roll & brush each coat									
Slow	300	350	91.40	3.50	.99	26.11	9.49	6.02	46.11
Medium	325	325	80.00	5.15	1.81	24.62	7.73	4.32	43.63
Fast	350	300	68.60	6.43	2.69	22.87	5.92	2.27	40.18

Measurements are based on the square feet of wall coated. These figures assume paint products are being applied over a rough finish, sand finish, or orange peel texture finish. For heights above 8 feet, use the High Time Difficulty Factors on page 137. "Slow" work is based on a $10.50 hourly wage, "Medium" work on a $16.75 hourly wage, and "Fast" work on a $22.50 hourly wage. Other qualifications that apply to this table are on page 9.

	Labor SF per manhour	Material coverage SF/gallon	Material cost per gallon	Labor cost per 100 SF	Labor burden 100 SF	Material cost per 100 SF	Overhead per 100 SF	Profit per 100 SF	Total price per 100 SF

Walls, plaster, interior, rough texture, brush, per 100 square feet of wall

	Labor SF per manhour	Material coverage SF/gallon	Material cost per gallon	Labor cost per 100 SF	Labor burden 100 SF	Material cost per 100 SF	Overhead per 100 SF	Profit per 100 SF	Total price per 100 SF
Flat latex, water base (material #5)									
Brush 1st coat									
Slow	115	300	17.10	9.13	2.60	5.70	5.40	3.42	26.25
Medium	140	275	15.00	11.96	4.17	5.45	5.29	2.96	29.83
Fast	165	250	12.90	13.64	5.69	5.16	4.53	1.74	30.76
Brush 2nd coat									
Slow	125	325	17.10	8.40	2.38	5.26	4.98	3.15	24.17
Medium	153	300	15.00	10.95	3.82	5.00	4.84	2.71	27.32
Fast	180	275	12.90	12.50	5.23	4.69	4.15	1.59	28.16
Brush 3rd or additional coats									
Slow	135	350	17.10	7.78	2.21	4.89	4.61	2.92	22.41
Medium	168	325	15.00	9.97	3.47	4.62	4.42	2.47	24.95
Fast	200	300	12.90	11.25	4.70	4.30	3.75	1.44	25.44
Enamel, water base (material #9)									
Brush 1st finish coat									
Slow	100	300	24.50	10.50	2.98	8.17	6.71	4.26	32.62
Medium	125	275	21.40	13.40	4.67	7.78	6.33	3.54	35.72
Fast	150	250	18.40	15.00	6.27	7.36	5.29	2.03	35.95
Brush 2nd finish coat									
Slow	115	325	24.50	9.13	2.60	7.54	5.97	3.79	29.03
Medium	140	300	21.40	11.96	4.17	7.13	5.70	3.19	32.15
Fast	165	275	18.40	13.64	5.69	6.69	4.81	1.85	32.68
Brush 3rd or additional finish coats									
Slow	125	350	24.50	8.40	2.38	7.00	5.51	3.50	26.79
Medium	160	325	21.40	10.47	3.65	6.58	5.07	2.83	28.60
Fast	185	300	18.40	12.16	5.09	6.13	4.32	1.66	29.36
Enamel, oil base (material #10)									
Brush 1st finish coat									
Slow	100	300	26.60	10.50	2.98	8.87	6.93	4.39	33.67
Medium	125	288	23.30	13.40	4.67	8.09	6.41	3.58	36.15
Fast	150	275	20.00	15.00	6.27	7.27	5.28	2.03	35.85
Brush 2nd finish coat									
Slow	115	375	26.60	9.13	2.60	7.09	5.83	3.70	28.35
Medium	140	350	23.30	11.96	4.17	6.66	5.58	3.12	31.49
Fast	165	325	20.00	13.64	5.69	6.15	4.71	1.81	32.00

	Labor SF per manhour	Material coverage SF/gallon	Material cost per gallon	Labor cost per 100 SF	Labor burden 100 SF	Material cost per 100 SF	Overhead per 100 SF	Profit per 100 SF	Total price per 100 SF
Brush 3rd or additional finish coats									
Slow	125	400	26.60	8.40	2.38	6.65	5.41	3.43	26.27
Medium	160	375	23.30	10.47	3.65	6.21	4.98	2.78	28.09
Fast	185	350	20.00	12.16	5.09	5.71	4.25	1.63	28.84
Epoxy coating, 2 part system - white (material #52)									
Brush 1st coat									
Slow	125	375	61.70	8.40	2.38	16.45	8.44	5.35	41.02
Medium	160	363	54.00	10.47	3.65	14.88	7.11	3.97	40.08
Fast	180	350	46.30	12.50	5.23	13.23	5.73	2.20	38.89
Brush 2nd or additional coats									
Slow	135	400	61.70	7.78	2.21	15.43	7.88	5.00	38.30
Medium	175	388	54.00	9.57	3.33	13.92	6.57	3.67	37.06
Fast	200	375	46.30	11.25	4.70	12.35	5.24	2.01	35.55
Glazing & mottling over enamel (material #16)									
Brush each coat									
Slow	40	875	31.20	26.25	7.45	3.57	11.56	7.33	56.16
Medium	50	838	27.30	33.50	11.68	3.26	11.87	6.63	66.94
Fast	60	800	23.40	37.50	15.66	2.93	10.38	3.99	70.46
Stipple									
Slow	90	--	--	11.67	3.31	--	4.65	2.95	22.58
Medium	103	--	--	16.26	5.67	--	5.37	3.00	30.30
Fast	115	--	--	19.57	8.17	--	5.13	1.97	34.84

Measurements are based on the square feet of wall coated. These figures assume paint products are being applied over a rough finish. For heights above 8 feet, use the High Time Difficulty Factors on page 137. "Slow" work is based on a $10.50 hourly wage, "Medium" work on a $16.75 hourly wage, and "Fast" work on a $22.50 hourly wage. Other qualifications that apply to this table are on page 9.

	Labor SF per manhour	Material coverage SF/gallon	Material cost per gallon	Labor cost per 100 SF	Labor burden 100 SF	Material cost per 100 SF	Overhead per 100 SF	Profit per 100 SF	Total price per 100 SF

Walls, plaster, interior, rough texture, roll, per 100 square feet of wall

Flat latex, water base (material #5)									
Roll 1st coat									
Slow	200	250	17.10	5.25	1.49	6.84	4.21	2.67	20.46
Medium	413	238	15.00	4.06	1.41	6.30	2.88	1.61	16.26
Fast	625	225	12.90	3.60	1.50	5.73	2.00	.77	13.60
Roll 2nd coat									
Slow	225	300	17.10	4.67	1.32	5.70	3.63	2.30	17.62
Medium	438	288	15.00	3.82	1.33	5.21	3.21	2.04	15.61
Fast	650	275	12.90	3.46	1.45	4.69	1.77	.68	12.05
Roll 3rd or additional coats									
Slow	250	325	17.10	4.20	1.19	5.26	3.30	2.09	16.04
Medium	463	313	15.00	3.62	1.26	4.79	2.37	1.32	13.36
Fast	675	300	12.90	3.33	1.39	4.30	1.67	.64	11.33
Enamel, water base (material #9)									
Roll 1st finish coat									
Slow	175	250	24.50	6.00	1.70	9.80	5.43	3.44	26.37
Medium	388	238	21.40	4.32	1.51	8.99	3.63	2.03	20.48
Fast	600	225	18.40	3.75	1.58	8.18	2.50	.96	16.97
Roll 2nd finish coat									
Slow	200	325	24.50	5.25	1.49	7.54	4.43	2.81	21.52
Medium	413	313	21.40	4.06	1.41	6.84	3.02	1.69	17.02
Fast	625	300	18.40	3.60	1.50	6.13	2.08	.80	14.11
Roll 3rd or additional finish coats									
Slow	225	325	24.50	4.67	1.32	7.54	4.20	2.66	20.39
Medium	438	313	21.40	3.82	1.33	6.84	2.94	1.64	16.57
Fast	650	300	18.40	3.46	1.45	6.13	2.04	.78	13.86
Enamel, oil base (material #10)									
Roll 1st finish coat									
Slow	175	300	26.60	6.00	1.70	8.87	5.14	3.26	24.97
Medium	388	288	23.30	4.32	1.51	8.09	3.41	1.91	19.24
Fast	600	275	20.00	3.75	1.58	7.27	2.33	.90	15.83
Roll 2nd finish coat									
Slow	200	350	26.60	5.25	1.49	7.60	4.45	2.82	21.61
Medium	413	325	23.30	4.06	1.41	7.17	3.10	1.73	17.47
Fast	625	300	20.00	3.60	1.50	6.67	2.18	.84	14.79

	Labor SF per manhour	Material coverage SF/gallon	Material cost per gallon	Labor cost per 100 SF	Labor burden 100 SF	Material cost per 100 SF	Overhead per 100 SF	Profit per 100 SF	Total price per 100 SF
Roll 3rd or additional finish coats									
Slow	225	375	26.60	4.67	1.32	7.09	4.06	2.57	19.71
Medium	438	350	23.30	3.82	1.33	6.66	2.89	1.62	16.32
Fast	650	325	20.00	3.46	1.45	6.15	2.04	.79	13.89
Epoxy coating, 2 part system - white (material #52)									
Roll 1st coat									
Slow	225	350	61.70	4.67	1.32	17.63	7.33	4.64	35.59
Medium	438	325	54.00	3.82	1.33	16.62	5.33	2.98	30.08
Fast	650	300	46.30	3.46	1.45	15.43	3.76	1.45	25.55
Roll 2nd or additional coats									
Slow	250	400	61.70	4.20	1.19	15.43	6.45	4.09	31.36
Medium	463	375	54.00	3.62	1.26	14.40	4.72	2.64	26.64
Fast	675	350	46.30	3.33	1.39	13.23	3.32	1.28	22.55

Measurements are based on the square feet of wall coated. These figures assume paint products are being applied over a rough finish. For heights above 8 feet, use the High Time Difficulty Factors on page 137. "Slow" work is based on a $10.50 hourly wage, "Medium" work on a $16.75 hourly wage, and "Fast" work on a $22.50 hourly wage. Other qualifications that apply to this table are on page 9.

	Labor SF per manhour	Material coverage SF/gallon	Material cost per gallon	Labor cost per 100 SF	Labor burden 100 SF	Material cost per 100 SF	Overhead per 100 SF	Profit per 100 SF	Total price per 100 SF

Walls, plaster, interior, rough texture, spray, per 100 square feet of wall

Flat latex, water base (material #5)									
Spray 1st coat									
Slow	500	325	17.10	2.10	.60	5.26	2.47	1.56	11.99
Medium	600	288	15.00	2.79	.98	5.21	2.20	1.23	12.41
Fast	700	250	12.90	3.21	1.35	5.16	1.80	.69	12.21
Spray 2nd coat									
Slow	600	400	17.10	1.75	.50	4.28	2.02	1.28	9.83
Medium	700	350	15.00	2.39	.84	4.29	1.84	1.03	10.39
Fast	800	300	12.90	2.81	1.18	4.30	1.53	.59	10.41
Spray 3rd or additional coats									
Slow	700	425	17.10	1.50	.43	4.02	1.84	1.17	8.96
Medium	800	375	15.00	2.09	.73	4.00	1.67	.93	9.42
Fast	900	325	12.90	2.50	1.04	3.97	1.39	.53	9.43
Enamel, water base (material #9)									
Spray 1st finish coat									
Slow	450	325	24.50	2.33	.66	7.54	3.26	2.07	15.86
Medium	550	288	21.40	3.05	1.06	7.43	2.83	1.58	15.95
Fast	650	250	18.40	3.46	1.45	7.36	2.27	.87	15.41
Spray 2nd finish coat									
Slow	550	400	24.50	1.91	.54	6.13	2.66	1.69	12.93
Medium	650	350	21.40	2.58	.90	6.11	2.35	1.31	13.25
Fast	750	300	18.40	3.00	1.24	6.13	1.92	.74	13.03
Spray 3rd or additional finish coats									
Slow	650	425	24.50	1.62	.46	5.76	2.43	1.54	11.81
Medium	750	375	21.40	2.23	.77	5.71	2.14	1.19	12.04
Fast	850	325	18.40	2.65	1.11	5.66	1.74	.67	11.83
Enamel, oil base (material #10)									
Spray 1st finish coat									
Slow	450	325	26.60	2.33	.66	8.18	3.46	2.19	16.82
Medium	550	300	23.30	3.05	1.06	7.77	2.91	1.63	16.42
Fast	650	275	20.00	3.46	1.45	7.27	2.25	.87	15.30
Spray 2nd finish coat									
Slow	550	400	26.60	1.91	.54	6.65	2.82	1.79	13.71
Medium	650	362	23.30	2.58	.90	6.44	2.43	1.36	13.71
Fast	750	325	20.00	3.00	1.24	6.15	1.92	.74	13.05

	Labor SF per manhour	Material coverage SF/gallon	Material cost per gallon	Labor cost per 100 SF	Labor burden 100 SF	Material cost per 100 SF	Overhead per 100 SF	Profit per 100 SF	Total price per 100 SF
Spray 3rd or additional finish coats									
Slow	650	425	26.60	1.62	.46	6.26	2.59	1.64	12.57
Medium	750	388	23.30	2.23	.77	6.01	2.21	1.24	12.46
Fast	850	350	20.00	2.65	1.11	5.71	1.75	.67	11.89
Epoxy coating, 2 part system - white (material #52)									
Spray 1st coat									
Slow	525	325	61.70	2.00	.56	18.98	6.68	4.23	32.45
Medium	663	313	54.00	2.53	.88	17.25	5.06	2.83	28.55
Fast	800	300	46.30	2.81	1.18	15.43	3.59	1.38	24.39
Spray 2nd or additional coats									
Slow	575	375	61.70	1.83	.52	16.45	5.83	3.69	28.32
Medium	713	363	54.00	2.35	.81	14.88	4.42	2.47	24.93
Fast	850	350	46.30	2.65	1.11	13.23	3.14	1.21	21.34

Measurements are based on the square feet of wall coated. These figures assume paint products are being applied over a rough finish. For heights above 8 feet, use the High Time Difficulty Factors on page 137. "Slow" work is based on a $10.50 hourly wage, "Medium" work on a $16.75 hourly wage, and "Fast" work on a $22.50 hourly wage. Other qualifications that apply to this table are on page 9.

	Labor SF per manhour	Material coverage SF/gallon	Material cost per gallon	Labor cost per 100 SF	Labor burden 100 SF	Material cost per 100 SF	Overhead per 100 SF	Profit per 100 SF	Total price per 100 SF

Walls, plaster, interior, smooth finish, per 100 square feet of wall

	Labor SF per manhour	Material coverage SF/gallon	Material cost per gallon	Labor cost per 100 SF	Labor burden 100 SF	Material cost per 100 SF	Overhead per 100 SF	Profit per 100 SF	Total price per 100 SF
Anti-graffiti stain eliminator									
Water base primer and sealer (material #39)									
Roll & brush each coat									
Slow	375	425	25.40	2.80	.80	5.98	2.97	1.88	14.43
Medium	400	400	22.30	4.19	1.46	5.58	2.75	1.54	15.52
Fast	425	375	19.10	5.29	2.20	5.09	2.33	.90	15.81
Oil base primer and sealer (material #40)									
Roll & brush each coat									
Slow	375	400	25.70	2.80	.80	6.43	3.11	1.97	15.11
Medium	400	375	22.50	4.19	1.46	6.00	2.85	1.60	16.10
Fast	425	350	19.30	5.29	2.20	5.51	2.41	.93	16.34
Polyurethane 2 part system (material #41)									
Roll & brush each coat									
Slow	325	375	91.40	3.23	.92	24.37	8.84	5.60	42.96
Medium	350	350	80.00	4.79	1.67	22.86	7.18	4.02	40.52
Fast	375	325	68.60	6.00	2.51	21.11	5.48	2.11	37.21

Measurements are based on the square feet of wall coated. These figures assume paint products are being applied over a smooth finish. For heights above 8 feet, use the High Time Difficulty Factors on page 137. "Slow" work is based on a $10.50 hourly wage, "Medium" work on a $16.75 hourly wage, and "Fast" work on a $22.50 hourly wage. Other qualifications that apply to this table are on page 9.

	Labor SF per manhour	Material coverage SF/gallon	Material cost per gallon	Labor cost per 100 SF	Labor burden 100 SF	Material cost per 100 SF	Overhead per 100 SF	Profit per 100 SF	Total price per 100 SF
Walls, plaster, interior, smooth finish, brush, per 100 square feet of wall									
Flat latex, water base (material #5)									
Brush 1st coat									
Slow	150	350	17.10	7.00	1.99	4.89	4.30	2.73	20.91
Medium	175	325	15.00	9.57	3.33	4.62	4.29	2.40	24.21
Fast	200	300	12.90	11.25	4.70	4.30	3.75	1.44	25.44
Brush 2nd coat									
Slow	175	375	17.10	6.00	1.70	4.56	3.80	2.41	18.47
Medium	200	350	15.00	8.38	2.92	4.29	3.82	2.14	21.55
Fast	225	325	12.90	10.00	4.16	3.97	3.36	1.29	22.78
Brush 3rd or additional coats									
Slow	200	400	17.10	5.25	1.49	4.28	3.42	2.17	16.61
Medium	225	375	15.00	7.44	2.59	4.00	3.44	1.92	19.39
Fast	250	350	12.90	9.00	3.76	3.69	3.04	1.17	20.66
Enamel, water base (material #9)									
Brush 1st finish coat									
Slow	125	350	24.50	8.40	2.38	7.00	5.51	3.50	26.79
Medium	163	325	21.40	10.28	3.57	6.58	5.01	2.80	28.24
Fast	200	300	18.40	11.25	4.70	6.13	4.08	1.57	27.73
Brush 2nd finish coat									
Slow	150	375	24.50	7.00	1.99	6.53	4.81	3.05	23.38
Medium	175	350	21.40	9.57	3.33	6.11	4.66	2.60	26.27
Fast	200	325	18.40	11.25	4.70	5.66	4.00	1.54	27.15
Brush 3rd or additional finish coats									
Slow	175	400	24.50	6.00	1.70	6.13	4.29	2.72	20.84
Medium	200	375	21.40	8.38	2.92	5.71	4.17	2.33	23.51
Fast	225	350	18.40	10.00	4.16	5.26	3.60	1.38	24.40
Enamel, oil base (material #10)									
Brush 1st finish coat									
Slow	125	400	26.60	8.40	2.38	6.65	5.41	3.43	26.27
Medium	163	375	23.30	10.28	3.57	6.21	4.92	2.75	27.73
Fast	200	350	20.00	11.25	4.70	5.71	4.01	1.54	27.21
Brush 2nd finish coat									
Slow	150	425	24.50	7.00	1.99	5.76	4.57	2.90	22.22
Medium	175	400	21.40	9.57	3.33	5.35	4.47	2.50	25.22
Fast	200	375	18.40	11.25	4.70	4.91	3.86	1.48	26.20

	Labor SF per manhour	Material coverage SF/gallon	Material cost per gallon	Labor cost per 100 SF	Labor burden 100 SF	Material cost per 100 SF	Overhead per 100 SF	Profit per 100 SF	Total price per 100 SF
Brush 3rd or additional finish coats									
Slow	175	450	26.60	6.00	1.70	5.91	4.22	2.68	20.51
Medium	200	425	23.30	8.38	2.92	5.48	4.11	2.30	23.19
Fast	225	400	20.00	10.00	4.16	5.00	3.55	1.36	24.07
Stipple finish									
Slow	130	--	--	8.08	2.29	--	3.22	2.04	15.63
Medium	150	--	--	11.17	3.90	--	3.69	2.06	20.82
Fast	170	--	--	13.24	5.51	--	3.47	1.33	23.55
Epoxy coating, 2 part system - white (material #52)									
Brush 1st coat									
Slow	175	400	61.70	6.00	1.70	15.43	7.17	4.55	34.85
Medium	200	388	54.00	8.38	2.92	13.92	6.18	3.45	34.85
Fast	225	375	46.30	10.00	4.16	12.35	4.91	1.89	33.31
Brush 2nd or additional coats									
Slow	200	425	61.70	5.25	1.49	14.52	6.59	4.18	32.03
Medium	225	413	54.00	7.44	2.59	13.08	5.66	3.16	31.93
Fast	250	400	46.30	9.00	3.76	11.58	4.50	1.73	30.57
Glazing & mottling over enamel (material #16)									
Brush each coat									
Slow	75	900	31.20	14.00	3.97	3.47	6.65	4.22	32.31
Medium	98	850	27.30	17.09	5.95	3.21	6.43	3.60	36.28
Fast	120	800	23.40	18.75	7.81	2.93	5.46	2.10	37.05
Stipple finish									
Slow	100	--	--	10.50	2.98	--	4.18	2.65	20.31
Medium	123	--	--	13.62	4.75	--	4.50	2.52	25.39
Fast	135	--	--	16.67	6.96	--	4.37	1.68	29.68

Measurements are based on the square feet of wall coated. These figures assume paint products are being applied over a smooth finish. For heights above 8 feet, use the High Time Difficulty Factors on page 137. "Slow" work is based on a $10.50 hourly wage, "Medium" work on a $16.75 hourly wage, and "Fast" work on a $22.50 hourly wage. Other qualifications that apply to this table are on page 9.

	Labor SF per manhour	Material coverage SF/gallon	Material cost per gallon	Labor cost per 100 SF	Labor burden 100 SF	Material cost per 100 SF	Overhead per 100 SF	Profit per 100 SF	Total price per 100 SF

Walls, plaster, interior, smooth finish, roll, per 100 square feet of wall

	Labor SF per manhour	Material coverage SF/gallon	Material cost per gallon	Labor cost per 100 SF	Labor burden 100 SF	Material cost per 100 SF	Overhead per 100 SF	Profit per 100 SF	Total price per 100 SF
Flat latex, water base (material #5)									
Roll 1st coat									
Slow	260	350	17.10	4.04	1.15	4.89	3.12	1.98	15.18
Medium	430	325	15.00	3.90	1.36	4.62	2.42	1.35	13.65
Fast	640	300	12.90	3.52	1.45	4.30	1.72	.66	11.65
Roll 2nd coat									
Slow	300	375	17.10	3.50	.99	4.56	2.81	1.78	13.64
Medium	488	350	15.00	3.43	1.20	4.29	2.19	1.22	12.33
Fast	675	325	12.90	3.33	1.39	3.97	1.61	.62	10.92
Roll 3rd or additional coats									
Slow	325	400	17.10	3.23	.92	4.28	2.61	1.66	12.70
Medium	513	375	15.00	3.27	1.14	4.00	2.06	1.15	11.62
Fast	700	350	12.90	3.21	1.35	3.69	1.52	.59	10.36
Enamel, water base (material #9)									
Roll 1st finish coat									
Slow	235	350	24.50	4.47	1.27	7.00	3.95	2.50	19.19
Medium	423	325	21.40	3.96	1.37	6.58	2.92	1.63	16.46
Fast	615	300	18.40	3.66	1.54	6.13	2.09	.80	14.22
Roll 2nd finish coat									
Slow	275	375	24.50	3.82	1.09	6.53	3.55	2.25	17.24
Medium	453	350	21.40	3.70	1.29	6.11	2.72	1.52	15.34
Fast	630	325	18.40	3.57	1.50	5.66	1.98	.76	13.47
Roll 3rd or additional finish coats									
Slow	300	400	24.50	3.50	.99	6.13	3.30	2.09	16.01
Medium	475	375	21.40	3.53	1.24	5.71	2.57	1.43	14.48
Fast	650	350	18.40	3.46	1.45	5.26	1.88	.72	12.77
Enamel, oil base (material #10)									
Roll 1st finish coat									
Slow	235	375	26.60	4.47	1.27	7.09	3.98	2.52	19.33
Medium	423	350	23.30	3.96	1.37	6.66	2.94	1.64	16.57
Fast	615	325	20.00	3.66	1.54	6.15	2.10	.81	14.26
Roll 2nd finish coat									
Slow	275	400	26.60	3.82	1.09	6.65	3.58	2.27	17.41
Medium	453	375	23.30	3.70	1.29	6.21	2.74	1.53	15.47
Fast	630	350	20.00	3.57	1.50	5.71	1.99	.77	13.54
Roll 3rd or additional finish coats									
Slow	300	425	26.60	3.50	.99	6.26	3.34	2.12	16.21
Medium	475	400	23.30	3.53	1.24	5.83	2.59	1.45	14.64
Fast	650	375	20.00	3.46	1.45	5.33	1.89	.73	12.86

	Labor SF per manhour	Material coverage SF/gallon	Material cost per gallon	Labor cost per 100 SF	Labor burden 100 SF	Material cost per 100 SF	Overhead per 100 SF	Profit per 100 SF	Total price per 100 SF
Epoxy coating, 2 part system - white (material #52)									
Roll 1st coat									
Slow	300	375	61.70	3.50	.99	16.45	6.49	4.12	31.55
Medium	488	350	54.00	3.43	1.20	15.43	4.91	2.75	27.72
Fast	675	325	46.30	3.33	1.39	14.25	3.51	1.35	23.83
Roll 2nd or additional coats									
Slow	325	425	61.70	3.23	.92	14.52	5.79	3.67	28.13
Medium	513	400	54.00	3.27	1.14	13.50	4.39	2.45	24.75
Fast	700	375	46.30	3.21	1.35	12.35	3.13	1.20	21.24

Measurements are based on the square feet of wall coated. These figures assume paint products are being applied over a smooth finish. For heights above 8 feet, use the High Time Difficulty Factors on page 137. "Slow" work is based on a $10.50 hourly wage, "Medium" work on a $16.75 hourly wage, and "Fast" work on a $22.50 hourly wage. Other qualifications that apply to this table are on page 9.

	Labor SF per manhour	Material coverage SF/gallon	Material cost per gallon	Labor cost per 100 SF	Labor burden 100 SF	Material cost per 100 SF	Overhead per 100 SF	Profit per 100 SF	Total price per 100 SF

Walls, plaster, interior, smooth finish, spray, per 100 square feet of wall

	Labor SF per manhour	Material coverage SF/gallon	Material cost per gallon	Labor cost per 100 SF	Labor burden 100 SF	Material cost per 100 SF	Overhead per 100 SF	Profit per 100 SF	Total price per 100 SF
Flat latex, water base (material #5)									
Spray 1st coat									
Slow	500	375	17.10	2.10	.60	4.56	2.25	1.43	10.94
Medium	625	338	15.00	2.68	.93	4.44	1.97	1.10	11.12
Fast	750	300	12.90	3.00	1.24	4.30	1.58	.61	10.73
Spray 2nd coat									
Slow	550	400	17.10	1.91	.54	4.28	2.09	1.32	10.14
Medium	700	363	15.00	2.39	.84	4.13	1.80	1.01	10.17
Fast	850	325	12.90	2.65	1.11	3.97	1.43	.55	9.71
Spray 3rd or additional coats									
Slow	600	425	17.10	1.75	.50	4.02	1.94	1.23	9.44
Medium	775	388	15.00	2.16	.75	3.87	1.66	.93	9.37
Fast	950	350	12.90	2.37	.98	3.69	1.30	.50	8.84
Enamel, water base (material #9)									
Spray 1st finish coat									
Slow	475	375	24.50	2.21	.63	6.53	2.90	1.84	14.11
Medium	600	338	21.40	2.79	.98	6.33	2.47	1.38	13.95
Fast	725	300	18.40	3.10	1.30	6.13	1.95	.75	13.23
Spray 2nd finish coat									
Slow	525	400	24.50	2.00	.56	6.13	2.70	1.71	13.10
Medium	675	388	21.40	2.48	.86	5.52	2.17	1.21	12.24
Fast	825	325	18.40	2.73	1.13	5.66	1.76	.68	11.96
Spray 3rd or additional finish coats									
Slow	575	425	24.50	1.83	.52	5.76	2.51	1.59	12.21
Medium	775	388	21.40	2.16	.75	5.52	2.07	1.16	11.66
Fast	925	350	18.40	2.43	1.01	5.26	1.61	.62	10.93
Enamel, oil base (material #10)									
Spray 1st finish coat									
Slow	475	425	26.60	2.21	.63	6.26	2.82	1.79	13.71
Medium	575	388	23.30	2.91	1.02	6.01	2.43	1.36	13.73
Fast	725	350	20.00	3.10	1.30	5.71	1.87	.72	12.70
Spray 2nd finish coat									
Slow	525	450	26.60	2.00	.56	5.91	2.63	1.67	12.77
Medium	675	413	23.30	2.48	.86	5.64	2.20	1.23	12.41
Fast	825	375	20.00	2.73	1.13	5.33	1.70	.65	11.54

	Labor SF per manhour	Material coverage SF/gallon	Material cost per gallon	Labor cost per 100 SF	Labor burden 100 SF	Material cost per 100 SF	Overhead per 100 SF	Profit per 100 SF	Total price per 100 SF
Spray 3rd or additional finish coats									
Slow	575	475	26.60	1.83	.52	5.60	2.46	1.56	11.97
Medium	750	438	23.30	2.23	.77	5.32	2.04	1.14	11.50
Fast	925	400	20.00	2.43	1.01	5.00	1.56	.60	10.60
Epoxy coating, 2 part system - white (material #52)									
Spray 1st coat									
Slow	550	325	61.70	1.91	.54	18.98	6.64	4.21	32.28
Medium	700	300	54.00	2.39	.84	18.00	5.20	2.91	29.34
Fast	850	275	46.30	2.65	1.11	16.84	3.81	1.46	25.87
Spray 2nd or additional coats									
Slow	600	375	61.70	1.75	.50	16.45	5.80	3.68	28.18
Medium	750	350	54.00	2.23	.77	15.43	4.52	2.53	25.48
Fast	900	325	46.30	2.50	1.04	14.25	3.29	1.26	22.34

Measurements are based on the square feet of wall coated. These figures assume paint products are being applied over a smooth finish. For heights above 8 feet, use the High Time Difficulty Factors on page 137. "Slow" work is based on a $10.50 hourly wage, "Medium" work on a $16.75 hourly wage, and "Fast" work on a $22.50 hourly wage. Other qualifications that apply to this table are on page 9.

	Labor SF per manhour	Material coverage SF/gallon	Material cost per gallon	Labor cost per 100 SF	Labor burden 100 SF	Material cost per 100 SF	Overhead per 100 SF	Profit per 100 SF	Total price per 100 SF
Walls, wood paneled, interior, paint grade, brush, per 100 square feet of wall									
Undercoat, water base (material #3)									
Brush, 1 coat									
Slow	65	300	21.10	16.15	4.58	7.03	8.61	5.46	41.83
Medium	75	288	18.50	22.33	7.78	6.42	8.95	5.00	50.48
Fast	85	275	15.80	26.47	11.03	5.75	8.00	3.08	54.33
Undercoat, oil base (material #4)									
Brush, 1 coat									
Slow	65	375	21.60	16.15	4.58	5.76	8.22	5.21	39.92
Medium	75	363	18.90	22.33	7.78	5.21	8.65	4.84	48.81
Fast	85	350	16.20	26.47	11.03	4.63	7.80	3.00	52.93
Split coat (1/2 undercoat + 1/2 enamel), water base (material #3 & #9)									
Brush, 1 coat									
Slow	55	300	22.80	19.09	5.42	7.60	9.96	6.31	48.38
Medium	65	288	19.95	25.77	8.97	6.93	10.21	5.71	57.59
Fast	75	275	17.10	30.00	12.51	6.22	9.02	3.47	61.22
Split coat (1/2 undercoat + 1/2 enamel), oil base (material #4 & #10)									
Brush, 1 coat									
Slow	55	375	24.10	19.09	5.42	6.43	9.59	6.08	46.61
Medium	65	363	21.10	25.77	8.97	5.81	9.94	5.56	56.05
Fast	75	350	18.10	30.00	12.51	5.17	8.82	3.39	59.89
Enamel, water base (material #9)									
Brush 1st finish coat									
Slow	80	350	24.50	13.13	3.72	7.00	7.40	4.69	35.94
Medium	95	338	21.40	17.63	6.16	6.33	7.37	4.12	41.61
Fast	110	325	18.40	20.45	8.54	5.66	6.41	2.46	43.52
Brush 2nd or additional finish coats									
Slow	100	375	24.50	10.50	2.98	6.53	6.21	3.93	30.15
Medium	110	363	21.40	15.23	5.30	5.90	6.48	3.62	36.53
Fast	120	350	18.40	18.75	7.81	5.26	5.89	2.26	39.97

	Labor SF per manhour	Material coverage SF/gallon	Material cost per gallon	Labor cost per 100 SF	Labor burden 100 SF	Material cost per 100 SF	Overhead per 100 SF	Profit per 100 SF	Total price per 100 SF
Enamel, oil base (material #10)									
Brush 1st finish coat									
Slow	80	400	26.60	13.13	3.72	6.65	7.29	4.62	35.41
Medium	95	388	23.30	17.63	6.16	6.01	7.30	4.08	41.18
Fast	110	375	20.00	20.45	8.54	5.33	6.35	2.44	43.11
Brush 2nd or additional finish coats									
Slow	100	425	26.60	10.50	2.98	6.26	6.12	3.88	29.74
Medium	110	413	23.30	15.23	5.30	5.64	6.41	3.58	36.16
Fast	120	400	20.00	18.75	7.81	5.00	5.84	2.25	39.65

These costs are based on painting interior tongue and groove, wood veneer or plain wainscot paneling. For heights above 8 feet, use the High Time Difficulty Factors on page 137. "Slow" work is based on a $10.50 hourly wage, "Medium" work on a $16.75 hourly wage, and "Fast" work on a $22.50 hourly wage. Other qualifications that apply to this table are on page 9.

	Labor SF per manhour	Material coverage SF/gallon	Material cost per gallon	Labor cost per 100 SF	Labor burden 100 SF	Material cost per 100 SF	Overhead per 100 SF	Profit per 100 SF	Total price per 100 SF
Walls, wood paneled, interior, paint grade, roll, per 100 square feet of wall									
Undercoat, water base (material #3)									
Roll, 1 coat									
Slow	200	275	21.10	5.25	1.49	7.67	4.47	2.83	21.71
Medium	300	263	18.50	5.58	1.94	7.03	3.56	1.99	20.10
Fast	400	250	15.80	5.63	2.34	6.32	2.65	1.02	17.96
Undercoat, oil base (material #4)									
Roll, 1 coat									
Slow	200	350	21.60	5.25	1.49	6.17	4.00	2.54	19.45
Medium	300	325	18.90	5.58	1.94	5.82	3.27	1.83	18.44
Fast	400	300	16.20	5.63	2.34	5.40	2.48	.95	16.80
Split coat (1/2 undercoat + 1/2 enamel), water base (material #3 & #9)									
Roll, 1 coat									
Slow	175	275	22.80	6.00	1.70	8.29	4.96	3.14	24.09
Medium	275	263	19.95	6.09	2.13	7.59	3.87	2.16	21.84
Fast	375	250	17.10	6.00	2.51	6.84	2.84	1.09	19.28
Split coat (1/2 undercoat + 1/2 enamel), oil base (material #4 & #10)									
Roll, 1 coat									
Slow	175	350	24.10	6.00	1.70	6.89	4.53	2.87	21.99
Medium	275	325	21.10	6.09	2.13	6.49	3.60	2.01	20.32
Fast	375	300	18.10	6.00	2.51	6.03	2.69	1.03	18.26
Enamel, water base (material #9)									
Roll 1st finish coat									
Slow	250	325	24.50	4.20	1.19	7.54	4.01	2.54	19.48
Medium	375	313	21.40	4.47	1.56	6.84	3.15	1.76	17.78
Fast	500	300	18.40	4.50	1.88	6.13	2.31	.89	15.71
Roll 2nd or additional finish coats									
Slow	300	350	24.50	3.50	.99	7.00	3.57	2.26	17.32
Medium	425	338	21.40	3.94	1.37	6.33	2.85	1.59	16.08
Fast	550	325	18.40	4.09	1.71	5.66	2.12	.81	14.39

	Labor SF per manhour	Material coverage SF/gallon	Material cost per gallon	Labor cost per 100 SF	Labor burden 100 SF	Material cost per 100 SF	Overhead per 100 SF	Profit per 100 SF	Total price per 100 SF
Enamel, oil base (material #10)									
Roll 1st finish coat									
Slow	250	375	26.60	4.20	1.19	7.09	3.87	2.45	18.80
Medium	375	363	23.30	4.47	1.56	6.42	3.05	1.71	17.21
Fast	500	350	20.00	4.50	1.88	5.71	2.24	.86	15.19
Roll 2nd or additional finish coats									
Slow	300	400	26.60	3.50	.99	6.65	3.46	2.19	16.79
Medium	425	388	23.30	3.94	1.37	6.01	2.77	1.55	15.64
Fast	550	375	20.00	4.09	1.71	5.33	2.06	.79	13.98

These costs are based on painting interior tongue and groove, wood veneer or plain wainscot paneling. For heights above 8 feet, use the High Time Difficulty Factors on page 137. "Slow" work is based on a $10.50 hourly wage, "Medium" work on a $16.75 hourly wage, and "Fast" work on a $22.50 hourly wage. Other qualifications that apply to this table are on page 9.

	Labor SF per manhour	Material coverage SF/gallon	Material cost per gallon	Labor cost per 100 SF	Labor burden 100 SF	Material cost per 100 SF	Overhead per 100 SF	Profit per 100 SF	Total price per 100 SF
Walls, wood paneled, interior, paint grade, spray, per 100 square feet of wall									
Undercoat, water base (material #3)									
Spray, 1 coat									
Slow	350	175	21.10	3.00	.86	12.06	4.93	3.13	23.98
Medium	425	150	18.50	3.94	1.37	12.33	4.32	2.42	24.38
Fast	500	125	15.80	4.50	1.88	12.64	3.52	1.35	23.89
Undercoat, oil base (material #4)									
Spray, 1 coat									
Slow	350	200	21.60	3.00	.86	10.80	4.54	2.88	22.08
Medium	425	188	18.90	3.94	1.37	10.05	3.76	2.10	21.22
Fast	500	175	16.20	4.50	1.88	9.26	2.89	1.11	19.64
Split coat (1/2 undercoat + 1/2 enamel), water base (material #3 & #9)									
Spray, 1 coat									
Slow	325	175	22.80	3.23	.92	13.03	5.33	3.38	25.89
Medium	400	150	19.95	4.19	1.46	13.30	4.64	2.59	26.18
Fast	475	125	17.10	4.74	1.99	13.68	3.77	1.45	25.63
Split coat (1/2 undercoat + 1/2 enamel), oil base (material #4 & #10)									
Spray, 1 coat									
Slow	325	200	24.10	3.23	.92	12.05	5.02	3.18	24.40
Medium	400	188	21.10	4.19	1.46	11.22	4.13	2.31	23.31
Fast	475	175	18.10	4.74	1.99	10.34	3.16	1.21	21.44
Enamel, water base (material #9)									
Spray 1st finish coat									
Slow	500	250	24.50	2.10	.60	9.80	3.88	2.46	18.84
Medium	550	225	21.40	3.05	1.06	9.51	3.34	1.87	18.83
Fast	600	200	18.40	3.75	1.58	9.20	2.69	1.03	18.25
Spray 2nd or additional finish coats									
Slow	600	350	24.50	1.75	.50	7.00	2.87	1.82	13.94
Medium	650	325	21.40	2.58	.90	6.58	2.46	1.38	13.90
Fast	700	300	18.40	3.21	1.35	6.13	1.98	.76	13.43

	Labor SF per manhour	Material coverage SF/gallon	Material cost per gallon	Labor cost per 100 SF	Labor burden 100 SF	Material cost per 100 SF	Overhead per 100 SF	Profit per 100 SF	Total price per 100 SF
Enamel, oil base (material #10)									
Spray 1st finish coat									
Slow	500	300	26.60	2.10	.60	8.87	3.59	2.27	17.43
Medium	550	275	23.30	3.05	1.06	8.47	3.08	1.72	17.38
Fast	600	250	20.00	3.75	1.58	8.00	2.46	.95	16.74
Spray 2nd or additional finish coats									
Slow	600	400	26.60	1.75	.50	6.65	2.76	1.75	13.41
Medium	650	375	23.30	2.58	.90	6.21	2.37	1.33	13.39
Fast	700	350	20.00	3.21	1.35	5.71	1.90	.73	12.90

These costs are based on painting interior tongue and groove, wood veneer or plain wainscot paneling. For heights above 8 feet, use the High Time Difficulty Factors on page 137. "Slow" work is based on a $10.50 hourly wage, "Medium" work on a $16.75 hourly wage, and "Fast" work on a $22.50 hourly wage. Other qualifications that apply to this table are on page 9.

	Labor SF per manhour	Material coverage SF/gallon	Material cost per gallon	Labor cost per 100 SF	Labor burden 100 SF	Material cost per 100 SF	Overhead per 100 SF	Profit per 100 SF	Total price per 100 SF

Walls, wood paneled, interior, stain grade, per 100 square feet of wall

Solid body or semi-transparent stain, water or oil base (material #18 or #19 or #20 or #21)

Roll & brush each coat									
Slow	225	250	23.23	4.67	1.32	9.29	4.74	3.00	23.02
Medium	263	238	20.33	6.37	2.21	8.54	4.20	2.35	23.67
Fast	300	225	17.40	7.50	3.12	7.73	3.40	1.31	23.06

Solid body or semi-transparent stain, water or oil base (material #18 or #19 or #20 or #21)

Spray each coat									
Slow	350	300	23.23	3.00	.86	7.74	3.59	2.28	17.47
Medium	400	250	20.33	4.19	1.46	8.13	3.38	1.89	19.05
Fast	450	200	17.40	5.00	2.08	8.70	2.92	1.12	19.82

Use these figures for quantities greater than 100 square feet. For quantities less than 100 square feet, use Fireplace siding. These costs are based on painting interior tongue and groove, wood veneer or plain wainscot paneling. Do not deduct for openings less than 100 square feet. For heights above 8 feet, use the High Time Difficulty Factors on page 137. "Slow" work is based on a $10.50 hourly wage, "Medium" work on a $16.75 hourly wage, and "Fast" work on a $22.50 hourly wage. Other qualifications that apply to this table are on page 9.

	Labor SF per manhour	Material coverage SF/gallon	Material cost per gallon	Labor cost per 100 SF	Labor burden 100 SF	Material cost per 100 SF	Overhead per 100 SF	Profit per 100 SF	Total price per 100 SF

Walls, wood paneled, interior, stain grade, per 100 square feet of wall

Stain, seal and 2 coat lacquer system (7 step process)

	Labor SF per manhour	Material coverage SF/gallon	Material cost per gallon	Labor cost per 100 SF	Labor burden 100 SF	Material cost per 100 SF	Overhead per 100 SF	Profit per 100 SF	Total price per 100 SF
STEP 1: Sand & putty									
Slow	175	--	--	6.00	1.70	--	2.39	1.52	11.61
Medium	200	--	--	8.38	2.92	--	2.77	1.55	15.62
Fast	225	--	--	10.00	4.16	--	2.62	1.01	17.79
STEP 2 & 3: Wiping stain, oil base (material #11a) & wipe									
Brush 1 coat & wipe									
Slow	100	400	30.90	10.50	2.98	7.73	6.58	4.17	31.96
Medium	125	375	27.00	13.40	4.67	7.20	6.19	3.46	34.92
Fast	150	350	23.10	15.00	6.27	6.60	5.15	1.98	35.00
Spray 1 coat & wipe									
Slow	350	175	30.90	3.00	.86	17.66	6.67	4.23	32.42
Medium	425	150	27.00	3.94	1.37	18.00	5.71	3.19	32.21
Fast	500	125	23.10	4.50	1.88	18.48	4.60	1.77	31.23
STEP 4 & 5: Sanding sealer (material #11b) & sand lightly									
Brush 1 coat & wipe									
Slow	200	450	17.20	5.25	1.49	3.82	3.27	2.07	15.90
Medium	220	425	15.00	7.61	2.67	3.53	3.38	1.89	19.08
Fast	240	400	12.90	9.38	3.92	3.23	3.06	1.18	20.77
Spray 1 coat & wipe									
Slow	400	175	17.20	2.63	.74	9.83	4.10	2.60	19.90
Medium	500	150	15.00	3.35	1.17	10.00	3.56	1.99	20.07
Fast	600	125	12.90	3.75	1.58	10.32	2.89	1.11	19.65
STEP 6 & 7: Lacquer (material #11c)									
Brush 1st coat									
Slow	175	375	19.30	6.00	1.70	5.15	3.99	2.53	19.37
Medium	225	350	16.80	7.44	2.59	4.80	3.63	2.03	20.49
Fast	300	325	14.40	7.50	3.12	4.43	2.79	1.07	18.91
Brush 2nd coat									
Slow	225	400	19.30	4.67	1.32	4.83	3.36	2.13	16.31
Medium	288	388	16.80	5.82	2.02	4.33	2.98	1.67	16.82
Fast	350	375	14.40	6.43	2.69	3.84	2.40	.92	16.28
Spray 1st coat									
Slow	450	150	19.30	2.33	.66	12.87	4.92	3.12	23.90
Medium	550	125	16.80	3.05	1.06	13.44	4.30	2.40	24.25
Fast	650	100	14.40	3.46	1.45	14.40	3.57	1.37	24.25

	Labor SF per manhour	Material coverage SF/gallon	Material cost per gallon	Labor cost per 100 SF	Labor burden 100 SF	Material cost per 100 SF	Overhead per 100 SF	Profit per 100 SF	Total price per 100 SF
Spray 2nd coat									
Slow	450	225	19.30	2.33	.66	8.58	3.59	2.27	17.43
Medium	550	200	16.80	3.05	1.06	8.40	3.06	1.71	17.28
Fast	650	175	14.40	3.46	1.45	8.23	2.43	.93	16.50
Complete 7 step stain, seal & 2 coat lacquer system (material #11)									
Brush all coats									
Slow	40	150	21.60	26.25	7.45	14.40	14.92	9.46	72.48
Medium	45	138	19.60	37.22	12.97	14.20	15.78	8.82	88.99
Fast	50	125	16.20	45.00	18.78	12.96	14.20	5.46	96.40
Spray all coats									
Slow	70	50	21.60	15.00	4.26	43.20	19.37	12.28	94.11
Medium	80	40	19.60	20.94	7.30	49.00	18.92	10.58	106.74
Fast	90	30	16.20	25.00	10.43	54.00	16.55	6.36	112.34
Penetrating stain wax (material #14)									
Brush each coat									
Slow	300	500	27.20	3.50	.99	5.44	3.08	1.95	14.96
Medium	350	475	23.80	4.79	1.67	5.01	2.81	1.57	15.85
Fast	400	450	20.40	5.63	2.34	4.53	2.31	.89	15.70

These costs are based on painting interior tongue and groove, wood veneer or plain wainscot paneling. Do not deduct for openings less than 100 square feet. For heights above 8 feet, use the High Time Difficulty Factors on page 137. "Slow" work is based on a $10.50 hourly wage, "Medium" work on a $16.75 hourly wage, and "Fast" work on a $22.50 hourly wage. Other qualifications that apply to this table are on page 9.

	Frames per manhour	Frames per gallon	Material cost per gallon	Labor cost per frame	Labor burden frame	Material cost per frame	Overhead per frame	Profit per frame	Total price per frame
Window screen frames, paint grade, per frame (15 square feet)									
Undercoat, water or oil base (material #3 or #4)									
Brush 1 coat									
Slow	5	50	21.35	2.10	.60	.43	.97	.62	4.72
Medium	6	45	18.70	2.79	.98	.42	1.02	.57	5.78
Fast	7	40	16.00	3.21	1.35	.40	.92	.35	6.23
Split coat (1/2 undercoat + 1/2 enamel), water or oil base (material #3 or #4 or #9 or #10)									
Brush 1 coat									
Slow	8	60	23.45	1.31	.38	.39	.64	.41	3.13
Medium	9	58	20.53	1.86	.65	.35	.70	.39	3.95
Fast	10	55	17.60	2.25	.94	.32	.65	.25	4.41
Enamel, water or oil base (material #9 or #10)									
Brush each finish coat									
Slow	6	55	25.55	1.75	.50	.46	.84	.53	4.08
Medium	7	53	22.35	2.39	.84	.42	.89	.50	5.04
Fast	8	50	19.20	2.81	1.18	.38	.81	.31	5.49

These figures will apply when painting all sides of wood window screens up to 15 square feet (length times width). Add: Preparation time for protecting adjacent surfaces with masking tape and paper as required. "Slow" work is based on a $10.50 hourly wage, "Medium" work on a $16.75 hourly wage, and "Fast" work on a $22.50 hourly wage. Other qualifications that apply to this table are on page 9.

	Labor SF per manhour	Material coverage SF/gallon	Material cost per gallon	Labor cost per 100 SF	Labor burden 100 SF	Material cost per 100 SF	Overhead per 100 SF	Profit per 100 SF	Total price per 100 SF

Window seats, wood, paint grade, per 100 square feet coated

Undercoat, water or oil base (material #3 or #4)									
Brush 1 coat									
Slow	20	45	21.35	52.50	14.90	47.44	35.61	22.57	173.02
Medium	25	43	18.70	67.00	23.36	43.49	32.79	18.33	184.97
Fast	30	40	16.00	75.00	31.29	40.00	27.07	10.40	183.76
Split coat (1/2 undercoat + 1/2 enamel), water or oil base (material #3 or #4 or #9 or #10)									
Brush 1 coat									
Slow	30	60	23.45	35.00	9.93	39.08	26.05	16.51	126.57
Medium	35	58	20.53	47.86	16.68	35.40	24.49	13.69	138.12
Fast	40	55	17.60	56.25	23.48	32.00	20.67	7.94	140.34
Enamel, water or oil base (material #9 or #10)									
Brush each finish coat									
Slow	25	55	25.55	42.00	11.92	46.45	31.12	19.73	151.22
Medium	30	53	22.35	55.83	19.46	42.17	28.78	16.09	162.33
Fast	35	50	19.20	64.29	26.82	38.40	23.96	9.21	162.68

Measurements are based on square feet of surface area of each window seat. "Slow" work is based on a $10.50 hourly wage, "Medium" work on a $16.75 hourly wage, and "Fast" work on a $22.50 hourly wage. Other qualifications that apply to this table are on page 9.

	Labor LF per manhour	Material coverage LF/gallon	Material cost per gallon	Labor cost per 100 LF	Labor burden 100 LF	Material cost per 100 LF	Overhead per 100 LF	Profit per 100 LF	Total price per 100 LF

Window sills, wood, paint grade, per 100 linear feet coated

Undercoat, water or oil base (material #3 or #4)									
Brush 1 coat									
Slow	40	140	21.35	26.25	7.45	15.25	15.18	9.62	73.75
Medium	50	130	18.70	33.50	11.68	14.38	14.59	8.16	82.31
Fast	60	120	16.00	37.50	15.66	13.33	12.30	4.73	83.52
Split coat (1/2 undercoat + 1/2 enamel), water or oil base (material #3 or #4 or #9 or #10)									
Brush 1 coat									
Slow	60	180	23.45	17.50	4.97	13.03	11.01	6.98	53.49
Medium	70	170	20.53	23.93	8.35	12.08	10.87	6.07	61.30
Fast	80	160	17.60	28.13	11.73	11.00	9.41	3.62	63.89
Enamel, water or oil base (material #9 or #10)									
Brush each finish coat									
Slow	50	175	25.55	21.00	5.96	14.60	12.89	8.17	62.62
Medium	60	163	22.35	27.92	9.74	13.71	12.58	7.03	70.98
Fast	70	150	19.20	32.14	13.43	12.80	10.80	4.15	73.32

Measurements are based on linear feet of window sill. Add: Preparation time for protecting adjacent surfaces with masking tape and paper as required. "Slow" work is based on a $10.50 hourly wage, "Medium" work on a $16.75 hourly wage, and "Fast" work on a $22.50 hourly wage. Other qualifications that apply to this table are on page 9.

	Frames per manhour	Frames per gallon	Material cost per gallon	Labor cost per frame	Labor burden frame	Material cost per frame	Overhead per frame	Profit per frame	Total price per frame
Window storm sash, paint grade, per 15 square feet painted									
Undercoat, water or oil base (material #3 or #4)									
Brush 1 coat									
Slow	3	25	21.35	3.50	.99	.85	1.66	1.05	8.05
Medium	4	24	18.70	4.19	1.46	.78	1.58	.88	8.89
Fast	5	22	16.00	4.50	1.88	.73	1.32	.51	8.94
Split coat (1/2 undercoat + 1/2 enamel), water or oil base (material #3 or #4 or #9 or #10)									
Brush 1 coat									
Slow	5	35	23.45	2.10	.60	.67	1.04	.66	5.07
Medium	6	33	20.53	2.79	.98	.62	1.07	.60	6.06
Fast	7	30	17.60	3.21	1.35	.59	.95	.37	6.47
Enamel, water or oil base (material #9 or #10)									
Brush each finish coat									
Slow	4	30	25.55	2.63	.74	.85	1.31	.83	6.36
Medium	5	28	22.35	3.35	1.17	.80	1.30	.73	7.35
Fast	6	25	19.20	3.75	1.58	.77	1.13	.43	7.66

These figures will apply when painting all sides of two-lite wood storm sash measuring up to 15 square feet overall (length times width). Add: Preparation time for protecting adjacent surfaces with window protective coating or masking tape and paper as required. "Slow" work is based on a $10.50 hourly wage, "Medium" work on a $16.75 hourly wage, and "Fast" work on a $22.50 hourly wage. Other qualifications that apply to this table are on page 9.

Windows, exterior

The estimates that follow include the time and material needed to paint the sash (mullions or muntins), trim, frames, jambs, sill and apron on the outside only. Manhour and material rates are the same for painting either the inside or outside of windows. These estimates are per window painted, and are based on a window size of 15 square feet overall (length times width).

	Labor SF per manhour	Material coverage SF/gallon	Material cost per gallon	Labor cost per 100 SF	Labor burden 100 SF	Material cost per 100 SF	Overhead per 100 SF	Profit per 100 SF	Total price per 100 SF
Windows, wood, exterior, paint grade, square foot basis, brush									
Undercoat, water or oil base (material #3 or #4)									
Brush 1 coat									
Slow	150	460	21.35	7.00	1.99	4.64	4.23	2.68	20.54
Medium	165	450	18.70	10.15	3.54	4.16	4.37	2.44	24.66
Fast	180	440	16.00	12.50	5.23	3.64	3.95	1.52	26.84
Split coat (1/2 undercoat + 1/2 enamel), water or oil base (material #3 or #4 or #9 or #10)									
Brush 1 coat									
Slow	120	520	23.45	8.75	2.48	4.51	4.88	3.09	23.71
Medium	135	500	20.53	12.41	4.33	4.11	5.11	2.85	28.81
Fast	150	480	17.60	15.00	6.27	3.67	4.61	1.77	31.32
Enamel, water or oil base (material #9 or #10)									
Brush each finish coat									
Slow	100	500	25.55	10.50	2.98	5.11	5.77	3.66	28.02
Medium	113	480	22.35	14.82	5.17	4.66	6.04	3.37	34.06
Fast	125	460	19.20	18.00	7.51	4.17	5.49	2.11	37.28

Use this table to estimate the costs for paint grade windows larger than 15 square feet (width times height). When calculating the window area, add 1 foot in each direction to both dimensions before multiplying width by the height. For example, a 4' x 4' window would have a calculated surface area of 36 square feet: Width of 4' + 1' + 1' = 6' times height of 4' + 1' + 1' = 6'. The total is 6' x 6', or 36 square feet per side. Add an extra 2 square feet to the calculated area for each window lite. This allows for the extra time needed to finish muntins, mullions and sash. For windows that are stained, sealed and varnished, use the previous tables based on window lites. Add: Preparation time to protect adjacent surfaces with window protective coating (wax) or masking tape and paper as required. For heights above 8 feet, use the High Time Difficulty Factors on page 137. "Slow" work is based on a $10.50 hourly wage, "Medium" work on a $16.75 hourly wage, and "Fast" work on a $22.50 hourly wage. Other qualifications that apply to this table are on page 9.

	Manhours per window	Windows per gallon	Material cost per gallon	Labor cost per window	Labor burden window	Material cost per window	Overhead per window	Profit per window	Total price per window

Windows, wood, exterior, 1, 2 and 3 lite, brush

	Manhours per window	Windows per gallon	Material cost per gallon	Labor cost per window	Labor burden window	Material cost per window	Overhead per window	Profit per window	Total price per window
Undercoat, water or oil base (material #3 or #4)									
Brush 1 coat									
Slow	.25	15.0	21.35	2.63	.74	1.42	1.49	.94	7.22
Medium	.20	14.5	18.70	3.35	1.17	1.29	1.42	.80	8.03
Fast	.15	14.0	16.00	3.38	1.40	1.14	1.10	.42	7.44
Split coat (1/2 undercoat + 1/2 enamel), water or oil base (material #3 or #4 or #9 or #10)									
Brush 1 coat									
Slow	.30	17.0	23.45	3.15	.89	1.38	1.68	1.07	8.17
Medium	.25	16.5	20.53	4.19	1.46	1.24	1.69	.94	9.52
Fast	.20	16.0	17.60	4.50	1.88	1.10	1.38	.53	9.39
Enamel, water or oil base (material #9 or #10)									
Brush each finish coat									
Slow	.35	16.0	25.55	3.68	1.04	1.60	1.96	1.24	9.52
Medium	.30	15.5	22.35	5.03	1.75	1.44	2.01	1.13	11.36
Fast	.25	15.0	19.20	5.63	2.34	1.28	1.71	.66	11.62
Stain, seal & 1 coat varnish system (material #30)									
Brush each finish coat									
Slow	.70	11.0	33.10	7.35	2.09	3.01	3.86	2.45	18.76
Medium	.65	10.0	29.00	10.89	3.79	2.90	4.31	2.41	24.30
Fast	.60	9.0	24.90	13.50	5.63	2.77	4.05	1.56	27.51
Varnish (material #30c)									
Brush additional coats of varnish									
Slow	.25	22.0	35.40	2.63	.74	1.61	1.55	.98	7.51
Medium	.23	20.0	31.00	3.85	1.35	1.55	1.65	.92	9.32
Fast	.20	18.0	26.60	4.50	1.88	1.48	1.45	.56	9.87
Preparation for varnish application (material - minimal)									
Steel wool buff									
Slow	.25	--	--	2.63	.74	--	1.05	.66	5.08
Medium	.20	--	--	3.35	1.17	--	1.11	.62	6.25
Fast	.15	--	--	3.38	1.40	--	.89	.34	6.01
Wax application (material - minimal)									
Slow	.25	--	--	2.63	.74	--	1.05	.66	5.08
Medium	.20	--	--	3.35	1.17	--	1.11	.62	6.25
Fast	.15	--	--	3.38	1.40	--	.89	.34	6.01

The stain, seal and 1 coat varnish system includes 1 coat of stain, sanding sealer, light sanding, putty and 1 coat of varnish. Minimum preparation for varnish usually includes a steel wool buff and wax operation with minimum material usage. Add: Preparation time for protecting adjacent surfaces with window protective coating (wax) or masking tape and paper as required. For heights above 8 feet, use the High Time Difficulty Factors on page 137. "Slow" work is based on a $10.50 hourly wage, "Medium" work on a $16.75 hourly wage, and "Fast" work on a $22.50 hourly wage. Other qualifications that apply to this table are on page 9.

	Manhours per window	Windows per gallon	Material cost per gallon	Labor cost per window	Labor burden window	Material cost per window	Overhead per window	Profit per window	Total price per window
Windows, wood, exterior, 4 to 6 lite, brush									
Undercoat, water or oil base (material #3 or #4)									
Brush 1 coat									
Slow	.35	14.0	21.35	3.68	1.04	1.53	1.94	1.23	9.42
Medium	.30	13.5	18.70	5.03	1.75	1.39	2.00	1.12	11.29
Fast	.25	13.0	16.00	5.63	2.34	1.23	1.70	.65	11.55
Split coat (1/2 undercoat + 1/2 enamel), water or oil base (material #3 or #4 or #9 or #10)									
Brush 1 coat									
Slow	.40	16.0	23.45	4.20	1.19	1.47	2.13	1.35	10.34
Medium	.35	15.5	20.53	5.86	2.05	1.32	2.26	1.26	12.75
Fast	.30	15.0	17.60	6.75	2.82	1.17	1.99	.76	13.49
Enamel, water or oil base (material #9 or #10)									
Brush each finish coat									
Slow	.50	15.0	25.55	5.25	1.49	1.70	2.62	1.66	12.72
Medium	.45	14.5	22.35	7.54	2.63	1.54	2.87	1.60	16.18
Fast	.40	14.0	19.20	9.00	3.76	1.37	2.61	1.00	17.74
Stain, seal & 1 coat varnish system (material #30)									
Brush each finish coat									
Slow	.83	10.0	33.10	8.72	2.47	3.31	4.50	2.85	21.85
Medium	.73	9.0	29.00	12.23	4.26	3.22	4.83	2.70	27.24
Fast	.63	8.0	24.90	14.18	5.91	3.11	4.29	1.65	29.14
Varnish (material #30c)									
Brush additional coats of varnish									
Slow	.33	21.0	35.40	3.47	.98	1.69	1.91	1.21	9.26
Medium	.27	19.0	31.00	4.52	1.58	1.63	1.89	1.06	10.68
Fast	.22	17.0	26.60	4.95	2.07	1.56	1.59	.61	10.78
Preparation for varnish application (material - minimal)									
Steel wool buff									
Slow	.30	--	--	3.15	.89	--	1.26	.80	6.10
Medium	.25	--	--	4.19	1.46	--	1.38	.77	7.80
Fast	.20	--	--	4.50	1.88	--	1.18	.45	8.01
Wax application (material - minimal)									
Slow	.30	--	--	3.15	.89	--	1.26	.80	6.10
Medium	.25	--	--	4.19	1.46	--	1.38	.77	7.80
Fast	.20	--	--	4.50	1.88	--	1.18	.45	8.01

The stain, seal and 1 coat varnish system includes 1 coat of stain, sanding sealer, light sanding, putty and 1 coat of varnish. Minimum preparation for varnish usually includes a steel wool buff and wax operation with minimum material usage. Add: Preparation time for protecting adjacent surfaces with window protective coating (wax) or masking tape and paper as required. For heights above 8 feet, use the High Time Difficulty Factors on page 137. "Slow" work is based on a $10.50 hourly wage, "Medium" work on a $16.75 hourly wage, and "Fast" work on a $22.50 hourly wage. Other qualifications that apply to this table are on page 9.

	Manhours per window	Windows per gallon	Material cost per gallon	Labor cost per window	Labor burden window	Material cost per window	Overhead per window	Profit per window	Total price per window

Windows, wood, exterior, 7 to 8 lite, brush

	Manhours per window	Windows per gallon	Material cost per gallon	Labor cost per window	Labor burden window	Material cost per window	Overhead per window	Profit per window	Total price per window
Undercoat, water or oil base (material #3 or #4)									
Brush 1 coat									
Slow	.45	13	21.35	4.73	1.34	1.64	2.39	1.52	11.62
Medium	.40	13	18.70	6.70	2.34	1.44	2.57	1.43	14.48
Fast	.35	12	16.00	7.88	3.28	1.33	2.31	.89	15.69
Split coat (1/2 undercoat + 1/2 enamel), water or oil base (material #3 or #4 or #9 or #10)									
Brush 1 coat									
Slow	.55	15	23.45	5.78	1.63	1.56	2.78	1.76	13.51
Medium	.50	15	20.53	8.38	2.92	1.37	3.10	1.73	17.50
Fast	.45	14	17.60	10.13	4.22	1.26	2.89	1.11	19.61
Enamel, water or oil base (material #9 or #10)									
Brush each finish coat									
Slow	.65	14	25.55	6.83	1.93	1.83	3.29	2.08	15.96
Medium	.60	14	22.35	10.05	3.50	1.60	3.71	2.07	20.93
Fast	.55	13	19.20	12.38	5.16	1.48	3.52	1.35	23.89
Stain, seal & 1 coat varnish system (material #30)									
Brush each finish coat									
Slow	.90	9	33.10	9.45	2.68	3.68	4.90	3.11	23.82
Medium	.80	8	29.00	13.40	4.67	3.63	5.32	2.97	29.99
Fast	.70	7	24.90	15.75	6.57	3.56	4.79	1.84	32.51
Varnish (material #30c)									
Brush additional coats of varnish									
Slow	.38	20	35.40	3.99	1.13	1.77	2.14	1.36	10.39
Medium	.33	18	31.00	5.53	1.92	1.72	2.25	1.26	12.68
Fast	.27	16	26.60	6.08	2.53	1.66	1.90	.73	12.90
Preparation for varnish application (material - minimal)									
Steel wool buff									
Slow	.35	--	--	3.68	1.04	--	1.47	.93	7.12
Medium	.30	--	--	5.03	1.75	--	1.66	.93	9.37
Fast	.25	--	--	5.63	2.34	--	1.48	.57	10.02
Wax application (material - minimal)									
Slow	.35	--	--	3.68	1.04	--	1.47	.93	7.12
Medium	.30	--	--	5.03	1.75	--	1.66	.93	9.37
Fast	.25	--	--	5.63	2.34	--	1.48	.57	10.02

The stain, seal and 1 coat varnish system includes 1 coat of stain, sanding sealer, light sanding, putty and 1 coat of varnish. Minimum preparation for varnish usually includes a steel wool buff and wax operation with minimum material usage. Add: Preparation time for protecting adjacent surfaces with window protective coating (wax) or masking tape and paper as required. For heights above 8 feet, use the High Time Difficulty Factors on page 137. "Slow" work is based on a $10.50 hourly wage, "Medium" work on a $16.75 hourly wage, and "Fast" work on a $22.50 hourly wage. Other qualifications that apply to this table are on page 9.

	Manhours per window	Windows per gallon	Material cost per gallon	Labor cost per window	Labor burden window	Material cost per window	Overhead per window	Profit per window	Total price per window
Windows, wood, exterior, 9 to 11 lite, brush									
Undercoat, water or oil base (material #3 or #4)									
Brush 1 coat									
Slow	.55	12	21.35	5.78	1.63	1.78	2.85	1.81	13.85
Medium	.50	12	18.70	8.38	2.92	1.56	3.15	1.76	17.77
Fast	.45	11	16.00	10.13	4.22	1.45	2.92	1.12	19.84
Split coat (1/2 undercoat + 1/2 enamel), water or oil base (material #3 or #4 or #9 or #10)									
Brush 1 coat									
Slow	.67	14	23.45	7.04	1.99	1.68	3.32	2.11	16.14
Medium	.62	14	20.53	10.39	3.62	1.47	3.79	2.12	21.39
Fast	.57	13	17.60	12.83	5.35	1.35	3.61	1.39	24.53
Enamel, water or oil base (material #9 or #10)									
Brush each finish coat									
Slow	.80	13	25.55	8.40	2.38	1.97	3.96	2.51	19.22
Medium	.73	13	22.35	12.23	4.26	1.72	4.46	2.49	25.16
Fast	.65	12	19.20	14.63	6.10	1.60	4.13	1.59	28.05
Stain, seal & 1 coat varnish system (material #30)									
Brush each finish coat									
Slow	1.00	7	33.10	10.50	2.98	4.73	5.65	3.58	27.44
Medium	.90	6	29.00	15.08	5.25	4.83	6.17	3.45	34.78
Fast	.80	5	24.90	18.00	7.51	4.98	5.64	2.17	38.30
Varnish (material #30c)									
Brush additional coats of varnish									
Slow	.45	19	35.40	4.73	1.34	1.86	2.46	1.56	11.95
Medium	.40	17	31.00	6.70	2.34	1.82	2.66	1.49	15.01
Fast	.35	15	26.60	7.88	3.28	1.77	2.39	.92	16.24
Preparation for varnish application (material - minimal)									
Steel wool buff									
Slow	.45	--	--	4.73	1.34	--	1.88	1.19	9.14
Medium	.40	--	--	6.70	2.34	--	2.21	1.24	12.49
Fast	.35	--	--	7.88	3.28	--	2.07	.79	14.02
Wax application (material - minimal)									
Slow	.40	--	--	4.20	1.19	--	1.67	1.06	8.12
Medium	.35	--	--	5.86	2.05	--	1.94	1.08	10.93
Fast	.30	--	--	6.75	2.82	--	1.77	.68	12.02

The stain, seal and 1 coat varnish system includes 1 coat of stain, sanding sealer, light sanding, putty and 1 coat of varnish. Minimum preparation for varnish usually includes a steel wool buff and wax operation with minimum material usage. Add: Preparation time for protecting adjacent surfaces with window protective coating (wax) or masking tape and paper as required. For heights above 8 feet, use the High Time Difficulty Factors on page 137. "Slow" work is based on a $10.50 hourly wage, "Medium" work on a $16.75 hourly wage, and "Fast" work on a $22.50 hourly wage. Other qualifications that apply to this table are on page 9.

	Manhours per window	Windows per gallon	Material cost per gallon	Labor cost per window	Labor burden window	Material cost per window	Overhead per window	Profit per window	Total price per window

Windows, wood, exterior, 12 lite, brush

	Manhours per window	Windows per gallon	Material cost per gallon	Labor cost per window	Labor burden window	Material cost per window	Overhead per window	Profit per window	Total price per window
Undercoat, water or oil base (material #3 or #4)									
Brush 1 coat									
Slow	.67	11	21.35	7.04	1.99	1.94	3.40	2.16	16.53
Medium	.62	11	18.70	10.39	3.62	1.70	3.85	2.15	21.71
Fast	.57	10	16.00	12.83	5.35	1.60	3.66	1.41	24.85
Split coat (1/2 undercoat + 1/2 enamel), water or oil base (material #3 or #4 or #9 or #10)									
Brush 1 coat									
Slow	.72	13	23.45	7.56	2.15	1.80	3.57	2.26	17.34
Medium	.67	13	20.53	11.22	3.92	1.58	4.09	2.29	23.10
Fast	.62	12	17.60	13.95	5.82	1.47	3.93	1.51	26.68
Enamel, water or oil base (material #9 or #10)									
Brush each finish coat									
Slow	.85	12	25.55	8.93	2.53	2.13	4.22	2.67	20.48
Medium	.80	12	22.35	13.40	4.67	1.86	4.88	2.73	27.54
Fast	.75	11	19.20	16.88	7.04	1.75	4.75	1.83	32.25
Stain, seal & 1 coat varnish system (material #30)									
Brush each finish coat									
Slow	1.10	6	33.10	11.55	3.28	5.52	6.31	4.00	30.66
Medium	1.00	5	29.00	16.75	5.84	5.80	6.96	3.89	39.24
Fast	.90	4	24.90	20.25	8.45	6.23	6.46	2.48	43.87
Varnish (material #30c)									
Brush additional coats of varnish									
Slow	.45	18	35.40	4.73	1.34	1.97	2.50	1.58	12.12
Medium	.40	16	31.00	6.70	2.34	1.94	2.69	1.50	15.17
Fast	.35	14	26.60	7.88	3.28	1.90	2.42	.93	16.41
Preparation for varnish application (material - minimal)									
Steel wool buff									
Slow	.45	--	--	4.73	1.34	--	1.88	1.19	9.14
Medium	.40	--	--	6.70	2.34	--	2.21	1.24	12.49
Fast	.35	--	--	7.88	3.28	--	2.07	.79	14.02
Wax application (material - minimal)									
Slow	.45	--	--	4.73	1.34	--	1.88	1.19	9.14
Medium	.40	--	--	6.70	2.34	--	2.21	1.24	12.49
Fast	.35	--	--	7.88	3.28	--	2.07	.79	14.02

The stain, seal and 1 coat varnish system includes 1 coat of stain, sanding sealer, light sanding, putty and 1 coat of varnish. Minimum preparation for varnish usually includes a steel wool buff and wax operation with minimum material usage. Add: Preparation time for protecting adjacent surfaces with window protective coating (wax) or masking tape and paper as required. For heights above 8 feet, use the High Time Difficulty Factors on page 137. "Slow" work is based on a $10.50 hourly wage, "Medium" work on a $16.75 hourly wage, and "Fast" work on a $22.50 hourly wage. Other qualifications that apply to this table are on page 9.

Windows, interior

The figures that follow include costs for painting the sash (mullions or muntins), trim, frames, jambs, sill and apron on the inside only. Manhour and material rates are the same for painting either the inside or outside of windows. These estimates are per window painted, and are based on a window size of 15 square feet overall (length times width).

	Manhour per window	Windows per gallon	Material cost per gallon	Labor cost per window	Labor burden window	Material cost per window	Overhead per window	Profit per window	Total price per window
Windows, wood, interior, 1, 2 or 3 lite, brush									
Undercoat, water or oil base (material #3 or #4)									
Brush 1 coat									
Slow	.25	15	21.35	2.63	.74	1.42	1.49	.94	7.22
Medium	.20	15	18.70	3.35	1.17	1.25	1.41	.79	7.97
Fast	.15	14	16.00	3.38	1.40	1.14	1.10	.42	7.44
Split coat (1/2 undercoat + 1/2 enamel), water or oil base (material #3 or #4 or #9 or #10)									
Brush 1 coat									
Slow	.30	17	23.45	3.15	.89	1.38	1.68	1.07	8.17
Medium	.25	17	20.53	4.19	1.46	1.21	1.68	.94	9.48
Fast	.20	16	17.60	4.50	1.88	1.10	1.38	.53	9.39
Enamel, water or oil base (material #9 or #10)									
Brush each finish coat									
Slow	.35	16	25.55	3.68	1.04	1.60	1.96	1.24	9.52
Medium	.30	16	22.35	5.03	1.75	1.40	2.00	1.12	11.30
Fast	.25	15	19.20	5.63	2.34	1.28	1.71	.66	11.62
Stain, seal & 2 coat lacquer system, (material #11)									
Brush each coat									
Slow	.67	11	21.60	7.04	1.99	1.96	3.41	2.16	16.56
Medium	.62	10	19.60	10.39	3.62	1.96	3.91	2.19	22.07
Fast	.57	9	16.20	12.83	5.35	1.80	3.70	1.42	25.10

The stain, seal and 2 coat lacquer system includes 1 coat of stain, sanding sealer, light sanding, putty and 2 coats of lacquer. Add: Preparation time for protecting adjacent surfaces with window protective coating (wax) or masking tape and paper as required. "Slow" work is based on a $10.50 hourly wage, "Medium" work on a $16.75 hourly wage, and "Fast" work on a $22.50 hourly wage. Other qualifications that apply to this table are on page 9.

	Manhours per window	Windows per gallon	Material cost per gallon	Labor cost per window	Labor burden window	Material cost per window	Overhead per window	Profit per window	Total price per window

Windows, wood, interior, 4 to 6 lite, brush

	Manhours per window	Windows per gallon	Material cost per gallon	Labor cost per window	Labor burden window	Material cost per window	Overhead per window	Profit per window	Total price per window
Undercoat, water or oil base (material #3 or #4)									
Brush 1 coat									
Slow	.35	14	21.35	3.68	1.04	1.53	1.94	1.23	9.42
Medium	.30	14	18.70	5.03	1.75	1.34	1.99	1.11	11.22
Fast	.25	13	16.00	5.63	2.34	1.23	1.70	.65	11.55
Split coat (1/2 undercoat + 1/2 enamel), water or oil base (material #3 or #4 or #9 or #10)									
Brush 1 coat									
Slow	.40	16	23.45	4.20	1.19	1.47	2.13	1.35	10.34
Medium	.35	16	20.53	5.86	2.05	1.28	2.25	1.26	12.70
Fast	.30	15	17.60	6.75	2.82	1.17	1.99	.76	13.49
Enamel, water or oil base (material #9 or #10)									
Brush each finish coat									
Slow	.50	15	25.55	5.25	1.49	1.70	2.62	1.66	12.72
Medium	.45	15	22.35	7.54	2.63	1.49	2.86	1.60	16.12
Fast	.40	14	19.20	9.00	3.76	1.37	2.61	1.00	17.74
Stain, seal & 2 coat lacquer system, (material #11)									
Brush each coat									
Slow	.80	10	21.60	8.40	2.38	2.16	4.01	2.54	19.49
Medium	.73	9	19.60	12.23	4.26	2.18	4.57	2.56	25.80
Fast	.65	8	16.20	14.63	6.10	2.03	4.21	1.62	28.59

The stain, seal and 2 coat lacquer system includes 1 coat of stain, sanding sealer, light sanding, putty and 2 coats of lacquer. Add: Preparation time for protecting adjacent surfaces with window protective coating (wax) or masking tape and paper as required. "Slow" work is based on a $10.50 hourly wage, "Medium" work on a $16.75 hourly wage, and "Fast" work on a $22.50 hourly wage. Other qualifications that apply to this table are on page 9.

	Manhours per window	Windows per gallon	Material cost per gallon	Labor cost per window	Labor burden window	Material cost per window	Overhead per window	Profit per window	Total price per window
Windows, wood, interior, 7 to 8 lite, brush									
Undercoat, water or oil base (material #3 or #4)									
Brush 1 coat									
Slow	.45	13	21.35	4.73	1.34	1.64	2.39	1.52	11.62
Medium	.40	13	18.70	6.70	2.34	1.44	2.57	1.43	14.48
Fast	.35	12	16.00	7.88	3.28	1.33	2.31	.89	15.69
Split coat (1/2 undercoat + 1/2 enamel), water or oil base (material #3 or #4 or #9 or #10)									
Brush 1 coat									
Slow	.55	15	23.45	5.78	1.63	1.56	2.78	1.76	13.51
Medium	.50	15	20.53	8.38	2.92	1.37	3.10	1.73	17.50
Fast	.45	14	17.60	10.13	4.22	1.26	2.89	1.11	19.61
Enamel, water or oil base (material #9 or #10)									
Brush each finish coat									
Slow	.67	14	25.55	7.04	1.99	1.83	3.37	2.14	16.37
Medium	.62	14	22.35	10.39	3.62	1.60	3.82	2.14	21.57
Fast	.57	13	19.20	12.83	5.35	1.48	3.64	1.40	24.70
Stain, seal & 2 coat lacquer system, (material #11)									
Brush each coat									
Slow	.90	9	21.60	9.45	2.68	2.40	4.51	2.86	21.90
Medium	.80	8	19.60	13.40	4.67	2.45	5.03	2.81	28.36
Fast	.70	7	16.20	15.75	6.57	2.31	4.56	1.75	30.94

The stain, seal and 2 coat lacquer system includes 1 coat of stain, sanding sealer, light sanding, putty and 2 coats of lacquer. Add: Preparation time for protecting adjacent surfaces with window protective coating (wax) or masking tape and paper as required. "Slow" work is based on a $10.50 hourly wage, "Medium" work on a $16.75 hourly wage, and "Fast" work on a $22.50 hourly wage. Other qualifications that apply to this table are on page 9.

	Manhours per window	Windows per gallon	Material cost per gallon	Labor cost per window	Labor burden window	Material cost per window	Overhead per window	Profit per window	Total price per window
Windows, wood, interior, 9 to 11 lite, brush									
Undercoat, water or oil base (material #3 or #4)									
Brush 1 coat									
Slow	.55	12	21.35	5.78	1.63	1.78	2.85	1.81	13.85
Medium	.50	12	18.70	8.38	2.92	1.56	3.15	1.76	17.77
Fast	.45	11	16.00	10.13	4.22	1.45	2.92	1.12	19.84
Split coat (1/2 undercoat + 1/2 enamel), water or oil base (material #3 or #4 or #9 or #10)									
Brush 1 coat									
Slow	.65	14	23.45	6.83	1.93	1.68	3.24	2.05	15.73
Medium	.60	14	20.53	10.05	3.50	1.47	3.68	2.06	20.76
Fast	.55	13	17.60	12.38	5.16	1.35	3.50	1.34	23.73
Enamel, water or oil base (material #9 or #10)									
Brush each finish coat									
Slow	.75	13	25.55	7.88	2.23	1.97	3.75	2.38	18.21
Medium	.70	13	22.35	11.73	4.08	1.72	4.30	2.40	24.23
Fast	.65	12	19.20	14.63	6.10	1.60	4.13	1.59	28.05
Stain, seal & 2 coat lacquer system, (material #11)									
Brush each coat									
Slow	1.00	8	21.60	10.50	2.98	2.70	5.02	3.18	24.38
Medium	.90	7	19.60	15.08	5.25	2.80	5.67	3.17	31.97
Fast	.80	6	16.20	18.00	7.51	2.70	5.22	2.01	35.44

The stain, seal and 2 coat lacquer system includes 1 coat of stain, sanding sealer, light sanding, putty and 2 coats of lacquer. Add: Preparation time for protecting adjacent surfaces with window protective coating (wax) or masking tape and paper as required. "Slow" work is based on a $10.50 hourly wage, "Medium" work on a $16.75 hourly wage, and "Fast" work on a $22.50 hourly wage. Other qualifications that apply to this table are on page 9.

	Manhours per window	Windows per gallon	Material cost per gallon	Labor cost per window	Labor burden window	Material cost per window	Overhead per window	Profit per window	Total price per window

Windows, wood, interior, 12 lite, brush

	Manhours per window	Windows per gallon	Material cost per gallon	Labor cost per window	Labor burden window	Material cost per window	Overhead per window	Profit per window	Total price per window
Undercoat, water or oil base (material #3 or #4)									
Brush 1 coat									
Slow	.65	11	21.35	6.83	1.93	1.94	3.32	2.10	16.12
Medium	.60	11	18.70	10.05	3.50	1.70	3.74	2.09	21.08
Fast	.55	10	16.00	12.38	5.16	1.60	3.54	1.36	24.04
Split coat (1/2 undercoat + 1/2 enamel), water or oil base (material #3 or #4 or #9 or #10)									
Brush 1 coat									
Slow	.70	13	23.45	7.35	2.09	1.80	3.48	2.21	16.93
Medium	.65	13	20.53	10.89	3.79	1.58	3.99	2.23	22.48
Fast	.60	12	17.60	13.50	5.63	1.47	3.81	1.47	25.88
Enamel, water or oil base (material #9 or #10)									
Brush each finish coat									
Slow	.80	12	25.55	8.40	2.38	2.13	4.01	2.54	19.46
Medium	.75	12	22.35	12.56	4.38	1.86	4.61	2.58	25.99
Fast	.70	11	19.20	15.75	6.57	1.75	4.45	1.71	30.23
Stain, seal & 2 coat lacquer system, (material #11)									
Brush each coat									
Slow	1.10	7	21.60	11.55	3.28	3.09	5.56	3.52	27.00
Medium	1.00	6	19.60	16.75	5.84	3.27	6.34	3.54	35.74
Fast	.90	5	16.20	20.25	8.45	3.24	5.91	2.27	40.12

The stain, seal and 2 coat lacquer system includes 1 coat of stain, sanding sealer, light sanding, putty and 2 coats of lacquer. Add: Preparation time for protecting adjacent surfaces with window protective coating (wax) or masking tape and paper as required. "Slow" work is based on a $10.50 hourly wage, "Medium" work on a $16.75 hourly wage, and "Fast" work on a $22.50 hourly wage. Other qualifications that apply to this table are on page 9.

	Labor SF per manhour	Material coverage SF/gallon	Material cost per gallon	Labor cost per 100 SF	Labor burden 100 SF	Material cost per 100 SF	Overhead per 100 SF	Profit per 100 SF	Total price per 100 SF
Windows, wood, interior, paint grade, square foot basis, brush									
Undercoat, water or oil base (material #3 or #4)									
Brush, 1 coat									
Slow	140	460	21.35	7.50	2.12	4.64	4.42	2.80	21.48
Medium	165	450	18.70	10.15	3.54	4.16	4.37	2.44	24.66
Fast	190	440	16.00	11.84	4.93	3.64	3.78	1.45	25.64
Split coat (1/2 undercoat + 1/2 enamel), water base (material #3 or #4 or #9 or#10)									
Brush 1 coat									
Slow	120	520	23.45	8.75	2.48	4.51	4.88	3.09	23.71
Medium	135	500	20.53	12.41	4.33	4.11	5.11	2.85	28.81
Fast	150	480	17.60	15.00	6.27	3.67	4.61	1.77	31.32
Enamel, water or oil base (material #9 or #10)									
Brush each finish coat									
Slow	95	500	25.55	11.05	3.14	5.11	5.98	3.79	29.07
Medium	110	480	22.35	15.23	5.30	4.66	6.17	3.45	34.81
Fast	125	460	19.20	18.00	7.51	4.17	5.49	2.11	37.28

Use this table to estimate the costs for paint grade windows larger than 15 square feet (width times height). When calculating the window area, add 1 foot in each direction to both dimensions before multiplying width by the height. For example, a 4' x 4' window would have a calculated surface area of 36 square feet: Width of 4' + 1' + 1' = 6' times height of 4' + 1' + 1' = 6'. The total is 6' x 6', or 36 square feet per side. Add an extra 2 square feet to the calculated area for each window lite. This allows for the extra time needed to finish muntins, mullions and sash. For windows that are stained, sealed and varnished, use the previous tables based on window lites. Add: Preparation time to protect adjacent surfaces with window protective coating (wax) or masking tape and paper as required. "Slow" work is based on a $10.50 hourly wage, "Medium" work on a $16.75 hourly wage, and "Fast" work on a $22.50 hourly wage. Other qualifications that apply to this table are on page 9.

Window conversion factors

The table below gives you a faster way to figure the costs for painting windows by using the square foot basis tables above. First, calculate the square feet of window area. For example, assume you're undercoating a 3 by 5 foot window with 4 lites. Multiply length times width to find square feet. Then multiply the area by the manhour conversion factor below for a 4 lite window. Then divide the calculated square feet by the medium manhour rate from the previous table. Example: Area of a 3.0 x 5.0 window is 15 square feet, times the conversion factor for 4 lites (3.0) equals 45 square feet. Divide that by the medium rate from the above table for undercoating, (165) for a total time of .27 hours.

To find the material needed to paint the same window, multiply the 15 square feet by the factor in the right column below for a 4 lite window (2.2), for a total of 33 square feet. Now divide that figure by the light coverage rate, 460 square feet per gallon. It will take .072 gallons to undercoat that window. Divide 1 gallon by .072 to find that you can undercoat 13.9 windows with 1 gallon of undercoat material.

Lites	Manhours per SF conversion factor	Material per SF conversion factor
1, 2 or 3 lites	L x W x 2.0	L x W x 2.0
4 to 6 lites	L x W x 3.0	L x W x 2.2
7 to 8 lites	L x W x 4.0	L x W x 2.4
9 to 11 lites	L x W x 5.0	L x W x 2.6
12 lites	L x W x 6.0	L x W x 2.9

	Labor SF per manhour	Material coverage SF/gallon	Material cost per gallon	Labor cost per 100 SF	Labor burden 100 SF	Material cost per 100 SF	Overhead per 100 SF	Profit per 100 SF	Total price per 100 SF
Wine racks, paint grade, spray, square feet of face									
Undercoat, water base (material #3)									
Spray 1 coat									
Slow	50	100	21.10	21.00	5.96	21.10	14.90	9.45	72.41
Medium	65	88	18.50	25.77	8.97	21.02	13.66	7.64	77.06
Fast	80	75	15.80	28.13	11.73	21.07	11.27	4.33	76.53
Undercoat, oil base (material #4)									
Spray 1 coat									
Slow	50	100	21.60	21.00	5.96	21.60	15.06	9.54	73.16
Medium	65	88	18.90	25.77	8.97	21.48	13.78	7.70	77.70
Fast	80	75	16.20	28.13	11.73	21.60	11.37	4.37	77.20
Split coat (1/2 undercoat + 1/2 enamel), water base (material #3 or #9)									
Spray 1 coat									
Slow	75	150	22.80	14.00	3.97	15.20	10.29	6.52	49.98
Medium	90	138	19.95	18.61	6.49	14.46	9.69	5.42	54.67
Fast	125	125	17.10	18.00	7.51	13.68	7.25	2.79	49.23
Split coat (1/2 undercoat + 1/2 enamel), oil base (material #4 or #10)									
Spray 1 coat									
Slow	75	150	24.10	14.00	3.97	16.07	10.56	6.69	51.29
Medium	90	138	21.10	18.61	6.49	15.29	9.90	5.53	55.82
Fast	125	125	18.10	18.00	7.51	14.48	7.40	2.84	50.23
Enamel, water base (material #9)									
Spray 1st finish coat									
Slow	65	125	24.50	16.15	4.58	19.60	12.51	7.93	60.77
Medium	88	113	21.40	19.03	6.63	18.94	10.93	6.11	61.64
Fast	100	100	18.40	22.50	9.39	18.40	9.30	3.58	63.17
Spray additional finish coats									
Slow	75	150	24.50	14.00	3.97	16.33	10.64	6.74	51.68
Medium	100	138	21.40	16.75	5.84	15.51	9.33	5.22	52.65
Fast	125	125	18.40	18.00	7.51	14.72	7.44	2.86	50.53
Enamel, oil base (material #10)									
Spray 1st finish coat									
Slow	65	125	26.60	16.15	4.58	21.28	13.03	8.26	63.30
Medium	88	113	23.30	19.03	6.63	20.62	11.34	6.34	63.96
Fast	100	100	20.00	22.50	9.39	20.00	9.60	3.69	65.18
Spray additional finish coats									
Slow	75	150	26.60	14.00	3.97	17.73	11.07	7.02	53.79
Medium	100	138	23.30	16.75	5.84	16.88	9.67	5.41	54.55
Fast	125	125	20.00	18.00	7.51	16.00	7.68	2.95	52.14

These figures include coating all interior and exterior surfaces and are based on overall dimensions (length times width) of the wine rack face. For heights above 8 feet, use the High Time Difficulty Factors on page 137. "Slow" work is based on a $10.50 hourly wage, "Medium" work on a $16.75 hourly wage, and "Fast" work on a $22.50 hourly wage. Other qualifications that apply to this table are on page 9.

	Labor SF per manhour	Material coverage SF/gallon	Material cost per gallon	Labor cost per 100 SF	Labor burden 100 SF	Material cost per 100 SF	Overhead per 100 SF	Profit per 100 SF	Total price per 100 SF
Wine racks, stain grade, per square feet of face									
Stain, seal and 2 coat lacquer system (7 step process)									
STEP 1: Sand & putty									
Slow	50	--	--	21.00	5.96	--	8.36	5.30	40.62
Medium	75	--	--	22.33	7.78	--	7.38	4.12	41.61
Fast	100	--	--	22.50	9.39	--	5.90	2.27	40.06
STEP 2 & 3: Wiping stain, oil base (material #11a) & wipe									
Spray 1 coat & wipe									
Slow	100	150	30.90	10.50	2.98	20.60	10.57	6.70	51.35
Medium	175	113	27.00	9.57	3.33	23.89	9.02	5.04	50.85
Fast	225	75	23.10	10.00	4.16	30.80	8.32	3.20	56.48
STEP 4: Sanding sealer (material #11b)									
Spray 1 coat & sand									
Slow	130	150	17.20	8.08	2.29	11.47	6.77	4.29	32.90
Medium	208	113	15.00	8.05	2.82	13.27	5.91	3.30	33.35
Fast	275	75	12.90	8.18	3.43	17.20	5.33	2.05	36.19
STEP 5: Sand lightly									
Slow	75	--	--	14.00	3.97	--	5.57	3.53	27.07
Medium	100	--	--	16.75	5.84	--	5.53	3.09	31.21
Fast	125	--	--	18.00	7.51	--	4.72	1.81	32.04
STEP 6 & 7: Lacquer (material #11c)									
Spray 1st coat									
Slow	100	100	19.30	10.50	2.98	19.30	10.16	6.44	49.38
Medium	200	75	16.80	8.38	2.92	22.40	8.26	4.62	46.58
Fast	300	50	14.40	7.50	3.12	28.80	7.29	2.80	49.51
Spray 2nd coat									
Slow	175	100	19.30	6.00	1.70	19.30	8.37	5.31	40.68
Medium	313	75	16.80	5.35	1.86	22.40	7.25	4.05	40.91
Fast	450	50	14.40	5.00	2.08	28.80	6.64	2.55	45.07
Complete 7 step stain, seal & 2 coat lacquer system (material #11)									
Spray all coats									
Slow	20	25	21.60	52.50	14.90	86.40	47.69	30.23	231.72
Medium	28	20	19.60	59.82	20.85	98.00	43.77	24.47	246.91
Fast	35	15	16.20	64.29	26.82	108.00	36.84	14.16	250.11

These figures include coating all interior and exterior surfaces and are based on overall dimensions (length times width) of the wine rack face. For heights above 8 feet, use the High Time Difficulty Factors on page 137. "Slow" work is based on a $10.50 hourly wage, "Medium" work on a $16.75 hourly wage, and "Fast" work on a $22.50 hourly wage. Other qualifications that apply to this table are on page 9.

Part II

Preparation Costs

	Labor LF per manhour	Material coverage LF/gallon	Material cost per gallon	Labor cost per 100 LF	Labor burden 100 LF	Material cost per 100 LF	Overhead per 100 LF	Profit per 100 LF	Total price per 100 LF
Acid wash gutters & downspouts									
Acid wash, muriatic acid (material #49)									
Brush or mitt 1 coat									
Slow	80	450	5.40	13.13	3.72	1.20	5.60	3.55	27.20
Medium	95	425	4.70	17.63	6.16	1.11	6.10	3.41	34.41
Fast	110	400	4.00	20.45	8.54	1.00	5.55	2.13	37.67

"Slow" work is based on a $10.50 hourly wage, "Medium" work on a $16.75 hourly wage, and "Fast" work on a $22.50 hourly wage. Other qualifications that apply to this table are on page 9.

	Labor SF per manhour	Material coverage SF/gallon	Material cost per gallon	Labor cost per 100 SF	Labor burden 100 SF	Material cost per 100 SF	Overhead per 100 SF	Profit per 100 SF	Total price per 100 SF
Airblast, compressed air									
Average production									
Slow	150	--	--	7.00	1.99	--	2.79	1.77	13.55
Medium	175	--	--	9.57	3.33	--	3.16	1.77	17.83
Fast	200	--	--	11.25	4.70	--	2.95	1.13	20.03

All widths less than 12", consider as 1 square foot per linear foot. Add equipment rental costs with Overhead and Profit. "Slow" work is based on a $10.50 hourly wage, "Medium" work on a $16.75 hourly wage, and "Fast" work on a $22.50 hourly wage. Other qualifications that apply to this table are on page 9.

	Labor SF per manhour	Material coverage SF/gallon	Material cost per gallon	Labor cost per 100 SF	Labor burden 100 SF	Material cost per 100 SF	Overhead per 100 SF	Profit per 100 SF	Total price per 100 SF
Burn off paint									
Exterior:									
Exterior trim									
Slow	20	--	--	52.50	14.90	--	20.91	13.25	101.56
Medium	25	--	--	67.00	23.36	--	22.14	12.37	124.87
Fast	30	--	--	75.00	31.29	--	19.67	7.56	133.52
Plain surfaces									
Slow	35	--	--	30.00	8.51	--	11.95	7.57	58.03
Medium	45	--	--	37.22	12.97	--	12.30	6.87	69.36
Fast	55	--	--	40.91	17.07	--	10.73	4.12	72.83
Beveled wood siding									
Slow	25	--	--	42.00	11.92	--	16.72	10.60	81.24
Medium	35	--	--	47.86	16.68	--	15.81	8.84	89.19
Fast	45	--	--	50.00	20.86	--	13.11	5.04	89.01
Interior:									
Interior trim									
Slow	15	--	--	70.00	19.87	--	27.88	17.67	135.42
Medium	20	--	--	83.75	29.20	--	27.67	15.47	156.09
Fast	25	--	--	90.00	37.56	--	23.60	9.07	160.23
Plain surfaces									
Slow	20	--	--	52.50	14.90	--	20.91	13.25	101.56
Medium	30	--	--	55.83	19.46	--	18.45	10.31	104.05
Fast	40	--	--	56.25	23.48	--	14.75	5.67	100.15

All widths less than 12", consider as 1 square foot per linear foot. Note: Because surfaces and the material being removed vary widely, it's best to quote prices for burning-off existing finishes on a Time and Material or Cost Plus Fee basis at a preset hourly rate. "Slow" work is based on a $10.50 hourly wage, "Medium" work on a $16.75 hourly wage, and "Fast" work on a $22.50 hourly wage. Other qualifications that apply to this table are on page 9.

	Labor LF per manhour	Material LF/fluid oz ounce	Material cost per ounce	Labor cost per 100 LF	Labor burden 100 LF	Material cost per 100 LF	Overhead per 100 LF	Profit per 100 LF	Total price per 100 LF
Caulk									
1/8" gap (material #42)									
Slow	60	14	.21	17.50	4.97	1.50	7.43	4.71	36.11
Medium	65	13	.20	25.77	8.97	1.54	8.89	4.97	50.14
Fast	70	12	.20	32.14	13.43	1.67	8.74	3.36	59.34
1/4" gap (material #42)									
Slow	50	3.5	.21	21.00	5.96	6.00	10.22	6.48	49.66
Medium	55	3.3	.20	30.45	10.62	6.06	11.54	6.45	65.12
Fast	60	3.0	.20	37.50	15.66	6.67	11.07	4.25	75.15
3/8" gap (material #42)									
Slow	40	1.5	.21	26.25	7.45	14.00	14.79	9.38	71.87
Medium	45	1.4	.20	37.22	12.97	14.29	15.80	8.83	89.11
Fast	50	1.3	.20	45.00	18.78	15.38	14.65	5.63	99.44
1/2" gap (material #42)									
Slow	33	1.0	.21	31.82	9.02	21.00	19.18	12.16	93.18
Medium	38	0.9	.20	44.08	15.38	22.22	20.01	11.18	112.87
Fast	43	0.8	.20	52.33	21.85	25.00	18.35	7.05	124.58

Caulking that's part of normal surface preparation is included in the painting cost tables. When extra caulking is required, use this cost guide. It's based on oil or latex base, silicone or urethane caulk in 10 ounce tubes. "Slow" work is based on a $10.50 hourly wage, "Medium" work on a $16.75 hourly wage, and "Fast" work on a $22.50 hourly wage. Other qualifications that apply to this table are on page 9.

	Labor SF per manhour	Material coverage SF/gallon	Material cost per gallon	Labor cost per 100 SF	Labor burden 100 SF	Material cost per 100 SF	Overhead per 100 SF	Profit per 100 SF	Total price per 100 SF
Cut cracks									
Varnish or hard oil and repair cracks									
Slow	120	--	--	8.75	2.48	--	3.48	2.21	16.92
Medium	130	--	--	12.88	4.49	--	4.26	2.38	24.01
Fast	140	--	--	16.07	6.70	--	4.21	1.62	28.60
Gloss painted walls and fix cracks									
Slow	125	--	--	8.40	2.38	--	3.34	2.12	16.24
Medium	135	--	--	12.41	4.33	--	4.10	2.29	23.13
Fast	145	--	--	15.52	6.48	--	4.07	1.56	27.63

All widths less than 12", consider as 1 square foot per linear foot. "Slow" work is based on a $10.50 hourly wage, "Medium" work on a $16.75 hourly wage, and "Fast" work on a $22.50 hourly wage. Other qualifications that apply to this table are on page 9.

	Labor SF per manhour	Material coverage SF/gallon	Material cost per gallon	Labor cost per 100 SF	Labor burden 100 SF	Material cost per 100 SF	Overhead per 100 SF	Profit per 100 SF	Total price per 100 SF
Fill wood floors									
Fill and wipe wood floors (material #48)									
Slow	45	155	32.90	23.33	6.62	21.23	15.87	10.06	77.11
Medium	60	145	28.80	27.92	9.74	19.86	14.09	7.88	79.49
Fast	75	135	24.70	30.00	12.51	18.30	11.25	4.32	76.38

All widths less than 12", consider as 1 square foot per linear foot. "Slow" work is based on a $10.50 hourly wage, "Medium" work on a $16.75 hourly wage, and "Fast" work on a $22.50 hourly wage. Other qualifications that apply to this table are on page 9.

	Labor SF per manhour	Material coverage SF/pound	Material cost per pound	Labor cost per 100 SF	Labor burden 100 SF	Material cost per 100 SF	Overhead per 100 SF	Profit per 100 SF	Total price per 100 SF
Putty application									
Good condition, 1 coat (material #45)									
Slow	60	150	7.00	17.50	4.97	4.67	8.42	5.34	40.90
Medium	90	135	6.10	18.61	6.49	4.52	7.26	4.06	40.94
Fast	120	120	5.20	18.75	7.81	4.33	5.72	2.20	38.81
Average condition, 1 coat (material #45)									
Slow	35	90	7.00	30.00	8.51	7.78	14.36	9.10	69.75
Medium	65	75	6.10	25.77	8.97	8.13	10.51	5.87	59.25
Fast	95	60	5.20	23.68	9.90	8.67	7.81	3.00	53.06
Poor conditions, 1 coat (material #45)									
Slow	15	40	7.00	70.00	19.87	17.50	33.30	21.11	161.78
Medium	30	30	6.10	55.83	19.46	20.33	23.43	13.10	132.15
Fast	45	20	5.20	50.00	20.86	26.00	17.92	6.89	121.67

These figures apply to either spackle or Swedish putty. All widths less than 12", consider as 1 square foot per linear foot. For flat trim or sash: Estimate 1 linear foot of trim as 1 square foot of surface. "Slow" work is based on a $10.50 hourly wage, "Medium" work on a $16.75 hourly wage, and "Fast" work on a $22.50 hourly wage. Other qualifications that apply to this table are on page 9.

	Labor SF per manhour	Material coverage SF/gallon	Material cost per gallon	Labor cost per 100 SF	Labor burden 100 SF	Material cost per 100 SF	Overhead per 100 SF	Profit per 100 SF	Total price per 100 SF
Sand, medium (before first coat)									
Interior flatwall areas									
Slow	275	--	--	3.82	1.09	--	1.20	.92	7.03
Medium	300	--	--	5.58	1.94	--	1.84	1.03	10.39
Fast	325	--	--	6.92	2.90	--	1.81	.70	12.33
Interior enamel areas									
Slow	250	--	--	4.20	1.19	--	1.67	1.06	8.12
Medium	275	--	--	6.09	2.13	--	2.01	1.12	11.35
Fast	300	--	--	7.50	3.12	--	1.97	.76	13.35

All widths less than 12", consider as 1 square foot per linear foot. "Slow" work is based on a $10.50 hourly wage, "Medium" work on a $16.75 hourly wage, and "Fast" work on a $22.50 hourly wage. Other qualifications that apply to this table are on page 9.

	Labor SF per manhour	Material coverage SF/gallon	Material cost per gallon	Labor cost per 100 SF	Labor burden 100 SF	Material cost per 100 SF	Overhead per 100 SF	Profit per 100 SF	Total price per 100 SF
Sand & putty (before second coat)									
Interior flatwall areas									
Slow	190	--	--	5.53	1.56	--	2.20	1.40	10.69
Medium	200	--	--	8.38	2.92	--	2.77	1.55	15.62
Fast	210	--	--	10.71	4.47	--	2.81	1.08	19.07
Interior enamel areas									
Slow	110	--	--	9.55	2.70	--	3.80	2.41	18.46
Medium	125	--	--	13.40	4.67	--	4.43	2.48	24.98
Fast	140	--	--	16.07	6.70	--	4.21	1.62	28.60
Exterior siding & trim-plain									
Slow	180	--	--	5.83	1.66	--	2.32	1.47	11.28
Medium	200	--	--	8.38	2.92	--	2.77	1.55	15.62
Fast	220	--	--	10.23	4.28	--	2.68	1.03	18.22
Exterior trim only									
Slow	100	--	--	10.50	2.98	--	4.18	2.65	20.31
Medium	110	--	--	15.23	5.30	--	5.03	2.81	28.37
Fast	120	--	--	18.75	7.81	--	4.92	1.89	33.37
Bookshelves									
Slow	100	--	--	10.50	2.98	--	4.18	2.65	20.31
Medium	125	--	--	13.40	4.67	--	4.43	2.48	24.98
Fast	150	--	--	15.00	6.27	--	3.93	1.51	26.71
Cabinets									
Slow	125	--	--	8.40	2.38	--	3.34	2.12	16.24
Medium	150	--	--	11.17	3.90	--	3.69	2.06	20.82
Fast	175	--	--	12.86	5.35	--	3.37	1.30	22.88

Consider all trim which is less than 12" wide to be 12" wide. High grade work - Use the manhours equal to 1 coat of paint. Medium grade work - Use half (50%) of the manhours for 1 coat of paint. All widths less than 12", consider as 1 square foot per linear foot. "Slow" work is based on a $10.50 hourly wage, "Medium" work on a $16.75 hourly wage, and "Fast" work on a $22.50 hourly wage. Other qualifications that apply to this table are on page 9.

	Labor SF per manhour	Material coverage SF/gallon	Material cost per gallon	Labor cost per 100 SF	Labor burden 100 SF	Material cost per 100 SF	Overhead per 100 SF	Profit per 100 SF	Total price per 100 SF
Sand, light (before third coat)									
Interior flatwall areas									
Slow	335	--	--	3.13	.90	--	1.25	.79	6.07
Medium	345	--	--	4.86	1.69	--	1.60	.90	9.05
Fast	355	--	--	6.34	2.65	--	1.66	.64	11.29
Interior enamel areas									
Slow	130	--	--	8.08	2.29	--	3.22	2.04	15.63
Medium	140	--	--	11.96	4.17	--	3.95	2.21	22.29
Fast	150	--	--	15.00	6.27	--	3.93	1.51	26.71
Exterior siding & trim-plain									
Slow	250	--	--	4.20	1.19	--	1.67	1.06	8.12
Medium	275	--	--	6.09	2.13	--	2.01	1.12	11.35
Fast	300	--	--	7.50	3.12	--	1.97	.76	13.35
Exterior trim only									
Slow	150	--	--	7.00	1.99	--	2.79	1.77	13.55
Medium	175	--	--	9.57	3.33	--	3.16	1.77	17.83
Fast	200	--	--	11.25	4.70	--	2.95	1.13	20.03
Bookshelves									
Slow	175	--	--	6.00	1.70	--	2.39	1.52	11.61
Medium	225	--	--	7.44	2.59	--	2.46	1.37	13.86
Fast	275	--	--	8.18	3.43	--	2.15	.83	14.59
Cabinets									
Slow	200	--	--	5.25	1.49	--	2.09	1.32	10.15
Medium	250	--	--	6.70	2.34	--	2.21	1.24	12.49
Fast	300	--	--	7.50	3.12	--	1.97	.76	13.35

For heights above 8 feet, use the High Time Difficulty Factors on page 137. "Slow" work is based on a $10.50 hourly wage, "Medium" work on a $16.75 hourly wage, and "Fast" work on a $22.50 hourly wage. Other qualifications that apply to this table are on page 9.

	Labor SF per manhour	Material coverage SF/gallon	Material cost per gallon	Labor cost per 100 SF	Labor burden 100 SF	Material cost per 100 SF	Overhead per 100 SF	Profit per 100 SF	Total price per 100 SF
Sand, extra fine, flat surfaces, varnish									
Sand or steel wool									
Slow	50	--	--	21.00	5.96	--	8.36	5.30	40.62
Medium	88	--	--	19.03	6.63	--	6.29	3.51	35.46
Fast	125	--	--	18.00	7.51	--	4.72	1.81	32.04

All widths less than 12", consider as 1 square foot per linear foot. "Slow" work is based on a $10.50 hourly wage, "Medium" work on a $16.75 hourly wage, and "Fast" work on a $22.50 hourly wage. Other qualifications that apply to this table are on page 9.

Sandblast, general

Sandblasting production rates may vary widely. Use the following figures as a reference for estimating and to establish performance data for your company. The abrasive material used in sandblasting is usually white silica sand although slags have recently gained in popularity. Material consumption varies with several factors:

1) Type of finish required
2) Condition of the surface
3) Quality of abrasive material (sharpness, cleanliness and hardness)
4) Nozzle size
5) Equipment arrangement and placement
6) Operator skill

All material consumption values are based on three uses of a 25 to 35 mesh white silica sand abrasive at a cost of $40 to $60 per ton. (Check the current price in your area.) Note: See the Structural steel conversion table at Figure 23 on pages 387 through 395 for converting linear feet of structural steel to square feet.

	Labor SF per manhour	Material coverage pounds/SF	Material cost per pound	Labor cost per 100 SF	Labor burden 100 SF	Material cost per 100 SF	Overhead per 100 SF	Profit per 100 SF	Total price per 100 SF
Sandblast, brush-off blast									
Surface condition basis - large projects & surface areas (material #46)									
Remove cement base paint									
Slow	150	2.0	.03	7.00	1.99	6.00	4.65	2.95	22.59
Medium	175	2.5	.03	9.57	3.33	7.50	5.00	2.80	28.20
Fast	200	3.0	.02	11.25	4.70	6.00	4.06	1.56	27.57
Remove oil or latex base paint									
Slow	100	3.0	.03	10.50	2.98	9.00	6.97	4.42	33.87
Medium	125	3.5	.03	13.40	4.67	10.50	7.00	3.91	39.48
Fast	150	4.0	.02	15.00	6.27	8.00	5.41	2.08	36.76
Surface area basis - large projects & surface areas (material #46)									
Pipe up to 12" O/D									
Slow	125	4.0	.03	8.40	2.38	12.00	7.06	4.48	34.32
Medium	150	4.5	.03	11.17	3.90	13.50	7.00	3.91	39.48
Fast	175	5.0	.02	12.86	5.35	10.00	5.22	2.01	35.44
Structural steel									
Sizes up to 2 SF/LF									
Slow	150	4.0	.03	7.00	1.99	12.00	6.51	4.13	31.63
Medium	175	4.5	.03	9.57	3.33	13.50	6.47	3.62	36.49
Fast	200	5.0	.02	11.25	4.70	10.00	4.80	1.85	32.60
Sizes from 2 to 5 SF/LF									
Slow	200	3.0	.03	5.25	1.49	9.00	4.88	3.09	23.71
Medium	225	3.5	.03	7.44	2.59	10.50	5.03	2.81	28.37
Fast	250	4.0	.02	9.00	3.76	8.00	3.84	1.48	26.08
Sizes over 5 SF/LF									
Slow	250	2.0	.03	4.20	1.19	6.00	3.53	2.24	17.16
Medium	275	3.0	.03	6.09	2.13	9.00	4.22	2.36	23.80
Fast	300	4.0	.02	7.50	3.12	8.00	3.45	1.32	23.39
Tanks and vessels									
Sizes up to 12'0" O/D									
Slow	200	3.0	.03	5.25	1.49	9.00	4.88	3.09	23.71
Medium	225	3.5	.03	7.44	2.59	10.50	5.03	2.81	28.37
Fast	250	4.0	.02	9.00	3.76	8.00	3.84	1.48	26.08
Sizes over 12'0" O/D									
Slow	250	2.0	.03	4.20	1.19	6.00	3.53	2.24	17.16
Medium	275	3.0	.03	6.09	2.13	9.00	4.22	2.36	23.80
Fast	300	4.0	.02	7.50	3.12	8.00	3.45	1.32	23.39

For heights above 8 feet, use the High Time Difficulty Factors on page 137. "Slow" work is based on a $10.50 hourly wage, "Medium" work on a $16.75 hourly wage, and "Fast" work on a $22.50 hourly wage. Other qualifications that apply to this table are on page 9.

	Labor SF per manhour	Material coverage pounds/SF	Material cost per pound	Labor cost per 100 SF	Labor burden 100 SF	Material cost per 100 SF	Overhead per 100 SF	Profit per 100 SF	Total price per 100 SF
Sandblast, commercial blast (67% white)									
Surface condition basis - large projects & surface areas (material #46)									
Loose mill scale & fine powder rust									
Slow	150	4.0	.03	7.00	1.99	12.00	6.51	4.13	31.63
Medium	175	4.5	.03	9.57	3.33	13.50	6.47	3.62	36.49
Fast	200	5.0	.02	11.25	4.70	10.00	4.80	1.85	32.60
Tight mill scale & little or no rust									
Slow	125	5.0	.03	8.40	2.38	15.00	7.99	5.07	38.84
Medium	150	5.5	.03	11.17	3.90	16.50	7.73	4.32	43.62
Fast	175	6.0	.02	12.86	5.35	12.00	5.59	2.15	37.95
Hard scale, blistered, rusty surface									
Slow	75	6.0	.03	14.00	3.97	18.00	11.15	7.07	54.19
Medium	100	7.0	.03	16.75	5.84	21.00	10.68	5.97	60.24
Fast	125	8.0	.02	18.00	7.51	16.00	7.68	2.95	52.14
Rust nodules and pitted surface									
Slow	50	8.0	.03	21.00	5.96	24.00	15.80	10.02	76.78
Medium	60	9.5	.03	27.92	9.74	28.50	16.21	9.06	91.43
Fast	70	11.0	.02	32.14	13.43	22.00	12.50	4.80	84.87
Surface area basis - large projects & surface areas (material #46)									
Pipe up to 12" O/D									
Slow	45	5.0	.03	23.33	6.62	15.00	13.94	8.84	67.73
Medium	60	6.0	.03	27.92	9.74	18.00	13.63	7.62	76.91
Fast	75	7.0	.02	30.00	12.51	14.00	10.46	4.02	70.99
Structural steel									
Sizes up to 2 SF/LF									
Slow	70	5.0	.03	15.00	4.26	15.00	10.62	6.73	51.61
Medium	85	6.0	.03	19.71	6.86	18.00	10.92	6.11	61.60
Fast	100	7.0	.02	22.50	9.39	14.00	8.49	3.26	57.64
Sizes from 2 to 5 SF/LF									
Slow	80	5.0	.03	13.13	3.72	15.00	9.88	6.26	47.99
Medium	95	5.5	.03	17.63	6.16	16.50	9.87	5.52	55.68
Fast	110	6.0	.02	20.45	8.54	12.00	7.58	2.91	51.48
Sizes over 5 SF/LF									
Slow	85	5.0	.03	12.35	3.50	15.00	9.57	6.06	46.48
Medium	100	5.5	.03	16.75	5.84	16.50	9.58	5.35	54.02
Fast	115	6.0	.02	19.57	8.17	12.00	7.35	2.83	49.92

	Labor SF per manhour	Material coverage pounds/SF	Material cost per pound	Labor cost per 100 SF	Labor burden 100 SF	Material cost per 100 SF	Overhead per 100 SF	Profit per 100 SF	Total price per 100 SF
Tanks and vessels									
Sizes up to 12'0" O/D									
Slow	80	6.0	.03	13.13	3.72	18.00	10.81	6.85	52.51
Medium	95	6.5	.03	17.63	6.16	19.50	10.60	5.93	59.82
Fast	110	7.0	.02	20.45	8.54	14.00	7.95	3.06	54.00
Sizes over 12'0" O/D									
Slow	75	6.0	.03	14.00	3.97	18.00	11.15	7.07	54.19
Medium	100	6.3	.03	16.75	5.84	18.90	10.17	5.68	57.34
Fast	125	6.5	.02	18.00	7.51	13.00	7.13	2.74	48.38

For heights above 8 feet, use the High Time Difficulty Factors on page 137. "Slow" work is based on a $10.50 hourly wage, "Medium" work on a $16.75 hourly wage, and "Fast" work on a $22.50 hourly wage. Other qualifications that apply to this table are on page 9.

	Labor SF per manhour	Material coverage pounds/SF	Material cost per pound	Labor cost per 100 SF	Labor burden 100 SF	Material cost per 100 SF	Overhead per 100 SF	Profit per 100 SF	Total price per 100 SF
Sandblast, near white blast (95% white)									
Surface condition basis - large projects & surface areas (material #46)									
Loose mill scale & fine powder rust									
Slow	125	5.0	.03	8.40	2.38	15.00	7.99	5.07	38.84
Medium	150	6.0	.03	11.17	3.90	18.00	8.10	4.53	45.70
Fast	175	7.0	.02	12.86	5.35	14.00	5.96	2.29	40.46
Tight mill scale & little or no rust									
Slow	75	7.0	.03	14.00	3.97	21.00	12.08	7.66	58.71
Medium	100	8.0	.03	16.75	5.84	24.00	11.41	6.38	64.38
Fast	125	9.0	.02	18.00	7.51	18.00	8.05	3.09	54.65
Hard scale, blistered, rusty surface									
Slow	50	9.0	.03	21.00	5.96	27.00	16.73	10.61	81.30
Medium	75	11.0	.03	22.33	7.78	33.00	15.46	8.64	87.21
Fast	100	13.0	.02	22.50	9.39	26.00	10.71	4.12	72.72
Rust nodules and pitted surface									
Slow	35	12.0	.03	30.00	8.51	36.00	23.11	14.65	112.27
Medium	50	14.5	.03	33.50	11.68	43.50	21.72	12.14	122.54
Fast	65	17.0	.02	34.62	14.43	34.00	15.37	5.91	104.33
Surface area basis - large projects & surface areas (material #46)									
Pipe up to 12" O/D									
Slow	30	8.0	.03	35.00	9.93	24.00	21.38	13.55	103.86
Medium	45	9.0	.03	37.22	12.97	27.00	18.91	10.57	106.67
Fast	60	10.0	.02	37.50	15.66	20.00	13.53	5.20	91.89
Structural steel									
Sizes up to 2 SF/LF									
Slow	40	7.0	.03	26.25	7.45	21.00	16.96	10.75	82.41
Medium	55	8.5	.03	30.45	10.62	25.50	16.31	9.12	92.00
Fast	70	10.0	.02	32.14	13.43	20.00	12.13	4.66	82.36
Sizes from 2 to 5 SF/LF									
Slow	45	7.0	.03	23.33	6.62	21.00	15.80	10.02	76.77
Medium	60	8.0	.03	27.92	9.74	24.00	15.10	8.44	85.20
Fast	75	9.0	.02	30.00	12.51	18.00	11.20	4.30	76.01
Sizes over 5 SF/LF									
Slow	55	8.0	.03	19.09	5.42	24.00	15.04	9.53	73.08
Medium	70	8.5	.03	23.93	8.35	25.50	14.15	7.91	79.84
Fast	85	9.0	.02	26.47	11.03	18.00	10.27	3.95	69.72

	Labor SF per manhour	Material coverage pounds/SF	Material cost per pound	Labor cost per 100 SF	Labor burden 100 SF	Material cost per 100 SF	Overhead per 100 SF	Profit per 100 SF	Total price per 100 SF
Tanks and vessels									
Sizes up to 12'0" O/D									
Slow	65	6.0	.03	16.15	4.58	18.00	12.01	7.61	58.35
Medium	80	7.0	.03	20.94	7.30	21.00	12.06	6.74	68.04
Fast	95	8.0	.02	23.68	9.90	16.00	9.17	3.52	62.27
Sizes over 12'0" O/D									
Slow	70	6.0	.03	15.00	4.26	18.00	11.55	7.32	56.13
Medium	85	7.0	.03	19.71	6.86	21.00	11.66	6.52	65.75
Fast	100	8.0	.02	22.50	9.39	16.00	8.86	3.41	60.16

For heights above 8 feet, use the High Time Difficulty Factors on page 137. "Slow" work is based on a $10.50 hourly wage, "Medium" work on a $16.75 hourly wage, and "Fast" work on a $22.50 hourly wage. Other qualifications that apply to this table are on page 9.

	Labor SF per manhour	Material coverage pounds/SF	Material cost per pound	Labor cost per 100 SF	Labor burden 100 SF	Material cost per 100 SF	Overhead per 100 SF	Profit per 100 SF	Total price per 100 SF
Sandblast, white blast (100% uniform white stage)									
Surface condition basis - large projects & surface areas (material #46)									
Loose mill scale & fine powder rust									
Slow	50	7.0	.03	21.00	5.96	21.00	14.87	9.43	72.26
Medium	75	8.5	.03	22.33	7.78	25.50	13.62	7.62	76.85
Fast	100	10.0	.02	22.50	9.39	20.00	9.60	3.69	65.18
Tight mill scale & little or no rust									
Slow	40	8.0	.03	26.25	7.45	24.00	17.89	11.34	86.93
Medium	60	9.5	.03	27.92	9.74	28.50	16.21	9.06	91.43
Fast	80	11.0	.02	28.13	11.73	22.00	11.45	4.40	77.71
Hard scale, blistered, rusty surface									
Slow	30	10.0	.03	35.00	9.93	30.00	23.24	14.73	112.90
Medium	45	12.5	.03	37.22	12.97	37.50	21.48	12.01	121.18
Fast	60	15.0	.02	37.50	15.66	30.00	15.38	5.91	104.45
Rust nodules and pitted surface									
Slow	25	15.0	.03	42.00	11.92	45.00	30.67	19.44	149.03
Medium	35	17.5	.03	47.86	16.68	52.50	28.67	16.03	161.74
Fast	45	20.0	.02	50.00	20.86	40.00	20.51	7.88	139.25
Surface area basis - large projects & surface areas (material #46)									
Pipe up to 12" O/D									
Slow	30	10.0	.03	35.00	9.93	30.00	23.24	14.73	112.90
Medium	40	11.5	.03	41.88	14.60	34.50	22.29	12.46	125.73
Fast	50	13.0	.02	45.00	18.78	26.00	16.61	6.38	112.77
Structural steel									
Sizes up to 2 SF/LF									
Slow	40	9.0	.03	26.25	7.45	27.00	18.82	11.93	91.45
Medium	50	10.5	.03	33.50	11.68	31.50	18.78	10.50	105.96
Fast	60	12.0	.02	37.50	15.66	24.00	14.27	5.49	96.92
Sizes from 2 to 5 SF/LF									
Slow	45	8.0	.03	23.33	6.62	24.00	16.73	10.61	81.29
Medium	55	9.5	.03	30.45	10.62	28.50	17.04	9.53	96.14
Fast	65	11.0	.02	34.62	14.43	22.00	13.15	5.05	89.25
Sizes over 5 SF/LF									
Slow	50	8.0	.03	21.00	5.96	24.00	15.80	10.02	76.78
Medium	60	9.5	.03	27.92	9.74	28.50	16.21	9.06	91.43
Fast	70	11.0	.02	32.14	13.43	22.00	12.50	4.80	84.87

	Labor SF per manhour	Material coverage pounds/SF	Material cost per pound	Labor cost per 100 SF	Labor burden 100 SF	Material cost per 100 SF	Overhead per 100 SF	Profit per 100 SF	Total price per 100 SF
Tanks and vessels									
Sizes up to 12'0" O/D									
Slow	60	8.0	.03	17.50	4.97	24.00	14.41	9.13	70.01
Medium	70	9.5	.03	23.93	8.35	28.50	14.89	8.32	83.99
Fast	80	11.0	.02	28.13	11.73	22.00	11.45	4.40	77.71
Sizes over 12'0" O/D									
Slow	70	8.0	.03	15.00	4.26	24.00	13.41	8.50	65.17
Medium	80	9.5	.03	20.94	7.30	28.50	13.90	7.77	78.41
Fast	90	11.0	.02	25.00	10.43	22.00	10.63	4.08	72.14

For heights above 8 feet, use the High Time Difficulty Factors on page 137. "Slow" work is based on a $10.50 hourly wage, "Medium" work on a $16.75 hourly wage, and "Fast" work on a $22.50 hourly wage. Other qualifications that apply to this table are on page 9.

	Labor SF per manhour	Material coverage SF/gallon	Material cost per gallon	Labor cost per 100 SF	Labor burden 100 SF	Material cost per 100 SF	Overhead per 100 SF	Profit per 100 SF	Total price per 100 SF
Strip, remove, or bleach									
Remove wallcover by hand									
Slow	50	--	--	21.00	5.96	--	8.36	5.30	40.62
Medium	70	--	--	23.93	8.35	--	7.91	4.42	44.61
Fast	90	--	--	25.00	10.43	--	6.56	2.52	44.51
Stripping flat, vertical, varnished surfaces									
Light duty liquid remover (material #43)									
Slow	25	175	19.10	42.00	11.92	10.91	15.89	12.11	92.83
Medium	35	158	16.80	47.86	16.68	10.63	18.42	10.29	103.88
Fast	45	140	14.40	50.00	20.86	10.29	15.02	5.77	101.94
Heavy duty liquid remover (material #44)									
Slow	20	150	23.90	52.50	14.90	15.93	20.43	15.57	119.33
Medium	30	138	20.90	55.83	19.46	15.14	22.16	12.38	124.97
Fast	40	125	17.90	56.25	23.48	14.32	17.40	6.69	118.14
Stripping flat, horizontal floor surfaces									
Paint removal with light duty liquid remover (material #43)									
Slow	20	180	19.10	52.50	14.90	10.61	19.12	14.58	111.71
Medium	30	175	16.80	55.83	19.46	9.60	20.80	11.63	117.32
Fast	40	170	14.40	56.25	23.48	8.47	16.32	6.27	110.79
Paint removal with heavy duty liquid remover (material #44)									
Slow	15	170	23.90	70.00	19.87	14.06	25.48	19.42	148.83
Medium	25	160	20.90	67.00	23.36	13.06	25.34	14.16	142.92
Fast	35	150	17.90	64.29	26.82	11.93	19.07	7.33	129.44
Varnish removal with light duty liquid remover (material #43)									
Slow	30	185	19.10	35.00	9.93	10.32	13.54	10.32	79.11
Medium	40	180	16.80	41.88	14.60	9.33	16.12	9.01	90.94
Fast	50	175	14.40	45.00	18.78	8.23	13.32	5.12	90.45
Varnish removal with heavy duty liquid remover (material #44)									
Slow	25	180	23.90	42.00	11.92	13.28	16.47	12.56	96.23
Medium	35	170	20.90	47.86	16.68	12.29	18.82	10.52	106.17
Fast	45	160	17.90	50.00	20.86	11.19	15.18	5.84	103.07

All widths less than 12", consider as 1 square foot per linear foot. For heights above 8 feet, use the High Time Difficulty Factors on page 137. "Slow" work is based on a $10.50 hourly wage, "Medium" work on a $16.75 hourly wage, and "Fast" work on a $22.50 hourly wage. Other qualifications that apply to this table are on page 9.

Unstick windows

On repaint jobs, test all windows during your estimating walk-through and allow approximately 15 minutes (.25 hours) for each stuck window. But don't price yourself out of the job with this extra time.

	Labor SF per manhour	Material coverage SF/gallon	Material cost per gallon	Labor cost per 100 SF	Labor burden 100 SF	Material cost per 100 SF	Overhead per 100 SF	Profit per 100 SF	Total price per 100 SF
Wash									
Interior flatwall (smooth surfaces)									
Wash only									
Slow	175	--	--	6.00	1.70	--	2.39	1.52	11.61
Medium	200	--	--	8.38	2.92	--	2.77	1.55	15.62
Fast	225	--	--	10.00	4.16	--	2.62	1.01	17.79
Wash & touchup									
Slow	135	--	--	7.78	2.21	--	3.10	1.96	15.05
Medium	160	--	--	10.47	3.65	--	3.46	1.93	19.51
Fast	185	--	--	12.16	5.09	--	3.19	1.23	21.67
Interior flatwall (rough surfaces)									
Wash only									
Slow	125	--	--	8.40	2.38	--	3.34	2.12	16.24
Medium	150	--	--	11.17	3.90	--	3.69	2.06	20.82
Fast	175	--	--	12.86	5.35	--	3.37	1.30	22.88
Wash & touchup									
Slow	55	--	--	19.09	5.42	--	7.60	4.82	36.93
Medium	95	--	--	17.63	6.16	--	5.82	3.25	32.86
Fast	135	--	--	16.67	6.96	--	4.37	1.68	29.68
Interior enamel (wall surfaces)									
Wash only									
Slow	190	--	--	5.53	1.56	--	2.20	1.40	10.69
Medium	215	--	--	7.79	2.71	--	2.57	1.44	14.51
Fast	240	--	--	9.38	3.92	--	2.46	.95	16.71
Wash & touchup									
Slow	85	--	--	12.35	3.50	--	4.92	3.12	23.89
Medium	113	--	--	14.82	5.17	--	4.90	2.74	27.63
Fast	140	--	--	16.07	6.70	--	4.21	1.62	28.60
Interior enamel trim									
Wash only									
Slow	100	--	--	10.50	2.98	--	4.18	2.65	20.31
Medium	150	--	--	11.17	3.90	--	3.69	2.06	20.82
Fast	200	--	--	11.25	4.70	--	2.95	1.13	20.03
Wash & touchup									
Slow	90	--	--	11.67	3.31	--	4.65	2.95	22.58
Medium	120	--	--	13.96	4.86	--	4.61	2.58	26.01
Fast	150	--	--	15.00	6.27	--	3.93	1.51	26.71

	Labor SF per manhour	Material coverage SF/gallon	Material cost per gallon	Labor cost per 100 SF	Labor burden 100 SF	Material cost per 100 SF	Overhead per 100 SF	Profit per 100 SF	Total price per 100 SF
Interior varnish trim									
Wash only									
Slow	150	--	--	7.00	1.99	--	2.79	1.77	13.55
Medium	195	--	--	8.59	3.00	--	2.84	1.59	16.02
Fast	240	--	--	9.38	3.92	--	2.46	.95	16.71
Wash & touchup									
Slow	120	--	--	8.75	2.48	--	3.48	2.21	16.92
Medium	140	--	--	11.96	4.17	--	3.95	2.21	22.29
Fast	160	--	--	14.06	5.87	--	3.69	1.42	25.04
Interior varnish floors									
Wash only									
Slow	160	--	--	6.56	1.87	--	2.61	1.66	12.70
Medium	210	--	--	7.98	2.77	--	2.64	1.47	14.86
Fast	260	--	--	8.65	3.63	--	2.27	.87	15.42
Wash & touchup									
Slow	130	--	--	8.08	2.29	--	3.22	2.04	15.63
Medium	153	--	--	10.95	3.82	--	3.62	2.02	20.41
Fast	175	--	--	12.86	5.35	--	3.37	1.30	22.88
Interior plaster (smooth)									
Wash only									
Slow	150	--	--	7.00	1.99	--	2.79	1.77	13.55
Medium	175	--	--	9.57	3.33	--	3.16	1.77	17.83
Fast	200	--	--	11.25	4.70	--	2.95	1.13	20.03
Wash & touchup									
Slow	125	--	--	8.40	2.38	--	3.34	2.12	16.24
Medium	140	--	--	11.96	4.17	--	3.95	2.21	22.29
Fast	155	--	--	14.52	6.05	--	3.81	1.46	25.84
Interior plaster (sand finish)									
Wash only									
Slow	110	--	--	9.55	2.70	--	3.80	2.41	18.46
Medium	135	--	--	12.41	4.33	--	4.10	2.29	23.13
Fast	160	--	--	14.06	5.87	--	3.69	1.42	25.04
Wash & touchup									
Slow	85	--	--	12.35	3.50	--	4.92	3.12	23.89
Medium	110	--	--	15.23	5.30	--	5.03	2.81	28.37
Fast	140	--	--	16.07	6.70	--	4.21	1.62	28.60

Consider all trim which is less than 12" wide to be 12" wide and figured as 1 square foot per linear foot. For heights above 8 feet, use the High Time Difficulty Factors on page 137. Note: Because the type of surface and type of material being removed will alter rates, it is best to wash surfaces with calcium deposits or other unusual surface markings or debris on a Time and Material or a Cost Plus Fee basis at a pre-set hourly rate. "Slow" work is based on a $10.50 hourly wage, "Medium" work on a $16.75 hourly wage, and "Fast" work on a $22.50 hourly wage. Other qualifications that apply to this table are on page 9.

	Labor SF per manhour	Material coverage SF/gallon	Material cost per gallon	Labor cost per 100 SF	Labor burden 100 SF	Material cost per 100 SF	Overhead per 100 SF	Profit per 100 SF	Total price per 100 SF
Waterblast									
Waterblast									
Slow	450	--	--	2.33	.66	--	.93	.59	4.51
Medium	500	--	--	3.35	1.17	--	1.11	.62	6.25
Fast	550	--	--	4.09	1.71	--	1.07	.41	7.28

Use an hourly or daily rate depending on the quantity of work involved. For heights above 8 feet, use the High Time Difficulty Factors on page 137. Waterblast to remove deteriorated, cracked, flaking paint from accessible concrete, block, brick, wood, plaster or stucco surfaces. Rates assume a 1.4" diameter nozzle with 100 lbs. pressure. "Slow" work is based on a $10.50 hourly wage, "Medium" work on a $16.75 hourly wage, and "Fast" work on a $22.50 hourly wage. Other qualifications that apply to this table are on page 9.

	Labor SF per manhour	Material coverage SF/gallon	Material cost per gallon	Labor cost per 100 SF	Labor burden 100 SF	Material cost per 100 SF	Overhead per 100 SF	Profit per 100 SF	Total price per 100 SF
Window protective coating (paraffin wax)									
Window protective coating - wax (material #47)									
Hand application									
Slow	90	350	15.00	11.67	3.31	4.29	5.98	3.79	29.04
Medium	100	325	13.10	16.75	5.84	4.03	6.52	3.65	36.79
Fast	110	300	11.20	20.45	8.54	3.73	6.05	2.33	41.10
Spray application									
Slow	350	250	15.00	3.00	.86	6.00	3.05	1.94	14.85
Medium	375	225	13.10	4.47	1.56	5.82	2.90	1.62	16.37
Fast	400	200	11.20	5.63	2.34	5.60	2.51	.97	17.05

For heights above 8 feet, use the High Time Difficulty Factors on page 137. "Slow" work is based on a $10.50 hourly wage, "Medium" work on a $16.75 hourly wage, and "Fast" work on a $22.50 hourly wage. Other qualifications that apply to this table are on page 9.

	Labor SF per manhour	Material coverage SF/gallon	Material cost per gallon	Labor cost per 100 SF	Labor burden 100 SF	Material cost per 100 SF	Overhead per 100 SF	Profit per 100 SF	Total price per 100 SF
Wire brush									
Surface area basis - large projects & surface areas									
Pipe up to 12" O/D									
Slow	50	--	--	21.00	5.96	--	8.36	5.30	40.62
Medium	75	--	--	22.33	7.78	--	7.38	4.12	41.61
Fast	100	--	--	22.50	9.39	--	5.90	2.27	40.06
Structural steel									
Sizes up to 2 SF/LF									
Slow	90	--	--	11.67	3.31	--	4.65	2.95	22.58
Medium	110	--	--	15.23	5.30	--	5.03	2.81	28.37
Fast	125	--	--	18.00	7.51	--	4.72	1.81	32.04
Sizes from 2 to 5 SF/LF									
Slow	100	--	--	10.50	2.98	--	4.18	2.65	20.31
Medium	120	--	--	13.96	4.86	--	4.61	2.58	26.01
Fast	140	--	--	16.07	6.70	--	4.21	1.62	28.60
Sizes over 5 SF/LF									
Slow	110	--	--	9.55	2.70	--	3.80	2.41	18.46
Medium	130	--	--	12.88	4.49	--	4.26	2.38	24.01
Fast	150	--	--	15.00	6.27	--	3.93	1.51	26.71
Tanks and vessels									
Sizes up to 12'0" O/D									
Slow	110	--	--	9.55	2.70	--	3.80	2.41	18.46
Medium	130	--	--	12.88	4.49	--	4.26	2.38	24.01
Fast	150	--	--	15.00	6.27	--	3.93	1.51	26.71
Sizes over 12'0" O/D									
Slow	120	--	--	8.75	2.48	--	3.48	2.21	16.92
Medium	140	--	--	11.96	4.17	--	3.95	2.21	22.29
Fast	160	--	--	14.06	5.87	--	3.69	1.42	25.04

For heights above 8 feet, use the High Time Difficulty Factors on page 137. "Slow" work is based on a $10.50 hourly wage, "Medium" work on a $16.75 hourly wage, and "Fast" work on a $22.50 hourly wage. Other qualifications that apply to this table are on page 9.

- INDUSTRIAL
- INSTITUTIONAL
- HEAVY COMMERCIAL

Painting Costs

Conduit, electric, brush application

	Labor SF per manhour	Material coverage SF/gallon	Material cost per gallon	Labor cost per 100 SF	Labor burden 100 SF	Material cost per 100 SF	Overhead per 100 SF	Profit per 100 SF	Total price per 100 SF
Acid wash coat, muriatic acid (material #49)									
Brush each coat									
Slow	60	700	5.40	17.50	4.97	.77	7.21	4.57	35.02
Medium	80	650	4.70	20.94	7.30	.72	7.10	3.97	40.03
Fast	100	600	4.00	22.50	9.39	.67	6.02	2.31	40.89
Metal primer, rust inhibitor - clean metal (material #35)									
Brush prime coat									
Slow	60	450	29.90	17.50	4.97	6.64	9.03	5.72	43.86
Medium	80	425	26.20	20.94	7.30	6.16	8.43	4.71	47.54
Fast	100	400	22.40	22.50	9.39	5.60	6.94	2.67	47.10
Metal primer, rust inhibitor - rusty metal (material #36)									
Brush prime coat									
Slow	60	400	36.60	17.50	4.97	9.15	9.81	6.22	47.65
Medium	80	375	32.10	20.94	7.30	8.56	9.02	5.04	50.86
Fast	100	350	27.50	22.50	9.39	7.86	7.35	2.83	49.93
Industrial enamel, oil base, high gloss - light colors (material #56)									
Brush 1st or additional finish coats									
Slow	100	450	29.60	10.50	2.98	6.58	6.22	3.94	30.22
Medium	125	425	25.90	13.40	4.67	6.09	5.92	3.31	33.39
Fast	150	400	22.20	15.00	6.27	5.55	4.96	1.91	33.69
Industrial enamel, oil base, high gloss - dark (OSHA) colors (material #57)									
Brush 1st or additional finish coats									
Slow	100	500	37.00	10.50	2.98	7.40	6.48	4.11	31.47
Medium	125	475	32.40	13.40	4.67	6.82	6.10	3.41	34.40
Fast	150	450	27.70	15.00	6.27	6.16	5.07	1.95	34.45
Epoxy coating, 2 part system - clear (material #51)									
Brush 1st coat									
Slow	60	425	58.30	17.50	4.97	13.72	11.22	7.11	54.52
Medium	80	400	51.00	20.94	7.30	12.75	10.04	5.61	56.64
Fast	100	375	43.70	22.50	9.39	11.65	8.05	3.10	54.69
Brush 2nd or additional finish coats									
Slow	100	475	58.30	10.50	2.98	12.27	7.99	5.06	38.80
Medium	125	450	51.00	13.40	4.67	11.33	7.20	4.03	40.63
Fast	150	425	43.70	15.00	6.27	10.28	5.83	2.24	39.62

	Labor SF per manhour	Material coverage SF/gallon	Material cost per gallon	Labor cost per 100 SF	Labor burden 100 SF	Material cost per 100 SF	Overhead per 100 SF	Profit per 100 SF	Total price per 100 SF
Epoxy coating, 2 part system - white (material #52)									
Brush 1st coat									
Slow	60	425	61.70	17.50	4.97	14.52	11.47	7.27	55.73
Medium	80	400	54.00	20.94	7.30	13.50	10.23	5.72	57.69
Fast	100	375	46.30	22.50	9.39	12.35	8.18	3.15	55.57
Brush 2nd or additional finish coats									
Slow	100	475	61.70	10.50	2.98	12.99	8.21	5.20	39.88
Medium	125	450	54.00	13.40	4.67	12.00	7.37	4.12	41.56
Fast	150	425	46.30	15.00	6.27	10.89	5.95	2.29	40.40

This table is based on square feet of conduit surface area. See Figure 21 on page 321 to convert from linear feet of various conduit or pipe sizes to square feet of surface area. For heights above 8 feet, use the High Time Difficulty Factors on page 137. Note: A two coat system, prime and finish, using oil base material is recommended for any metal surface. Although water base material is often used, it may cause oxidation, corrosion and rust. One coat of oil base solid body stain is often used on exterior metal but it may crack, peel or chip without the proper prime coat application. "Slow" work is based on a $10.50 hourly wage, "Medium" work on a $16.75 hourly wage, and "Fast" work on a $22.50 hourly wage. Other qualifications that apply to this table are on page 9.

	Labor SF per manhour	Material coverage SF/gallon	Material cost per gallon	Labor cost per 100 SF	Labor burden 100 SF	Material cost per 100 SF	Overhead per 100 SF	Profit per 100 SF	Total price per 100 SF

Conduit, electric, roll application

	Labor SF per manhour	Material coverage SF/gallon	Material cost per gallon	Labor cost per 100 SF	Labor burden 100 SF	Material cost per 100 SF	Overhead per 100 SF	Profit per 100 SF	Total price per 100 SF
Acid wash coat, muriatic acid (material #49)									
Roll each coat									
Slow	175	700	5.40	6.00	1.70	.77	2.63	1.67	12.77
Medium	200	650	4.70	8.38	2.92	.72	2.94	1.65	16.61
Fast	225	600	4.00	10.00	4.16	.67	2.75	1.06	18.64
Metal primer, rust inhibitor - clean metal (material #35)									
Roll prime coat									
Slow	175	425	29.90	6.00	1.70	7.04	4.57	2.90	22.21
Medium	200	400	26.20	8.38	2.92	6.55	4.37	2.44	24.66
Fast	225	375	22.40	10.00	4.16	5.97	3.73	1.43	25.29
Metal primer, rust inhibitor - rusty metal (material #36)									
Roll prime coat									
Slow	175	375	36.60	6.00	1.70	9.76	5.42	3.43	26.31
Medium	200	350	32.10	8.38	2.92	9.17	5.02	2.80	28.29
Fast	225	325	27.50	10.00	4.16	8.46	4.19	1.61	28.42
Industrial enamel, oil base, high gloss - light colors (material #56)									
Roll 1st or additional finish coats									
Slow	225	425	29.60	4.67	1.32	6.96	4.02	2.55	19.52
Medium	250	400	25.90	6.70	2.34	6.48	3.80	2.12	21.44
Fast	275	375	22.20	8.18	3.43	5.92	3.24	1.25	22.02
Industrial enamel, oil base, high gloss - dark (OSHA) colors (material #57)									
Roll 1st or additional finish coats									
Slow	225	475	37.00	4.67	1.32	7.79	4.27	2.71	20.76
Medium	250	450	32.40	6.70	2.34	7.20	3.98	2.22	22.44
Fast	275	425	27.70	8.18	3.43	6.52	3.35	1.29	22.77
Epoxy coating, 2 part system - clear (material #51)									
Roll 1st coat									
Slow	175	400	58.30	6.00	1.70	14.58	6.91	4.38	33.57
Medium	200	375	51.00	8.38	2.92	13.60	6.10	3.41	34.41
Fast	225	350	43.70	10.00	4.16	12.49	4.93	1.90	33.48
Roll 2nd or additional finish coats									
Slow	225	450	58.30	4.67	1.32	12.96	5.88	3.73	28.56
Medium	250	425	51.00	6.70	2.34	12.00	5.15	2.88	29.07
Fast	275	400	43.70	8.18	3.43	10.93	4.17	1.60	28.31

	Labor SF per manhour	Material coverage SF/gallon	Material cost per gallon	Labor cost per 100 SF	Labor burden 100 SF	Material cost per 100 SF	Overhead per 100 SF	Profit per 100 SF	Total price per 100 SF
Epoxy coating, 2 part system - white (material #52)									
Roll 1st coat									
Slow	175	400	61.70	6.00	1.70	15.43	7.17	4.55	34.85
Medium	200	375	54.00	8.38	2.92	14.40	6.30	3.52	35.52
Fast	225	350	46.30	10.00	4.16	13.23	5.07	1.95	34.41
Roll 2nd or additional finish coats									
Slow	225	450	61.70	4.67	1.32	13.71	6.11	3.87	29.68
Medium	250	425	54.00	6.70	2.34	12.71	5.33	2.98	30.06
Fast	275	400	46.30	8.18	3.43	11.58	4.29	1.65	29.13

This table is based on square feet of conduit surface area. See Figure 21 on page 321 to convert from linear feet of various conduit or pipe sizes to square feet of surface area. For heights above 8 feet, use the High Time Difficulty Factors on page 137. Note: A two coat system, prime and finish, using oil base material is recommended for any metal surface. Although water base material is often used, it may cause oxidation, corrosion and rust. One coat of oil base solid body stain is often used on exterior metal but it may crack, peel or chip without the proper prime coat application. "Slow" work is based on a $10.50 hourly wage, "Medium" work on a $16.75 hourly wage, and "Fast" work on a $22.50 hourly wage. Other qualifications that apply to this table are on page 9.

	Labor SF per manhour	Material coverage SF/gallon	Material cost per gallon	Labor cost per 100 SF	Labor burden 100 SF	Material cost per 100 SF	Overhead per 100 SF	Profit per 100 SF	Total price per 100 SF

Conduit, electric, spray application

	Labor SF per manhour	Material coverage SF/gallon	Material cost per gallon	Labor cost per 100 SF	Labor burden 100 SF	Material cost per 100 SF	Overhead per 100 SF	Profit per 100 SF	Total price per 100 SF
Acid wash coat, muriatic acid (material #49)									
Spray each coat									
Slow	350	350	5.40	3.00	.86	1.54	1.67	1.06	8.13
Medium	400	325	4.70	4.19	1.46	1.45	1.74	.97	9.81
Fast	450	300	4.00	5.00	2.08	1.33	1.56	.60	10.57
Metal primer, rust inhibitor - clean metal (material #35)									
Spray prime coat									
Slow	350	275	29.90	3.00	.86	10.87	4.56	2.89	22.18
Medium	400	263	26.20	4.19	1.46	9.96	3.82	2.14	21.57
Fast	450	250	22.40	5.00	2.08	8.96	2.97	1.14	20.15
Metal primer, rust inhibitor - rusty metal (material #36)									
Spray prime coat									
Slow	350	225	36.60	3.00	.86	16.27	6.24	3.95	30.32
Medium	400	213	32.10	4.19	1.46	15.07	5.08	2.84	28.64
Fast	450	200	27.50	5.00	2.08	13.75	3.86	1.48	26.17
Industrial enamel, oil base, high gloss - light colors (material #56)									
Spray 1st or additional finish coats									
Slow	450	275	29.60	2.33	.66	10.76	4.26	2.70	20.71
Medium	500	263	25.90	3.35	1.17	9.85	3.52	1.97	19.86
Fast	550	250	22.20	4.09	1.71	8.88	2.72	1.04	18.44
Industrial enamel, oil base, high gloss - dark (OSHA) colors (material #57)									
Spray 1st or additional finish coats									
Slow	450	300	37.00	2.33	.66	12.33	4.75	3.01	23.08
Medium	500	288	32.40	3.35	1.17	11.25	3.86	2.16	21.79
Fast	550	275	27.70	4.09	1.71	10.07	2.94	1.13	19.94
Epoxy coating, 2 part system - clear (material #51)									
Spray 1st coat									
Slow	350	250	58.30	3.00	.86	23.32	8.42	5.34	40.94
Medium	400	238	51.00	4.19	1.46	21.43	6.63	3.71	37.42
Fast	450	225	43.70	5.00	2.08	19.42	4.90	1.88	33.28
Spray 2nd or additional finish coats									
Slow	450	285	58.30	2.33	.66	20.46	7.27	4.61	35.33
Medium	500	273	51.00	3.35	1.17	18.68	5.68	3.18	32.06
Fast	550	260	43.70	4.09	1.71	16.81	4.18	1.61	28.40

	Labor SF per manhour	Material coverage SF/gallon	Material cost per gallon	Labor cost per 100 SF	Labor burden 100 SF	Material cost per 100 SF	Overhead per 100 SF	Profit per 100 SF	Total price per 100 SF
Epoxy coating, 2 part system - white (material #52)									
Spray 1st coat									
Slow	350	250	61.70	3.00	.86	24.68	8.84	5.61	42.99
Medium	400	238	54.00	4.19	1.46	22.69	6.94	3.88	39.16
Fast	450	225	46.30	5.00	2.08	20.58	5.12	1.97	34.75
Spray 2nd or additional finish coats									
Slow	450	285	61.70	2.33	.66	21.65	7.64	4.84	37.12
Medium	500	273	54.00	3.35	1.17	19.78	5.95	3.33	33.58
Fast	550	260	46.30	4.09	1.71	17.81	4.37	1.68	29.66

This table is based on square feet of conduit surface area. See Figure 21 on page 321 to convert from linear feet of various conduit or pipe sizes to square feet of surface area. For heights above 8 feet, use the High Time Difficulty Factors on page 137. Note: A two coat system, prime and finish, using oil base material is recommended for any metal surface. Although water base material is often used, it may cause oxidation, corrosion and rust. One coat of oil base solid body stain is often used on exterior metal but it may crack, peel or chip without the proper prime coat application. "Slow" work is based on a $10.50 hourly wage, "Medium" work on a $16.75 hourly wage, and "Fast" work on a $22.50 hourly wage. Other qualifications that apply to this table are on page 9.

	Labor SF per manhour	Material coverage SF/gallon	Material cost per gallon	Labor cost per 100 SF	Labor burden 100 SF	Material cost per 100 SF	Overhead per 100 SF	Profit per 100 SF	Total price per 100 SF
Conduit, electric, mitt or glove application									
Acid wash coat, muriatic acid (material #49)									
Mitt or glove each coat									
Slow	175	700	5.40	6.00	1.70	.77	2.63	1.67	12.77
Medium	200	650	4.70	8.38	2.92	.72	2.94	1.65	16.61
Fast	225	600	4.00	10.00	4.16	.67	2.75	1.06	18.64
Metal primer, rust inhibitor - clean metal (material #35)									
Mitt or glove prime coat									
Slow	175	450	29.90	6.00	1.70	6.64	4.45	2.82	21.61
Medium	200	438	26.20	8.38	2.92	5.98	4.23	2.37	23.88
Fast	225	425	22.40	10.00	4.16	5.27	3.60	1.38	24.41
Metal primer, rust inhibitor - rusty metal (material #36)									
Mitt or glove prime coat									
Slow	175	400	36.60	6.00	1.70	9.15	5.23	3.31	25.39
Medium	200	388	32.10	8.38	2.92	8.27	4.79	2.68	27.04
Fast	225	375	27.50	10.00	4.16	7.33	3.98	1.53	27.00
Industrial enamel, oil base, high gloss - light colors (material #56)									
Mitt or glove 1st or additional finish coats									
Slow	225	450	29.60	4.67	1.32	6.58	3.90	2.47	18.94
Medium	250	438	25.90	6.70	2.34	5.91	3.66	2.05	20.66
Fast	275	425	22.20	8.18	3.43	5.22	3.11	1.20	21.14
Industrial enamel, oil base, high gloss - dark (OSHA) colors (material #57)									
Mitt or glove 1st or additional finish coats									
Slow	225	500	37.00	4.67	1.32	7.40	4.15	2.63	20.17
Medium	250	488	32.40	6.70	2.34	6.64	3.84	2.15	21.67
Fast	275	475	27.70	8.18	3.43	5.83	3.22	1.24	21.90
Epoxy coating, 2 part system - clear (material #51)									
Mitt or glove 1st coat									
Slow	175	425	58.30	6.00	1.70	13.72	6.64	4.21	32.27
Medium	200	413	51.00	8.38	2.92	12.35	5.79	3.24	32.68
Fast	225	400	43.70	10.00	4.16	10.93	4.65	1.79	31.53
Mitt or glove 2nd or additional finish coats									
Slow	225	475	58.30	4.67	1.32	12.27	5.66	3.59	27.51
Medium	250	463	51.00	6.70	2.34	11.02	4.91	2.75	27.72
Fast	275	450	43.70	8.18	3.43	9.71	3.94	1.52	26.78

	Labor SF per manhour	Material coverage SF/gallon	Material cost per gallon	Labor cost per 100 SF	Labor burden 100 SF	Material cost per 100 SF	Overhead per 100 SF	Profit per 100 SF	Total price per 100 SF
Epoxy coating, 2 part system - white (material #52)									
Mitt or glove 1st coat									
Slow	175	425	61.70	6.00	1.70	14.52	6.89	4.37	33.48
Medium	200	413	54.00	8.38	2.92	13.08	5.97	3.34	33.69
Fast	225	400	46.30	10.00	4.16	11.58	4.77	1.83	32.34
Mitt or glove 2nd or additional finish coats									
Slow	225	475	61.70	4.67	1.32	12.99	5.89	3.73	28.60
Medium	250	463	54.00	6.70	2.34	11.66	5.07	2.83	28.60
Fast	275	450	46.30	8.18	3.43	10.29	4.05	1.56	27.51

This table is based on square feet of conduit surface area. See Figure 21 below to convert from linear feet of various conduit or pipe sizes to square feet of surface area. For heights above 8 feet, use the High Time Difficulty Factors on page 137. Note: A two coat system, prime and finish, using oil base material is recommended for any metal surface. Although water base material is often used, it may cause oxidation, corrosion and rust. One coat of oil base solid body stain is often used on exterior metal but it may crack, peel or chip without the proper prime coat application. "Slow" work is based on a $10.50 hourly wage, "Medium" work on a $16.75 hourly wage, and "Fast" work on a $22.50 hourly wage. Other qualifications that apply to this table are on page 9.

Pipe O/D (inches)	Conversion Factor (SF per measured LF)
1" to 3"	1 SF for each 1 LF
4" to 7"	2 SF for each 1 LF
8" to 11"	3 SF for each 1 LF
12" to 15"	4 SF for each 1 LF
16" to 19"	5 SF for each 1 LF
20" to 22"	6 SF for each 1 LF
23" to 26"	7 SF for each 1 LF
27" to 30"	8 SF for each 1 LF

Figure 21
Conduit/pipe area conversions

	Labor SF per manhour	Material coverage SF/gallon	Material cost per gallon	Labor cost per 100 SF	Labor burden 100 SF	Material cost per 100 SF	Overhead per 100 SF	Profit per 100 SF	Total price per 100 SF
Decking and siding, metal, corrugated metal									
Acid wash coat, muriatic acid (material #49)									
Spray each coat									
Slow	700	500	5.40	1.50	.43	1.08	.93	.59	4.53
Medium	750	450	4.70	2.23	.77	1.04	.99	.55	5.58
Fast	800	400	4.00	2.81	1.18	1.00	.92	.35	6.26
Metal primer, rust inhibitor - clean metal (material #35)									
Spray prime coat									
Slow	700	325	29.90	1.50	.43	9.20	3.45	2.19	16.77
Medium	750	300	26.20	2.23	.77	8.73	2.88	1.61	16.22
Fast	800	275	22.40	2.81	1.18	8.15	2.24	.86	15.24
Metal primer, rust inhibitor - rusty metal (material #36)									
Spray prime coat									
Slow	700	275	36.60	1.50	.43	13.31	4.72	2.99	22.95
Medium	750	250	32.10	2.23	.77	12.84	3.88	2.17	21.89
Fast	800	225	27.50	2.81	1.18	12.22	3.00	1.15	20.36
Industrial enamel, oil base, high gloss - light colors (material #56)									
Spray 1st or additional finish coats									
Slow	850	325	29.60	1.24	.35	9.11	3.32	2.10	16.12
Medium	900	300	25.90	1.86	.65	8.63	2.73	1.53	15.40
Fast	950	275	22.20	2.37	.98	8.07	2.11	.81	14.34
Industrial enamel, oil base, high gloss - dark (OSHA) colors (material #57)									
Spray 1st or additional finish coats									
Slow	850	425	37.00	1.24	.35	8.71	3.19	2.02	15.51
Medium	900	400	32.40	1.86	.65	8.10	2.60	1.45	14.66
Fast	950	375	27.70	2.37	.98	7.39	1.99	.76	13.49
Epoxy coating, 2 part system - clear (material #51)									
Spray 1st coat									
Slow	700	300	58.30	1.50	.43	19.43	6.62	4.20	32.18
Medium	750	275	51.00	2.23	.77	18.55	5.28	2.95	29.78
Fast	800	250	43.70	2.81	1.18	17.48	3.97	1.53	26.97
Spray 2nd or additional finish coats									
Slow	850	400	58.30	1.24	.35	14.58	5.01	3.18	24.36
Medium	900	375	51.00	1.86	.65	13.60	3.95	2.21	22.27
Fast	950	350	43.70	2.37	.98	12.49	2.93	1.13	19.90

	Labor SF per manhour	Material coverage SF/gallon	Material cost per gallon	Labor cost per 100 SF	Labor burden 100 SF	Material cost per 100 SF	Overhead per 100 SF	Profit per 100 SF	Total price per 100 SF
Epoxy coating, 2 part system - white (material #52)									
Spray 1st coat									
Slow	700	300	61.70	1.50	.43	20.57	6.98	4.42	33.90
Medium	750	275	54.00	2.23	.77	19.64	5.55	3.10	31.29
Fast	800	250	46.30	2.81	1.18	18.52	4.16	1.60	28.27
Spray 2nd or additional finish coats									
Slow	850	400	61.70	1.24	.35	15.43	5.28	3.35	25.65
Medium	900	375	54.00	1.86	.65	14.40	4.14	2.32	23.37
Fast	950	350	46.30	2.37	.98	13.23	3.07	1.18	20.83

The figures in the table above are based on overall dimensions (length times width). But all decking and siding has corrugations, peaks and valleys that increase the surface that has to be painted. For example, corrugated siding with 2-1/2" center to center corrugations has a surface area 10% greater than the width times length dimension. For corrugated siding with 1-1/4" center to center corrugations, increase the surface area by 15%. For square corner decking, Figure 22 below shows how much area must be added to allow for peaks and valleys. For heights above 8 feet, use the High Time Difficulty Factors on page 137. Note: A two coat system, prime and finish, using oil base material is recommended for any metal surface. Although water base material is often used, it may cause oxidation, corrosion and rust. One coat of oil base solid body stain is often used on exterior metal but it may crack, peel or chip without the proper prime coat application. "Slow" work is based on a $10.50 hourly wage, "Medium" work on a $16.75 hourly wage, and "Fast" work on a $22.50 hourly wage. Other qualifications that apply to this table are on page 9.

C	2C	D	PW	VW	Factor
--	12"	4-1/2"	3"	2"	2.50
--	12"	1-1/2"	3-1/8"	2"	1.50
--	12"	1-1/2"	5-1/16"	1"	1.45
12"	--	3"	9-5/8"	1"	1.50
12"	--	4-1/2"	9-5/8"	1"	1.75
24"	--	4-1/2"	12"	12"	1.60
24"	--	6"	12"	12"	1.75
24"	--	8"	12"	12"	1.95

For square corner decking, calculate the overall (length times width) deck area. Then measure the peaks and valleys on the deck. Select the row in the table above that most nearly matches the deck you're painting. Multiply the overall deck area by the number listed in the column headed "factor" to find the actual area you're painting. Use this actual area when calculating labor and material requirements. In the table above, figures in the column headed C show the distance between the center of the peaks. Column 2C shows the distance between every second center of peak (2 centers). Column D shows the depth of corrugation. Column PW shows the peak width. Column VW shows the valley width. If the deck you're painting doesn't match any deck listed in this table, use the factor for the most similar deck in the table.

Figure 22
Square corner decking factors

	Labor SF per manhour	Material coverage SF/gallon	Material cost per gallon	Labor cost per 100 SF	Labor burden 100 SF	Material cost per 100 SF	Overhead per 100 SF	Profit per 100 SF	Total price per 100 SF

Decking and siding, metal, flat pan metal

Acid wash coat, muriatic acid (material #49)

Spray each coat									
Slow	800	600	5.40	1.31	.38	.90	.80	.51	3.90
Medium	850	550	4.70	1.97	.70	.85	.86	.48	4.86
Fast	900	500	4.00	2.50	1.04	.80	.80	.31	5.45

Metal primer, rust inhibitor - clean metal (material #35)

Spray prime coat									
Slow	800	375	29.90	1.31	.38	7.97	2.99	1.90	14.55
Medium	850	350	26.20	1.97	.70	7.49	2.49	1.39	14.04
Fast	900	325	22.40	2.50	1.04	6.89	1.93	.74	13.10

Metal primer, rust inhibitor - rusty metal (material #36)

Spray prime coat									
Slow	800	300	36.60	1.31	.38	12.20	4.30	2.73	20.92
Medium	850	275	32.10	1.97	.70	11.67	3.51	1.96	19.81
Fast	900	250	27.50	2.50	1.04	11.00	2.69	1.03	18.26

Industrial enamel, oil base, high gloss - light colors (material #56)

Spray 1st or additional finish coats									
Slow	1000	375	29.60	1.05	.30	7.89	2.86	1.82	13.92
Medium	1050	350	25.90	1.60	.55	7.40	2.34	1.31	13.20
Fast	1100	325	22.20	2.05	.85	6.83	1.80	.69	12.22

Industrial enamel, oil base, high gloss - dark (OSHA) colors (material #57)

Spray 1st or additional finish coats									
Slow	1000	425	37.00	1.05	.30	8.71	3.12	1.98	15.16
Medium	1050	400	32.40	1.60	.55	8.10	2.51	1.40	14.16
Fast	1100	375	27.70	2.05	.85	7.39	1.91	.73	12.93

Epoxy coating, 2 part system - clear (material #51)

Spray 1st coat									
Slow	800	350	58.30	1.31	.38	16.66	5.69	3.60	27.64
Medium	850	325	51.00	1.97	.70	15.69	4.50	2.51	25.37
Fast	900	300	43.70	2.50	1.04	14.57	3.35	1.29	22.75

Spray 2nd or additional finish coats									
Slow	1000	400	58.30	1.05	.30	14.58	4.94	3.13	24.00
Medium	1050	375	51.00	1.60	.55	13.60	3.86	2.16	21.77
Fast	1100	350	43.70	2.05	.85	12.49	2.85	1.10	19.34

	Labor SF per manhour	Material coverage SF/gallon	Material cost per gallon	Labor cost per 100 SF	Labor burden 100 SF	Material cost per 100 SF	Overhead per 100 SF	Profit per 100 SF	Total price per 100 SF
Epoxy coating, 2 part system - white (material #52)									
Spray 1st coat									
Slow	800	350	61.70	1.31	.38	17.63	5.99	3.80	29.11
Medium	850	325	54.00	1.97	.70	16.62	4.72	2.64	26.65
Fast	900	300	46.30	2.50	1.04	15.43	3.51	1.35	23.83
Spray 2nd or additional coats									
Slow	1000	400	61.70	1.05	.30	15.43	5.20	3.30	25.28
Medium	1050	375	54.00	1.60	.55	14.40	4.06	2.27	22.88
Fast	1100	350	46.30	2.05	.85	13.23	2.99	1.15	20.27

The figures in table above are based on overall dimensions (length times width). But all decking and siding has corrugations, peaks and valleys that increase the surface that has to be painted. For example, corrugated siding with 2-1/2" center to center corrugations has a surface area 10% greater than the width times length dimension. For corrugated siding with 1-1/4" center to center corrugations, increase the surface area by 15%. For square corner decking, Figure 22 on page 323 shows how much area must be added to allow for peaks and valleys. For heights above 8 feet, use the High Time Difficulty Factors on page 137. Note: A two coat system, prime and finish, using oil base material is recommended for any metal surface. Although water base material is often used, it may cause oxidation, corrosion and rust. One coat of oil base solid body stain is often used on exterior metal but it may crack, peel or chip without the proper prime coat application. "Slow" work is based on a $10.50 hourly wage, "Medium" work on a $16.75 hourly wage, and "Fast" work on a $22.50 hourly wage. Other qualifications that apply to this table are on page 9.

	Labor SF per manhour	Material coverage SF/gallon	Material cost per gallon	Labor cost per 100 SF	Labor burden 100 SF	Material cost per 100 SF	Overhead per 100 SF	Profit per 100 SF	Total price per 100 SF

Doors, hollow metal, brush application, square foot basis

	Labor SF per manhour	Material coverage SF/gallon	Material cost per gallon	Labor cost per 100 SF	Labor burden 100 SF	Material cost per 100 SF	Overhead per 100 SF	Profit per 100 SF	Total price per 100 SF
Metal primer, rust inhibitor - clean metal (material #35)									
Roll and brush prime coat									
Slow	160	450	29.90	6.56	1.87	6.64	4.67	2.96	22.70
Medium	180	438	26.20	9.31	3.25	5.98	4.54	2.54	25.62
Fast	200	425	22.40	11.25	4.70	5.27	3.93	1.51	26.66
Metal primer, rust inhibitor - rusty metal (material #36)									
Roll and brush prime coat									
Slow	160	425	36.60	6.56	1.87	8.61	5.28	3.35	25.67
Medium	180	413	32.10	9.31	3.25	7.77	4.98	2.78	28.09
Fast	200	400	27.50	11.25	4.70	6.88	4.22	1.62	28.67
Metal finish - synthetic enamel, gloss, interior or exterior - off white (material #37)									
Roll and brush 1st or additional finish coats									
Slow	175	450	27.90	6.00	1.70	6.20	4.31	2.73	20.94
Medium	195	438	24.40	8.59	3.00	5.57	4.20	2.35	23.71
Fast	215	425	20.90	10.47	4.36	4.92	3.66	1.41	24.82
Metal finish - synthetic enamel, gloss, interior or exterior - colors, except orange & red (material #38)									
Roll and brush 1st or additional finish coats									
Slow	175	475	30.00	6.00	1.70	6.32	4.35	2.76	21.13
Medium	195	463	26.30	8.59	3.00	5.68	4.23	2.36	23.86
Fast	215	450	22.50	10.47	4.36	5.00	3.67	1.41	24.91

To calculate the cost per door, figure the square feet for each side based on a 3 x 7 door. Add 3 feet to the width and 1 foot to the top of the door to allow for the time and material necessary to finish the door edges, frames and jambs. The result is a 6 x 8 door or 48 square feet per side, times 2 is 96 square feet per door. The cost per door is simply the 96 square feet times the cost per square foot indicated in the table above. Typically, hollow metal doors are pre-primed by the manufacturer and shipped ready to install and paint. These doors usually require only one coat of finish paint to cover. The metal finish figures include minor touchup of the prime coat. If off white or other light colored finish paint is specified, make sure the prime coat is a light color also, or more than one finish coat will be necessary. Note: A two coat system, prime and finish, using oil base material is recommended for any metal surface. Although water base material is often used, it may cause oxidation, corrosion and rust. One coat of oil base solid body stain is often used on exterior metal but it may crack, peel or chip without the proper prime coat application. "Slow" work is based on a $10.50 hourly wage, "Medium" work on a $16.75 hourly wage, and "Fast" work on a $22.50 hourly wage. Other qualifications that apply to this table are on page 9.

	Labor SF per manhour	Material coverage SF/gallon	Material cost per gallon	Labor cost per 100 SF	Labor burden 100 SF	Material cost per 100 SF	Overhead per 100 SF	Profit per 100 SF	Total price per 100 SF
Ductwork, bare duct, brush application									
Acid wash coat, muriatic acid (material #49)									
Brush each coat									
Slow	80	750	5.40	13.13	3.72	.72	5.45	3.46	26.48
Medium	100	700	4.70	16.75	5.84	.67	5.70	3.19	32.15
Fast	120	650	4.00	18.75	7.81	.62	5.03	1.93	34.14
Metal primer, rust inhibitor - clean metal (material #35)									
Brush prime coat									
Slow	80	400	29.90	13.13	3.72	7.48	7.55	4.79	36.67
Medium	100	375	26.20	16.75	5.84	6.99	7.25	4.05	40.88
Fast	120	350	22.40	18.75	7.81	6.40	6.10	2.34	41.40
Metal primer, rust inhibitor - rusty metal (material #36)									
Brush prime coat									
Slow	80	350	36.60	13.13	3.72	10.46	8.47	5.37	41.15
Medium	100	325	32.10	16.75	5.84	9.88	7.96	4.45	44.88
Fast	120	300	27.50	18.75	7.81	9.17	6.61	2.54	44.88
Industrial enamel, oil base, high gloss - light colors (material #56)									
Brush 1st or additional finish coats									
Slow	90	400	29.60	11.67	3.31	7.40	6.94	4.40	33.72
Medium	115	375	25.90	14.57	5.08	6.91	6.51	3.64	36.71
Fast	140	350	22.20	16.07	6.70	6.34	5.39	2.07	36.57
Industrial enamel, oil base, high gloss - dark (OSHA) colors (material #57)									
Brush 1st or additional finish coats									
Slow	90	450	37.00	11.67	3.31	8.22	7.20	4.56	34.96
Medium	115	425	32.40	14.57	5.08	7.62	6.68	3.73	37.68
Fast	140	400	27.70	16.07	6.70	6.93	5.50	2.11	37.31
Epoxy coating, 2 part system - clear (material #51)									
Brush 1st coat									
Slow	80	350	58.30	13.13	3.72	16.66	10.39	6.59	50.49
Medium	100	325	51.00	16.75	5.84	15.69	9.38	5.24	52.90
Fast	120	300	43.70	18.75	7.81	14.57	7.61	2.93	51.67
Brush 2nd or additional finish coats									
Slow	90	425	58.30	11.67	3.31	13.72	8.90	5.64	43.24
Medium	115	400	51.00	14.57	5.08	12.75	7.94	4.44	44.78
Fast	140	375	43.70	16.07	6.70	11.65	6.37	2.45	43.24

	Labor SF per manhour	Material coverage SF/gallon	Material cost per gallon	Labor cost per 100 SF	Labor burden 100 SF	Material cost per 100 SF	Overhead per 100 SF	Profit per 100 SF	Total price per 100 SF
Epoxy coating, 2 part system - white (material #52)									
Brush 1st coat									
Slow	80	350	61.70	13.13	3.72	17.63	10.70	6.78	51.96
Medium	100	325	54.00	16.75	5.84	16.62	9.61	5.37	54.19
Fast	120	300	46.30	18.75	7.81	15.43	7.77	2.99	52.75
Brush 2nd or additional finish coats									
Slow	90	425	61.70	11.67	3.31	14.52	9.15	5.80	44.45
Medium	115	400	54.00	14.57	5.08	13.50	8.12	4.54	45.81
Fast	140	375	46.30	16.07	6.70	12.35	6.50	2.50	44.12

For heights over 8 feet, use the High Time Difficulty Factors on page 137. Note: A two coat system, prime and finish, using oil base material is recommended for any metal surface. Although water base material is often used, it may cause oxidation, corrosion and rust. One coat of oil base solid body stain is often used on exterior metal but it may crack, peel or chip without the proper prime coat application. "Slow" work is based on a $10.50 hourly wage, "Medium" work on a $16.75 hourly wage, and "Fast" work on a $22.50 hourly wage. Other qualifications that apply to this table are on page 9.

	Labor SF per manhour	Material coverage SF/gallon	Material cost per gallon	Labor cost per 100 SF	Labor burden 100 SF	Material cost per 100 SF	Overhead per 100 SF	Profit per 100 SF	Total price per 100 SF
Ductwork, bare duct, roll application									
Acid wash coat, muriatic acid (material #49)									
Roll each coat									
Slow	225	700	5.40	4.67	1.32	.77	2.10	1.33	10.19
Medium	250	650	4.70	6.70	2.34	.72	2.39	1.34	13.49
Fast	275	600	4.00	8.18	3.43	.67	2.27	.87	15.42
Metal primer, rust inhibitor - clean metal (material #35)									
Roll prime coat									
Slow	225	375	29.90	4.67	1.32	7.97	4.33	2.75	21.04
Medium	250	350	26.20	6.70	2.34	7.49	4.05	2.26	22.84
Fast	275	325	22.40	8.18	3.43	6.89	3.42	1.31	23.23
Metal primer, rust inhibitor - rusty metal (material #36)									
Roll prime coat									
Slow	225	325	36.60	4.67	1.32	11.26	5.35	3.39	25.99
Medium	250	300	32.10	6.70	2.34	10.70	4.83	2.70	27.27
Fast	275	275	27.50	8.18	3.43	10.00	4.00	1.54	27.15
Industrial enamel, oil base, high gloss - light colors (material #56)									
Roll 1st or additional finish coats									
Slow	275	425	29.60	3.82	1.09	6.96	3.68	2.33	17.88
Medium	300	400	25.90	5.58	1.94	6.48	3.43	1.92	19.35
Fast	325	375	22.20	6.92	2.90	5.92	2.91	1.12	19.77
Industrial enamel, oil base, high gloss - dark (OSHA) colors (material #57)									
Roll 1st or additional finish coats									
Slow	275	475	37.00	3.82	1.09	7.79	3.94	2.50	19.14
Medium	300	450	32.40	5.58	1.94	7.20	3.61	2.02	20.35
Fast	325	425	27.70	6.92	2.90	6.52	3.02	1.16	20.52
Epoxy coating, 2 part system - clear (material #51)									
Roll 1st coat									
Slow	225	350	58.30	4.67	1.32	16.66	7.02	4.45	34.12
Medium	250	325	51.00	6.70	2.34	15.69	6.06	3.39	34.18
Fast	275	300	43.70	8.18	3.43	14.57	4.84	1.86	32.88
Roll 2nd or additional finish coats									
Slow	275	450	58.30	3.82	1.09	12.96	5.54	3.51	26.92
Medium	300	425	51.00	5.58	1.94	12.00	4.78	2.67	26.97
Fast	325	400	43.70	6.92	2.90	10.93	3.84	1.47	26.06

	Labor SF per manhour	Material coverage SF/gallon	Material cost per gallon	Labor cost per 100 SF	Labor burden 100 SF	Material cost per 100 SF	Overhead per 100 SF	Profit per 100 SF	Total price per 100 SF
Epoxy coating, 2 part system - white (material #52)									
Roll 1st coat									
Slow	225	350	61.70	4.67	1.32	17.63	7.33	4.64	35.59
Medium	250	325	54.00	6.70	2.34	16.62	6.28	3.51	35.45
Fast	275	300	46.30	8.18	3.43	15.43	5.00	1.92	33.96
Roll 2nd or additional finish coats									
Slow	275	450	61.70	3.82	1.09	13.71	5.77	3.66	28.05
Medium	300	425	54.00	5.58	1.94	12.71	4.96	2.77	27.96
Fast	325	400	46.30	6.92	2.90	11.58	3.96	1.52	26.88

For heights over 8 feet, use the High Time Difficulty Factors on page 137. Note: A two coat system, prime and finish, using oil base material is recommended for any metal surface. Although water base material is often used, it may cause oxidation, corrosion and rust. One coat of oil base solid body stain is often used on exterior metal but it may crack, peel or chip without the proper prime coat application. "Slow" work is based on a $10.50 hourly wage, "Medium" work on a $16.75 hourly wage, and "Fast" work on a $22.50 hourly wage. Other qualifications that apply to this table are on page 9.

	Labor SF per manhour	Material coverage SF/gallon	Material cost per gallon	Labor cost per 100 SF	Labor burden 100 SF	Material cost per 100 SF	Overhead per 100 SF	Profit per 100 SF	Total price per 100 SF
Ductwork, bare duct, spray application									
Acid wash coat, muriatic acid (material #49)									
Spray each coat									
Slow	550	450	5.40	1.91	.54	1.20	1.13	.72	5.50
Medium	600	400	4.70	2.79	.98	1.18	1.21	.68	6.84
Fast	650	350	4.00	3.46	1.45	1.14	1.12	.43	7.60
Metal primer, rust inhibitor - clean metal (material #35)									
Spray prime coat									
Slow	550	250	29.90	1.91	.54	11.96	4.47	2.83	21.71
Medium	600	225	26.20	2.79	.98	11.64	3.77	2.11	21.29
Fast	650	200	22.40	3.46	1.45	11.20	2.98	1.14	20.23
Metal primer, rust inhibitor - rusty metal (material #36)									
Spray prime coat									
Slow	550	200	36.60	1.91	.54	18.30	6.43	4.08	31.26
Medium	600	188	32.10	2.79	.98	17.07	5.10	2.85	28.79
Fast	650	175	27.50	3.46	1.45	15.71	3.81	1.47	25.90
Industrial enamel, oil base, high gloss - light colors (material #56)									
Spray 1st or additional finish coats									
Slow	700	225	29.60	1.50	.43	13.16	4.68	2.97	22.74
Medium	750	213	25.90	2.23	.77	12.16	3.72	2.08	20.96
Fast	800	200	22.20	2.81	1.18	11.10	2.79	1.07	18.95
Industrial enamel, oil base, high gloss - dark (OSHA) colors (material #57)									
Spray 1st or additional finish coats									
Slow	700	275	37.00	1.50	.43	13.45	4.77	3.02	23.17
Medium	750	250	32.40	2.23	.77	12.96	3.91	2.19	22.06
Fast	800	225	27.70	2.81	1.18	12.31	3.01	1.16	20.47
Epoxy coating, 2 part system - clear (material #51)									
Spray 1st coat									
Slow	550	225	58.30	1.91	.54	25.91	8.79	5.57	42.72
Medium	600	213	51.00	2.79	.98	23.94	6.79	3.79	38.29
Fast	650	200	43.70	3.46	1.45	21.85	4.95	1.90	33.61
Spray 2nd or additional finish coats									
Slow	700	250	58.30	1.50	.43	23.32	7.83	4.96	38.04
Medium	750	238	51.00	2.23	.77	21.43	5.99	3.35	33.77
Fast	800	225	43.70	2.81	1.18	19.42	4.33	1.66	29.40

	Labor SF per manhour	Material coverage SF/gallon	Material cost per gallon	Labor cost per 100 SF	Labor burden 100 SF	Material cost per 100 SF	Overhead per 100 SF	Profit per 100 SF	Total price per 100 SF
Epoxy coating, 2 part system - white (material #52)									
Spray 1st coat									
Slow	550	225	61.70	1.91	.54	27.42	9.26	5.87	45.00
Medium	600	213	54.00	2.79	.98	25.35	7.13	3.99	40.24
Fast	650	200	46.30	3.46	1.45	23.15	5.19	1.99	35.24
Spray 2nd or additional finish coats									
Slow	700	250	61.70	1.50	.43	24.68	8.25	5.23	40.09
Medium	750	238	54.00	2.23	.77	22.69	6.30	3.52	35.51
Fast	800	225	46.30	2.81	1.18	20.58	4.54	1.75	30.86

For heights over 8 feet, use the High Time Difficulty Factors on page 137. Note: A two coat system, prime and finish, using oil base material is recommended for any metal surface. Although water base material is often used, it may cause oxidation, corrosion and rust. One coat of oil base solid body stain is often used on exterior metal but it may crack, peel or chip without the proper prime coat application. "Slow" work is based on a $10.50 hourly wage, "Medium" work on a $16.75 hourly wage, and "Fast" work on a $22.50 hourly wage. Other qualifications that apply to this table are on page 9.

	Labor SF per manhour	Material coverage SF/gallon	Material cost per gallon	Labor cost per 100 SF	Labor burden 100 SF	Material cost per 100 SF	Overhead per 100 SF	Profit per 100 SF	Total price per 100 SF
Ductwork, bare duct, mitt or glove application									
Acid wash coat, muriatic acid (material #49)									
Mitt or glove each coat									
Slow	175	700	5.40	6.00	1.70	.77	2.63	1.67	12.77
Medium	200	650	4.70	8.38	2.92	.72	2.94	1.65	16.61
Fast	225	600	4.00	10.00	4.16	.67	2.75	1.06	18.64
Metal primer, rust inhibitor - clean metal (material #35)									
Mitt or glove prime coat									
Slow	175	400	29.90	6.00	1.70	7.48	4.71	2.99	22.88
Medium	200	388	26.20	8.38	2.92	6.75	4.42	2.47	24.94
Fast	225	375	22.40	10.00	4.16	5.97	3.73	1.43	25.29
Metal primer, rust inhibitor - rusty metal (material #36)									
Mitt or glove prime coat									
Slow	175	400	36.60	6.00	1.70	9.15	5.23	3.31	25.39
Medium	200	350	32.10	8.38	2.92	9.17	5.02	2.80	28.29
Fast	225	325	27.50	10.00	4.16	8.46	4.19	1.61	28.42
Industrial enamel, oil base, high gloss - light colors (material #56)									
Mitt or glove 1st or additional finish coats									
Slow	225	425	29.60	4.67	1.32	6.96	4.02	2.55	19.52
Medium	250	400	25.90	6.70	2.34	6.48	3.80	2.12	21.44
Fast	275	375	22.20	8.18	3.43	5.92	3.24	1.25	22.02
Industrial enamel, oil base, high gloss - dark (OSHA) colors (material #57)									
Mitt or glove 1st or additional finish coats									
Slow	225	475	37.00	4.67	1.32	7.79	4.27	2.71	20.76
Medium	250	450	32.40	6.70	2.34	7.20	3.98	2.22	22.44
Fast	275	425	27.70	8.18	3.43	6.52	3.35	1.29	22.77
Epoxy coating, 2 part system - clear (material #51)									
Mitt or glove 1st coat									
Slow	175	400	58.30	6.00	1.70	14.58	6.91	4.38	33.57
Medium	200	375	51.00	8.38	2.92	13.60	6.10	3.41	34.41
Fast	225	350	43.70	10.00	4.16	12.49	4.93	1.90	33.48
Mitt or glove 2nd or additional finish coats									
Slow	225	425	58.30	4.67	1.32	13.72	6.11	3.87	29.69
Medium	250	400	51.00	6.70	2.34	12.75	5.34	2.98	30.11
Fast	275	375	43.70	8.18	3.43	11.65	4.30	1.65	29.21

	Labor SF per manhour	Material coverage SF/gallon	Material cost per gallon	Labor cost per 100 SF	Labor burden 100 SF	Material cost per 100 SF	Overhead per 100 SF	Profit per 100 SF	Total price per 100 SF
Epoxy coating, 2 part system - white (material #52)									
Mitt or glove 1st coat									
Slow	175	400	61.70	6.00	1.70	15.43	7.17	4.55	34.85
Medium	200	375	54.00	8.38	2.92	14.40	6.30	3.52	35.52
Fast	225	350	46.30	10.00	4.16	13.23	5.07	1.95	34.41
Mitt or glove 2nd or additional finish coats									
Slow	225	425	61.70	4.67	1.32	14.52	6.36	4.03	30.90
Medium	250	400	54.00	6.70	2.34	13.50	5.52	3.09	31.15
Fast	275	375	46.30	8.18	3.43	12.35	4.43	1.70	30.09

For heights over 8 feet, use the High Time Difficulty Factors on page 137. Note: A two coat system, prime and finish, using oil base material is recommended for any metal surface. Although water base material is often used, it may cause oxidation, corrosion and rust. One coat of oil base solid body stain is often used on exterior metal but it may crack, peel or chip without the proper prime coat application. "Slow" work is based on a $10.50 hourly wage, "Medium" work on a $16.75 hourly wage, and "Fast" work on a $22.50 hourly wage. Other qualifications that apply to this table are on page 9.

	Labor SF per manhour	Material coverage SF/gallon	Material cost per gallon	Labor cost per 100 SF	Labor burden 100 SF	Material cost per 100 SF	Overhead per 100 SF	Profit per 100 SF	Total price per 100 SF
Ductwork, canvas insulated, brush application									
Flat latex, water base (material #5)									
Brush 1st coat									
Slow	60	250	17.10	17.50	4.97	6.84	9.09	5.76	44.16
Medium	75	238	15.00	22.33	7.78	6.30	8.92	4.99	50.32
Fast	90	225	12.90	25.00	10.43	5.73	7.62	2.93	51.71
Brush 2nd coat									
Slow	85	275	17.10	12.35	3.50	6.22	6.84	4.34	33.25
Medium	105	238	15.00	15.95	5.56	6.30	6.81	3.81	38.43
Fast	125	250	12.90	18.00	7.51	5.16	5.68	2.18	38.53
Brush 3rd or additional coats									
Slow	100	325	17.10	10.50	2.98	5.26	5.81	3.68	28.23
Medium	125	313	15.00	13.40	4.67	4.79	5.60	3.13	31.59
Fast	150	300	12.90	15.00	6.27	4.30	4.73	1.82	32.12
Sealer, off white, water base (material #1)									
Brush 1 coat									
Slow	60	250	16.30	17.50	4.97	6.52	8.99	5.70	43.68
Medium	75	238	14.30	22.33	7.78	6.01	8.85	4.95	49.92
Fast	90	225	12.30	25.00	10.43	5.47	7.57	2.91	51.38
Sealer, off white, oil base (material #2)									
Brush 1 coat									
Slow	60	275	22.90	17.50	4.97	8.33	9.55	6.05	46.40
Medium	75	263	20.10	22.33	7.78	7.64	9.25	5.17	52.17
Fast	90	250	17.20	25.00	10.43	6.88	7.83	3.01	53.15
Enamel, water base (material #9)									
Brush 1st finish coat									
Slow	85	275	24.50	12.35	3.50	8.91	7.68	4.87	37.31
Medium	105	238	21.40	15.95	5.56	8.99	7.47	4.18	42.15
Fast	125	250	18.40	18.00	7.51	7.36	6.08	2.34	41.29
Brush 2nd or additional finish coats									
Slow	100	325	24.50	10.50	2.98	7.54	6.52	4.13	31.67
Medium	125	313	21.40	13.40	4.67	6.84	6.10	3.41	34.42
Fast	150	300	18.40	15.00	6.27	6.13	5.07	1.95	34.42
Industrial enamel, oil base, high gloss - light colors (material #56)									
Brush 1st finish coat									
Slow	85	300	29.60	12.35	3.50	9.87	7.98	5.06	38.76
Medium	105	288	25.90	15.95	5.56	8.99	7.47	4.18	42.15
Fast	125	275	22.20	18.00	7.51	8.07	6.21	2.39	42.18

	Labor SF per manhour	Material coverage SF/gallon	Material cost per gallon	Labor cost per 100 SF	Labor burden 100 SF	Material cost per 100 SF	Overhead per 100 SF	Profit per 100 SF	Total price per 100 SF
Brush 2nd or additional finish coats									
Slow	100	350	29.60	10.50	2.98	8.46	6.80	4.31	33.05
Medium	125	338	25.90	13.40	4.67	7.66	6.30	3.52	35.55
Fast	150	325	22.20	15.00	6.27	6.83	5.20	2.00	35.30
Industrial enamel, oil base, high gloss - dark (OSHA) colors (material #57)									
Brush 1st finish coat									
Slow	85	300	37.00	12.35	3.50	12.33	8.74	5.54	42.46
Medium	105	288	32.40	15.95	5.56	11.25	8.03	4.49	45.28
Fast	125	275	27.70	18.00	7.51	10.07	6.58	2.53	44.69
Brush 2nd or additional finish coats									
Slow	100	350	37.00	10.50	2.98	10.57	7.46	4.73	36.24
Medium	125	338	32.40	13.40	4.67	9.59	6.78	3.79	38.23
Fast	150	325	27.70	15.00	6.27	8.52	5.51	2.12	37.42
Epoxy coating, 2 part system - clear (material #51)									
Brush 1st coat									
Slow	60	260	58.30	17.50	4.97	22.42	13.92	8.82	67.63
Medium	75	243	51.00	22.33	7.78	20.99	12.52	7.00	70.62
Fast	90	235	43.70	25.00	10.43	18.60	10.00	3.84	67.87
Brush 2nd coat									
Slow	85	285	58.30	12.35	3.50	20.46	11.26	7.14	54.71
Medium	105	273	51.00	15.95	5.56	18.68	9.85	5.50	55.54
Fast	125	260	43.70	18.00	7.51	16.81	7.83	3.01	53.16
Brush 3rd or additional coats									
Slow	100	335	58.30	10.50	2.98	17.40	9.58	6.07	46.53
Medium	125	323	51.00	13.40	4.67	15.79	8.30	4.64	46.80
Fast	150	310	43.70	15.00	6.27	14.10	6.54	2.51	44.42

	Labor SF per manhour	Material coverage SF/gallon	Material cost per gallon	Labor cost per 100 SF	Labor burden 100 SF	Material cost per 100 SF	Overhead per 100 SF	Profit per 100 SF	Total price per 100 SF
Epoxy coating, 2 part system - white (material #52)									
Brush 1st coat									
Slow	60	260	61.70	17.50	4.97	23.73	14.33	9.08	69.61
Medium	75	243	54.00	22.33	7.78	22.22	12.82	7.17	72.32
Fast	90	235	46.30	25.00	10.43	19.70	10.20	3.92	69.25
Brush 2nd coat									
Slow	85	285	61.70	12.35	3.50	21.65	11.63	7.37	56.50
Medium	100	273	54.00	16.75	5.84	19.78	10.38	5.80	58.55
Fast	125	260	46.30	18.00	7.51	17.81	8.02	3.08	54.42
Brush 3rd or additional coats									
Slow	100	335	61.70	10.50	2.98	18.42	9.89	6.27	48.06
Medium	125	323	54.00	13.40	4.67	16.72	8.52	4.76	48.07
Fast	150	310	46.30	15.00	6.27	14.94	6.70	2.57	45.48

For heights over 8 feet, use the High Time Difficulty Factors on page 137. "Slow" work is based on a $10.50 hourly wage, "Medium" work on a $16.75 hourly wage, and "Fast" work on a $22.50 hourly wage. Other qualifications that apply to this table are on page 9.

	Labor SF per manhour	Material coverage SF/gallon	Material cost per gallon	Labor cost per 100 SF	Labor burden 100 SF	Material cost per 100 SF	Overhead per 100 SF	Profit per 100 SF	Total price per 100 SF

Ductwork, canvas insulated, roll application

	Labor SF per manhour	Material coverage SF/gallon	Material cost per gallon	Labor cost per 100 SF	Labor burden 100 SF	Material cost per 100 SF	Overhead per 100 SF	Profit per 100 SF	Total price per 100 SF
Flat latex, water base (material #5)									
Roll 1st coat									
Slow	125	250	17.10	8.40	2.38	6.84	5.47	3.47	26.56
Medium	150	238	15.00	11.17	3.90	6.30	5.23	2.92	29.52
Fast	175	225	12.90	12.86	5.35	5.73	4.43	1.70	30.07
Roll 2nd coat									
Slow	175	275	17.10	6.00	1.70	6.22	4.32	2.74	20.98
Medium	200	263	15.00	8.38	2.92	5.70	4.17	2.33	23.50
Fast	225	250	12.90	10.00	4.16	5.16	3.58	1.38	24.28
Roll 3rd or additional coats									
Slow	225	400	17.10	4.67	1.32	4.28	3.19	2.02	15.48
Medium	250	388	15.00	6.70	2.34	3.87	3.16	1.77	17.84
Fast	275	375	12.90	8.18	3.43	3.44	2.78	1.07	18.90
Sealer, off white, water base (material #1)									
Roll 1 coat									
Slow	125	250	16.30	8.40	2.38	6.52	5.37	3.40	26.07
Medium	150	238	14.30	11.17	3.90	6.01	5.16	2.89	29.13
Fast	175	225	12.30	12.86	5.35	5.47	4.38	1.68	29.74
Sealer, off white, oil base (material #2)									
Roll 1 coat									
Slow	125	275	22.90	8.40	2.38	8.33	5.93	3.76	28.80
Medium	150	263	20.10	11.17	3.90	7.64	5.56	3.11	31.38
Fast	175	250	17.20	12.86	5.35	6.88	4.65	1.79	31.53
Enamel, water base (material #9)									
Roll 1st finish coat									
Slow	175	350	24.50	6.00	1.70	7.00	4.56	2.89	22.15
Medium	200	325	21.40	8.38	2.92	6.58	4.38	2.45	24.71
Fast	225	300	18.40	10.00	4.16	6.13	3.76	1.44	25.49
Roll 2nd or additional finish coats									
Slow	225	400	24.50	4.67	1.32	6.13	3.76	2.38	18.26
Medium	250	388	21.40	6.70	2.34	5.52	3.56	1.99	20.11
Fast	275	375	18.40	8.18	3.43	4.91	3.05	1.17	20.74

	Labor SF per manhour	Material coverage SF/gallon	Material cost per gallon	Labor cost per 100 SF	Labor burden 100 SF	Material cost per 100 SF	Overhead per 100 SF	Profit per 100 SF	Total price per 100 SF
Industrial enamel, oil base, high gloss - light colors (material #56)									
Roll 1st finish coat									
Slow	175	300	29.60	6.00	1.70	9.87	5.45	3.45	26.47
Medium	200	288	25.90	8.38	2.92	8.99	4.97	2.78	28.04
Fast	225	275	22.20	10.00	4.16	8.07	4.12	1.58	27.93
Roll 2nd or additional finish coats									
Slow	225	450	29.60	4.67	1.32	6.58	3.90	2.47	18.94
Medium	250	438	25.90	6.70	2.34	5.91	3.66	2.05	20.66
Fast	275	425	22.20	8.18	3.43	5.22	3.11	1.20	21.14
Industrial enamel, oil base, high gloss - dark (OSHA) colors (material #57)									
Roll 1st finish coat									
Slow	175	300	37.00	6.00	1.70	12.33	6.21	3.94	30.18
Medium	200	288	32.40	8.38	2.92	11.25	5.52	3.09	31.16
Fast	225	275	27.70	10.00	4.16	10.07	4.49	1.72	30.44
Roll 2nd or additional finish coats									
Slow	225	450	37.00	4.67	1.32	8.22	4.41	2.79	21.41
Medium	250	438	32.40	6.70	2.34	7.40	4.03	2.25	22.72
Fast	275	425	27.70	8.18	3.43	6.52	3.35	1.29	22.77
Epoxy coating, 2 part system - clear (material #51)									
Roll 1st coat									
Slow	125	250	58.30	8.40	2.38	23.32	10.57	6.70	51.37
Medium	150	238	51.00	11.17	3.90	21.43	8.94	5.00	50.44
Fast	175	225	43.70	12.86	5.35	19.42	6.97	2.68	47.28
Roll 2nd coat									
Slow	175	375	58.30	6.00	1.70	15.55	7.21	4.57	35.03
Medium	200	338	51.00	8.38	2.92	15.09	6.47	3.61	36.47
Fast	225	300	43.70	10.00	4.16	14.57	5.32	2.04	36.09
Roll 3rd or additional coats									
Slow	225	425	58.30	4.67	1.32	13.72	6.11	3.87	29.69
Medium	250	413	51.00	6.70	2.34	12.35	5.24	2.93	29.56
Fast	275	400	43.70	8.18	3.43	10.93	4.17	1.60	28.31

	Labor SF per manhour	Material coverage SF/gallon	Material cost per gallon	Labor cost per 100 SF	Labor burden 100 SF	Material cost per 100 SF	Overhead per 100 SF	Profit per 100 SF	Total price per 100 SF
Epoxy coating, 2 part system - white (material #52)									
Roll 1st coat									
Slow	125	250	61.70	8.40	2.38	24.68	11.00	6.97	53.43
Medium	150	238	54.00	11.17	3.90	22.69	9.25	5.17	52.18
Fast	175	225	46.30	12.86	5.35	20.58	7.18	2.76	48.73
Roll 2nd coat									
Slow	175	375	61.70	6.00	1.70	16.45	7.49	4.75	36.39
Medium	200	338	54.00	8.38	2.92	15.98	6.68	3.74	37.70
Fast	225	300	46.30	10.00	4.16	15.43	5.48	2.11	37.18
Roll 3rd or additional coats									
Slow	225	425	61.70	4.67	1.32	14.52	6.36	4.03	30.90
Medium	250	413	54.00	6.70	2.34	13.08	5.42	3.03	30.57
Fast	275	400	46.30	8.18	3.43	11.58	4.29	1.65	29.13

For heights over 8 feet, use the High Time Difficulty Factors on page 137. "Slow" work is based on a $10.50 hourly wage, "Medium" work on a $16.75 hourly wage, and "Fast" work on a $22.50 hourly wage. Other qualifications that apply to this table are on page 9.

	Labor SF per manhour	Material coverage SF/gallon	Material cost per gallon	Labor cost per 100 SF	Labor burden 100 SF	Material cost per 100 SF	Overhead per 100 SF	Profit per 100 SF	Total price per 100 SF
Ductwork, canvas insulated, spray application									
Flat latex, water base (material #5)									
Spray prime coat									
Slow	450	200	17.10	2.33	.66	8.55	3.58	2.27	17.39
Medium	500	188	15.00	3.35	1.17	7.98	3.06	1.71	17.27
Fast	550	175	12.90	4.09	1.71	7.37	2.44	.94	16.55
Spray 1st finish coat									
Slow	550	225	17.10	1.91	.54	7.60	3.12	1.98	15.15
Medium	625	213	15.00	2.68	.93	7.04	2.61	1.46	14.72
Fast	700	200	12.90	3.21	1.35	6.45	2.04	.78	13.83
Spray 2nd or additional finish coats									
Slow	700	250	17.10	1.50	.43	6.84	2.72	1.72	13.21
Medium	750	238	15.00	2.23	.77	6.30	2.28	1.27	12.85
Fast	800	225	12.90	2.81	1.18	5.73	1.80	.69	12.21
Sealer, off white, water base (material #1)									
Spray 1 coat									
Slow	450	200	16.30	2.33	.66	8.15	3.45	2.19	16.78
Medium	500	188	14.30	3.35	1.17	7.61	2.97	1.66	16.76
Fast	550	175	12.30	4.09	1.71	7.03	2.37	.91	16.11
Sealer, off white, oil base (material #2)									
Spray 1 coat									
Slow	450	225	22.90	2.33	.66	10.18	4.08	2.59	19.84
Medium	500	213	20.10	3.35	1.17	9.44	3.42	1.91	19.29
Fast	550	200	17.20	4.09	1.71	8.60	2.66	1.02	18.08
Enamel, water base (material #9)									
Spray 1st finish coat									
Slow	550	225	24.50	1.91	.54	10.89	4.14	2.62	20.10
Medium	625	213	21.40	2.68	.93	10.05	3.35	1.87	18.88
Fast	700	200	18.40	3.21	1.35	9.20	2.54	.98	17.28
Spray 2nd or additional finish coats									
Slow	700	250	24.50	1.50	.43	9.80	3.64	2.31	17.68
Medium	750	238	21.40	2.23	.77	8.99	2.94	1.64	16.57
Fast	800	225	18.40	2.81	1.18	8.18	2.25	.86	15.28

	Labor SF per manhour	Material coverage SF/gallon	Material cost per gallon	Labor cost per 100 SF	Labor burden 100 SF	Material cost per 100 SF	Overhead per 100 SF	Profit per 100 SF	Total price per 100 SF
Industrial enamel, oil base, high gloss - light colors (material #56)									
Spray 1st finish coat									
Slow	700	250	29.60	1.50	.43	11.84	4.27	2.71	20.75
Medium	625	238	25.90	2.68	.93	10.88	3.55	1.98	20.02
Fast	550	225	22.20	4.09	1.71	9.87	2.90	1.11	19.68
Spray 2nd or additional finish coats									
Slow	800	275	29.60	1.31	.38	10.76	3.86	2.45	18.76
Medium	750	263	25.90	2.23	.77	9.85	3.15	1.76	17.76
Fast	700	250	22.20	3.21	1.35	8.88	2.48	.95	16.87
Industrial enamel, oil base, high gloss - dark (OSHA) colors (material #57)									
Spray 1st finish coat									
Slow	700	250	37.00	1.50	.43	14.80	5.19	3.29	25.21
Medium	625	238	32.40	2.68	.93	13.61	4.22	2.36	23.80
Fast	550	225	27.70	4.09	1.71	12.31	3.35	1.29	22.75
Spray 2nd or additional finish coats									
Slow	800	275	37.00	1.31	.38	13.45	4.69	2.97	22.80
Medium	750	263	32.40	2.23	.77	12.32	3.76	2.10	21.18
Fast	700	250	27.70	3.21	1.35	11.08	2.89	1.11	19.64
Epoxy coating, 2 part system - clear (material #51)									
Spray prime coat									
Slow	450	200	58.30	2.33	.66	29.15	9.96	6.32	48.42
Medium	500	188	51.00	3.35	1.17	27.13	7.75	4.33	43.73
Fast	550	175	43.70	4.09	1.71	24.97	5.69	2.19	38.65
Spray 1st finish coat									
Slow	550	235	58.30	1.91	.54	24.81	8.45	5.36	41.07
Medium	625	225	51.00	2.68	.93	22.67	6.44	3.60	36.32
Fast	700	215	43.70	3.21	1.35	20.33	4.60	1.77	31.26
Spray 2nd or additional finish coats									
Slow	700	265	58.30	1.50	.43	22.00	7.42	4.70	36.05
Medium	750	253	51.00	2.23	.77	20.16	5.68	3.17	32.01
Fast	800	240	43.70	2.81	1.18	18.21	4.11	1.58	27.89

	Labor SF per manhour	Material coverage SF/gallon	Material cost per gallon	Labor cost per 100 SF	Labor burden 100 SF	Material cost per 100 SF	Overhead per 100 SF	Profit per 100 SF	Total price per 100 SF
Epoxy coating, 2 part system - white (material #52)									
Spray prime coat									
Slow	450	200	61.70	2.33	.66	30.85	10.49	6.65	50.98
Medium	500	188	54.00	3.35	1.17	28.72	8.14	4.55	45.93
Fast	550	175	46.30	4.09	1.71	26.46	5.97	2.29	40.52
Spray 1st finish coat									
Slow	550	235	61.70	1.91	.54	26.26	8.90	5.64	43.25
Medium	625	225	54.00	2.68	.93	24.00	6.76	3.78	38.15
Fast	700	215	46.30	3.21	1.35	21.53	4.82	1.85	32.76
Spray 2nd or additional finish coats									
Slow	700	265	61.70	1.50	.43	23.28	7.82	4.95	37.98
Medium	750	253	54.00	2.23	.77	21.34	5.97	3.34	33.65
Fast	800	240	46.30	2.81	1.18	19.29	4.30	1.65	29.23

For heights over 8 feet, use the High Time Difficulty Factors on page 137. "Slow" work is based on a $10.50 hourly wage, "Medium" work on a $16.75 hourly wage, and "Fast" work on a $22.50 hourly wage. Other qualifications that apply to this table are on page 9.

	Manhours per flight	Flights per gallon	Material cost per gallon	Labor cost per flight	Labor burden flight	Material cost per flight	Overhead per flight	Profit per flight	Total price per flight
Fire escapes									
Solid (plain) deck									
Spray each coat									
Metal primer, rust inhibitor - clean metal (material #35)									
Slow	2.0	1.25	29.90	21.00	5.96	23.92	15.78	10.00	76.66
Medium	1.5	1.00	26.20	25.13	8.76	26.20	14.72	8.23	83.04
Fast	1.0	0.75	22.40	22.50	9.39	29.87	11.43	4.39	77.58
Metal primer, rust inhibitor - rusty metal (material #36)									
Slow	2.0	1.25	36.60	21.00	5.96	29.28	17.44	11.05	84.73
Medium	1.5	1.00	32.10	25.13	8.76	32.10	16.17	9.04	91.20
Fast	1.0	0.75	27.50	22.50	9.39	36.67	12.68	4.87	86.11
Industrial enamel, oil base, high gloss - light colors (material #56)									
Slow	2.25	1.50	29.60	23.63	6.70	19.73	15.52	9.84	75.42
Medium	1.75	1.25	25.90	29.31	10.22	20.72	14.76	8.25	83.26
Fast	1.25	1.00	22.20	28.13	11.73	22.20	11.48	4.41	77.95
Industrial enamel, oil base, high gloss - dark (OSHA) colors (material #57)									
Slow	2.25	1.50	37.00	23.63	6.70	24.67	17.06	10.81	82.87
Medium	1.75	1.25	32.40	29.31	10.22	25.92	16.03	8.96	90.44
Fast	1.25	1.00	27.70	28.13	11.73	27.70	12.50	4.80	84.86
Grating deck									
Spray each coat									
Metal primer, rust inhibitor - clean metal (material #35)									
Slow	3.0	1.75	29.90	31.50	8.94	17.09	17.84	11.31	86.68
Medium	2.5	1.50	26.20	41.88	14.60	17.47	18.12	10.13	102.20
Fast	2.0	1.25	22.40	45.00	18.78	17.92	15.12	5.81	102.63
Metal primer, rust inhibitor - rusty metal (material #36)									
Slow	3.0	1.75	36.60	31.50	8.94	20.91	19.02	12.06	92.43
Medium	2.5	1.50	32.10	41.88	14.60	21.40	19.08	10.67	107.63
Fast	2.0	1.25	27.50	45.00	18.78	22.00	15.87	6.10	107.75

	Manhours per flight	Flights per gallon	Material cost per gallon	Labor cost per flight	Labor burden flight	Material cost per flight	Overhead per flight	Profit per flight	Total price per flight
Industrial enamel, oil base, high gloss - light colors (material #56)									
Slow	3.25	2.00	29.60	34.13	9.68	14.80	18.18	11.52	88.31
Medium	2.75	1.75	25.90	46.06	16.06	14.80	18.84	10.53	106.29
Fast	2.25	1.50	22.20	50.63	21.12	14.80	16.02	6.16	108.73
Industrial enamel, oil base, high gloss - dark (OSHA) colors (material #57)									
Slow	3.25	2.00	37.00	34.13	9.68	18.50	19.33	12.25	93.89
Medium	2.75	1.75	32.40	46.06	16.06	18.51	19.75	11.04	111.42
Fast	2.25	1.50	27.70	50.63	21.12	18.47	16.69	6.42	113.33

Fire escapes can also be estimated by the square foot. Calculate the actual area to be coated. For continuous solid (plain) deck, use the rates listed under Decking and siding. For continuous grating deck, use the rates listed under Grates, steel. Note: A two coat system, prime and finish, using oil base material is recommended for any metal surface. Although water base material is often used, it may cause oxidation, corrosion and rust. One coat of oil base solid body stain is often used on exterior metal but it may crack, peel or chip without the proper prime coat application. "Slow" work is based on a $10.50 hourly wage, "Medium" work on a $16.75 hourly wage, and "Fast" work on a $22.50 hourly wage. Other qualifications that apply to this table are on page 9.

Fire sprinkler systems

Use the costs listed for 1" to 4" pipe. For painting sprinkler heads at 12 feet on center at a ceiling height of 12 feet, figure 3 minutes per head (20 per hour). Very little paint is needed. Your material estimate for the sprinkler pipe will include enough to cover the heads. Note: A two coat system, prime and finish, using oil base material is recommended for any metal surface. Although water base material is often used, it may cause oxidation, corrosion and rust. One coat of oil base solid body stain is often used on exterior metal but it may crack, peel or chip without the proper prime coat application.

	Labor SF per manhour	Material coverage SF/gallon	Material cost per gallon	Labor cost per 100 SF	Labor burden 100 SF	Material cost per 100 SF	Overhead per 100 SF	Profit per 100 SF	Total price per 100 SF
Grates, steel, over 1" thick									
Without supports									
Brush each coat									
Metal primer, rust inhibitor - clean metal (material #35)									
Slow	60	175	29.90	17.50	4.97	.17	7.02	4.45	34.11
Medium	85	150	26.20	19.71	6.86	.17	6.55	3.66	36.95
Fast	110	125	22.40	20.45	8.54	.18	5.40	2.07	36.64
Metal primer, rust inhibitor - rusty metal (material #36)									
Slow	60	175	36.60	17.50	4.97	.21	7.03	4.46	34.17
Medium	85	150	32.10	19.71	6.86	.21	6.56	3.67	37.01
Fast	110	125	27.50	20.45	8.54	.22	5.40	2.08	36.69
Industrial enamel, oil base, high gloss - light colors (material #56)									
Slow	75	225	29.60	14.00	3.97	.13	5.61	3.56	27.27
Medium	100	200	25.90	16.75	5.84	.13	5.57	3.11	31.40
Fast	125	175	22.20	18.00	7.51	.13	4.75	1.82	32.21
Industrial enamel, oil base, high gloss - dark (OSHA) colors (material #57)									
Slow	75	225	37.00	14.00	3.97	.16	5.62	3.56	27.31
Medium	100	200	32.40	16.75	5.84	.16	5.57	3.12	31.44
Fast	125	175	27.70	18.00	7.51	.16	4.75	1.83	32.25
Spray each coat									
Metal primer, rust inhibitor - clean metal (material #35)									
Slow	190	125	29.90	5.53	1.56	.24	2.28	1.44	11.05
Medium	208	113	26.20	8.05	2.82	.23	2.72	1.52	15.34
Fast	225	100	22.40	10.00	4.16	.22	2.66	1.02	18.06
Metal primer, rust inhibitor - rusty metal (material #36)									
Slow	190	125	36.60	5.53	1.56	.29	2.29	1.45	11.12
Medium	208	113	32.10	8.05	2.82	.28	2.73	1.53	15.41
Fast	225	100	27.50	10.00	4.16	.28	2.68	1.03	18.15
Industrial enamel, oil base, high gloss - light colors (material #56)									
Slow	215	160	29.60	4.88	1.39	.19	2.00	1.27	9.73
Medium	233	148	25.90	7.19	2.50	.18	2.42	1.35	13.64
Fast	250	135	22.20	9.00	3.76	.16	2.39	.92	16.23
Industrial enamel, oil base, high gloss - dark (OSHA) colors (material #57)									
Slow	215	160	37.00	4.88	1.39	.23	2.02	1.28	9.80
Medium	233	148	32.40	7.19	2.50	.22	2.43	1.36	13.70
Fast	250	135	27.70	9.00	3.76	.21	2.40	.92	16.29

	Labor SF per manhour	Material coverage SF/gallon	Material cost per gallon	Labor cost per 100 SF	Labor burden 100 SF	Material cost per 100 SF	Overhead per 100 SF	Profit per 100 SF	Total price per 100 SF
Including typical supports									
Brush each coat									
Metal primer, rust inhibitor - clean metal (material #35)									
Slow	40	125	29.90	26.25	7.45	.24	10.53	6.67	51.14
Medium	55	113	26.20	30.45	10.62	.23	10.12	5.66	57.08
Fast	70	100	22.40	32.14	13.43	.22	8.47	3.26	57.52
Metal primer, rust inhibitor - rusty metal (material #36)									
Slow	40	125	36.60	26.25	7.45	.29	10.54	6.68	51.21
Medium	55	113	32.10	30.45	10.62	.28	10.13	5.66	57.14
Fast	70	100	27.50	32.14	13.43	.28	8.48	3.26	57.59
Industrial enamel, oil base, high gloss - light colors (material #56)									
Slow	50	160	29.60	21.00	5.96	.19	8.42	5.34	40.91
Medium	75	148	25.90	22.33	7.78	.18	7.42	4.15	41.86
Fast	100	135	22.20	22.50	9.39	.16	5.93	2.28	40.26
Industrial enamel, oil base, high gloss - dark (OSHA) colors (material #57)									
Slow	50	160	37.00	21.00	5.96	.23	8.43	5.34	40.96
Medium	75	148	32.40	22.33	7.78	.22	7.43	4.15	41.91
Fast	100	135	27.70	22.50	9.39	.21	5.94	2.28	40.32
Spray each coat									
Metal primer, rust inhibitor - clean metal (material #35)									
Slow	120	100	29.90	8.75	2.48	.30	3.58	2.27	17.38
Medium	135	88	26.20	12.41	4.33	.30	4.17	2.33	23.54
Fast	150	75	22.40	15.00	6.27	.30	3.99	1.53	27.09
Metal primer, rust inhibitor - rusty metal (material #36)									
Slow	120	100	36.60	8.75	2.48	.37	3.60	2.28	17.48
Medium	135	88	32.10	12.41	4.33	.36	4.19	2.34	23.63
Fast	150	75	27.50	15.00	6.27	.37	4.00	1.54	27.18
Industrial enamel, oil base, high gloss - light colors (material #56)									
Slow	150	125	29.60	7.00	1.99	.24	2.86	1.81	13.90
Medium	163	113	25.90	10.28	3.57	.23	3.45	1.93	19.46
Fast	175	100	22.20	12.86	5.35	.22	3.41	1.31	23.15
Industrial enamel, oil base, high gloss - dark (OSHA) colors (material #57)									
Slow	150	125	37.00	7.00	1.99	.30	2.88	1.83	14.00
Medium	163	113	32.40	10.28	3.57	.29	3.47	1.94	19.55
Fast	175	100	27.70	12.86	5.35	.28	3.42	1.32	23.23

Use these figures when estimating steel grates over 1" thick. The figures will apply when both sides are painted with oil or water base paint. Square feet calculations for grates are based on overall (length times width) dimensions. For grills under 1" thick, see the following table. Note: A two coat system, prime and finish, using oil base material is recommended for any metal surface. Although water base material is often used, it may cause oxidation, corrosion and rust. Using one coat of oil base paint on exterior metal may result in cracking, peeling, or chipping without the proper prime coat application. "Slow" work is based on a $10.50 hourly wage, "Medium" work on a $16.75 hourly wage, and "Fast" work on a $22.50 hourly wage. Other qualifications that apply to this table are on page 9.

	Labor SF per manhour	Material coverage SF/gallon	Material cost per gallon	Labor cost per 100 SF	Labor burden 100 SF	Material cost per 100 SF	Overhead per 100 SF	Profit per 100 SF	Total price per 100 SF

Grates, steel, under 1" thick

Without supports

Brush each coat

	Labor SF per manhour	Material coverage SF/gallon	Material cost per gallon	Labor cost per 100 SF	Labor burden 100 SF	Material cost per 100 SF	Overhead per 100 SF	Profit per 100 SF	Total price per 100 SF
Metal primer, rust inhibitor, clean metal (material #35)									
Slow	175	200	29.90	6.00	1.70	14.95	7.02	4.45	34.12
Medium	200	175	26.20	8.38	2.92	14.97	6.44	3.60	36.31
Fast	225	150	22.40	10.00	4.16	14.93	5.39	2.07	36.55
Metal primer, rust inhibitor, rusty metal (material #36)									
Slow	175	200	36.60	6.00	1.70	18.30	8.06	5.11	39.17
Medium	200	175	32.10	8.38	2.92	18.34	7.26	4.06	40.96
Fast	225	150	27.50	10.00	4.16	18.33	6.01	2.31	40.81
Industrial enamel, oil base, high gloss, light colors (material #56)									
Slow	200	250	29.60	5.25	1.49	11.84	5.76	3.65	27.99
Medium	225	225	25.90	7.44	2.59	11.51	5.28	2.95	29.77
Fast	250	200	22.20	9.00	3.76	11.10	4.41	1.70	29.97
Industrial enamel, oil base, high gloss, dark (OSHA) colors (material #57)									
Slow	200	250	37.00	5.25	1.49	14.80	6.68	4.23	32.45
Medium	225	225	32.40	7.44	2.59	14.40	5.99	3.35	33.77
Fast	250	200	27.70	9.00	3.76	13.85	4.92	1.89	33.42

Spray each coat

	Labor SF per manhour	Material coverage SF/gallon	Material cost per gallon	Labor cost per 100 SF	Labor burden 100 SF	Material cost per 100 SF	Overhead per 100 SF	Profit per 100 SF	Total price per 100 SF
Metal primer, rust inhibitor, clean metal (material #35)									
Slow	400	150	29.90	2.63	.74	19.93	7.23	4.58	35.11
Medium	450	138	26.20	3.72	1.29	18.99	5.88	3.29	33.17
Fast	500	125	22.40	4.50	1.88	17.92	4.50	1.73	30.53
Metal primer, rust inhibitor, rusty metal (material #36)									
Slow	400	150	36.60	2.63	.74	24.40	8.61	5.46	41.84
Medium	450	138	32.10	3.72	1.29	23.26	6.93	3.87	39.07
Fast	500	125	27.50	4.50	1.88	22.00	5.25	2.02	35.65
Industrial enamel, oil base, high gloss, light colors (material #56)									
Slow	425	175	29.60	2.47	.70	16.91	6.22	3.95	30.25
Medium	475	163	25.90	3.53	1.24	15.89	5.06	2.83	28.55
Fast	525	150	22.20	4.29	1.77	14.80	3.86	1.48	26.20
Industrial enamel, oil base, high gloss, dark (OSHA) colors (material #57)									
Slow	425	175	37.00	2.47	.70	21.14	7.54	4.78	36.63
Medium	475	163	32.40	3.53	1.24	19.88	6.04	3.37	34.06
Fast	525	150	27.70	4.29	1.77	18.47	4.54	1.75	30.82

	Labor SF per manhour	Material coverage SF/gallon	Material cost per gallon	Labor cost per 100 SF	Labor burden 100 SF	Material cost per 100 SF	Overhead per 100 SF	Profit per 100 SF	Total price per 100 SF
Including typical supports									
Brush each coat									
Metal primer, rust inhibitor, clean metal (material #35)									
Slow	125	150	29.90	8.40	2.38	19.93	9.52	6.04	46.27
Medium	150	125	26.20	11.17	3.90	20.96	8.82	4.93	49.78
Fast	175	100	22.40	12.86	5.35	22.40	7.52	2.89	51.02
Metal primer, rust inhibitor, rusty metal (material #36)									
Slow	125	150	36.60	8.40	2.38	24.40	10.91	6.92	53.01
Medium	150	125	32.10	11.17	3.90	25.68	9.98	5.58	56.31
Fast	175	100	27.50	12.86	5.35	27.50	8.46	3.25	57.42
Industrial enamel, oil base, high gloss, light colors (material #56)									
Slow	150	200	29.60	7.00	1.99	14.80	7.37	4.67	35.83
Medium	175	175	25.90	9.57	3.33	14.80	6.79	3.80	38.29
Fast	200	150	22.20	11.25	4.70	14.80	5.69	2.19	38.63
Industrial enamel, oil base, high gloss, dark (OSHA) colors (material #57)									
Slow	150	200	37.00	7.00	1.99	18.50	8.52	5.40	41.41
Medium	175	175	32.40	9.57	3.33	18.51	7.70	4.30	43.41
Fast	200	150	27.70	11.25	4.70	18.47	6.37	2.45	43.24
Spray each coat									
Metal primer, rust inhibitor, clean metal (material #35)									
Slow	325	100	29.90	3.23	.92	29.90	10.56	6.69	51.30
Medium	375	88	26.20	4.47	1.56	29.77	8.77	4.90	49.47
Fast	425	75	22.40	5.29	2.20	29.87	6.91	2.66	46.93
Metal primer, rust inhibitor, rusty metal (material #36)									
Slow	325	100	36.60	3.23	.92	36.60	12.63	8.01	61.39
Medium	375	88	32.10	4.47	1.56	36.48	10.41	5.82	58.74
Fast	425	75	27.50	5.29	2.20	36.67	8.17	3.14	55.47
Industrial enamel, oil base, high gloss, light colors (material #56)									
Slow	350	125	29.60	3.00	.86	23.68	8.53	5.41	41.48
Medium	400	113	25.90	4.19	1.46	22.92	7.00	3.91	39.48
Fast	450	100	22.20	5.00	2.08	22.20	5.42	2.08	36.78
Industrial enamel, oil base, high gloss, dark (OSHA) colors (material #57)									
Slow	350	125	37.00	3.00	.86	29.60	10.37	6.57	50.40
Medium	400	113	32.40	4.19	1.46	28.67	8.41	4.70	47.43
Fast	450	100	27.70	5.00	2.08	27.70	6.44	2.47	43.69

Use these figures to estimate steel grates under 1" thick. The figures will apply when both sides are painted with oil or water base paint. Square foot calculations for grates are based on overall (length times width) dimensions. For grills over 1" thick, see the previous table. Note: A two coat system, prime and finish, using oil base material is recommended for any metal surface. Although water base material is often used, it may cause oxidation, corrosion and rust. Using one coat of oil base paint on exterior metal may result in cracking, peeling, or chipping without the proper prime coat application. "Slow" work is based on a $10.50 hourly wage, "Medium" work on a $16.75 hourly wage, and "Fast" work on a $22.50 hourly wage. Other qualifications that apply to this table are on page 9.

Ladders

Measure the length of the ladder rungs and vertical members. Then multiply by a difficulty factor of 1.5 (Length x 1.5) to allow for limited access to the back of the ladder. Then use the rates in the Bare pipe tables to figure the labor and material costs.

	Labor SF per manhour	Material coverage SF/gallon	Material cost per gallon	Labor cost per 100 SF	Labor burden 100 SF	Material cost per 100 SF	Overhead per 100 SF	Profit per 100 SF	Total price per 100 SF
Masonry, Concrete Masonry Units (CMU), rough, porous surface									
Industrial bonding & penetrating oil paint (material #55)									
Brush 1st coat									
Slow	225	200	27.80	4.67	1.32	13.90	6.17	3.91	29.97
Medium	250	188	24.30	6.70	2.34	12.93	5.38	3.01	30.36
Fast	275	175	20.80	8.18	3.43	11.89	4.35	1.67	29.52
Brush 2nd coat									
Slow	230	275	27.80	4.57	1.29	10.11	4.95	3.14	24.06
Medium	260	250	24.30	6.44	2.26	9.72	4.51	2.52	25.45
Fast	290	225	20.80	7.76	3.24	9.24	3.74	1.44	25.42
Industrial waterproofing (material #58)									
Brush 1st coat									
Slow	65	65	24.80	16.15	4.58	38.15	18.26	11.57	88.71
Medium	90	55	21.70	18.61	6.49	39.45	15.81	8.84	89.20
Fast	115	45	18.60	19.57	8.17	41.33	12.78	4.91	86.76
Brush 2nd or additional coats									
Slow	90	150	24.80	11.67	3.31	16.53	9.77	6.19	47.47
Medium	115	125	21.70	14.57	5.08	17.36	9.07	5.07	51.15
Fast	145	100	18.60	15.52	6.48	18.60	7.51	2.89	51.00
Roll 1st coat									
Slow	100	125	24.80	10.50	2.98	19.84	10.33	6.55	50.20
Medium	125	108	21.70	13.40	4.67	20.09	9.35	5.23	52.74
Fast	150	90	18.60	15.00	6.27	20.67	7.76	2.98	52.68
Roll 2nd or additional coats									
Slow	150	175	24.80	7.00	1.99	14.17	7.18	4.55	34.89
Medium	180	150	21.70	9.31	3.25	14.47	6.62	3.70	37.35
Fast	210	125	18.60	10.71	4.47	14.88	5.56	2.14	37.76

Use these figures for Concrete Masonry Units (CMU) where the block surfaces are rough, porous or unfilled, with joints struck to average depth. The more porous the surface, the rougher the texture, the more time and material will be required. For heights above 8 feet, use the High Time Difficulty Factors on page 137. Also see Masonry in the "General Painting Operations" section of this book. "Slow" work is based on a $10.50 hourly wage, "Medium" work on a $16.75 hourly wage, "Fast" work on a $22.50 hourly wage. Other qualifications that apply to this table are on page 9.

	Labor SF per manhour	Material coverage SF/gallon	Material cost per gallon	Labor cost per 100 SF	Labor burden 100 SF	Material cost per 100 SF	Overhead per 100 SF	Profit per 100 SF	Total price per 100 SF
Masonry, Concrete Masonry Units (CMU), smooth surface									
Industrial bonding & penetrating oil paint (material #55)									
Brush 1st coat									
Slow	325	240	27.80	3.23	.92	11.58	4.88	3.09	23.70
Medium	350	230	24.30	4.79	1.67	10.57	4.17	2.33	23.53
Fast	375	220	20.80	6.00	2.51	9.45	3.32	1.28	22.56
Brush 2nd coat									
Slow	340	300	27.80	3.09	.87	9.27	4.10	2.60	19.93
Medium	370	275	24.30	4.53	1.57	8.84	3.66	2.05	20.65
Fast	400	250	20.80	5.63	2.34	8.32	3.02	1.16	20.47
Industrial waterproofing (material #58)									
Brush 1st coat									
Slow	75	100	24.80	14.00	3.97	24.80	13.26	8.41	64.44
Medium	100	95	21.70	16.75	5.84	22.84	11.13	6.22	62.78
Fast	125	90	18.60	18.00	7.51	20.67	8.55	3.28	58.01
Brush 2nd or additional coats									
Slow	100	150	24.80	10.50	2.98	16.53	9.31	5.90	45.22
Medium	125	138	21.70	13.40	4.67	15.72	8.28	4.63	46.70
Fast	150	125	18.60	15.00	6.27	14.88	6.69	2.57	45.41
Roll 1st coat									
Slow	125	150	24.80	8.40	2.38	16.53	8.47	5.37	41.15
Medium	150	138	21.70	11.17	3.90	15.72	7.54	4.22	42.55
Fast	175	125	18.60	12.86	5.35	14.88	6.13	2.35	41.57
Roll 2nd or additional coats									
Slow	175	200	24.80	6.00	1.70	12.40	6.23	3.95	30.28
Medium	200	175	21.70	8.38	2.92	12.40	5.81	3.25	32.76
Fast	225	150	18.60	10.00	4.16	12.40	4.92	1.89	33.37

Use these figures for Concrete Masonry Units (CMU), precision block, filled block or slump stone with joints struck to average depth. The more porous the surface, the rougher the texture, the more time and material will be required. For heights above 8 feet, use the High Time Difficulty Factors on page 137. Also see Masonry in the "General Painting Operations" section of this book. "Slow" work is based on a $10.50 hourly wage, "Medium" work on a $16.75 hourly wage, "Fast" work on a $22.50 hourly wage. Other qualifications that apply to this table are on page 9.

	Labor SF per manhour	Material coverage SF/gallon	Material cost per gallon	Labor cost per 100 SF	Labor burden 100 SF	Material cost per 100 SF	Overhead per 100 SF	Profit per 100 SF	Total price per 100 SF

Mechanical equipment

Brush each coat

Metal primer, rust inhibitor, clean metal (material #35)

Slow	175	275	29.90	6.00	1.70	10.87	5.76	3.65	27.98
Medium	200	263	26.20	8.38	2.92	9.96	5.21	2.91	29.38
Fast	225	250	22.40	10.00	4.16	8.96	4.28	1.65	29.05

Metal primer, rust inhibitor, rusty metal (material #36)

Slow	175	275	36.60	6.00	1.70	13.31	6.52	4.13	31.66
Medium	200	263	32.10	8.38	2.92	12.21	5.76	3.22	32.49
Fast	225	250	27.50	10.00	4.16	11.00	4.66	1.79	31.61

Industrial enamel, oil base, high gloss, light colors (material #56)

Slow	200	375	29.60	5.25	1.49	7.89	4.54	2.88	22.05
Medium	225	363	25.90	7.44	2.59	7.13	4.20	2.35	23.71
Fast	250	350	22.20	9.00	3.76	6.34	3.53	1.36	23.99

Industrial enamel, oil base, high gloss, dark (OSHA) colors (material #57)

Slow	200	375	37.00	5.25	1.49	9.87	5.15	3.26	25.02
Medium	225	363	32.40	7.44	2.59	8.93	4.65	2.60	26.21
Fast	250	350	27.70	9.00	3.76	7.91	3.82	1.47	25.96

Spray each coat

Metal primer, rust inhibitor, clean metal (material #35)

Slow	350	200	29.90	3.00	.86	14.95	5.83	3.69	28.33
Medium	375	175	26.20	4.47	1.56	14.97	5.15	2.88	29.03
Fast	400	150	22.40	5.63	2.34	14.93	4.24	1.63	28.77

Metal primer, rust inhibitor, rusty metal (material #36)

Slow	350	200	36.60	3.00	.86	18.30	6.87	4.35	33.38
Medium	375	175	32.10	4.47	1.56	18.34	5.97	3.34	33.68
Fast	400	150	27.50	5.63	2.34	18.33	4.87	1.87	33.04

Industrial enamel, oil base, high gloss, light colors (material #56)

Slow	375	275	29.60	2.80	.80	10.76	4.45	2.82	21.63
Medium	400	250	25.90	4.19	1.46	10.36	3.92	2.19	22.12
Fast	425	225	22.20	5.29	2.20	9.87	3.21	1.23	21.80

Industrial enamel, oil base, high gloss, dark (OSHA) colors (material #57)

Slow	375	275	37.00	2.80	.80	13.45	5.29	3.35	25.69
Medium	400	250	32.40	4.19	1.46	12.96	4.56	2.55	25.72
Fast	425	225	27.70	5.29	2.20	12.31	3.66	1.41	24.87

Use these figures to estimate the cost of painting mechanical equipment (such as compressors and mixing boxes). Measurements are based on square feet of surface area covered. For heights above 8 feet, use the High Time Difficulty Factors on page 137. Note: A two coat system, prime and finish, using oil base material is recommended for any metal surface. Although water base material is often used, it may cause oxidation, corrosion and rust. Using one coat of oil base paint on exterior metal may result in cracking, peeling or chipping without the proper prime coat application. "Slow" work is based on a $10.50 hourly wage, "Medium" work on a $16.75 hourly wage, "Fast" work on a $22.50 hourly wage. Other qualifications

Mechanical equipment, boiler room

Don't bother figuring the exact area of boiler room equipment that has to be coated. Instead, take the area as equal to 1/2 the wall height times the ceiling area in the room. Figure a painter will coat 125 square feet per hour and a gallon of paint will cover 300 square feet. This rate does not include time needed to paint walls, ceiling or floor around mechanical equipment. In any case, you'll need to rely on judgment when painting boiler room equipment.

	Labor SF per manhour	Material coverage SF/gallon	Material cost per gallon	Labor cost per 100 SF	Labor burden 100 SF	Material cost per 100 SF	Overhead per 100 SF	Profit per 100 SF	Total price per 100 SF
Piping, bare pipe, brush application									
Metal primer, rust inhibitor, clean metal (material #35)									
Brush prime coat									
Slow	75	360	29.90	14.00	3.97	8.31	8.15	5.17	39.60
Medium	100	335	26.20	16.75	5.84	7.82	7.45	4.16	42.02
Fast	125	310	22.40	18.00	7.51	7.23	6.06	2.33	41.13
Metal primer, rust inhibitor, rusty metal (material #36)									
Brush prime coat									
Slow	75	360	36.60	14.00	3.97	10.17	8.73	5.53	42.40
Medium	100	335	32.10	16.75	5.84	9.58	7.88	4.41	44.46
Fast	125	310	27.50	18.00	7.51	8.87	6.36	2.45	43.19
Industrial enamel, oil base, high gloss, light colors (material #56)									
Brush 1st finish coat									
Slow	90	450	29.60	11.67	3.31	6.58	6.69	4.24	32.49
Medium	115	425	25.90	14.57	5.08	6.09	6.31	3.53	35.58
Fast	140	400	22.20	16.07	6.70	5.55	5.24	2.01	35.57
Brush 2nd or additional finish coats									
Slow	125	500	29.60	8.40	2.38	5.92	5.18	3.28	25.16
Medium	150	475	25.90	11.17	3.90	5.45	5.02	2.81	28.35
Fast	175	450	22.20	12.86	5.35	4.93	4.28	1.65	29.07
Industrial enamel, oil base, high gloss, dark (OSHA) colors (material #57)									
Brush 1st finish coat									
Slow	90	450	37.00	11.67	3.31	8.22	7.20	4.56	34.96
Medium	115	425	32.40	14.57	5.08	7.62	6.68	3.73	37.68
Fast	140	400	27.70	16.07	6.70	6.93	5.50	2.11	37.31
Brush 2nd or additional finish coats									
Slow	125	500	37.00	8.40	2.38	7.40	5.64	3.57	27.39
Medium	150	475	32.40	11.17	3.90	6.82	5.36	3.00	30.25
Fast	175	450	27.70	12.86	5.35	6.16	4.51	1.73	30.61

	Labor SF per manhour	Material coverage SF/gallon	Material cost per gallon	Labor cost per 100 SF	Labor burden 100 SF	Material cost per 100 SF	Overhead per 100 SF	Profit per 100 SF	Total price per 100 SF
Epoxy coating, 2 part system, clear (material #51)									
Brush 1st coat									
Slow	75	375	58.30	14.00	3.97	15.55	10.39	6.59	50.50
Medium	100	350	51.00	16.75	5.84	14.57	9.10	5.09	51.35
Fast	125	325	43.70	18.00	7.51	13.45	7.21	2.77	48.94
Brush 2nd coat									
Slow	90	425	58.30	11.67	3.31	13.72	8.90	5.64	43.24
Medium	115	400	51.00	14.57	5.08	12.75	7.94	4.44	44.78
Fast	140	375	43.70	16.07	6.70	11.65	6.37	2.45	43.24
Brush 3rd or additional coats									
Slow	125	475	58.30	8.40	2.38	12.27	7.15	4.53	34.73
Medium	150	450	51.00	11.17	3.90	11.33	6.47	3.61	36.48
Fast	175	425	43.70	12.86	5.35	10.28	5.27	2.03	35.79
Epoxy coating, 2 part system, white (material #52)									
Brush 1st coat									
Slow	75	375	61.70	14.00	3.97	16.45	10.67	6.77	51.86
Medium	100	350	54.00	16.75	5.84	15.43	9.31	5.21	52.54
Fast	125	325	46.30	18.00	7.51	14.25	7.36	2.83	49.95
Brush 2nd coat									
Slow	90	425	61.70	11.67	3.31	14.52	9.15	5.80	44.45
Medium	115	400	54.00	14.57	5.08	13.50	8.12	4.54	45.81
Fast	140	375	46.30	16.07	6.70	12.35	6.50	2.50	44.12
Brush 3rd or additional coats									
Slow	125	475	61.70	8.40	2.38	12.99	7.37	4.67	35.81
Medium	150	450	54.00	11.17	3.90	12.00	6.63	3.71	37.41
Fast	175	425	46.30	12.86	5.35	10.89	5.39	2.07	36.56
Aluminum base paint (material #50)									
Brush each coat									
Slow	50	600	40.10	21.00	5.96	6.68	10.43	6.61	50.68
Medium	75	575	35.10	22.33	7.78	6.10	8.87	4.96	50.04
Fast	100	550	30.10	22.50	9.39	5.47	6.91	2.66	46.93

	Labor SF per manhour	Material coverage SF/gallon	Material cost per gallon	Labor cost per 100 SF	Labor burden 100 SF	Material cost per 100 SF	Overhead per 100 SF	Profit per 100 SF	Total price per 100 SF
Heat resistant enamel, 800 to 1200 degree range (material #53)									
Brush each coat									
Slow	50	600	71.30	21.00	5.96	11.88	12.04	7.63	58.51
Medium	75	575	62.40	22.33	7.78	10.85	10.04	5.61	56.61
Fast	100	550	53.50	22.50	9.39	9.73	7.70	2.96	52.28
Heat resistant enamel, 300 to 800 degree range (material #54)									
Brush each coat									
Slow	50	600	54.10	21.00	5.96	9.02	11.16	7.07	54.21
Medium	75	575	47.30	22.33	7.78	8.23	9.39	5.25	52.98
Fast	100	550	40.50	22.50	9.39	7.36	7.26	2.79	49.30

Use the pipe conversion factors in Figure 21 on page 321 to convert linear feet of pipe to square feet of surface. Vertical pipe runs require 2 to 3 times the manhours plus 10% more material. Solid color coded piping requires 15% to 25% more labor and material. For color bands on piping at 10' to 15' intervals, add the cost of an additional 1st coat. For heights above 8 feet, use the High Time Difficulty Factors on page 137. Note: A two coat system, prime and finish, using oil base material is recommended for any metal surface. Although water base material is often used, it may cause oxidation, corrosion and rust. One coat of oil base solid body stain is often used on interior or exterior metal but it may crack, peel or chip without the proper prime coat application. "Slow" work is based on a $10.50 hourly wage, "Medium" work on a $16.75 hourly wage, "Fast" work on a $22.50 hourly wage. Other qualifications that apply to this table are on page 9.

	Labor SF per manhour	Material coverage SF/gallon	Material cost per gallon	Labor cost per 100 SF	Labor burden 100 SF	Material cost per 100 SF	Overhead per 100 SF	Profit per 100 SF	Total price per 100 SF
Piping, bare pipe, roll application									
Metal primer, rust inhibitor, clean metal (material #35)									
Roll prime coat									
Slow	175	400	29.90	6.00	1.70	7.48	4.71	2.99	22.88
Medium	200	375	26.20	8.38	2.92	6.99	4.48	2.50	25.27
Fast	225	350	22.40	10.00	4.16	6.40	3.81	1.46	25.83
Metal primer, rust inhibitor, rusty metal (material #36)									
Roll prime coat									
Slow	175	400	36.60	6.00	1.70	9.15	5.23	3.31	25.39
Medium	200	375	32.10	8.38	2.92	8.56	4.87	2.72	27.45
Fast	225	350	27.50	10.00	4.16	7.86	4.08	1.57	27.67
Industrial enamel, oil base, high gloss, light colors (material #56)									
Roll 1st finish coat									
Slow	200	450	29.60	5.25	1.49	6.58	4.13	2.62	20.07
Medium	225	425	25.90	7.44	2.59	6.09	3.95	2.21	22.28
Fast	250	400	22.20	9.00	3.76	5.55	3.39	1.30	23.00
Roll 2nd or additional finish coats									
Slow	275	500	29.60	3.82	1.09	5.92	3.36	2.13	16.32
Medium	300	475	25.90	5.58	1.94	5.45	3.18	1.78	17.93
Fast	325	450	22.20	6.92	2.90	4.93	2.73	1.05	18.53
Industrial enamel, oil base, high gloss, dark (OSHA) colors (material #57)									
Roll 1st finish coat									
Slow	200	450	37.00	5.25	1.49	8.22	4.64	2.94	22.54
Medium	225	425	32.40	7.44	2.59	7.62	4.32	2.42	24.39
Fast	250	400	27.70	9.00	3.76	6.93	3.64	1.40	24.73
Roll 2nd or additional finish coats									
Slow	275	500	37.00	3.82	1.09	7.40	3.82	2.42	18.55
Medium	300	475	32.40	5.58	1.94	6.82	3.51	1.96	19.81
Fast	325	450	27.70	6.92	2.90	6.16	2.95	1.14	20.07

Use the pipe conversion factors in Figure 21 on page 321 to convert linear feet of pipe to square feet of surface. Vertical pipe runs require 2 to 3 times the manhours plus 10% more material. Solid color coded piping requires 15% to 25% more labor and material. For color bands on piping at 10' to 15' intervals, add the cost of an additional 1st coat. For heights above 8 feet, use the High Time Difficulty Factors on page 137. Note: A two coat system, prime and finish, using oil base material is recommended for any metal surface. Although water base material is often used, it may cause oxidation, corrosion and rust. One coat of oil base solid body stain is often used on interior or exterior metal but it may crack, peel or chip without the proper prime coat application. "Slow" work is based on a $10.50 hourly wage, "Medium" work on a $16.75 hourly wage, "Fast" work on a $22.50 hourly wage. Other qualifications that apply to this table are on page 9.

	Labor SF per manhour	Material coverage SF/gallon	Material cost per gallon	Labor cost per 100 SF	Labor burden 100 SF	Material cost per 100 SF	Overhead per 100 SF	Profit per 100 SF	Total price per 100 SF

Piping, bare pipe, spray application

	Labor SF per manhour	Material coverage SF/gallon	Material cost per gallon	Labor cost per 100 SF	Labor burden 100 SF	Material cost per 100 SF	Overhead per 100 SF	Profit per 100 SF	Total price per 100 SF
Metal primer, rust inhibitor, clean metal (material #35)									
Spray prime coat									
Slow	300	175	29.90	3.50	.99	17.09	6.69	4.24	32.51
Medium	350	163	26.20	4.79	1.67	16.07	5.52	3.09	31.14
Fast	400	150	22.40	5.63	2.34	14.93	4.24	1.63	28.77
Metal primer, rust inhibitor, rusty metal (material #36)									
Spray prime coat									
Slow	300	175	36.60	3.50	.99	20.91	7.88	4.99	38.27
Medium	350	163	32.10	4.79	1.67	19.69	6.41	3.58	36.14
Fast	400	150	27.50	5.63	2.34	18.33	4.87	1.87	33.04
Industrial enamel, oil base, high gloss, light colors (material #56)									
Spray 1st finish coat									
Slow	375	210	29.60	2.80	.80	14.10	5.49	3.48	26.67
Medium	425	198	25.90	3.94	1.37	13.08	4.51	2.52	25.42
Fast	475	185	22.20	4.74	1.99	12.00	3.46	1.33	23.52
Spray 2nd or additional finish coats									
Slow	425	300	29.60	2.47	.70	9.87	4.04	2.56	19.64
Medium	475	288	25.90	3.53	1.24	8.99	3.37	1.88	19.01
Fast	525	275	22.20	4.29	1.77	8.07	2.62	1.01	17.76
Industrial enamel, oil base, high gloss, dark (OSHA) colors (material #57)									
Spray 1st finish coat									
Slow	375	210	37.00	2.80	.80	17.62	6.58	4.17	31.97
Medium	425	198	32.40	3.94	1.37	16.36	5.31	2.97	29.95
Fast	475	185	27.70	4.74	1.99	14.97	4.01	1.54	27.25
Spray 2nd or additional finish coats									
Slow	425	300	37.00	2.47	.70	12.33	4.81	3.05	23.36
Medium	475	288	32.40	3.53	1.24	11.25	3.92	2.19	22.13
Fast	525	275	27.70	4.29	1.77	10.07	2.99	1.15	20.27
Epoxy coating, 2 part system, clear (material #51)									
Spray 1st coat									
Slow	300	185	58.30	3.50	.99	31.51	11.16	7.08	54.24
Medium	350	173	51.00	4.79	1.67	29.48	8.81	4.92	49.67
Fast	400	160	43.70	5.63	2.34	27.31	6.53	2.51	44.32
Spray 2nd coat									
Slow	375	200	58.30	2.80	.80	29.15	10.15	6.44	49.34
Medium	425	188	51.00	3.94	1.37	27.13	7.95	4.44	44.83
Fast	475	175	43.70	4.74	1.99	24.97	5.86	2.25	39.81

	Labor SF per manhour	Material coverage SF/gallon	Material cost per gallon	Labor cost per 100 SF	Labor burden 100 SF	Material cost per 100 SF	Overhead per 100 SF	Profit per 100 SF	Total price per 100 SF
Spray 3rd or additional coats									
Slow	425	275	58.30	2.47	.70	21.20	7.55	4.79	36.71
Medium	475	263	51.00	3.53	1.24	19.39	5.92	3.31	33.39
Fast	525	250	43.70	4.29	1.77	17.48	4.36	1.68	29.58
Epoxy coating, 2 part system, white (material #52)									
Spray 1st coat									
Slow	300	185	61.70	3.50	.99	33.35	11.73	7.44	57.01
Medium	350	173	54.00	4.79	1.67	31.21	9.23	5.16	52.06
Fast	400	160	46.30	5.63	2.34	28.94	6.83	2.63	46.37
Spray 2nd coat									
Slow	375	200	61.70	2.80	.80	30.85	10.68	6.77	51.90
Medium	425	188	54.00	3.94	1.37	28.72	8.34	4.66	47.03
Fast	475	175	46.30	4.74	1.99	26.46	6.14	2.36	41.69
Spray 3rd or additional coats									
Slow	425	275	61.70	2.47	.70	22.44	7.94	5.03	38.58
Medium	475	263	54.00	3.53	1.24	20.53	6.20	3.46	34.96
Fast	525	250	46.30	4.29	1.77	18.52	4.55	1.75	30.88

Use the pipe conversion factors in Figure 21 on page 321 to convert linear feet of pipe to square feet of surface. Vertical pipe runs require 2 to 3 times the manhours plus 10% more material. Solid color coded piping requires 15% to 25% more labor and material. For color bands on piping at 10' to 15' intervals, add the cost of an additional 1st coat. For heights above 8 feet, use the High Time Difficulty Factors on page 137. Note: A two coat system, prime and finish, using oil base material is recommended for any metal surface. Although water base material is often used, it may cause oxidation, corrosion and rust. One coat of oil base solid body stain is often used on interior or exterior metal but it may crack, peel or chip without the proper prime coat application. "Slow" work is based on a $10.50 hourly wage, "Medium" work on a $16.75 hourly wage, "Fast" work on a $22.50 hourly wage. Other qualifications that apply to this table are on page 9.

	Labor SF per manhour	Material coverage SF/gallon	Material cost per gallon	Labor cost per 100 SF	Labor burden 100 SF	Material cost per 100 SF	Overhead per 100 SF	Profit per 100 SF	Total price per 100 SF

Piping, bare pipe, mitt or glove application

Metal primer, rust inhibitor, clean metal (material #35)

Mitt or glove prime coat									
Slow	175	325	29.90	6.00	1.70	9.20	5.24	3.32	25.46
Medium	200	313	26.20	8.38	2.92	8.37	4.82	2.69	27.18
Fast	225	300	22.40	10.00	4.16	7.47	4.01	1.54	27.18

Metal primer, rust inhibitor, rusty metal (material #36)

Mitt or glove prime coat									
Slow	175	325	36.60	6.00	1.70	11.26	5.88	3.73	28.57
Medium	200	313	32.10	8.38	2.92	10.26	5.28	2.95	29.79
Fast	225	300	27.50	10.00	4.16	9.17	4.32	1.66	29.31

Industrial enamel, oil base, high gloss, light colors (material #56)

Mitt or glove 1st finish coat									
Slow	200	375	29.60	5.25	1.49	7.89	4.54	2.88	22.05
Medium	225	363	25.90	7.44	2.59	7.13	4.20	2.35	23.71
Fast	250	350	22.20	9.00	3.76	6.34	3.53	1.36	23.99

Mitt or glove 2nd or additional finish coats									
Slow	275	400	29.60	3.82	1.09	7.40	3.82	2.42	18.55
Medium	300	388	25.90	5.58	1.94	6.68	3.48	1.94	19.62
Fast	325	375	22.20	6.92	2.90	5.92	2.91	1.12	19.77

Industrial enamel, oil base, high gloss, dark (OSHA) colors (material #57)

Mitt or glove 1st finish coat									
Slow	200	375	37.00	5.25	1.49	9.87	5.15	3.26	25.02
Medium	225	363	32.40	7.44	2.59	8.93	4.65	2.60	26.21
Fast	250	350	27.70	9.00	3.76	7.91	3.82	1.47	25.96

Mitt or glove 2nd or additional finish coats									
Slow	275	400	37.00	3.82	1.09	9.25	4.39	2.78	21.33
Medium	300	388	32.40	5.58	1.94	8.35	3.89	2.17	21.93
Fast	325	375	27.70	6.92	2.90	7.39	3.18	1.22	21.61

Epoxy coating, 2 part system, clear (material #51)

Mitt or glove 1st coat									
Slow	175	360	58.30	6.00	1.70	16.19	7.41	4.70	36.00
Medium	200	348	51.00	8.38	2.92	14.66	6.36	3.56	35.88
Fast	225	335	43.70	10.00	4.16	13.04	5.04	1.94	34.18

Mitt or glove 2nd coat									
Slow	200	375	58.30	5.25	1.49	15.55	6.91	4.38	33.58
Medium	225	363	51.00	7.44	2.59	14.05	5.90	3.30	33.28
Fast	250	350	43.70	9.00	3.76	12.49	4.67	1.80	31.72

	Labor SF per manhour	Material coverage SF/gallon	Material cost per gallon	Labor cost per 100 SF	Labor burden 100 SF	Material cost per 100 SF	Overhead per 100 SF	Profit per 100 SF	Total price per 100 SF
Mitt or glove 3rd or additional coats									
Slow	275	390	58.30	3.82	1.09	14.95	6.16	3.90	29.92
Medium	300	378	51.00	5.58	1.94	13.49	5.15	2.88	29.04
Fast	325	365	43.70	6.92	2.90	11.97	4.03	1.55	27.37
Epoxy coating, 2 part system, white (material #52)									
Mitt or glove 1st coat									
Slow	175	360	61.70	6.00	1.70	17.14	7.70	4.88	37.42
Medium	200	348	54.00	8.38	2.92	15.52	6.57	3.67	37.06
Fast	225	335	46.30	10.00	4.16	13.82	5.18	1.99	35.15
Mitt or glove 2nd coat									
Slow	200	375	61.70	5.25	1.49	16.45	7.19	4.56	34.94
Medium	225	363	54.00	7.44	2.59	14.88	6.10	3.41	34.42
Fast	250	350	46.30	9.00	3.76	13.23	4.81	1.85	32.65
Mitt or glove 3rd or additional coats									
Slow	275	390	61.70	3.82	1.09	15.82	6.43	4.07	31.23
Medium	300	378	54.00	5.58	1.94	14.29	5.34	2.99	30.14
Fast	325	365	46.30	6.92	2.90	12.68	4.16	1.60	28.26

Use the pipe conversion factors in Figure 21 on page 321 to convert linear feet of pipe to square feet of surface. Vertical pipe runs require 2 to 3 times the manhours plus 10% more material. Solid color coded piping requires 15% to 25% more labor and material. For color bands on piping at 10' to 15' intervals, add the cost of an additional 1st coat. For heights above 8 feet, use the High Time Difficulty Factors on page 137. Note: A two coat system, prime and finish, using oil base material is recommended for any metal surface. Although water base material is often used, it may cause oxidation, corrosion and rust. One coat of oil base solid body stain is often used on interior or exterior metal but it may crack, peel or chip without the proper prime coat application. "Slow" work is based on a $10.50 hourly wage, "Medium" work on a $16.75 hourly wage, "Fast" work on a $22.50 hourly wage. Other qualifications that apply to this table are on page 9.

	Labor SF per manhour	Material coverage SF/gallon	Material cost per gallon	Labor cost per 100 SF	Labor burden 100 SF	Material cost per 100 SF	Overhead per 100 SF	Profit per 100 SF	Total price per 100 SF
Piping, insulated, canvas jacket, brush									
Flat latex, water base (material #5)									
Brush 1st coat									
Slow	60	150	17.10	17.50	4.97	11.40	10.50	6.66	51.03
Medium	80	138	15.00	20.94	7.30	10.87	9.58	5.36	54.05
Fast	100	125	12.90	22.50	9.39	10.32	7.81	3.00	53.02
Brush 2nd coat									
Slow	75	300	17.10	14.00	3.97	5.70	7.34	4.65	35.66
Medium	100	288	15.00	16.75	5.84	5.21	6.81	3.81	38.42
Fast	125	275	12.90	18.00	7.51	4.69	5.59	2.15	37.94
Brush 3rd or additional coats									
Slow	100	400	17.10	10.50	2.98	4.28	5.51	3.49	26.76
Medium	138	375	15.00	12.14	4.24	4.00	4.99	2.79	28.16
Fast	175	350	12.90	12.86	5.35	3.69	4.06	1.56	27.52
Sealer, off white, water base (material #1)									
Brush 1 coat									
Slow	60	150	16.30	17.50	4.97	10.87	10.34	6.55	50.23
Medium	80	138	14.30	20.94	7.30	10.36	9.46	5.29	53.35
Fast	100	125	12.30	22.50	9.39	9.84	7.72	2.97	52.42
Sealer, off white, oil base (material #2)									
Brush 1 coat									
Slow	60	200	22.90	17.50	4.97	11.45	10.52	6.67	51.11
Medium	80	188	20.10	20.94	7.30	10.69	9.54	5.33	53.80
Fast	100	175	17.20	22.50	9.39	9.83	7.72	2.97	52.41
Enamel, water base latex (material #9)									
Brush 1st finish coat									
Slow	75	300	24.50	14.00	3.97	8.17	8.11	5.14	39.39
Medium	100	288	21.40	16.75	5.84	7.43	7.35	4.11	41.48
Fast	125	275	18.40	18.00	7.51	6.69	5.96	2.29	40.45
Brush 2nd or additional finish coats									
Slow	100	400	24.50	10.50	2.98	6.13	6.08	3.86	29.55
Medium	138	375	21.40	12.14	4.24	5.71	5.41	3.02	30.52
Fast	175	350	18.40	12.86	5.35	5.26	4.35	1.67	29.49
Industrial enamel, oil base, high gloss, light colors (material #56)									
Brush 1st finish coat									
Slow	75	335	29.60	14.00	3.97	8.84	8.31	5.27	40.39
Medium	100	323	25.90	16.75	5.84	8.02	7.50	4.19	42.30
Fast	125	310	22.20	18.00	7.51	7.16	6.05	2.32	41.04

	Labor SF per manhour	Material coverage SF/gallon	Material cost per gallon	Labor cost per 100 SF	Labor burden 100 SF	Material cost per 100 SF	Overhead per 100 SF	Profit per 100 SF	Total price per 100 SF
Brush 2nd or additional finish coats									
Slow	100	450	29.60	10.50	2.98	6.58	6.22	3.94	30.22
Medium	150	438	25.90	11.17	3.90	5.91	5.14	2.87	28.99
Fast	175	425	22.20	12.86	5.35	5.22	4.34	1.67	29.44
Industrial enamel, oil base, high gloss, dark (OSHA) colors (material #57)									
Brush 1st finish coat									
Slow	75	335	37.00	14.00	3.97	11.04	9.00	5.70	43.71
Medium	100	323	32.40	16.75	5.84	10.03	7.99	4.47	45.08
Fast	125	310	27.70	18.00	7.51	8.94	6.38	2.45	43.28
Brush 2nd or additional finish coats									
Slow	100	450	37.00	10.50	2.98	8.22	6.73	4.27	32.70
Medium	150	438	32.40	11.17	3.90	7.40	5.50	3.08	31.05
Fast	175	425	27.70	12.86	5.35	6.52	4.58	1.76	31.07
Epoxy coating, 2 part system, clear (material #51)									
Brush 1st coat									
Slow	75	325	58.30	14.00	3.97	17.94	11.14	7.06	54.11
Medium	100	313	51.00	16.75	5.84	16.29	9.53	5.33	53.74
Fast	125	300	43.70	18.00	7.51	14.57	7.42	2.85	50.35
Brush 2nd or additional coats									
Slow	100	450	58.30	10.50	2.98	12.96	8.20	5.20	39.84
Medium	150	438	51.00	11.17	3.90	11.64	6.54	3.66	36.91
Fast	175	425	43.70	12.86	5.35	10.28	5.27	2.03	35.79
Epoxy coating, 2 part system, white (material #52)									
Brush 1st coat									
Slow	75	325	61.70	14.00	3.97	18.98	11.46	7.26	55.67
Medium	100	313	54.00	16.75	5.84	17.25	9.76	5.46	55.06
Fast	125	300	46.30	18.00	7.51	15.43	7.58	2.91	51.43
Brush 2nd or additional coats									
Slow	100	450	61.70	10.50	2.98	13.71	8.43	5.34	40.96
Medium	150	438	54.00	11.17	3.90	12.33	6.71	3.75	37.86
Fast	175	425	46.30	12.86	5.35	10.89	5.39	2.07	36.56

Use the pipe conversion factors in Figure 21 on page 321 to convert linear feet of pipe to square feet of surface. Vertical pipe runs require 2 to 3 times the manhours plus 10% more material. Solid color coded piping requires 15% to 25% more labor and material. For color bands on piping at 10' to 15' intervals, add the cost of an additional 1st coat. For heights above 8 feet, use the High Time Difficulty Factors on page 137. "Slow" work is based on a $10.50 hourly wage, "Medium" work on a $16.75 hourly wage, "Fast" work on a $22.50 hourly wage. Other qualifications that apply to this table are on page 9.

	Labor SF per manhour	Material coverage SF/gallon	Material cost per gallon	Labor cost per 100 SF	Labor burden 100 SF	Material cost per 100 SF	Overhead per 100 SF	Profit per 100 SF	Total price per 100 SF
Piping, insulated, canvas jacket, roll									
Flat latex, water base (material #5)									
Roll 1st coat									
Slow	135	150	17.10	7.78	2.21	11.40	6.63	4.20	32.22
Medium	160	138	15.00	10.47	3.65	10.87	6.12	3.42	34.53
Fast	185	125	12.90	12.16	5.09	10.32	5.10	1.96	34.63
Roll 2nd coat									
Slow	175	300	17.10	6.00	1.70	5.70	4.16	2.64	20.20
Medium	200	288	15.00	8.38	2.92	5.21	4.04	2.26	22.81
Fast	225	275	12.90	10.00	4.16	4.69	3.49	1.34	23.68
Roll 3rd or additional coats									
Slow	275	400	17.10	3.82	1.09	4.28	2.85	1.81	13.85
Medium	300	388	15.00	5.58	1.94	3.87	2.79	1.56	15.74
Fast	325	375	12.90	6.92	2.90	3.44	2.45	.94	16.65
Sealer, off white, water base (material #1)									
Roll 1 coat									
Slow	135	150	16.30	7.78	2.21	10.87	6.47	4.10	31.43
Medium	160	138	14.30	10.47	3.65	10.36	6.00	3.35	33.83
Fast	185	125	12.30	12.16	5.09	9.84	5.01	1.93	34.03
Sealer, off white, oil base (material #2)									
Roll 1 coat									
Slow	135	225	22.90	7.78	2.21	10.18	6.25	3.96	30.38
Medium	160	213	20.10	10.47	3.65	9.44	5.77	3.23	32.56
Fast	185	200	17.20	12.16	5.09	8.60	4.78	1.84	32.47
Enamel, water base latex (material #9)									
Roll 1st finish coat									
Slow	175	300	24.50	6.00	1.70	8.17	4.92	3.12	23.91
Medium	200	288	21.40	8.38	2.92	7.43	4.59	2.57	25.89
Fast	225	275	18.40	10.00	4.16	6.69	3.86	1.48	26.19
Roll 2nd or additional finish coats									
Slow	275	400	24.50	3.82	1.09	6.13	3.42	2.17	16.63
Medium	300	388	21.40	5.58	1.94	5.52	3.19	1.79	18.02
Fast	325	375	18.40	6.92	2.90	4.91	2.72	1.05	18.50
Industrial enamel, oil base, high gloss, light colors (material #56)									
Roll 1st finish coat									
Slow	175	335	29.60	6.00	1.70	8.84	5.13	3.25	24.92
Medium	200	323	25.90	8.38	2.92	8.02	4.73	2.65	26.70
Fast	225	310	22.20	10.00	4.16	7.16	3.95	1.52	26.79

	Labor SF per manhour	Material coverage SF/gallon	Material cost per gallon	Labor cost per 100 SF	Labor burden 100 SF	Material cost per 100 SF	Overhead per 100 SF	Profit per 100 SF	Total price per 100 SF
Roll 2nd or additional finish coats									
Slow	275	450	29.60	3.82	1.09	6.58	3.56	2.26	17.31
Medium	300	438	25.90	5.58	1.94	5.91	3.29	1.84	18.56
Fast	325	425	22.20	6.92	2.90	5.22	2.78	1.07	18.89
Industrial enamel, oil base, high gloss, dark (OSHA) colors (material #57)									
Roll 1st finish coat									
Slow	175	335	37.00	6.00	1.70	11.04	5.81	3.68	28.23
Medium	200	323	32.40	8.38	2.92	10.03	5.23	2.92	29.48
Fast	225	310	27.70	10.00	4.16	8.94	4.28	1.64	29.02
Roll 2nd or additional finish coats									
Slow	275	450	37.00	3.82	1.09	8.22	4.07	2.58	19.78
Medium	300	438	32.40	5.58	1.94	7.40	3.66	2.04	20.62
Fast	325	425	27.70	6.92	2.90	6.52	3.02	1.16	20.52
Epoxy coating, 2 part system, clear (material #51)									
Roll 1st coat									
Slow	175	325	58.30	6.00	1.70	17.94	7.95	5.04	38.63
Medium	200	313	51.00	8.38	2.92	16.29	6.76	3.78	38.13
Fast	225	300	43.70	10.00	4.16	14.57	5.32	2.04	36.09
Roll 2nd or additional coats									
Slow	275	425	58.30	3.82	1.09	13.72	5.78	3.66	28.07
Medium	300	413	51.00	5.58	1.94	12.35	4.87	2.72	27.46
Fast	325	400	43.70	6.92	2.90	10.93	3.84	1.47	26.06
Epoxy coating, 2 part system, white (material #52)									
Roll 1st coat									
Slow	175	325	61.70	6.00	1.70	18.98	8.27	5.24	40.19
Medium	200	313	54.00	8.38	2.92	17.25	6.99	3.91	39.45
Fast	225	300	46.30	10.00	4.16	15.43	5.48	2.11	37.18
Roll 2nd or additional coats									
Slow	275	425	61.70	3.82	1.09	14.52	6.02	3.82	29.27
Medium	300	413	54.00	5.58	1.94	13.08	5.05	2.82	28.47
Fast	325	400	46.30	6.92	2.90	11.58	3.96	1.52	26.88

Use the pipe conversion factors in Figure 21 on page 321 to convert linear feet of pipe to square feet of surface. Vertical pipe runs require 2 to 3 times the manhours plus 10% more material. Solid color coded piping requires 15% to 25% more labor and material. For color bands on piping at 10' to 15' intervals, add the cost of an additional 1st coat. For heights above 8 feet, use the High Time Difficulty Factors on page 137. "Slow" work is based on a $10.50 hourly wage, "Medium" work on a $16.75 hourly wage, "Fast" work on a $22.50 hourly wage. Other qualifications that apply to this table are on page 9.

	Labor SF per manhour	Material coverage SF/gallon	Material cost per gallon	Labor cost per 100 SF	Labor burden 100 SF	Material cost per 100 SF	Overhead per 100 SF	Profit per 100 SF	Total price per 100 SF
Piping, insulated, canvas jacket, spray									
Flat latex, water base (material #5)									
Spray 1st coat									
Slow	225	100	17.10	4.67	1.32	17.10	7.16	4.54	34.79
Medium	250	88	15.00	6.70	2.34	17.05	6.39	3.57	36.05
Fast	275	75	12.90	8.18	3.43	17.20	5.33	2.05	36.19
Spray 2nd coat									
Slow	275	200	17.10	3.82	1.09	8.55	4.17	2.64	20.27
Medium	300	188	15.00	5.58	1.94	7.98	3.80	2.12	21.42
Fast	325	175	12.90	6.92	2.90	7.37	3.18	1.22	21.59
Spray 3rd or additional coats									
Slow	375	300	17.10	2.80	.80	5.70	2.88	1.83	14.01
Medium	400	288	15.00	4.19	1.46	5.21	2.66	1.49	15.01
Fast	425	275	12.90	5.29	2.20	4.69	2.26	.87	15.31
Sealer, off white, water base (material #1)									
Spray 1 coat									
Slow	225	100	16.30	4.67	1.32	16.30	6.91	4.38	33.58
Medium	250	88	14.30	6.70	2.34	16.25	6.19	3.46	34.94
Fast	275	75	12.30	8.18	3.43	16.40	5.18	1.99	35.18
Sealer, off white, oil base (material #2)									
Spray 1 coat									
Slow	225	150	22.90	4.67	1.32	15.27	6.59	4.18	32.03
Medium	250	138	20.10	6.70	2.34	14.57	5.78	3.23	32.62
Fast	275	125	17.20	8.18	3.43	13.76	4.69	1.80	31.86
Enamel, water base latex (material #9)									
Spray 1st finish coat									
Slow	275	200	24.50	3.82	1.09	12.25	5.32	3.37	25.85
Medium	300	188	21.40	5.58	1.94	11.38	4.63	2.59	26.12
Fast	325	175	18.40	6.92	2.90	10.51	3.76	1.44	25.53
Spray 2nd or additional finish coats									
Slow	375	300	24.50	2.80	.80	8.17	3.65	2.31	17.73
Medium	400	288	21.40	4.19	1.46	7.43	3.20	1.79	18.07
Fast	425	275	18.40	5.29	2.20	6.69	2.63	1.01	17.82
Industrial enamel, oil base, high gloss, light colors (material #56)									
Spray 1st finish coat									
Slow	275	235	29.60	3.82	1.09	12.60	5.43	3.44	26.38
Medium	300	223	25.90	5.58	1.94	11.61	4.69	2.62	26.44
Fast	325	210	22.20	6.92	2.90	10.57	3.77	1.45	25.61

	Labor SF per manhour	Material coverage SF/gallon	Material cost per gallon	Labor cost per 100 SF	Labor burden 100 SF	Material cost per 100 SF	Overhead per 100 SF	Profit per 100 SF	Total price per 100 SF
Spray 2nd or additional finish coats									
Slow	375	335	29.60	2.80	.80	8.84	3.86	2.45	18.75
Medium	400	323	25.90	4.19	1.46	8.02	3.35	1.87	18.89
Fast	425	310	22.20	5.29	2.20	7.16	2.71	1.04	18.40
Industrial enamel, oil base, high gloss, dark (OSHA) colors (material #57)									
Spray 1st finish coat									
Slow	275	235	37.00	3.82	1.09	15.74	6.40	4.06	31.11
Medium	300	223	32.40	5.58	1.94	14.53	5.40	3.02	30.47
Fast	325	210	27.70	6.92	2.90	13.19	4.26	1.64	28.91
Spray 2nd or additional finish coats									
Slow	375	335	37.00	2.80	.80	11.04	4.54	2.88	22.06
Medium	400	323	32.40	4.19	1.46	10.03	3.84	2.15	21.67
Fast	425	310	27.70	5.29	2.20	8.94	3.04	1.17	20.64
Epoxy coating, 2 part system, clear (material #51)									
Spray 1st coat									
Slow	275	225	58.30	3.82	1.09	25.91	9.55	6.06	46.43
Medium	300	213	51.00	5.58	1.94	23.94	7.71	4.31	43.48
Fast	325	200	43.70	6.92	2.90	21.85	5.86	2.25	39.78
Spray 2nd or additional coats									
Slow	375	325	58.30	2.80	.80	17.94	6.68	4.23	32.45
Medium	400	313	51.00	4.19	1.46	16.29	5.38	3.01	30.33
Fast	425	300	43.70	5.29	2.20	14.57	4.08	1.57	27.71
Epoxy coating, 2 part system, white (material #52)									
Spray 1st coat									
Slow	275	225	61.70	3.82	1.09	27.42	10.02	6.35	48.70
Medium	300	213	54.00	5.58	1.94	25.35	8.05	4.50	45.42
Fast	325	200	46.30	6.92	2.90	23.15	6.10	2.34	41.41
Spray 2nd or additional coats									
Slow	375	325	61.70	2.80	.80	18.98	7.00	4.44	34.02
Medium	400	313	54.00	4.19	1.46	17.25	5.61	3.14	31.65
Fast	425	300	46.30	5.29	2.20	15.43	4.24	1.63	28.79

Use the pipe conversion factors in Figure 21 on page 321 to convert linear feet of pipe to square feet of surface. Vertical pipe runs require 2 to 3 times the manhours plus 10% more material. Solid color coded piping requires 15% to 25% more labor and material. For color bands on piping at 10' to 15' intervals, add the cost of an additional 1st coat. For heights above 8 feet, use the High Time Difficulty Factors on page 137. "Slow" work is based on a $10.50 hourly wage, "Medium" work on a $16.75 hourly wage, "Fast" work on a $22.50 hourly wage. Other qualifications that apply to this table are on page 9.

	Labor SF per manhour	Material coverage SF/gallon	Material cost per gallon	Labor cost per 100 SF	Labor burden 100 SF	Material cost per 100 SF	Overhead per 100 SF	Profit per 100 SF	Total price per 100 SF
Radiators									
Brush each coat									
Metal primer, rust inhibitor, clean metal (material #35)									
Slow	50	100	29.90	21.00	5.96	29.90	17.63	11.18	85.67
Medium	70	95	26.20	23.93	8.35	27.58	14.66	8.20	82.72
Fast	90	90	22.40	25.00	10.43	24.89	11.16	4.29	75.77
Metal primer, rust inhibitor, rusty metal (material #36)									
Slow	50	100	36.60	21.00	5.96	36.60	19.71	12.49	95.76
Medium	70	95	32.10	23.93	8.35	33.79	16.18	9.05	91.30
Fast	90	90	27.50	25.00	10.43	30.56	12.21	4.69	82.89
Industrial enamel, oil base, high gloss, light colors (material #56)									
Slow	60	150	29.60	17.50	4.97	19.73	13.09	8.30	63.59
Medium	80	138	25.90	20.94	7.30	18.77	11.52	6.44	64.97
Fast	100	125	22.20	22.50	9.39	17.76	9.19	3.53	62.37
Industrial enamel, oil base, high gloss, dark (OSHA) colors (material #57)									
Slow	60	150	37.00	17.50	4.97	24.67	14.62	9.27	71.03
Medium	80	138	32.40	20.94	7.30	23.48	12.67	7.08	71.47
Fast	100	125	27.70	22.50	9.39	22.16	10.00	3.84	67.89
Spray each coat									
Metal primer, rust inhibitor, clean metal (material #35)									
Slow	225	90	29.90	4.67	1.32	33.22	12.16	7.71	59.08
Medium	250	83	26.20	6.70	2.34	31.57	9.95	5.56	56.12
Fast	275	75	22.40	8.18	3.43	29.87	7.67	2.95	52.10
Metal primer, rust inhibitor, rusty metal (material #36)									
Slow	225	90	36.60	4.67	1.32	40.67	14.47	9.17	70.30
Medium	250	83	32.10	6.70	2.34	38.67	11.69	6.53	65.93
Fast	275	75	27.50	8.18	3.43	36.67	8.93	3.43	60.64
Industrial enamel, oil base, high gloss, light colors (material #56)									
Slow	250	110	29.60	4.20	1.19	26.91	10.01	6.35	48.66
Medium	275	100	25.90	6.09	2.13	25.90	8.36	4.67	47.15
Fast	300	90	22.20	7.50	3.12	24.67	6.53	2.51	44.33
Industrial enamel, oil base, high gloss, dark (OSHA) colors (material #57)									
Slow	250	110	37.00	4.20	1.19	33.64	12.10	7.67	58.80
Medium	275	100	32.40	6.09	2.13	32.40	9.95	5.56	56.13
Fast	300	90	27.70	7.50	3.12	30.78	7.66	2.94	52.00

Use these figures to estimate the cost of painting both sides of 6" to 18" deep hot water or steam radiators with oil or water base paint. Measurements are per square foot of area measured, one side (length times width). For heights above 8 feet, use the High Time Difficulty Factors on page 137. Note: A two coat system, prime and finish, using oil base material is recommended for any metal surface. Although water base material is often used, it may cause oxidation, corrosion and rust. One coat of oil base solid body stain is often used on interior or exterior metal but it may crack, peel or chip without the proper prime coat application. "Slow" work is based on a $10.50 hourly wage, "Medium" work on a $16.75 hourly wage, "Fast" work on a $22.50 hourly wage. Other qualifications that apply to this table are on page 9.

Structural steel

Fabrication and erection estimates are usually based on weight of the steel in tons. As a paint estimator you need to convert tons of steel to square feet of surface. Of course, the conversion factor depends on the size of the steel members. On larger jobs and where accuracy is essential, use the Structural Steel conversion table in Figure 23 on pages 387 through 395 at the end of this Structural steel section to make exact conversions. On smaller jobs, a rule of thumb is that there are 225 square feet of paintable surface per ton of steel.

	Labor SF per manhour	Material coverage SF/gallon	Material cost per gallon	Labor cost per 100 SF	Labor burden 100 SF	Material cost per 100 SF	Overhead per 100 SF	Profit per 100 SF	Total price per 100 SF
Structural steel, heavy, brush application									
Metal primer, rust inhibitor - clean metal (material #35)									
Brush prime coat									
Slow	130	400	29.90	8.08	2.29	7.48	5.54	3.51	26.90
Medium	150	388	26.20	11.17	3.90	6.75	5.34	2.99	30.15
Fast	170	375	22.40	13.24	5.51	5.97	4.58	1.76	31.06
Metal primer, rust inhibitor - rusty metal (material #36)									
Brush prime coat									
Slow	130	400	36.60	8.08	2.29	9.15	6.05	3.84	29.41
Medium	150	388	32.10	11.17	3.90	8.27	5.72	3.20	32.26
Fast	170	375	27.50	13.24	5.51	7.33	4.83	1.86	32.77
Industrial enamel, oil base, high gloss - light colors (material #56)									
Brush 1st or additional finish coats									
Slow	175	425	29.60	6.00	1.70	6.96	4.55	2.88	22.09
Medium	200	413	25.90	8.38	2.92	6.27	4.30	2.41	24.28
Fast	225	400	22.20	10.00	4.16	5.55	3.65	1.40	24.76
Industrial enamel, oil base, high gloss - dark (OSHA) colors (material #57)									
Brush 1st or additional finish coats									
Slow	175	475	37.00	6.00	1.70	7.79	4.81	3.05	23.35
Medium	200	463	32.40	8.38	2.92	7.00	4.48	2.51	25.29
Fast	225	450	27.70	10.00	4.16	6.16	3.76	1.45	25.53
Epoxy coating, 2 part system, clear (material #51)									
Brush 1st coat									
Slow	130	425	58.30	8.08	2.29	13.72	7.47	4.74	36.30
Medium	150	413	51.00	11.17	3.90	12.35	6.72	3.75	37.89
Fast	170	400	43.70	13.24	5.51	10.93	5.49	2.11	37.28
Brush 2nd or additional coats									
Slow	175	450	58.30	6.00	1.70	12.96	6.41	4.06	31.13
Medium	200	438	51.00	8.38	2.92	11.64	5.62	3.14	31.70
Fast	225	425	43.70	10.00	4.16	10.28	4.53	1.74	30.71

	Labor SF per manhour	Material coverage SF/gallon	Material cost per gallon	Labor cost per 100 SF	Labor burden 100 SF	Material cost per 100 SF	Overhead per 100 SF	Profit per 100 SF	Total price per 100 SF
Epoxy coating, 2 part system, white (material #52)									
Brush 1st coat									
Slow	130	425	61.70	8.08	2.29	14.52	7.72	4.89	37.50
Medium	150	413	54.00	11.17	3.90	13.08	6.89	3.85	38.89
Fast	170	400	46.30	13.24	5.51	11.58	5.61	2.16	38.10
Brush 2nd or additional coats									
Slow	175	450	61.70	6.00	1.70	13.71	6.64	4.21	32.26
Medium	200	438	54.00	8.38	2.92	12.33	5.79	3.24	32.66
Fast	225	425	46.30	10.00	4.16	10.89	4.64	1.78	31.47

For field painting at heights above 8 feet, use the High Time Difficulty Factors on page 137. Heavy structural steel has from 100 to 150 square feet of surface area per ton. Extra heavy structural steel has from 50 to 100 square feet of surface area per ton. Rule of thumb: When coatings are applied by brush, a journeyman painter will apply a first coat on 6 to 7 tons per 8 hour day. When coatings are applied by spray, figure output at 0.2 hours per ton and material use at about 0.2 gallons per ton. Use Figure 23 on pages 387 to 395 to convert structural steel linear feet or tonnage to surface area. Note: A two coat system, prime and finish, using oil base material is recommended for any metal surface. Although water base material is often used, it may cause oxidation, corrosion and rust. One coat of oil base solid body stain is often used on interior or exterior metal but it may crack, peel or chip without the proper prime coat application. "Slow" work is based on a $10.50 hourly wage, "Medium" work on a $16.75 hourly wage, and "Fast" work on a $22.50 hourly wage. Other qualifications that apply to this table are on page 9.

	Labor SF per manhour	Material coverage SF/gallon	Material cost per gallon	Labor cost per 100 SF	Labor burden 100 SF	Material cost per 100 SF	Overhead per 100 SF	Profit per 100 SF	Total price per 100 SF
Structural steel, heavy, roll application									
Metal primer, rust inhibitor - clean metal (material #35)									
Roll prime coat									
Slow	250	390	29.90	4.20	1.19	7.67	4.05	2.57	19.68
Medium	275	378	26.20	6.09	2.13	6.93	3.71	2.07	20.93
Fast	300	365	22.40	7.50	3.12	6.14	3.10	1.19	21.05
Metal primer, rust inhibitor - rusty metal (material #36)									
Roll prime coat									
Slow	250	380	36.60	4.20	1.19	9.63	4.66	2.95	22.63
Medium	275	363	32.10	6.09	2.13	8.84	4.18	2.34	23.58
Fast	300	355	27.50	7.50	3.12	7.75	3.40	1.31	23.08
Industrial enamel, oil base, high gloss - light colors (material #56)									
Roll 1st or additional finish coats									
Slow	275	425	29.60	3.82	1.09	6.96	3.68	2.33	17.88
Medium	300	413	25.90	5.58	1.94	6.27	3.38	1.89	19.06
Fast	325	400	22.20	6.92	2.90	5.55	2.84	1.09	19.30
Industrial enamel, oil base, high gloss - dark (OSHA) colors (material #57)									
Roll 1st or additional finish coats									
Slow	275	440	37.00	3.82	1.09	8.41	4.13	2.62	20.07
Medium	300	428	32.40	5.58	1.94	7.57	3.70	2.07	20.86
Fast	325	415	27.70	6.92	2.90	6.67	3.05	1.17	20.71
Epoxy coating, 2 part system, clear (material #51)									
Roll 1st coat									
Slow	250	400	58.30	4.20	1.19	14.58	6.19	3.92	30.08
Medium	275	388	51.00	6.09	2.13	13.14	5.23	2.92	29.51
Fast	300	375	43.70	7.50	3.12	11.65	4.12	1.58	27.97
Roll 2nd or additional coats									
Slow	275	425	58.30	3.82	1.09	13.72	5.78	3.66	28.07
Medium	300	413	51.00	5.58	1.94	12.35	4.87	2.72	27.46
Fast	325	400	43.70	6.92	2.90	10.93	3.84	1.47	26.06

	Labor SF per manhour	Material coverage SF/gallon	Material cost per gallon	Labor cost per 100 SF	Labor burden 100 SF	Material cost per 100 SF	Overhead per 100 SF	Profit per 100 SF	Total price per 100 SF
Epoxy coating, 2 part system, white (material #52)									
Roll 1st coat									
Slow	250	400	61.70	4.20	1.19	15.43	6.45	4.09	31.36
Medium	275	388	54.00	6.09	2.13	13.92	5.42	3.03	30.59
Fast	300	375	46.30	7.50	3.12	12.35	4.25	1.63	28.85
Roll 2nd or additional coats									
Slow	275	425	61.70	3.82	1.09	14.52	6.02	3.82	29.27
Medium	300	413	54.00	5.58	1.94	13.08	5.05	2.82	28.47
Fast	325	400	46.30	6.92	2.90	11.58	3.96	1.52	26.88

For field painting at heights above 8 feet, use the High Time Difficulty Factors on page 137. Heavy structural steel has from 100 to 150 square feet of surface area per ton. Extra heavy structural steel has from 50 to 100 square feet of surface area per ton. Rule of thumb: When coatings are applied by brush, a journeyman painter will apply a first coat on from 6 to 7 tons per 8 hour day. When coatings are applied by spray, figure output at 0.2 hours per ton and material use at about 0.2 gallons per ton. Use Figure 23 on pages 387 to 395 to convert structural steel linear feet or tonnage to surface area. Note: A two coat system, prime and finish, using oil base material is recommended for any metal surface. Although water base material is often used, it may cause oxidation, corrosion and rust. One coat of oil base solid body stain is often used on interior or exterior metal but it may crack, peel or chip without the proper prime coat application. "Slow" work is based on a $10.50 hourly wage, "Medium" work on a $16.75 hourly wage, and "Fast" work on a $22.50 hourly wage. Other qualifications that apply to this table are on page 9.

	Labor SF per manhour	Material coverage SF/gallon	Material cost per gallon	Labor cost per 100 SF	Labor burden 100 SF	Material cost per 100 SF	Overhead per 100 SF	Profit per 100 SF	Total price per 100 SF
Structural steel, heavy, spray application									
Metal primer, rust inhibitor - clean metal (material #35)									
Spray prime coat									
Slow	650	325	29.90	1.62	.46	9.20	3.50	2.22	17.00
Medium	750	313	26.20	2.23	.77	8.37	2.79	1.56	15.72
Fast	850	300	22.40	2.65	1.11	7.47	2.08	.80	14.11
Metal primer, rust inhibitor - rusty metal (material #36)									
Spray prime coat									
Slow	650	300	36.60	1.62	.46	12.20	4.43	2.81	21.52
Medium	750	288	32.10	2.23	.77	11.15	3.47	1.94	19.56
Fast	850	275	27.50	2.65	1.11	10.00	2.55	.98	17.29
Industrial enamel, oil base, high gloss - light colors (material #56)									
Spray 1st or additional finish coats									
Slow	750	340	29.60	1.40	.39	8.71	3.26	2.07	15.83
Medium	850	325	25.90	1.97	.70	7.97	2.60	1.46	14.70
Fast	950	310	22.20	2.37	.98	7.16	1.95	.75	13.21
Industrial enamel, oil base, high gloss - dark (OSHA) colors (material #57)									
Spray 1st or additional finish coats									
Slow	750	365	37.00	1.40	.39	10.14	3.70	2.35	17.98
Medium	850	350	32.40	1.97	.70	9.26	2.92	1.63	16.48
Fast	950	335	27.70	2.37	.98	8.27	2.15	.83	14.60
Epoxy coating, 2 part system, clear (material #51)									
Spray 1st coat									
Slow	650	325	58.30	1.62	.46	17.94	6.21	3.93	30.16
Medium	750	313	51.00	2.23	.77	16.29	4.73	2.64	26.66
Fast	850	300	43.70	2.65	1.11	14.57	3.39	1.30	23.02
Spray 2nd or additional coats									
Slow	750	350	58.30	1.40	.39	16.66	5.72	3.63	27.80
Medium	850	338	51.00	1.97	.70	15.09	4.35	2.43	24.54
Fast	950	325	43.70	2.37	.98	13.45	3.11	1.20	21.11

	Labor SF per manhour	Material coverage SF/gallon	Material cost per gallon	Labor cost per 100 SF	Labor burden 100 SF	Material cost per 100 SF	Overhead per 100 SF	Profit per 100 SF	Total price per 100 SF
Epoxy coating, 2 part system, white (material #52)									
Spray 1st coat									
Slow	650	325	61.70	1.62	.46	18.98	6.53	4.14	31.73
Medium	750	313	54.00	2.23	.77	17.25	4.96	2.77	27.98
Fast	850	300	46.30	2.65	1.11	15.43	3.55	1.36	24.10
Spray 2nd or additional coats									
Slow	750	350	61.70	1.40	.39	17.63	6.02	3.82	29.26
Medium	850	338	54.00	1.97	.70	15.98	4.57	2.55	25.77
Fast	950	325	46.30	2.37	.98	14.25	3.26	1.25	22.11

For field painting at heights above 8 feet, use the High Time Difficulty Factors on page 137. Heavy structural steel has from 100 to 150 square feet of surface area per ton. Extra heavy structural steel has from 50 to 100 square feet of surface area per ton. Rule of thumb: When coatings are applied by brush, a journeyman painter will apply a first coat on from 6 to 7 tons per 8 hour day. When coatings are applied by spray, figure output at 0.2 hours per ton and material use at about 0.2 gallons per ton. Use Figure 23 on pages 387 to 395 to convert structural steel linear feet or tonnage to surface area. Note: A two coat system, prime and finish, using oil base material is recommended for any metal surface. Although water base material is often used, it may cause oxidation, corrosion and rust. One coat of oil base solid body stain is often used on interior or exterior metal but it may crack, peel or chip without the proper prime coat application. "Slow" work is based on a $10.50 hourly wage, "Medium" work on a $16.75 hourly wage, and "Fast" work on a $22.50 hourly wage. Other qualifications that apply to this table are on page 9.

	Labor SF per manhour	Material coverage SF/gallon	Material cost per gallon	Labor cost per 100 SF	Labor burden 100 SF	Material cost per 100 SF	Overhead per 100 SF	Profit per 100 SF	Total price per 100 SF
Structural steel, light, brush application									
Metal primer, rust inhibitor, clean metal (material #35)									
Brush prime coat									
Slow	60	425	29.90	17.50	4.97	7.04	9.15	5.80	44.46
Medium	80	413	26.20	20.94	7.30	6.34	8.47	4.74	47.79
Fast	100	400	22.40	22.50	9.39	5.60	6.94	2.67	47.10
Metal primer, rust inhibitor, rusty metal (material #36)									
Brush prime coat									
Slow	60	400	36.60	17.50	4.97	9.15	9.81	6.22	47.65
Medium	80	388	32.10	20.94	7.30	8.27	8.94	5.00	50.45
Fast	100	375	27.50	22.50	9.39	7.33	7.26	2.79	49.27
Industrial enamel, oil base, high gloss, light colors (material #56)									
Brush 1st or additional finish coats									
Slow	80	425	29.60	13.13	3.72	6.96	7.39	4.68	35.88
Medium	100	413	25.90	16.75	5.84	6.27	7.07	3.95	39.88
Fast	120	400	22.20	18.75	7.81	5.55	5.94	2.28	40.33
Industrial enamel, oil base, high gloss, dark (OSHA) colors (material #57)									
Brush 1st or additional finish coats									
Slow	80	475	37.00	13.13	3.72	7.79	7.64	4.85	37.13
Medium	100	463	32.40	16.75	5.84	7.00	7.25	4.05	40.89
Fast	120	450	27.70	18.75	7.81	6.16	6.06	2.33	41.11
Epoxy coating, 2 part system, clear (material #51)									
Brush 1st coat									
Slow	60	425	58.30	17.50	4.97	13.72	11.22	7.11	54.52
Medium	80	413	51.00	20.94	7.30	12.35	9.94	5.56	56.09
Fast	100	400	43.70	22.50	9.39	10.93	7.92	3.04	53.78
Brush 2nd or additional coats									
Slow	80	450	58.30	13.13	3.72	12.96	9.25	5.86	44.92
Medium	100	438	51.00	16.75	5.84	11.64	8.39	4.69	47.31
Fast	120	425	43.70	18.75	7.81	10.28	6.82	2.62	46.28

	Labor SF per manhour	Material coverage SF/gallon	Material cost per gallon	Labor cost per 100 SF	Labor burden 100 SF	Material cost per 100 SF	Overhead per 100 SF	Profit per 100 SF	Total price per 100 SF
Epoxy coating, 2 part system, white (material #52)									
Brush 1st coat									
Slow	60	425	61.70	17.50	4.97	14.52	11.47	7.27	55.73
Medium	80	413	54.00	20.94	7.30	13.08	10.12	5.66	57.10
Fast	100	400	46.30	22.50	9.39	11.58	8.04	3.09	54.60
Brush 2nd or additional coats									
Slow	80	450	61.70	13.13	3.72	13.71	9.48	6.01	46.05
Medium	100	438	54.00	16.75	5.84	12.33	8.56	4.78	48.26
Fast	120	425	46.30	18.75	7.81	10.89	6.93	2.66	47.04

For field painting at heights above 8 feet, use the High Time Difficulty Factors on page 137. Light structural steel has from 300 to 500 square feet of surface per ton. As a comparison, when coatings are applied by brush, a journeyman painter will apply a first coat on from 2 to 3 tons per 8 hour day. A second and subsequent coats can be applied on 3 to 4 tons per day. Use Figure 23 on pages 387 to 395 to convert structural steel linear feet or tonnage to surface area. Note: A two coat system, prime and finish, using oil base material is recommended for any metal surface. Although water base material is often used, it may cause oxidation, corrosion and rust. One coat of oil base solid body stain is often used on interior or exterior metal but it may crack, peel or chip without the proper prime coat application. "Slow" work is based on a $10.50 hourly wage, "Medium" work on a $16.75 hourly wage, "Fast" work on a $22.50 hourly wage. Other qualifications that apply to this table are on page 9.

	Labor SF per manhour	Material coverage SF/gallon	Material cost per gallon	Labor cost per 100 SF	Labor burden 100 SF	Material cost per 100 SF	Overhead per 100 SF	Profit per 100 SF	Total price per 100 SF
Structural steel, light, roll and brush application									
Metal primer, rust inhibitor, clean metal (material #35)									
Roll and brush prime coat									
Slow	125	390	29.90	8.40	2.38	7.67	5.72	3.63	27.80
Medium	150	378	26.20	11.17	3.90	6.93	5.39	3.01	30.40
Fast	175	365	22.40	12.86	5.35	6.14	4.51	1.73	30.59
Metal primer, rust inhibitor, rusty metal (material #36)									
Roll and brush prime coat									
Slow	125	380	36.60	8.40	2.38	9.63	6.33	4.01	30.75
Medium	150	368	32.10	11.17	3.90	8.72	5.83	3.26	32.88
Fast	175	355	27.50	12.86	5.35	7.75	4.81	1.85	32.62
Industrial enamel, oil base, high gloss, light colors (material #56)									
Roll and brush 1st or additional finish coats									
Slow	175	390	29.60	6.00	1.70	7.59	4.74	3.01	23.04
Medium	200	378	25.90	8.38	2.92	6.85	4.45	2.49	25.09
Fast	225	365	22.20	10.00	4.16	6.08	3.75	1.44	25.43
Industrial enamel, oil base, high gloss, dark (OSHA) colors (material #57)									
Roll and brush 1st or additional finish coats									
Slow	175	440	37.00	6.00	1.70	8.41	5.00	3.17	24.28
Medium	200	428	32.40	8.38	2.92	7.57	4.62	2.58	26.07
Fast	225	415	27.70	10.00	4.16	6.67	3.86	1.48	26.17
Epoxy coating, 2 part system, clear (material #51)									
Roll and brush 1st coat									
Slow	125	400	58.30	8.40	2.38	14.58	7.86	4.98	38.20
Medium	150	388	51.00	11.17	3.90	13.14	6.91	3.86	38.98
Fast	175	375	43.70	12.86	5.35	11.65	5.53	2.12	37.51
Roll and brush 2nd or additional coats									
Slow	175	425	58.30	6.00	1.70	13.72	6.64	4.21	32.27
Medium	200	413	51.00	8.38	2.92	12.35	5.79	3.24	32.68
Fast	225	400	43.70	10.00	4.16	10.93	4.65	1.79	31.53

	Labor SF per manhour	Material coverage SF/gallon	Material cost per gallon	Labor cost per 100 SF	Labor burden 100 SF	Material cost per 100 SF	Overhead per 100 SF	Profit per 100 SF	Total price per 100 SF
Epoxy coating, 2 part system, white (material #52)									
Roll and brush 1st coat									
Slow	125	400	61.70	8.40	2.38	15.43	8.13	5.15	39.49
Medium	150	388	54.00	11.17	3.90	13.92	7.10	3.97	40.06
Fast	175	375	46.30	12.86	5.35	12.35	5.66	2.17	38.39
Roll and brush 2nd or additional coats									
Slow	175	425	61.70	6.00	1.70	14.52	6.89	4.37	33.48
Medium	200	413	54.00	8.38	2.92	13.08	5.97	3.34	33.69
Fast	225	400	46.30	10.00	4.16	11.58	4.77	1.83	32.34

For field painting at heights above 8 feet, use the High Time Difficulty Factors on page 137. Light structural steel has from 300 to 500 square feet of surface per ton. As a comparison, when coatings are applied by brush, a journeyman painter will apply a first coat on from 2 to 3 tons per 8 hour day. A second and subsequent coats can be applied on 3 to 4 tons per day. Use Figure 23 on pages 387 to 395 to convert structural steel linear feet or tonnage to surface area. Note: A two coat system, prime and finish, using oil base material is recommended for any metal surface. Although water base material is often used, it may cause oxidation, corrosion and rust. One coat of oil base solid body stain is often used on interior or exterior metal but it may crack, peel or chip without the proper prime coat application. "Slow" work is based on a $10.50 hourly wage, "Medium" work on a $16.75 hourly wage, "Fast" work on a $22.50 hourly wage. Other qualifications that apply to this table are on page 9.

	Labor SF per manhour	Material coverage SF/gallon	Material cost per gallon	Labor cost per 100 SF	Labor burden 100 SF	Material cost per 100 SF	Overhead per 100 SF	Profit per 100 SF	Total price per 100 SF

Structural steel, light, spray application

Metal primer, rust inhibitor, clean metal (material #35)

	Labor SF per manhour	Material coverage SF/gallon	Material cost per gallon	Labor cost per 100 SF	Labor burden 100 SF	Material cost per 100 SF	Overhead per 100 SF	Profit per 100 SF	Total price per 100 SF
Spray prime coat									
Slow	400	325	29.90	2.63	.74	9.20	3.90	2.47	18.94
Medium	500	313	26.20	3.35	1.17	8.37	3.16	1.77	17.82
Fast	600	300	22.40	3.75	1.58	7.47	2.37	.91	16.08

Metal primer, rust inhibitor, rusty metal (material #36)

	Labor SF per manhour	Material coverage SF/gallon	Material cost per gallon	Labor cost per 100 SF	Labor burden 100 SF	Material cost per 100 SF	Overhead per 100 SF	Profit per 100 SF	Total price per 100 SF
Spray prime coat									
Slow	400	300	36.60	2.63	.74	12.20	4.83	3.06	23.46
Medium	500	288	32.10	3.35	1.17	11.15	3.84	2.15	21.66
Fast	600	275	27.50	3.75	1.58	10.00	2.83	1.09	19.25

Industrial enamel, oil base, high gloss, light colors (material #56)

	Labor SF per manhour	Material coverage SF/gallon	Material cost per gallon	Labor cost per 100 SF	Labor burden 100 SF	Material cost per 100 SF	Overhead per 100 SF	Profit per 100 SF	Total price per 100 SF
Spray 1st or additional finish coats									
Slow	500	325	29.60	2.10	.60	9.11	3.66	2.32	17.79
Medium	600	313	25.90	2.79	.98	8.27	2.95	1.65	16.64
Fast	700	300	22.20	3.21	1.35	7.40	2.21	.85	15.02

Industrial enamel, oil base, high gloss, dark (OSHA) colors (material #57)

	Labor SF per manhour	Material coverage SF/gallon	Material cost per gallon	Labor cost per 100 SF	Labor burden 100 SF	Material cost per 100 SF	Overhead per 100 SF	Profit per 100 SF	Total price per 100 SF
Spray 1st or additional finish coats									
Slow	500	365	37.00	2.10	.60	10.14	3.98	2.52	19.34
Medium	600	350	32.40	2.79	.98	9.26	3.19	1.78	18.00
Fast	700	335	27.70	3.21	1.35	8.27	2.37	.91	16.11

Epoxy coating, 2 part system, clear (material #51)

	Labor SF per manhour	Material coverage SF/gallon	Material cost per gallon	Labor cost per 100 SF	Labor burden 100 SF	Material cost per 100 SF	Overhead per 100 SF	Profit per 100 SF	Total price per 100 SF
Spray 1st coat									
Slow	400	325	58.30	2.63	.74	17.94	6.61	4.19	32.11
Medium	500	313	51.00	3.35	1.17	16.29	5.10	2.85	28.76
Fast	600	300	43.70	3.75	1.58	14.57	3.68	1.41	24.99
Spray 2nd or additional coats									
Slow	500	350	58.30	2.10	.60	16.66	6.00	3.80	29.16
Medium	600	338	51.00	2.79	.98	15.09	4.62	2.58	26.06
Fast	700	325	43.70	3.21	1.35	13.45	3.33	1.28	22.62

	Labor SF per manhour	Material coverage SF/gallon	Material cost per gallon	Labor cost per 100 SF	Labor burden 100 SF	Material cost per 100 SF	Overhead per 100 SF	Profit per 100 SF	Total price per 100 SF
Epoxy coating, 2 part system, white (material #52)									
Spray 1st coat									
Slow	400	325	61.70	2.63	.74	18.98	6.93	4.39	33.67
Medium	500	313	54.00	3.35	1.17	17.25	5.33	2.98	30.08
Fast	600	300	46.30	3.75	1.58	15.43	3.84	1.48	26.08
Spray 2nd or additional coats									
Slow	500	350	61.70	2.10	.60	17.63	6.30	3.99	30.62
Medium	600	338	54.00	2.79	.98	15.98	4.84	2.70	27.29
Fast	700	325	46.30	3.21	1.35	14.25	3.48	1.34	23.63

For field painting at heights above 8 feet, use the High Time Difficulty Factors on page 137. Light structural steel has from 300 to 500 square feet of surface per ton. The rule of thumb for labor output and material usage for spray application on light structural steel is shown in the table on the next page. Use Figure 23 on pages 387 to 395 to convert structural steel linear feet or tonnage to surface area. Note: A two coat system, prime and finish, using oil base material is recommended for any metal surface. Although water base material is often used, it may cause oxidation, corrosion and rust. One coat of oil base solid body stain is often used on interior or exterior metal but it may crack, peel or chip without the proper prime coat application. "Slow" work is based on a $10.50 hourly wage, "Medium" work on a $16.75 hourly wage, "Fast" work on a $22.50 hourly wage. Other qualifications that apply to this table are on page 9.

	Labor manhours per ton	Material gallons per ton	Material cost per gallon	Labor cost per ton	Labor burden per ton	Material cost per ton	Overhead per ton	Profit per ton	Total price per ton
Structural steel, light, coating rule of thumb, spray, per ton									
Metal primer, rust inhibitor, clean metal (material #35)									
Spray prime coat									
Slow	1.8	1.0	29.90	18.90	5.36	29.90	16.80	10.65	81.61
Medium	1.6	1.1	26.20	26.80	9.34	28.82	15.92	8.90	89.78
Fast	1.4	1.2	22.40	31.50	13.15	26.88	13.23	5.09	89.85
Metal primer, rust inhibitor, rusty metal (material #36)									
Spray prime coat									
Slow	1.8	1.0	36.60	18.90	5.36	36.60	18.87	11.96	91.69
Medium	1.6	1.1	32.10	26.80	9.34	35.31	17.51	9.79	98.75
Fast	1.4	1.2	27.50	31.50	13.15	33.00	14.37	5.52	97.54
Industrial enamel, oil base, high gloss, light colors (material #56)									
Spray 1st finish coat									
Slow	1.5	0.9	29.60	15.75	4.47	26.64	14.53	9.21	70.60
Medium	1.3	1.0	25.90	21.78	7.59	25.90	13.54	7.57	76.38
Fast	1.1	1.1	22.20	24.75	10.33	24.42	11.01	4.23	74.74
Spray 2nd or additional finish coats									
Slow	1.1	0.8	29.60	11.55	3.28	23.68	11.94	7.57	58.02
Medium	1.0	0.9	25.90	16.75	5.84	23.31	11.25	6.29	63.44
Fast	0.9	1.0	22.20	20.25	8.45	22.20	9.42	3.62	63.94
Industrial enamel, oil base, high gloss, dark (OSHA) colors (material #57)									
Spray 1st finish coat									
Slow	1.5	0.9	37.00	15.75	4.47	33.30	16.59	10.52	80.63
Medium	1.3	1.0	32.40	21.78	7.59	32.40	15.13	8.46	85.36
Fast	1.1	1.1	27.70	24.75	10.33	30.47	12.13	4.66	82.34
Spray 2nd or additional finish coats									
Slow	1.1	0.8	37.00	11.55	3.28	29.60	13.78	8.73	66.94
Medium	1.0	0.9	32.40	16.75	5.84	29.16	12.68	7.09	71.52
Fast	0.9	1.0	27.70	20.25	8.45	27.70	10.43	4.01	70.84

For field painting at heights above 8 feet, use the High Time Difficulty Factors on page 137. Note: A two coat system, prime and finish, using oil base material is recommended for any metal surface. Although water base material is often used, it may cause oxidation, corrosion and rust. One coat of oil base solid body stain is often used on interior or exterior metal but it may crack, peel or chip without the proper prime coat application. "Slow" work is based on a $10.50 hourly wage, "Medium" work on a $16.75 hourly wage, "Fast" work on a $22.50 hourly wage. Other qualifications that apply to this table are on page 9.

	Labor SF per manhour	Material coverage SF/gallon	Material cost per gallon	Labor cost per 100 SF	Labor burden 100 SF	Material cost per 100 SF	Overhead per 100 SF	Profit per 100 SF	Total price per 100 SF
Structural steel, medium, brush application									
Metal primer, rust inhibitor, clean metal (material #35)									
Brush prime coat									
Slow	80	425	29.90	13.13	3.72	7.04	7.41	4.70	36.00
Medium	100	413	26.20	16.75	5.84	6.34	7.09	3.96	39.98
Fast	120	400	22.40	18.75	7.81	5.60	5.95	2.29	40.40
Metal primer, rust inhibitor, rusty metal (material #36)									
Brush prime coat									
Slow	80	400	36.60	13.13	3.72	9.15	8.07	5.11	39.18
Medium	100	388	32.10	16.75	5.84	8.27	7.56	4.23	42.65
Fast	120	375	27.50	18.75	7.81	7.33	6.27	2.41	42.57
Industrial enamel, oil base, high gloss, light colors (material #56)									
Brush 1st or additional finish coats									
Slow	100	425	29.60	10.50	2.98	6.96	6.34	4.02	30.80
Medium	125	413	25.90	13.40	4.67	6.27	5.96	3.33	33.63
Fast	150	400	22.20	15.00	6.27	5.55	4.96	1.91	33.69
Industrial enamel, oil base, high gloss, dark (OSHA) colors (material #57)									
Brush 1st or additional finish coats									
Slow	100	475	37.00	10.50	2.98	7.79	6.60	4.18	32.05
Medium	125	463	32.40	13.40	4.67	7.00	6.14	3.43	34.64
Fast	150	450	27.70	15.00	6.27	6.16	5.07	1.95	34.45
Epoxy coating, 2 part system, clear (material #51)									
Brush 1st coat									
Slow	80	425	58.30	13.13	3.72	13.72	9.48	6.01	46.06
Medium	100	413	51.00	16.75	5.84	12.35	8.56	4.79	48.29
Fast	120	400	43.70	18.75	7.81	10.93	6.94	2.67	47.10
Brush 2nd or additional coats									
Slow	100	450	58.30	10.50	2.98	12.96	8.20	5.20	39.84
Medium	125	438	51.00	13.40	4.67	11.64	7.28	4.07	41.06
Fast	150	425	43.70	15.00	6.27	10.28	5.83	2.24	39.62

	Labor SF per manhour	Material coverage SF/gallon	Material cost per gallon	Labor cost per 100 SF	Labor burden 100 SF	Material cost per 100 SF	Overhead per 100 SF	Profit per 100 SF	Total price per 100 SF
Epoxy coating, 2 part system, white (material #52)									
Brush 1st coat									
Slow	80	425	61.70	13.13	3.72	14.52	9.73	6.17	47.27
Medium	100	413	54.00	16.75	5.84	13.08	8.74	4.89	49.30
Fast	120	400	46.30	18.75	7.81	11.58	7.06	2.71	47.91
Brush 2nd or additional coats									
Slow	100	450	61.70	10.50	2.98	13.71	8.43	5.34	40.96
Medium	125	438	54.00	13.40	4.67	12.33	7.45	4.16	42.01
Fast	150	425	46.30	15.00	6.27	10.89	5.95	2.29	40.40

For field painting at heights above 8 feet, use the High Time Difficulty Factors on page 137. Medium structural steel has from 150 to 300 square feet of surface per ton. As a comparison, when coatings are applied by brush, a journeyman painter will apply a first coat on 4 to 5 tons per 8 hour day. A second and subsequent coat can be applied on 5 to 6 tons per day. When coatings are applied by spray, figure output at 0.6 hours per ton and material use at about 0.6 gallons per ton. Use Figure 23 on pages 387 to 395 to convert structural steel linear feet or tonnage to surface area. Note: A two coat system, prime and finish, using oil base material is recommended for any metal surface. Although water base material is often used, it may cause oxidation, corrosion and rust. One coat of oil base solid body stain is often used on interior or exterior metal but it may crack, peel or chip without the proper prime coat application. "Slow" work is based on a $10.50 hourly wage, "Medium" work on a $16.75 hourly wage, "Fast" work on a $22.50 hourly wage. Other qualifications that apply to this table are on page 9.

	Labor SF per manhour	Material coverage SF/gallon	Material cost per gallon	Labor cost per 100 SF	Labor burden 100 SF	Material cost per 100 SF	Overhead per 100 SF	Profit per 100 SF	Total price per 100 SF

Structural steel, medium, roll and brush application

	Labor SF per manhour	Material coverage SF/gallon	Material cost per gallon	Labor cost per 100 SF	Labor burden 100 SF	Material cost per 100 SF	Overhead per 100 SF	Profit per 100 SF	Total price per 100 SF
Metal primer, rust inhibitor, clean metal (material #35)									
Roll and brush prime coat									
Slow	200	390	29.90	5.25	1.49	7.67	4.47	2.83	21.71
Medium	225	378	26.20	7.44	2.59	6.93	4.16	2.32	23.44
Fast	250	365	22.40	9.00	3.76	6.14	3.50	1.34	23.74
Metal primer, rust inhibitor, rusty metal (material #36)									
Roll and brush prime coat									
Slow	200	380	36.60	5.25	1.49	9.63	5.07	3.22	24.66
Medium	225	368	32.10	7.44	2.59	8.72	4.59	2.57	25.91
Fast	250	355	27.50	9.00	3.76	7.75	3.79	1.46	25.76
Industrial enamel, oil base, high gloss, light colors (material #56)									
Roll and brush 1st or additional finish coats									
Slow	225	390	29.60	4.67	1.32	7.59	4.21	2.67	20.46
Medium	250	378	25.90	6.70	2.34	6.85	3.89	2.17	21.95
Fast	275	365	22.20	8.18	3.43	6.08	3.27	1.26	22.22
Industrial enamel, oil base, high gloss, dark (OSHA) colors (material #57)									
Roll and brush 1st or additional finish coats									
Slow	225	440	37.00	4.67	1.32	8.41	4.47	2.83	21.70
Medium	250	428	32.40	6.70	2.34	7.57	4.07	2.27	22.95
Fast	275	415	27.70	8.18	3.43	6.67	3.38	1.30	22.96
Epoxy coating, 2 part system, clear (material #51)									
Roll and brush 1st coat									
Slow	200	400	58.30	5.25	1.49	14.58	6.61	4.19	32.12
Medium	225	388	51.00	7.44	2.59	13.14	5.68	3.17	32.02
Fast	250	375	43.70	9.00	3.76	11.65	4.52	1.74	30.67
Roll and brush 2nd or additional coats									
Slow	225	425	58.30	4.67	1.32	13.72	6.11	3.87	29.69
Medium	250	413	51.00	6.70	2.34	12.35	5.24	2.93	29.56
Fast	275	400	43.70	8.18	3.43	10.93	4.17	1.60	28.31

	Labor SF per manhour	Material coverage SF/gallon	Material cost per gallon	Labor cost per 100 SF	Labor burden 100 SF	Material cost per 100 SF	Overhead per 100 SF	Profit per 100 SF	Total price per 100 SF
Epoxy coating, 2 part system, white (material #52)									
Roll and brush 1st coat									
Slow	200	400	61.70	5.25	1.49	15.43	6.87	4.36	33.40
Medium	225	388	54.00	7.44	2.59	13.92	5.87	3.28	33.10
Fast	250	375	46.30	9.00	3.76	12.35	4.65	1.79	31.55
Roll and brush 2nd or additional coats									
Slow	225	425	61.70	4.67	1.32	14.52	6.36	4.03	30.90
Medium	250	413	54.00	6.70	2.34	13.08	5.42	3.03	30.57
Fast	275	400	46.30	8.18	3.43	11.58	4.29	1.65	29.13

For field painting at heights above 8 feet, use the High Time Difficulty Factors on page 137. Medium structural steel has from 150 to 300 square feet of surface per ton. As a comparison, when coatings are applied by brush, a journeyman painter will apply a first coat on 4 to 5 tons per 8 hour day. A second and subsequent coat can be applied on 5 to 6 tons per day. When coatings are applied by spray, figure output at 0.6 hours per ton and material use at about 0.6 gallons per ton. Use Figure 23 on pages 387 to 395 to convert structural steel linear feet or tonnage to surface area. Note: A two coat system, prime and finish, using oil base material is recommended for any metal surface. Although water base material is often used, it may cause oxidation, corrosion and rust. One coat of oil base solid body stain is often used on interior or exterior metal but it may crack, peel or chip without the proper prime coat application. "Slow" work is based on a $10.50 hourly wage, "Medium" work on a $16.75 hourly wage, "Fast" work on a $22.50 hourly wage. Other qualifications that apply to this table are on page 9.

	Labor SF per manhour	Material coverage SF/gallon	Material cost per gallon	Labor cost per 100 SF	Labor burden 100 SF	Material cost per 100 SF	Overhead per 100 SF	Profit per 100 SF	Total price per 100 SF

Structural steel, medium, spray application

Metal primer, rust inhibitor, clean metal (material #35)

Spray prime coat									
Slow	500	325	29.90	2.10	.60	9.20	3.69	2.34	17.93
Medium	600	313	26.20	2.79	.98	8.37	2.97	1.66	16.77
Fast	700	300	22.40	3.21	1.35	7.47	2.22	.85	15.10

Metal primer, rust inhibitor, rusty metal (material #36)

Spray prime coat									
Slow	500	300	36.60	2.10	.60	12.20	4.62	2.93	22.45
Medium	600	288	32.10	2.79	.98	11.15	3.65	2.04	20.61
Fast	700	275	27.50	3.21	1.35	10.00	2.69	1.03	18.28

Industrial enamel, oil base, high gloss, light colors (material #56)

Spray 1st or additional finish coats									
Slow	600	325	29.60	1.75	.50	9.11	3.52	2.23	17.11
Medium	700	313	25.90	2.39	.84	8.27	2.82	1.57	15.89
Fast	800	300	22.20	2.81	1.18	7.40	2.11	.81	14.31

Industrial enamel, oil base, high gloss, dark (OSHA) colors (material #57)

Spray 1st or additional finish coats									
Slow	600	365	37.00	1.75	.50	10.14	3.84	2.43	18.66
Medium	700	350	32.40	2.39	.84	9.26	3.06	1.71	17.26
Fast	800	335	27.70	2.81	1.18	8.27	2.27	.87	15.40

Epoxy coating, 2 part system, clear (material #51)

Spray 1st coat									
Slow	500	325	58.30	2.10	.60	17.94	6.40	4.06	31.10
Medium	600	313	51.00	2.79	.98	16.29	4.91	2.75	27.72
Fast	700	300	43.70	3.21	1.35	14.57	3.54	1.36	24.03

Spray 2nd or additional coats									
Slow	600	350	58.30	1.75	.50	16.66	5.86	3.72	28.49
Medium	700	338	51.00	2.39	.84	15.09	4.49	2.51	25.32
Fast	800	325	43.70	2.81	1.18	13.45	3.22	1.24	21.90

	Labor SF per manhour	Material coverage SF/gallon	Material cost per gallon	Labor cost per 100 SF	Labor burden 100 SF	Material cost per 100 SF	Overhead per 100 SF	Profit per 100 SF	Total price per 100 SF
Epoxy coating, 2 part system, white (material #52)									
Spray 1st coat									
Slow	500	325	61.70	2.10	.60	18.98	6.72	4.26	32.66
Medium	600	313	54.00	2.79	.98	17.25	5.15	2.88	29.05
Fast	700	300	46.30	3.21	1.35	15.43	3.70	1.42	25.11
Spray 2nd or additional coats									
Slow	600	350	61.70	1.75	.50	17.63	6.16	3.91	29.95
Medium	700	338	54.00	2.39	.84	15.98	4.70	2.63	26.54
Fast	800	325	46.30	2.81	1.18	14.25	3.37	1.30	22.91

For field painting at heights above 8 feet, use the High Time Difficulty Factors on page 137. Medium structural steel has from 150 to 300 square feet of surface per ton. As a comparison, when coatings are applied by brush, a journeyman painter will apply a first coat on 4 to 5 tons per 8 hour day. A second and subsequent coat can be applied on 5 to 6 tons per day. When coatings are applied by spray, figure output at 0.6 hours per ton and material use at about 0.6 gallons per ton. Use Figure 23 on pages 387 to 395 to convert structural steel linear feet or tonnage to surface area. Note: A two coat system, prime and finish, using oil base material is recommended for any metal surface. Although water base material is often used, it may cause oxidation, corrosion and rust. One coat of oil base solid body stain is often used on interior or exterior metal but it may crack, peel or chip without the proper prime coat application. "Slow" work is based on a $10.50 hourly wage, "Medium" work on a $16.75 hourly wage, "Fast" work on a $22.50 hourly wage. Other qualifications that apply to this table are on page 9.

Section designation		Square feet of surface area per foot of length: Minus one flange side	Square feet of surface area per foot of length: All around	Square feet of surface area per ton: Minus one flange side	Square feet of surface area per ton: All around
W 30	x 99	7.56	8.43	152.7	170.3
W 27	x 94	6.99	7.81	148.7	166.2
	x 84	6.99	7.79	166.4	185.5
W 24	x 100	7.00	8.00	140.0	160.0
	x 94	6.29	7.04	133.8	149.8
	x 84	6.27	7.02	149.3	167.1
	x 76	6.23	6.98	163.9	183.7
	x 68	6.21	6.96	182.6	204.7
	x 61	5.71	6.29	187.2	206.2
	x 55	5.67	6.25	206.2	227.2
W 21	x 96	5.77	6.52	120.2	135.8
	x 82	5.73	6.48	139.8	158.0
	x 73	5.60	6.29	153.4	172.3
	x 68	5.58	6.57	164.1	184.4
	x 62	5.56	6.25	179.4	201.6
	x 55	5.52	6.21	200.7	225.8
	x 49	5.10	5.65	208.2	230.6
	x 44	5.08	5.63	230.9	255.9
W 18	x 96	5.96	6.94	124.2	144.6
	x 85	5.28	6.02	124.2	141.6
	x 77	5.22	5.95	135.6	154.5
	x 70	5.19	5.92	148.3	169.1
	x 64	5.17	5.90	161.6	184.4
	x 60	4.92	5.54	164.0	184.7
	x 55	4.90	5.52	178.1	200.7
	x 50	4.88	5.50	195.2	220.0
	x 45	4.85	5.48	215.6	243.6
	x 40	4.48	4.98	224.0	249.0
	x 35	4.46	4.96	254.9	283.4
W 16	x 96	5.60	6.56	116.7	136.7
	x 88	5.57	6.53	126.6	148.4
	x 78	4.88	5.60	125.1	143.6
	x 71	4.82	5.53	135.8	155.8
	x 64	4.79	5.50	149.7	171.9
	x 58	4.77	5.48	164.5	189.0
	x 50	4.51	5.10	180.4	204.0
	x 45	4.45	5.03	197.8	223.6
	x 40	4.42	5.00	221.0	250.0
	x 36	4.40	4.98	244.4	276.7
	x 31	4.02	4.48	259.4	289.0
	x 26	3.98	4.44	306.2	341.5

Section designation		Square feet of surface area per foot of length: Minus one flange side	Square feet of surface area per foot of length: All around	Square feet of surface area per ton: Minus one flange side	Square feet of surface area per ton: All around
W 14	x 95	5.99	7.20	126.1	151.6
	x 87	5.96	7.17	137.0	164.8
	x 84	5.36	6.36	127.6	151.4
	x 78	5.33	6.33	136.7	162.3
	x 74	4.92	5.77	133.0	155.9
	x 68	4.83	5.67	142.1	166.8
	x 61	4.81	5.65	157.7	185.2
	x 53	4.33	5.00	163.4	188.7
	x 48	4.29	4.96	178.8	206.7
	x 43	4.27	4.94	198.6	229.8
	x 38	4.04	4.60	212.6	242.1
	x 34	4.02	4.58	236.5	269.4
	x 30	4.00	4.56	266.7	304.0
	x 26	3.56	3.98	273.8	306.2
	x 22	3.54	3.96	321.8	360.0
W 12	x 99	5.19	6.21	104.8	125.4
	x 92	5.14	6.15	111.7	133.7
	x 85	5.12	6.13	120.5	144.2
	x 79	5.09	6.10	128.9	154.4
	x 72	5.04	6.04	140.0	167.8
	x 65	5.02	6.02	154.5	185.2
	x 58	4.54	5.38	156.6	185.5
	x 53	4.50	5.33	169.8	201.1
	x 50	4.07	4.75	162.8	190.0
	x 45	4.00	4.67	177.8	207.6
	x 40	4.00	4.67	200.0	233.5
	x 36	3.70	4.25	205.6	236.1
	x 31	3.65	4.19	235.5	270.3
	x 27	3.63	4.17	268.9	308.9
	x 22	3.04	3.38	276.4	307.3
	x 19	3.02	3.35	317.9	352.6
	x 16.5	3.00	3.33	363.6	403.6
	x 14	2.98	3.31	425.7	472.9
W 10	x 100	4.45	5.31	89.0	106.2
	x 89	4.38	5.23	98.4	117.5
	x 77	4.33	5.19	112.5	134.8
	x 72	4.28	5.13	118.9	142.5
	x 66	4.26	5.10	129.1	154.5
	x 60	4.24	5.08	141.3	169.3

Figure 23

Structural steel conversion table

Section designation		Square feet of surface area per foot of length		Square feet of surface area per ton	
		Minus one flange side	All around	Minus one flange side	All around
W 10	x 54	4.19	5.02	155.2	185.9
	x 49	4.17	5.00	170.2	204.0
	x 45	3.69	4.35	164.0	193.3
	x 39	3.67	4.33	188.2	222.1
	x 33	3.63	4.29	220.0	260.0
	x 29	3.15	3.63	217.2	250.3
	x 25	3.13	3.60	250.4	288.0
	x 21	3.08	3.56	293.3	339.0
	x 19	2.71	3.04	285.3	320.0
	x 17	2.69	3.02	316.5	355.3
	x 15	2.67	3.00	356.0	400.0
	x 11.5	2.65	2.98	460.9	518.3
W 8	x 67	3.56	4.25	106.3	126.9
	x 58	3.52	4.21	121.4	145.2
	x 48	3.45	4.13	143.8	172.1
	x 40	3.41	4.08	170.5	204.0
	x 35	3.35	4.02	191.4	229.7
	x 31	3.33	4.00	214.8	258.1
	x 28	2.96	3.50	211.4	250.0
	x 24	2.94	3.48	245.0	290.0
	x 20	2.67	3.10	267.0	310.0
	x 17	2.65	3.08	311.8	362.4
	x 15	2.35	2.69	313.3	358.7
	x 13	2.33	2.67	358.5	410.8
	x 10	2.31	2.65	462.0	530.0
W 6	x 25	2.59	3.10	207.2	248.0
	x 20	2.54	3.04	254.0	304.0
	x 15.5	2.50	3.00	322.6	387.1
	x 16	2.04	2.38	255.0	297.5
	x 12	2.00	2.33	333.3	388.3
	x 8.5	1.98	2.31	465.9	543.5
W 5	x 18.5	2.10	2.52	227.0	272.4
	x 16	2.08	2.50	260.0	312.5
W 4	x 13	1.69	2.02	260.0	310.8
S-24	x 90	5.78	6.38	128.4	141.8
	x 79.9	5.75	6.33	143.9	158.4

Section designation		Square feet of surface area per foot of length		Square feet of surface area per ton	
		Minus one flange side	All around	Minus one flange side	All around
S-20	x 95	5.15	5.75	108.4	121.1
	x 85	5.08	5.67	119.5	133.4
	x 75	4.93	5.46	131.5	145.6
	x 65.4	4.90	5.42	149.8	165.7
S-18	x 70	4.56	5.08	130.3	145.1
	x 54.7	4.50	5.00	164.5	182.8
S-15	x 50	3.91	4.38	156.4	175.2
	x 42.9	3.88	4.33	180.9	201.9
S-12	x 50	3.38	3.83	135.2	153.2
	x 40.8	3.31	3.75	162.3	183.8
	x 35	3.28	3.71	187.4	212.0
	x 31.8	3.25	3.67	204.4	230.8
S-10	x 35	2.92	3.33	166.9	190.2
	x 25.4	2.82	3.21	222.0	252.8
S-8	x 23	2.36	2.71	205.2	235.7
	x 18.4	2.33	2.67	253.3	290.2
S-7	x 20	2.14	2.46	214.0	246.0
	x 15.3	2.07	2.38	270.6	311.1
S-6	x 17.25	1.91	2.21	221.4	256.2
	x 12.5	1.84	2.13	294.4	340.8
S-5	x 14.75	1.65	1.92	223.7	260.3
	x 10	1.58	1.83	316.0	366.0
S-4	x 9.5	1.35	1.58	284.2	332.6
	x 7.7	1.32	1.54	342.9	400.0
S-3	x 7.5	1.13	1.33	301.3	354.7
	x 5.7	1.09	1.29	382.5	452.6
Miscellaneous shape					
M-5	x 18.9	2.08	2.50	220.1	264.6

Figure 23 (cont'd)
Structural steel conversion table

Section designation		Square feet of surface area per foot of length		Square feet of surface area per ton	
		Minus one flange side	All around	Minus one flange side	All around
C-15	x 50	3.44	3.75	137.6	150.0
	x 40	3.38	3.67	169.0	183.5
	x 33.9	3.34	3.63	197.1	214.2
C-12	x 30	2.78	3.04	185.3	202.7
	x 25	2.75	3.00	220.0	240.0
	x 20.7	2.75	3.00	265.7	289.9
C-10	x 30	2.42	2.67	161.3	178.0
	x 25	2.39	2.63	191.2	210.4
	x 20	2.35	2.58	235.0	258.0
	x 15.3	2.32	2.54	305.3	334.2
C-9	x 20	2.16	2.38	216.0	238.0
	x 15	2.13	2.33	284.0	310.7
	x 13.4	2.09	2.29	311.9	341.8
C-8	x 18.75	1.96	2.17	209.1	231.5
	x 13.75	1.93	2.13	280.7	309.8
	x 11.5	1.90	2.08	330.4	361.7
C-7	x 14.75	1.73	1.92	234.6	260.3
	x 12.25	1.73	1.92	282.4	313.5
	x 9.8	1.70	1.88	346.9	383.7
C-6	x 13	1.53	1.71	235.4	263.1
	x 10.5	1.50	1.67	285.7	318.1
	x 8.2	1.47	1.63	358.5	397.6
C-5	x 9	1.30	1.46	288.9	324.4
	x 6.7	1.27	1.42	379.1	423.9
C-4	x 7.25	1.10	1.25	295.2	344.8
	x 5.4	1.07	1.21	396.3	448.1
C-3	x 6	.91	1.04	303.3	346.7
	x 5	.88	1.00	352.0	400.0
	x 4.1	.84	.96	409.8	468.3
MC-18	x 58	4.06	4.42	140.0	152.4
	x 51.9	4.03	4.38	155.3	168.8
	x 45.8	4.00	4.33	176.7	189.1
	x 42.7	4.00	4.33	187.4	202.8
MC-13	x 50	3.26	3.63	130.4	145.2
	x 40	3.20	3.54	160.0	177.0
	x 35	3.20	3.54	182.9	202.3
	x 31.8	3.17	3.50	199.4	220.1

Section designation		Square feet of surface area per foot of length		Square feet of surface area per ton	
		Minus one flange side	All around	Minus one flange side	All around
MC-12	x 50	3.03	3.38	121.2	135.2
	x 45	3.00	3.33	133.3	148.0
	x 40	2.97	3.29	148.5	164.5
	x 35	2.94	3.25	168.0	185.7
	x 37	2.91	3.21	157.3	173.5
	x 32.9	2.88	3.17	175.1	192.7
	x 30.9	2.88	3.17	186.4	205.2
MC-10	x 41.1	2.76	3.13	134.3	152.3
	x 33.6	2.70	3.04	160.7	181.0
	x 28.5	2.67	3.00	187.4	210.5
	x 28.3	2.54	2.83	179.5	200.0
	x 25.3	2.54	2.83	200.8	223.7
	x 24.9	2.51	2.79	201.6	224.1
	x 21.9	2.54	2.83	232.0	258.4
MC-9	x 25.4	2.38	2.67	187.4	210.2
	x 23.9	2.38	2.67	199.2	223.4
MC-8	x 22.8	2.21	2.50	193.9	219.3
	x 21.4	2.21	2.50	206.5	233.6
	x 20	2.08	2.33	208.0	233.0
	x 18.7	2.08	2.33	222.5	249.2
MC-7	x 22.7	2.07	2.38	182.4	209.7
	x 19.1	2.04	2.33	213.6	244.0
	x 17.6	1.92	2.17	218.2	246.6
MC-6	x 18	1.88	2.17	208.9	241.1
	x 15.3	1.88	2.17	245.8	283.7
	x 16.3	1.75	2.00	214.7	245.4
	x 15.1	1.75	2.00	231.8	264.9
	x 12	1.63	1.83	271.7	305.0
MC-3	x 9	1.03	1.21	228.9	268.9
	x 7.1	1.00	1.17	281.7	329.6

Figure 23 (cont'd)
Structural steel conversion table

Section designation		Surface area per foot of length	Surface area per ton
ST 18	x 97	5.06	104.3
	x 91	5.04	110.8
	x 85	5.02	118.1
	x 80	5.00	125.0
	x 75	4.98	132.8
	x 67.5	4.95	147.0
ST 16.5	x 76	4.72	124.2
	x 70.5	4.70	133.3
	x 65	4.38	144.0
	x 59	4.65	157.6
ST 15	x 95	5.02	105.7
	x 86	4.99	116.0
	x 66	4.28	129.7
	x 62	4.27	137.7
	x 58	4.25	146.6
	x 54	4.23	156.7
	x 49.5	4.21	170.1
ST 13.5	x 88.5	4.63	104.6
	x 80	4.59	114.8
	x 72.5	4.57	126.1
	x 57	3.95	138.6
	x 51	3.93	154.1
	x 47	3.91	166.4
	x 42	3.89	185.2
ST 12	x 80	4.41	110.3
	x 72.5	4.38	120.8
	x 65	4.36	134.1
	x 60	4.04	134.7
	x 55	4.02	146.2
	x 50	4.00	160.0
	x 47	3.54	150.6
	x 42	3.51	167.1
	x 38	3.49	183.7
	x 34	3.47	204.1
	x 30.5	3.15	206.6
	x 27.5	3.13	227.6

Section designation		Surface area per foot of length	Surface area per ton
ST 10.5	x 71	3.97	111.8
	x 63.5	3.95	124.4
	x 56	3.92	140.0
	x 48	3.27	136.3
	x 41	3.23	157.6
	x 36.5	3.15	172.6
	x 34	3.14	184.7
	x 31	3.12	201.3
	x 27.5	3.10	225.5
	x 24.5	2.82	230.2
	x 22	2.81	255.5
ST 9	x 57	3.51	123.2
	x 52.5	3.49	133.0
	x 48	3.47	144.6
	x 42.5	3.00	141.2
	x 38.5	2.98	154.8
	x 35	2.96	169.1
	x 32	2.94	183.8
	x 30	2.78	185.3
	x 27.5	2.77	201.5
	x 25	2.75	220.0
	x 22.5	2.73	242.7
	x 20	2.49	249.0
	x 17.5	2.48	283.4
ST 8	x 48	3.28	136.7
	x 44	3.26	148.2
	x 39	2.79	143.1
	x 35.5	2.77	156.1
	x 32	2.76	172.5
	x 29	2.73	188.3
	x 25	2.53	202.4
	x 22.5	2.52	224.0
	x 20	2.50	250.0
	x 18	2.49	276.7
	x 15.5	2.24	289.0
	x 13	2.22	341.5

Figure 23 (cont'd)
Structural steel conversion table

Section designation		Surface area per foot of length	Surface area per ton
ST 7	x 88	3.88	88.2
	x 83.5	3.86	92.5
	x 79	3.84	97.2
	x 75	3.83	102.1
	x 71	3.81	107.3
	x 68	3.69	108.5
	x 63.5	3.67	115.6
	x 59.5	3.65	122.7
	x 55.5	3.64	131.2
	x 51.5	3.62	140.6
	x 47.5	3.60	151.6
	x 43.5	3.58	164.6
	x 42	3.19	151.9
	x 39	3.17	162.6
	x 37	2.86	154.6
	x 34	2.85	167.6
	x 30.5	2.83	185.6
	x 26.5	2.51	189.4
	x 24	2.49	207.5
	x 21.5	2.47	229.8
	x 19	2.31	243.2
	x 17	2.29	269.4
	x 15	2.28	304.0
	x 13	2.00	307.7
	x 11	1.98	360.0

ST 6	x 95	3.31	69.7
	x 80.5	3.24	80.5
	x 66.5	3.18	97.8
	x 60	3.15	105.6
	x 53	3.11	117.4
	x 49.5	3.10	125.3
	x 46	3.08	133.9
	x 42.5	3.06	144.0
	x 39.5	3.05	154.4
	x 36	3.03	168.3
	x 32.5	3.01	185.2
	x 29	2.69	185.5
	x 26.5	2.67	201.5
	x 25	2.36	188.8
	x 22.5	2.35	208.9

Section designation		Surface area per foot of length	Surface area per ton
ST 6	x 20	2.33	233.0
	x 18	2.11	234.4
	x 15.5	2.10	271.0
	x 13.5	2.08	308.1
	x 11	1.70	309.1
	x 9.5	1.68	353.7
	x 8.25	1.67	404.8
	x 7	1.65	471.4
ST 5	x 56	2.68	95.7
	x 50	2.65	106.0
	x 44.5	2.62	117.8
	x 38.5	2.58	134.0
	x 36	2.57	142.8
	x 33	2.55	154.5
	x 30	2.53	168.7
	x 27	2.51	185.9
	x 24.5	2.50	204.1
	x 22.5	2.18	193.8
	x 19.5	2.16	221.5
	x 16.5	2.14	259.4
	x 14.5	1.82	251.0
	x 12.5	1.80	288.0
	x 10.5	1.78	339.0
	x 9.5	1.53	322.1
	x 8.5	1.51	355.3
	x 7.5	1.50	400.0
	x 5.75	1.48	514.8

ST 4	x 33.5	2.13	127.7
	x 29	2.10	144.8
	x 24	2.06	171.7
	x 20	2.03	203.0
	x 17.5	2.01	229.7
	x 15.5	2.00	258.1
	x 14	1.76	251.4
	x 12	1.75	291.7
	x 10	1.56	312.0
	x 8.5	1.54	362.4
	x 7.5	1.35	360.0
	x 6.5	1.33	409.2
	x 5	1.31	524.0

Figure 23 (cont'd)
Structural steel conversion table

Section designation		Surface area per foot of length	Surface area per ton
ST 3	x 12.5	1.55	248.0
	x 10	1.52	304.0
	x 7.75	1.50	387.1
	x 8	1.19	297.5
	x 6	1.17	390.0
	x 4.25	1.14	536.5
ST 2.5	x 9.25	1.26	272.4
	x 8	1.25	312.5
ST 2	x 6.5	1.02	313.8

Tees cut from American standard shapes

Section designation		Surface area per foot of length	Surface area per ton
ST 12	x 60	3.34	111.3
	x 52.95	3.31	125.0
	x 50	3.21	128.4
	x 45	3.19	141.8
	x 39.95	3.17	158.7
ST 10	x47.5	2.87	120.8
	x 42.5	2.84	133.6
	x 37.5	2.73	145.6
	x 32.7	2.70	165.1
ST 9	x 35	2.54	145.1
	x 27.35	2.50	182.8
ST 7.5	x 25	2.19	175.2
	x 21.45	2.17	202.3

Section designation		Surface area per foot of length	Surface area per ton
ST 6	x 25	1.91	152.8
	x 20.4	1.88	184.3
	x 17.5	1.85	211.4
	x 15.9	1.83	230.2
ST 5	x 17.5	1.66	189.7
	x 12.7	1.61	253.5
ST 4	x 11.5	1.36	236.5
	x 9.2	1.33	289.1
ST 3.5	x 10	1.23	246.0
	x 7.65	1.19	311.0
ST 3	x 8.625	1.09	252.8
	x 6.25	1.06	339.2
ST 2.5	x 7.375	.96	260.3
	x 5	.92	368.0
ST 2	x 4.75	.80	336.8
	x 3.85	.78	405.2

Miscellaneous tee (cut from M5 x 18.9)

Section designation		Surface area per foot of length	Surface area per ton
MT 2.5	x 9.45	.83	175.7

Figure 23 (cont'd)
Structural steel conversion table

Section designation			Surface area per foot of length	Surface area per ton
L 8	x 8	x 1-1/8	2.67	93.8
	x 8	x 1	2.67	104.7
	x 8	x 1	2.67	118.7
	x 8	x 7/8	2.67	118.7
	x 8	x 3/4	2.67	137.3
	x 8	x 5/8	2.67	163.3
	x 8	x 9/16	2.67	180.4
	x 8	x 1/2	2.67	202.3
L 6	x 6	x 1	2.00	107.0
	x 6	x 7/8	2.00	120.8
	x 6	x 3/4	2.00	139.4
	x 6	x 5/8	2.00	165.3
	x 6	x 9/16	2.00	182.6
	x 6	x 1/2	2.00	204.1
	x 6	x 7/16	2.00	232.6
	x 6	x 3/8	2.00	268.5
	x 6	x 5/16	2.00	322.6
L5	x 5	x 7/8	1.67	122.8
	x 5	x 3/4	1.67	141.5
	x 5	x 5/8	1.67	167.0
	x 5	x 1/2	1.67	206.2
	x 5	x 7/16	1.67	233.6
	x 5	x 3/8	1.67	271.5
	x 5	x 5/16	1.67	324.3
L 4	x 4	x 3/4	1.33	143.8
	x 4	x 5/8	1.33	169.4
	x 4	x 1/2	1.33	207.8
	x 4	x 7/16	1.33	235.4
	x 4	x 3/8	1.33	271.4
	x 4	x 5/16	1.33	324.4
	x 4	x 1/4	1.33	403.0
L 3-1/2	x 3/1/2	x 1/2	1.17	210.8
	x 3-1/2	x 7/16	1.17	238.8
	x 3-1/2	x 3/8	1.17	275.3
	x 3-1/2	x 5/16	1.17	325.0
	x 3-1/2	x 1/4	1.17	403.4

Section designation			Surface area per foot of length	Surface area per ton
L 3	x 3	x 1/2	1.00	212.8
	x 3	x 7/16	1.00	241.0
	x 3	x 3/8	1.00	277.8
	x 3	x 5/16	1.00	327.9
	x 3	x 1/4	1.00	408.2
	x 3	x 3/16	1.00	539.1
L 2-1/2	x 2-1/2	x 1/2	.83	215.6
	x 2-1/2	x 3/8	.83	281.4
	x 2-1/2	x 5/16	.83	332.0
	x 2-1/2	x 1/4	.83	404.9
	x 2-1/2	x 3/16	.83	540.7
L 2	x 2	x 3/8	.67	285.1
	x 2	x 5/16	.67	341.8
	x 2	x 1/4	.67	420.1
	x 2	x 3/16	.67	549.2
	x 2	x 1/8	.67	812.1
L 1-3/4	x 1-3/4	x 1/4	.58	418.8
	x 1-3/4	x 3/16	.58	547.2
	x 1-3/4	x 1/8	.58	805.6
L 1-1/2	x 1-1/2	x 1/4	.50	427.4
	x 1-1/2	x 3/16	.50	555.6
	x 1-1/2	x 5/32	.50	657.9
	x 1-1/2	x 1/8	.50	813.0
L 1-1/4	x 1-1/4	x 1/4	.42	437.5
	x 1-1/4	x 3/16	.42	567.6
	x 1-1/4	x 1/8	.42	831.7
L 1	x 1	x 1/4	.33	443.0
	x 1	x 3/16	.33	569.0
	x 1	x 1/8	.33	825.0

Figure 23 (cont'd)
Structural steel conversion table

Section designation			Surface area per foot of length	Surface area per ton
L 9	x 4	x 1	2.17	106.4
	x 4	x 7/8	2.17	120.2
	x 4	x 3/4	2.17	138.7
	x 4	x 5/8	2.17	165.0
	x 4	x 9/16	2.17	182.4
	x 4	x 1/2	2.17	203.8
L 8	x 6	x 1	2.33	105.4
	x 6	x 7/8	2.33	119.2
	x 6	x 3/4	2.33	137.9
	x 6	x 5/8	2.33	163.5
	x 6	x 9/16	2.33	181.3
	x 6	x 1/2	2.33	202.6
	x 6	x 7/16	2.33	230.7
L 8	x 4	x 1	2.00	107.0
	x 4	x 7/8	2.00	120.8
	x 4	x 3/4	2.00	139.4
	x 4	x 5/8	2.00	165.3
	x 4	x 9/16	2.00	182.6
	x 4	x 1/2	2.00	204.1
	x 4	x 7/16	2.00	232.6
L 7	x 4	x 7/8	1.83	121.2
	x 4	x 3/4	1.83	139.7
	x 4	x 5/8	1.83	165.6
	x 4	x 9/16	1.83	183.0
	x 4	x 1/2	1.83	204.5
	x 4	x 7/16	1.83	231.6
	x 4	x 3/8	1.83	269.1
L 6	x 4	x 7/8	1.67	122.8
	x 4	x 3/4	1.67	141.5
	x 4	x 5/8	1.67	167.0
	x 4	x 9/16	1.67	184.5
	x 4	x 1/2	1.67	206.2
	x 4	x 7/16	1.67	233.6
	x 4	x 3/8	1.67	271.5
	x 4	x 5/16	1.67	324.3
	x 4	x 1/4	1.67	402.4
L 6	x 3-1/2	x 1/2	1.58	206.5
	x 3-1/2	x 3/8	1.58	270.1
	x 3-1/2	x 5/16	1.58	322.4
	x 3-1/2	x 1/4	1.58	400.0

Section designation			Surface area per foot of length	Surface area per ton
L 5	x 3-1/2	x 3/4	1.42	143.4
	x 3-1/2	x 5/8	1.42	169.0
	x 3-1/2	x 1/2	1.42	208.8
	x 3-1/2	x 7/16	1.42	236.7
	x 3-1/2	x 3/8	1.42	273.1
	x 3-1/2	x 5/16	1.42	326.4
	x 3-1/2	x 1/4	1.42	405.7
L 5	x 3	x 1/2	1.33	207.8
	x 3	x 7/16	1.33	235.4
	x 3	x 3/8	1.33	271.4
	x 3	x 5/16	1.33	324.4
	x 3	x 1/4	1.33	403.0
L 4	x 3-1/2	x 5/8	1.25	170.1
	x 3-1/2	x 1/2	1.25	210.1
	x 3-1/2	x 7/16	1.25	235.8
	x 3-1/2	x 3/8	1.25	274.7
	x 3-1/2	x 5/16	1.25	324.7
	x 3-1/2	x 1/4	1.25	403.2
L 4	x 3	x 5/8	1.17	172.1
	x 3	x 1/2	1.17	210.8
	x 3	x 7/16	1.17	238.8
	x 3	x 3/8	1.17	275.3
	x 3	x 5/16	1.17	325.0
	x 3	x 1/4	1.17	403.4
L 3-1/2	x 3	x 1/2	1.08	211.8
	x 3	x 7/16	1.08	237.4
	x 3	x 3/8	1.08	273.4
	x 3	x 5/16	1.08	327.3
	x 3	x 1/4	1.08	400.0
L 3-1/2	x 2-1/2	x 1/2	1.00	212.8
	x 2-1/2	x 7/16	1.00	241.0
	x 2-1/2	x 3/8	1.00	277.8
	x 2-1/2	x 5/16	1.00	327.9
	x 2-1/2	x 1/4	1.00	408.2
L 3	x 2-1/2	x 1/2	.92	216.5
	x 2-1/2	x 7/16	.92	242.1
	x 2-1/2	x 3/8	.92	278.8
	x 2-1/2	x 5/16	.92	328.6
	x 2-1/2	x 1/4	.92	408.9
	x 2-1/2	x 3/16	.92	542.8

Figure 23 (cont'd)
Structural steel conversion table

Section designation			Surface area per foot of length	Surface area per ton
L 3	x 2	x 1/2	.83	215.6
	x 2	x 7/16	.83	244.1
	x 2	x 3/8	.83	281.4
	x 2	x 5/16	.83	332.0
	x 2	x 1/4	.83	404.9
	x 2	x 3/16	.83	540.7
L 2-1/2	x 2	x 3/8	.75	283.0
	x 2	x 5/16	.75	333.3
	x 2	x 1/4	.75	414.4
	x 2	x 3/16	.75	545.5
L 2-1/2	x 1-1/2	x 5/16	.67	341.8
	x 1-1/2	x 1/4	.67	420.1
	x 1-1/2	x 3/16	.67	549.2

Courtesy: Richardson Engineering Services, Inc.

Section designation			Surface area per foot of length	Surface area per ton
L 2	x 1-1/2	x 1/4	.58	418.8
	x 1-1/2	x 3/16	.58	547.2
	x 1-1/2	x 1/8	.58	805.6
L 2	x 1-1/4	x 1/4	.54	423.5
	x 1-1/4	x 3/16	.54	551.0
L 1-3/4	x 1-1/4	x 1/4	.50	427.4
	x 1-1/4	x 3/16	.50	555.6
	x 1-1/4	x 1/8	.50	813.0

Figure 23 (cont'd)
Structural steel conversion table

Diameter (in feet)	Area (SF)
10	314
15	707
20	1,257
25	1,963
30	2,827
35	3,848
40	5,027
45	6,362
50	7,854
55	9,503
60	11,310
65	13,273
70	15,394

Figure 24
Surface area of spheres

	Labor SF per manhour	Material coverage SF/gallon	Material cost per gallon	Labor cost per 100 SF	Labor burden 100 SF	Material cost per 100 SF	Overhead per 100 SF	Profit per 100 SF	Total price per 100 SF
Tank, silo, vessel, or hopper, brush, exterior walls only									
Metal primer, rust inhibitor - clean metal (material #35)									
Brush prime coat									
Slow	150	425	29.90	7.00	1.99	7.04	4.97	3.15	24.15
Medium	175	400	26.20	9.57	3.33	6.55	4.77	2.67	26.89
Fast	200	375	22.40	11.25	4.70	5.97	4.06	1.56	27.54
Metal primer, rust inhibitor - rusty metal (material #36)									
Brush prime coat									
Slow	150	400	36.60	7.00	1.99	9.15	5.62	3.56	27.32
Medium	175	375	32.10	9.57	3.33	8.56	5.26	2.94	29.66
Fast	200	350	27.50	11.25	4.70	7.86	4.40	1.69	29.90
Industrial enamel, oil base, high gloss - light colors (material #56)									
Brush 1st or additional finish coats									
Slow	200	450	29.60	5.25	1.49	6.58	4.13	2.62	20.07
Medium	225	425	25.90	7.44	2.59	6.09	3.95	2.21	22.28
Fast	250	400	22.20	9.00	3.76	5.55	3.39	1.30	23.00
Industrial enamel, oil base, high gloss - dark (OSHA) colors (material #57)									
Brush 1st or additional finish coats									
Slow	200	475	37.00	5.25	1.49	7.79	4.50	2.85	21.88
Medium	225	450	32.40	7.44	2.59	7.20	4.22	2.36	23.81
Fast	250	425	27.70	9.00	3.76	6.52	3.57	1.37	24.22
Epoxy coating, 2 part system, clear (material #51)									
Brush 1st coat									
Slow	150	425	58.30	7.00	1.99	13.72	7.04	4.46	34.21
Medium	175	400	51.00	9.57	3.33	12.75	6.29	3.51	35.45
Fast	200	375	43.70	11.25	4.70	11.65	5.11	1.96	34.67
Brush 2nd or additional coats									
Slow	200	450	58.30	5.25	1.49	12.96	6.11	3.87	29.68
Medium	225	425	51.00	7.44	2.59	12.00	5.40	3.02	30.45
Fast	250	400	43.70	9.00	3.76	10.93	4.38	1.68	29.75
Epoxy coating, 2 part system, white (material #52)									
Brush 1st coat									
Slow	150	425	61.70	7.00	1.99	14.52	7.29	4.62	35.42
Medium	175	400	54.00	9.57	3.33	13.50	6.47	3.62	36.49
Fast	200	375	46.30	11.25	4.70	12.35	5.24	2.01	35.55
Brush 2nd or additional coats									
Slow	200	450	61.70	5.25	1.49	13.71	6.34	4.02	30.81
Medium	225	425	54.00	7.44	2.59	12.71	5.57	3.11	31.42
Fast	250	400	46.30	9.00	3.76	11.58	4.50	1.73	30.57

	Labor SF per manhour	Material coverage SF/gallon	Material cost per gallon	Labor cost per 100 SF	Labor burden 100 SF	Material cost per 100 SF	Overhead per 100 SF	Profit per 100 SF	Total price per 100 SF
Vinyl coating (material #59)									
Brush 1st coat									
Slow	125	250	59.10	8.40	2.38	23.64	10.67	6.77	51.86
Medium	150	238	51.70	11.17	3.90	21.72	9.01	5.04	50.84
Fast	175	225	44.30	12.86	5.35	19.69	7.02	2.70	47.62
Brush 2nd or additional coats									
Slow	165	150	59.10	6.36	1.81	39.40	14.75	9.35	71.67
Medium	190	125	51.70	8.82	3.06	41.36	13.05	7.29	73.58
Fast	215	100	44.30	10.47	4.36	44.30	10.94	4.20	74.27

See Figure 24 on page 395 to find the surface area of a spherical vessel. Use this table when estimating walls only. The cost tables for painting steel tank, silo, vessel or hopper roofs follow those for painting walls. For heights above 8 feet, use the High Time Difficulty Factors on page 137. Note: A two coat system, prime and finish, using oil base material is recommended for any metal surface. Although water base material is often used, it may cause oxidation, corrosion and rust. One coat of oil base solid body stain is often used on interior or exterior metal but it may crack, peel or chip without the proper prime coat application. "Slow" work is based on a $10.50 hourly wage, "Medium" work on a $16.75 hourly wage, and "Fast" work on a $22.50 hourly wage. Other qualifications that apply to this table are on page 9.

	Labor SF per manhour	Material coverage SF/gallon	Material cost per gallon	Labor cost per 100 SF	Labor burden 100 SF	Material cost per 100 SF	Overhead per 100 SF	Profit per 100 SF	Total price per 100 SF
Tank, silo, vessel, or hopper, roll, exterior walls only									
Metal primer, rust inhibitor - clean metal (material #35)									
Roll prime coat									
Slow	275	400	29.90	3.82	1.09	7.48	3.84	2.43	18.66
Medium	300	375	26.20	5.58	1.94	6.99	3.55	1.99	20.05
Fast	325	350	22.40	6.92	2.90	6.40	3.00	1.15	20.37
Metal primer, rust inhibitor - rusty metal (material #36)									
Roll prime coat									
Slow	275	380	36.60	3.82	1.09	9.63	4.51	2.86	21.91
Medium	300	355	32.10	5.58	1.94	9.04	4.06	2.27	22.89
Fast	325	330	27.50	6.92	2.90	8.33	3.36	1.29	22.80
Industrial enamel, oil base, high gloss - light colors (material #56)									
Roll 1st or additional finish coats									
Slow	375	425	29.60	2.80	.80	6.96	3.27	2.07	15.90
Medium	400	400	25.90	4.19	1.46	6.48	2.97	1.66	16.76
Fast	425	375	22.20	5.29	2.20	5.92	2.48	.95	16.84
Industrial enamel, oil base, high gloss - dark (OSHA) colors (material #57)									
Roll 1st or additional finish coats									
Slow	375	450	37.00	2.80	.80	8.22	3.66	2.32	17.80
Medium	400	425	32.40	4.19	1.46	7.62	3.25	1.82	18.34
Fast	425	400	27.70	5.29	2.20	6.93	2.67	1.03	18.12
Epoxy coating, 2 part system, clear (material #51)									
Roll 1st coat									
Slow	275	425	58.30	3.82	1.09	13.72	5.78	3.66	28.07
Medium	300	400	51.00	5.58	1.94	12.75	4.97	2.78	28.02
Fast	325	375	43.70	6.92	2.90	11.65	3.97	1.53	26.97
Roll 2nd or additional coats									
Slow	375	450	58.30	2.80	.80	12.96	5.13	3.25	24.94
Medium	400	425	51.00	4.19	1.46	12.00	4.32	2.42	24.39
Fast	425	400	43.70	5.29	2.20	10.93	3.41	1.31	23.14

	Labor SF per manhour	Material coverage SF/gallon	Material cost per gallon	Labor cost per 100 SF	Labor burden 100 SF	Material cost per 100 SF	Overhead per 100 SF	Profit per 100 SF	Total price per 100 SF
Epoxy coating, 2 part system, white (material #52)									
Roll 1st coat									
Slow	275	425	61.70	3.82	1.09	14.52	6.02	3.82	29.27
Medium	300	400	54.00	5.58	1.94	13.50	5.15	2.88	29.05
Fast	325	375	46.30	6.92	2.90	12.35	4.10	1.58	27.85
Roll 2nd or additional coats									
Slow	375	450	61.70	2.80	.80	13.71	5.37	3.40	26.08
Medium	400	425	54.00	4.19	1.46	12.71	4.50	2.51	25.37
Fast	425	400	46.30	5.29	2.20	11.58	3.53	1.36	23.96

See Figure 24 on page 395 to find the surface area of a spherical vessel. Use this table when estimating walls only. The cost tables for painting steel tank, silo, vessel or hopper roofs follow those for painting walls. For heights above 8 feet, use the High Time Difficulty Factors on page 137. Note: A two coat system, prime and finish, using oil base material is recommended for any metal surface. Although water base material is often used, it may cause oxidation, corrosion and rust. One coat of oil base solid body stain is often used on interior or exterior metal but it may crack, peel or chip without the proper prime coat application. "Slow" work is based on a $10.50 hourly wage, "Medium" work on a $16.75 hourly wage, and "Fast" work on a $22.50 hourly wage. Other qualifications that apply to this table are on page 9.

	Labor SF per manhour	Material coverage SF/gallon	Material cost per gallon	Labor cost per 100 SF	Labor burden 100 SF	Material cost per 100 SF	Overhead per 100 SF	Profit per 100 SF	Total price per 100 SF
Tank, silo, vessel, or hopper, spray, exterior walls only									
Metal primer, rust inhibitor - clean metal (material #35)									
Spray prime coat									
Slow	700	325	29.90	1.50	.43	9.20	3.45	2.19	16.77
Medium	750	300	26.20	2.23	.77	8.73	2.88	1.61	16.22
Fast	800	275	22.40	2.81	1.18	8.15	2.24	.86	15.24
Metal primer, rust inhibitor - rusty metal (material #36)									
Spray prime coat									
Slow	700	300	36.60	1.50	.43	12.20	4.38	2.78	21.29
Medium	750	275	32.10	2.23	.77	11.67	3.60	2.01	20.28
Fast	800	250	27.50	2.81	1.18	11.00	2.77	1.07	18.83
Industrial enamel, oil base, high gloss - light colors (material #56)									
Spray 1st or additional finish coats									
Slow	850	350	29.60	1.24	.35	8.46	3.12	1.98	15.15
Medium	900	325	25.90	1.86	.65	7.97	2.57	1.44	14.49
Fast	950	300	22.20	2.37	.98	7.40	1.99	.77	13.51
Industrial enamel, oil base, high gloss - dark (OSHA) colors (material #57)									
Spray 1st or additional finish coats									
Slow	850	375	37.00	1.24	.35	9.87	3.55	2.25	17.26
Medium	900	350	32.40	1.86	.65	9.26	2.88	1.61	16.26
Fast	950	325	27.70	2.37	.98	8.52	2.20	.84	14.91
Epoxy coating, 2 part system, clear (material #51)									
Spray 1st coat									
Slow	700	325	58.30	1.50	.43	17.94	6.16	3.90	29.93
Medium	750	313	51.00	2.23	.77	16.29	4.73	2.64	26.66
Fast	800	300	43.70	2.81	1.18	14.57	3.43	1.32	23.31
Spray 2nd or additional coats									
Slow	850	350	58.30	1.24	.35	16.66	5.66	3.59	27.50
Medium	900	338	51.00	1.86	.65	15.09	4.31	2.41	24.32
Fast	950	325	43.70	2.37	.98	13.45	3.11	1.20	21.11
Epoxy coating, 2 part system, white (material #52)									
Spray 1st coat									
Slow	700	325	61.70	1.50	.43	18.98	6.48	4.11	31.50
Medium	750	313	54.00	2.23	.77	17.25	4.96	2.77	27.98
Fast	800	300	46.30	2.81	1.18	15.43	3.59	1.38	24.39
Spray 2nd or additional coats									
Slow	850	350	61.70	1.24	.35	17.63	5.96	3.78	28.96
Medium	900	338	54.00	1.86	.65	15.98	4.53	2.53	25.55
Fast	950	325	46.30	2.37	.98	14.25	3.26	1.25	22.11

	Labor SF per manhour	Material coverage SF/gallon	Material cost per gallon	Labor cost per 100 SF	Labor burden 100 SF	Material cost per 100 SF	Overhead per 100 SF	Profit per 100 SF	Total price per 100 SF
Vinyl coating (material #59)									
Spray 1st coat									
Slow	600	225	59.10	1.75	.50	26.27	8.84	5.60	42.96
Medium	625	213	51.70	2.68	.93	24.27	6.83	3.82	38.53
Fast	650	200	44.30	3.46	1.45	22.15	5.00	1.92	33.98
Spray 2nd or additional coats									
Slow	725	130	59.10	1.45	.41	45.46	14.67	9.30	71.29
Medium	750	105	51.70	2.23	.77	49.24	12.80	7.16	72.20
Fast	775	80	44.30	2.90	1.21	55.38	11.01	4.23	74.73

See Figure 24 on page 395 to find the surface area of a spherical vessel. Use this table when estimating walls only. The cost tables for painting steel tank, silo, vessel or hopper roofs follow those for painting walls. For heights above 8 feet, use the High Time Difficulty Factors on page 137. Note: A two coat system, prime and finish, using oil base material is recommended for any metal surface. Although water base material is often used, it may cause oxidation, corrosion and rust. One coat of oil base solid body stain is often used on interior or exterior metal but it may crack, peel or chip without the proper prime coat application. "Slow" work is based on a $10.50 hourly wage, "Medium" work on a $16.75 hourly wage, and "Fast" work on a $22.50 hourly wage. Other qualifications that apply to this table are on page 9.

	Labor SF per manhour	Material coverage SF/gallon	Material cost per gallon	Labor cost per 100 SF	Labor burden 100 SF	Material cost per 100 SF	Overhead per 100 SF	Profit per 100 SF	Total price per 100 SF
Tank, silo, vessel, or hopper, brush, exterior roof only									
Metal primer, rust inhibitor - clean metal (material #35)									
Brush prime coat									
Slow	175	425	29.90	6.00	1.70	7.04	4.57	2.90	22.21
Medium	200	400	26.20	8.38	2.92	6.55	4.37	2.44	24.66
Fast	225	375	22.40	10.00	4.16	5.97	3.73	1.43	25.29
Metal primer, rust inhibitor - rusty metal (material #36)									
Brush prime coat									
Slow	175	400	36.60	6.00	1.70	9.15	5.23	3.31	25.39
Medium	200	375	32.10	8.38	2.92	8.56	4.87	2.72	27.45
Fast	225	350	27.50	10.00	4.16	7.86	4.08	1.57	27.67
Industrial enamel, oil base, high gloss - light colors (material #56)									
Brush 1st or additional finish coats									
Slow	225	450	29.60	4.67	1.32	6.58	3.90	2.47	18.94
Medium	250	425	25.90	6.70	2.34	6.09	3.70	2.07	20.90
Fast	275	400	22.20	8.18	3.43	5.55	3.17	1.22	21.55
Industrial enamel, oil base, high gloss - dark (OSHA) colors (material #57)									
Brush 1st or additional finish coats									
Slow	225	475	37.00	4.67	1.32	7.79	4.27	2.71	20.76
Medium	250	450	32.40	6.70	2.34	7.20	3.98	2.22	22.44
Fast	275	425	27.70	8.18	3.43	6.52	3.35	1.29	22.77
Epoxy coating, 2 part system, clear (material #51)									
Brush 1st coat									
Slow	175	425	58.30	6.00	1.70	13.72	6.64	4.21	32.27
Medium	200	400	51.00	8.38	2.92	12.75	5.89	3.29	33.23
Fast	225	375	43.70	10.00	4.16	11.65	4.78	1.84	32.43
Brush 2nd or additional coats									
Slow	225	450	58.30	4.67	1.32	12.96	5.88	3.73	28.56
Medium	250	425	51.00	6.70	2.34	12.00	5.15	2.88	29.07
Fast	275	400	43.70	8.18	3.43	10.93	4.17	1.60	28.31
Epoxy coating, 2 part system, white (material #52)									
Brush 1st coat									
Slow	175	425	61.70	6.00	1.70	14.52	6.89	4.37	33.48
Medium	200	400	54.00	8.38	2.92	13.50	6.08	3.40	34.28
Fast	225	375	46.30	10.00	4.16	12.35	4.91	1.89	33.31
Brush 2nd or additional coats									
Slow	225	450	61.70	4.67	1.32	13.71	6.11	3.87	29.68
Medium	250	425	54.00	6.70	2.34	12.71	5.33	2.98	30.06
Fast	275	400	46.30	8.18	3.43	11.58	4.29	1.65	29.13

	Labor SF per manhour	Material coverage SF/gallon	Material cost per gallon	Labor cost per 100 SF	Labor burden 100 SF	Material cost per 100 SF	Overhead per 100 SF	Profit per 100 SF	Total price per 100 SF
Vinyl coating (material #59)									
Brush 1st coat									
Slow	125	250	59.10	8.40	2.38	23.64	10.67	6.77	51.86
Medium	150	238	51.70	11.17	3.90	21.72	9.01	5.04	50.84
Fast	175	225	44.30	12.86	5.35	19.69	7.02	2.70	47.62
Brush 2nd or additional coats									
Slow	200	150	59.10	5.25	1.49	39.40	14.30	9.07	69.51
Medium	225	125	51.70	7.44	2.59	41.36	12.59	7.04	71.02
Fast	250	100	44.30	9.00	3.76	44.30	10.56	4.06	71.68

Use these figures to estimate labor and material costs for painting the exterior surface of a flat roof on a steel tank, silo, vessel or hopper. Rule of thumb: For a vaulted, peaked or sloping roof, figure the roof area as though it were flat and add 5%. Note: A two coat system, prime and finish, using oil base material is recommended for any metal surface. Although water base material is often used, it may cause oxidation, corrosion and rust. One coat of oil base solid body stain is often used on interior or exterior metal but it may crack, peel or chip without the proper prime coat application. "Slow" work is based on a $10.50 hourly wage, "Medium" work on a $16.75 hourly wage, and "Fast" work on a $22.50 hourly wage. Other qualifications that apply to this table are on page 9.

	Labor SF per manhour	Material coverage SF/gallon	Material cost per gallon	Labor cost per 100 SF	Labor burden 100 SF	Material cost per 100 SF	Overhead per 100 SF	Profit per 100 SF	Total price per 100 SF
Tank, silo, vessel, or hopper, roll, exterior roof only									
Metal primer, rust inhibitor - clean metal (material #35)									
Roll prime coat									
Slow	325	400	29.90	3.23	.92	7.48	3.61	2.29	17.53
Medium	350	375	26.20	4.79	1.67	6.99	3.30	1.84	18.59
Fast	375	350	22.40	6.00	2.51	6.40	2.76	1.06	18.73
Metal primer, rust inhibitor - rusty metal (material #36)									
Roll prime coat									
Slow	325	380	36.60	3.23	.92	9.63	4.27	2.71	20.76
Medium	350	355	32.10	4.79	1.67	9.04	3.80	2.12	21.42
Fast	375	330	27.50	6.00	2.51	8.33	3.12	1.20	21.16
Industrial enamel, oil base, high gloss - light colors (material #56)									
Roll 1st or additional finish coats									
Slow	400	425	29.60	2.63	.74	6.96	3.21	2.03	15.57
Medium	425	400	25.90	3.94	1.37	6.48	2.89	1.61	16.29
Fast	450	375	22.20	5.00	2.08	5.92	2.41	.93	16.34
Industrial enamel, oil base, high gloss - dark (OSHA) colors (material #57)									
Roll 1st or additional finish coats									
Slow	400	450	37.00	2.63	.74	8.22	3.60	2.28	17.47
Medium	425	425	32.40	3.94	1.37	7.62	3.17	1.77	17.87
Fast	450	400	27.70	5.00	2.08	6.93	2.59	1.00	17.60
Epoxy coating, 2 part system, clear (material #51)									
Roll 1st coat									
Slow	325	425	58.30	3.23	.92	13.72	5.54	3.51	26.92
Medium	350	400	51.00	4.79	1.67	12.75	4.71	2.63	26.55
Fast	375	375	43.70	6.00	2.51	11.65	3.73	1.43	25.32
Roll 2nd or additional coats									
Slow	400	450	58.30	2.63	.74	12.96	5.07	3.21	24.61
Medium	425	425	51.00	3.94	1.37	12.00	4.24	2.37	23.92
Fast	450	400	43.70	5.00	2.08	10.93	3.33	1.28	22.62

	Labor SF per manhour	Material coverage SF/gallon	Material cost per gallon	Labor cost per 100 SF	Labor burden 100 SF	Material cost per 100 SF	Overhead per 100 SF	Profit per 100 SF	Total price per 100 SF
Epoxy coating, 2 part system, white (material #52)									
Roll 1st coat									
Slow	325	425	61.70	3.23	.92	14.52	5.79	3.67	28.13
Medium	350	400	54.00	4.79	1.67	13.50	4.89	2.73	27.58
Fast	375	375	46.30	6.00	2.51	12.35	3.86	1.48	26.20
Roll 2nd or additional coats									
Slow	400	450	61.70	2.63	.74	13.71	5.30	3.36	25.74
Medium	425	425	54.00	3.94	1.37	12.71	4.41	2.47	24.90
Fast	450	400	46.30	5.00	2.08	11.58	3.45	1.33	23.44

Use these figures to estimate labor and material costs for painting the exterior surface of a flat roof on a steel tank, silo, vessel or hopper. Rule of thumb: For a vaulted, peaked or sloping roof, figure the roof area as though it were flat and add 5%. Note: A two coat system, prime and finish, using oil base material is recommended for any metal surface. Although water base material is often used, it may cause oxidation, corrosion and rust. One coat of oil base solid body stain is often used on interior or exterior metal but it may crack, peel or chip without the proper prime coat application. "Slow" work is based on a $10.50 hourly wage, "Medium" work on a $16.75 hourly wage, and "Fast" work on a $22.50 hourly wage. Other qualifications that apply to this table are on page 9.

	Labor SF per manhour	Material coverage SF/gallon	Material cost per gallon	Labor cost per 100 SF	Labor burden 100 SF	Material cost per 100 SF	Overhead per 100 SF	Profit per 100 SF	Total price per 100 SF
Tank, silo, vessel, or hopper, spray, exterior roof only									
Metal primer, rust inhibitor - clean metal (material #35)									
Spray prime coat									
Slow	850	325	29.90	1.24	.35	9.20	3.34	2.12	16.25
Medium	900	300	26.20	1.86	.65	8.73	2.75	1.54	15.53
Fast	950	275	22.40	2.37	.98	8.15	2.13	.82	14.45
Metal primer, rust inhibitor - rusty metal (material #36)									
Spray prime coat									
Slow	850	300	36.60	1.24	.35	12.20	4.27	2.71	20.77
Medium	900	275	32.10	1.86	.65	11.67	3.47	1.94	19.59
Fast	950	250	27.50	2.37	.98	11.00	2.66	1.02	18.03
Industrial enamel, oil base, high gloss - light colors (material #56)									
Spray 1st or additional finish coats									
Slow	950	300	26.50	1.11	.31	8.83	3.18	2.02	15.45
Medium	1025	275	23.20	1.63	.58	8.44	2.61	1.46	14.72
Fast	1100	250	29.90	2.05	.85	11.96	2.75	1.06	18.67
Industrial enamel, oil base, high gloss - dark (OSHA) colors (material #57)									
Spray 1st or additional finish coats									
Slow	950	325	31.30	1.11	.31	9.63	3.43	2.17	16.65
Medium	1025	300	27.40	1.63	.58	9.13	2.78	1.55	15.67
Fast	1100	275	23.50	2.05	.85	8.55	2.12	.81	14.38
Epoxy coating, 2 part system, clear (material #51)									
Spray 1st coat									
Slow	850	325	58.30	1.24	.35	17.94	6.05	3.84	29.42
Medium	900	313	51.00	1.86	.65	16.29	4.61	2.58	25.99
Fast	950	300	43.70	2.37	.98	14.57	3.32	1.28	22.52
Spray 2nd or additional coats									
Slow	950	350	58.30	1.11	.31	16.66	5.61	3.56	27.25
Medium	1025	338	51.00	1.63	.58	15.09	4.24	2.37	23.91
Fast	1100	325	43.70	2.05	.85	13.45	3.03	1.16	20.54
Epoxy coating, 2 part system, white (material #52)									
Spray 1st coat									
Slow	850	325	61.70	1.24	.35	18.98	6.38	4.04	30.99
Medium	900	313	54.00	1.86	.65	17.25	4.84	2.71	27.31
Fast	950	300	46.30	2.37	.98	15.43	3.48	1.34	23.60
Spray 2nd or additional coats									
Slow	950	350	61.70	1.11	.31	17.63	5.91	3.75	28.71
Medium	1025	338	54.00	1.63	.58	15.98	4.45	2.49	25.13
Fast	1100	325	46.30	2.05	.85	14.25	3.17	1.22	21.54

	Labor SF per manhour	Material coverage SF/gallon	Material cost per gallon	Labor cost per 100 SF	Labor burden 100 SF	Material cost per 100 SF	Overhead per 100 SF	Profit per 100 SF	Total price per 100 SF
Vinyl coating (material #59)									
Spray 1st coat									
Slow	750	225	59.10	1.40	.39	26.27	8.70	5.52	42.28
Medium	775	213	51.70	2.16	.75	24.27	6.66	3.72	37.56
Fast	800	200	44.30	2.81	1.18	22.15	4.83	1.86	32.83
Spray 2nd or additional coats									
Slow	900	130	59.10	1.17	.33	45.46	14.56	9.23	70.75
Medium	950	105	51.70	1.76	.61	49.24	12.64	7.07	71.32
Fast	1000	80	44.30	2.25	.94	55.38	10.84	4.16	73.57

Use these figures to estimate labor and material costs for painting the exterior surface of a flat roof on a steel tank, silo, vessel or hopper. Rule of thumb: For a vaulted, peaked or sloping roof, figure the roof area as though it were flat and add 5%. Note: A two coat system, prime and finish, using oil base material is recommended for any metal surface. Although water base material is often used, it may cause oxidation, corrosion and rust. One coat of oil base solid body stain is often used on interior or exterior metal but it may crack, peel or chip without the proper prime coat application. "Slow" work is based on a $10.50 hourly wage, "Medium" work on a $16.75 hourly wage, and "Fast" work on a $22.50 hourly wage. Other qualifications that apply to this table are on page 9.

	Labor SF per manhour	Material coverage SF/gallon	Material cost per gallon	Labor cost per 100 SF	Labor burden 100 SF	Material cost per 100 SF	Overhead per 100 SF	Profit per 100 SF	Total price per 100 SF
Walls, concrete tilt-up, brush application									
Flat latex, water base (material #5)									
Brush 1st coat									
Slow	150	275	17.10	7.00	1.99	6.22	4.72	2.99	22.92
Medium	188	238	15.00	8.91	3.11	6.30	4.49	2.51	25.32
Fast	225	200	12.90	10.00	4.16	6.45	3.82	1.47	25.90
Brush 2nd or additional coats									
Slow	200	360	17.10	5.25	1.49	4.75	3.56	2.26	17.31
Medium	225	243	15.00	7.44	2.59	6.17	3.97	2.22	22.39
Fast	250	225	12.90	9.00	3.76	5.73	3.42	1.31	23.22
Enamel, water base (material #9)									
Brush 1st coat									
Slow	150	275	24.50	7.00	1.99	8.91	5.55	3.52	26.97
Medium	188	238	21.40	8.91	3.11	8.99	5.15	2.88	29.04
Fast	225	200	18.40	10.00	4.16	9.20	4.33	1.66	29.35
Brush 2nd or additional coats									
Slow	200	360	24.50	5.25	1.49	6.81	4.20	2.66	20.41
Medium	225	243	21.40	7.44	2.59	8.81	4.62	2.58	26.04
Fast	250	225	18.40	9.00	3.76	8.18	3.87	1.49	26.30
Enamel, oil base (material #10)									
Brush 1st coat									
Slow	150	300	26.60	7.00	1.99	8.87	5.54	3.51	26.91
Medium	188	250	23.30	8.91	3.11	9.32	5.23	2.92	29.49
Fast	225	200	20.00	10.00	4.16	10.00	4.47	1.72	30.35
Brush 2nd or additional coats									
Slow	200	400	26.60	5.25	1.49	6.65	4.15	2.63	20.17
Medium	225	325	23.30	7.44	2.59	7.17	4.21	2.36	23.77
Fast	250	250	20.00	9.00	3.76	8.00	3.84	1.48	26.08
Epoxy coating, 2 part system, clear (material #51)									
Brush 1st coat									
Slow	150	330	58.30	7.00	1.99	17.67	8.26	5.24	40.16
Medium	188	290	51.00	8.91	3.11	17.59	7.25	4.05	40.91
Fast	225	250	43.70	10.00	4.16	17.48	5.86	2.25	39.75
Brush 2nd or additional coats									
Slow	200	380	58.30	5.25	1.49	15.34	6.84	4.34	33.26
Medium	225	340	51.00	7.44	2.59	15.00	6.13	3.43	34.59
Fast	250	300	43.70	9.00	3.76	14.57	5.06	1.94	34.33

	Labor SF per manhour	Material coverage SF/gallon	Material cost per gallon	Labor cost per 100 SF	Labor burden 100 SF	Material cost per 100 SF	Overhead per 100 SF	Profit per 100 SF	Total price per 100 SF
Epoxy coating, 2 part system, white (material #52)									
Brush 1st coat									
Slow	150	330	61.70	7.00	1.99	18.70	8.58	5.44	41.71
Medium	188	290	54.00	8.91	3.11	18.62	7.51	4.20	42.35
Fast	225	250	46.30	10.00	4.16	18.52	6.05	2.33	41.06
Brush 2nd or additional coats									
Slow	200	380	61.70	5.25	1.49	16.24	7.12	4.52	34.62
Medium	225	340	54.00	7.44	2.59	15.88	6.35	3.55	35.81
Fast	250	300	46.30	9.00	3.76	15.43	5.22	2.00	35.41
Waterproofing, clear hydro sealer (material #34)									
Brush 1st coat									
Slow	150	160	17.80	7.00	1.99	11.13	6.24	3.95	30.31
Medium	175	140	15.60	9.57	3.33	11.14	5.89	3.29	33.22
Fast	200	120	13.30	11.25	4.70	11.08	5.00	1.92	33.95
Brush 2nd or additional coats									
Slow	230	200	17.80	4.57	1.29	8.90	4.58	2.90	22.24
Medium	275	188	15.60	6.09	2.13	8.30	4.04	2.26	22.82
Fast	295	175	13.30	7.63	3.18	7.60	3.41	1.31	23.13
Industrial waterproofing (material #58)									
Brush 1st coat									
Slow	90	100	24.80	11.67	3.31	24.80	12.33	7.82	59.93
Medium	100	95	21.70	16.75	5.84	22.84	11.13	6.22	62.78
Fast	110	90	18.60	20.45	8.54	20.67	9.19	3.53	62.38
Brush 2nd or additional coats									
Slow	150	200	24.80	7.00	1.99	12.40	6.63	4.20	32.22
Medium	180	188	21.70	9.31	3.25	11.54	5.90	3.30	33.30
Fast	195	175	18.60	11.54	4.82	10.63	4.99	1.92	33.90

Use these figures to estimate the costs for finishing concrete walls which have a smooth surface (trowel), rough texture, or exposed aggregate finish. For wall heights above 10', increase the computed area by 50%. "Slow" work is based on a $10.50 hourly wage, "Medium" work on a $16.75 hourly wage, and "Fast" work on a $22.50 hourly wage. Other qualifications that apply to this table are on page 9.

	Labor SF per manhour	Material coverage SF/gallon	Material cost per gallon	Labor cost per 100 SF	Labor burden 100 SF	Material cost per 100 SF	Overhead per 100 SF	Profit per 100 SF	Total price per 100 SF
Walls, concrete tilt-up, roll application									
Flat latex, water base (material #5)									
Roll 1st coat									
Slow	275	275	17.10	3.82	1.09	6.22	3.45	2.19	16.77
Medium	300	250	15.00	5.58	1.94	6.00	3.31	1.85	18.68
Fast	325	225	12.90	6.92	2.90	5.73	2.87	1.10	19.52
Roll 2nd or additional coats									
Slow	300	375	17.10	3.50	.99	4.56	2.81	1.78	13.64
Medium	338	325	15.00	4.96	1.73	4.62	2.77	1.55	15.63
Fast	375	275	12.90	6.00	2.51	4.69	2.44	.94	16.58
Enamel, water base (material #9)									
Roll 1st coat									
Slow	275	275	24.50	3.82	1.09	8.91	4.28	2.72	20.82
Medium	300	250	21.40	5.58	1.94	8.56	3.94	2.20	22.22
Fast	325	225	18.40	6.92	2.90	8.18	3.33	1.28	22.61
Roll 2nd or additional coats									
Slow	300	375	24.50	3.50	.99	6.53	3.42	2.17	16.61
Medium	338	325	21.40	4.96	1.73	6.58	3.25	1.82	18.34
Fast	375	275	18.40	6.00	2.51	6.69	2.81	1.08	19.09
Enamel, oil base (material #10)									
Roll 1st coat									
Slow	275	300	26.60	3.82	1.09	8.87	4.27	2.71	20.76
Medium	300	250	23.30	5.58	1.94	9.32	4.13	2.31	23.28
Fast	325	200	20.00	6.92	2.90	10.00	3.66	1.41	24.89
Roll 2nd or additional coats									
Slow	300	350	26.60	3.50	.99	7.60	3.75	2.38	18.22
Medium	338	300	23.30	4.96	1.73	7.77	3.54	1.98	19.98
Fast	375	250	20.00	6.00	2.51	8.00	3.05	1.17	20.73
Epoxy coating, 2 part system, clear (material #51)									
Roll 1st coat									
Slow	275	290	58.30	3.82	1.09	20.10	7.75	4.91	37.67
Medium	300	250	51.00	5.58	1.94	20.40	6.84	3.82	38.58
Fast	325	210	43.70	6.92	2.90	20.81	5.66	2.18	38.47
Roll 2nd or additional coats									
Slow	300	380	58.30	3.50	.99	15.34	6.15	3.90	29.88
Medium	338	340	51.00	4.96	1.73	15.00	5.31	2.97	29.97
Fast	375	300	43.70	6.00	2.51	14.57	4.27	1.64	28.99

	Labor SF per manhour	Material coverage SF/gallon	Material cost per gallon	Labor cost per 100 SF	Labor burden 100 SF	Material cost per 100 SF	Overhead per 100 SF	Profit per 100 SF	Total price per 100 SF
Epoxy coating, 2 part system, white (material #52)									
Roll 1st coat									
Slow	275	290	61.70	3.82	1.09	21.28	8.12	5.15	39.46
Medium	300	250	54.00	5.58	1.94	21.60	7.13	3.99	40.24
Fast	325	210	46.30	6.92	2.90	22.05	5.89	2.27	40.03
Roll 2nd or additional coats									
Slow	300	380	61.70	3.50	.99	16.24	6.43	4.08	31.24
Medium	338	340	54.00	4.96	1.73	15.88	5.53	3.09	31.19
Fast	375	300	46.30	6.00	2.51	15.43	4.43	1.70	30.07
Waterproofing, clear hydro sealer (material #34)									
Roll 1st coat									
Slow	170	200	17.80	6.18	1.75	8.90	5.22	3.31	25.36
Medium	200	165	15.60	8.38	2.92	9.45	5.08	2.84	28.67
Fast	245	130	13.30	9.18	3.83	10.23	4.30	1.65	29.19
Roll 2nd or additional coats									
Slow	275	325	17.80	3.82	1.09	5.48	3.22	2.04	15.65
Medium	300	275	15.60	5.58	1.94	5.67	3.23	1.81	18.23
Fast	325	225	13.30	6.92	2.90	5.91	2.91	1.12	19.76
Industrial waterproofing (material #58)									
Roll 1st coat									
Slow	100	125	24.80	10.50	2.98	19.84	10.33	6.55	50.20
Medium	113	113	21.70	14.82	5.17	19.20	9.60	5.37	54.16
Fast	125	100	18.60	18.00	7.51	18.60	8.16	3.14	55.41
Roll 2nd or additional coats									
Slow	180	200	24.80	5.83	1.66	12.40	6.17	3.91	29.97
Medium	198	188	21.70	8.46	2.95	11.54	5.62	3.14	31.71
Fast	215	175	18.60	10.47	4.36	10.63	4.71	1.81	31.98

Use these figures to estimate the costs for finishing concrete walls which have a smooth surface (trowel), rough texture, or exposed aggregate finish. For wall heights above 10', increase the computed area by 50%. "Slow" work is based on a $10.50 hourly wage, "Medium" work on a $16.75 hourly wage, and "Fast" work on a $22.50 hourly wage. Other qualifications that apply to this table are on page 9.

	Labor SF per manhour	Material coverage SF/gallon	Material cost per gallon	Labor cost per 100 SF	Labor burden 100 SF	Material cost per 100 SF	Overhead per 100 SF	Profit per 100 SF	Total price per 100 SF
Walls, concrete tilt-up, spray application									
Flat latex, water base (material #5)									
Spray 1st coat									
Slow	500	225	17.10	2.10	.60	7.60	3.19	2.02	15.51
Medium	700	188	15.00	2.39	.84	7.98	2.74	1.53	15.48
Fast	900	150	12.90	2.50	1.04	8.60	2.25	.86	15.25
Spray 2nd or additional coats									
Slow	600	275	17.10	1.75	.50	6.22	2.63	1.67	12.77
Medium	800	238	15.00	2.09	.73	6.30	2.23	1.25	12.60
Fast	1000	200	12.90	2.25	.94	6.45	1.78	.69	12.11
Enamel, water base (material #9)									
Spray 1st coat									
Slow	500	225	24.50	2.10	.60	10.89	4.21	2.67	20.47
Medium	700	188	21.40	2.39	.84	11.38	3.58	2.00	20.19
Fast	900	150	18.40	2.50	1.04	12.27	2.92	1.12	19.85
Spray 2nd or additional coats									
Slow	600	275	24.50	1.75	.50	8.91	3.46	2.19	16.81
Medium	800	238	21.40	2.09	.73	8.99	2.89	1.62	16.32
Fast	1000	200	18.40	2.25	.94	9.20	2.29	.88	15.56
Enamel, oil base (material #10)									
Spray 1st coat									
Slow	500	200	26.60	2.10	.60	13.30	4.96	3.14	24.10
Medium	700	163	23.30	2.39	.84	14.29	4.29	2.40	24.21
Fast	900	125	20.00	2.50	1.04	16.00	3.61	1.39	24.54
Spray 2nd or additional coats									
Slow	600	300	26.60	1.75	.50	8.87	3.45	2.19	16.76
Medium	800	243	23.30	2.09	.73	9.59	3.04	1.70	17.15
Fast	1000	175	20.00	2.25	.94	11.43	2.70	1.04	18.36
Epoxy coating, 2 part system, clear (material #51)									
Spray 1st coat									
Slow	500	270	58.30	2.10	.60	21.59	7.53	4.77	36.59
Medium	700	253	51.00	2.39	.84	20.16	5.73	3.20	32.32
Fast	900	235	43.70	2.50	1.04	18.60	4.10	1.57	27.81
Spray 2nd or additional coats									
Slow	600	375	58.30	1.75	.50	15.55	5.52	3.50	26.82
Medium	800	338	51.00	2.09	.73	15.09	4.39	2.45	24.75
Fast	1000	300	43.70	2.25	.94	14.57	3.29	1.26	22.31

	Labor SF per manhour	Material coverage SF/gallon	Material cost per gallon	Labor cost per 100 SF	Labor burden 100 SF	Material cost per 100 SF	Overhead per 100 SF	Profit per 100 SF	Total price per 100 SF
Epoxy coating, 2 part system, white (material #52)									
Spray 1st coat									
Slow	500	270	61.70	2.10	.60	22.85	7.92	5.02	38.49
Medium	700	253	54.00	2.39	.84	21.34	6.02	3.36	33.95
Fast	900	235	46.30	2.50	1.04	19.70	4.30	1.65	29.19
Spray 2nd or additional coats									
Slow	600	375	61.70	1.75	.50	16.45	5.80	3.68	28.18
Medium	800	338	54.00	2.09	.73	15.98	4.61	2.58	25.99
Fast	1000	300	46.30	2.25	.94	15.43	3.44	1.32	23.38
Waterproofing, clear hydro sealer (material #34)									
Spray 1st coat									
Slow	500	125	17.80	2.10	.60	14.24	5.25	3.33	25.52
Medium	700	113	15.60	2.39	.84	13.81	4.17	2.33	23.54
Fast	900	100	13.30	2.50	1.04	13.30	3.12	1.20	21.16
Spray 2nd or additional coats									
Slow	600	200	17.80	1.75	.50	8.90	3.46	2.19	16.80
Medium	800	170	15.60	2.09	.73	9.18	2.94	1.64	16.58
Fast	1000	140	13.30	2.25	.94	9.50	2.35	.90	15.94

Use these figures to estimate the costs for finishing concrete walls which have a smooth surface (trowel), rough texture, or exposed aggregate finish. For wall heights above 10', increase the computed area by 50%. "Slow" work is based on a $10.50 hourly wage, "Medium" work on a $16.75 hourly wage, and "Fast" work on a $22.50 hourly wage. Other qualifications that apply to this table are on page 9.

	Labor LF per manhour	Material coverage LF/gallon	Material cost per gallon	Labor cost per 100 LF	Labor burden 100 LF	Material cost per 100 LF	Overhead per 100 LF	Profit per 100 LF	Total price per 100 LF
Windows, steel factory sash, brush application									
Metal primer, rust inhibitor - clean metal (material #35)									
Brush prime coat									
Slow	100	850	29.90	10.50	2.98	3.52	5.27	3.34	25.61
Medium	125	800	26.20	13.40	4.67	3.28	5.23	2.92	29.50
Fast	150	750	22.40	15.00	6.27	2.99	4.49	1.72	30.47
Metal primer, rust inhibitor - rusty metal (material #36)									
Brush prime coat									
Slow	100	800	36.60	10.50	2.98	4.58	5.60	3.55	27.21
Medium	125	750	32.10	13.40	4.67	4.28	5.48	3.06	30.89
Fast	150	700	27.50	15.00	6.27	3.93	4.66	1.79	31.65
Metal finish - synthetic enamel, gloss, interior or exterior, off white (material #37)									
Brush 1st coat									
Slow	125	900	27.90	8.40	2.38	3.10	4.31	2.73	20.92
Medium	150	850	24.40	11.17	3.90	2.87	4.39	2.46	24.79
Fast	175	800	20.90	12.86	5.35	2.61	3.86	1.48	26.16
Brush 2nd or additional coats									
Slow	150	1000	27.90	7.00	1.99	2.79	3.65	2.31	17.74
Medium	175	950	24.40	9.57	3.33	2.57	3.79	2.12	21.38
Fast	200	900	20.90	11.25	4.70	2.32	3.38	1.30	22.95
Metal finish - synthetic enamel, gloss, interior or exterior, colors, except orange & red (material #38)									
Brush 1st coat									
Slow	125	950	30.00	8.40	2.38	3.16	4.32	2.74	21.00
Medium	150	900	26.30	11.17	3.90	2.92	4.41	2.46	24.86
Fast	175	850	22.50	12.86	5.35	2.65	3.86	1.48	26.20
Brush 2nd or additional coats									
Slow	150	1050	30.00	7.00	1.99	2.86	3.67	2.33	17.85
Medium	175	1000	26.30	9.57	3.33	2.63	3.81	2.13	21.47
Fast	200	950	22.50	11.25	4.70	2.37	3.39	1.30	23.01

These figures will apply when painting steel factory sash but do not include work on glazing or frame. For heights above 8 feet, use the High Time Difficulty Factors on page 137. Note: A two coat system, prime and finish, using oil base material is recommended for any metal surface. Although water base material is often used, it may cause oxidation, corrosion and rust. One coat of oil base solid body stain is often used on interior or exterior metal but it may crack, peel or chip without the proper prime coat application. "Slow" work is based on a $10.50 hourly wage, "Medium" work on a $16.75 hourly wage, and "Fast" work on a $22.50 hourly wage. Other qualifications that apply to this table are on page 9.

Part IV

Wallcovering Costs

Adhesive coverage, rolls to yards conversion

	American rolls per gallon	Linear yards per gallon	Adhesive cost per gallon
Ready-mix			
Light weight vinyl (material #60)			
Slow	8.0	20.0	7.10
Medium	7.5	18.5	6.30
Fast	7.0	17.0	5.40
Heavy weight vinyl (material #61)			
Slow	6.0	15.0	7.30
Medium	5.5	14.0	6.40
Fast	5.0	13.0	5.50
Cellulose (material #62)			
Slow	8.0	19.0	7.80
Medium	7.0	17.5	6.90
Fast	6.0	16.0	5.90
Vinyl to vinyl (material #63)			
Slow	7.0	17.0	16.80
Medium	6.5	16.0	14.70
Fast	6.0	15.0	12.60
Powdered cellulose (material #64)			
Slow	12.0	30.0	2.70
Medium	11.0	27.5	2.40
Fast	10.0	25.0	2.00
Powdered vinyl (material #65)			
Slow	12.0	30.0	3.20
Medium	11.0	27.5	2.80
Fast	10.0	25.0	2.40
Powdered wheat paste (material #66)			
Slow	13.0	33.0	2.70
Medium	12.0	30.5	2.40
Fast	11.0	28.0	2.00

These figures are based on rolls or yards per gallon of liquid paste. Vinyl to vinyl ready-mix is usually distributed in pint containers which have been converted to gallons. Powdered adhesive coverage is based on water added to powder to prepare gallon quantities.

	Labor SF per manhour	Material coverage SF/gallon	Material cost per gallon	Labor cost per 100 SF	Labor burden 100 SF	Material cost per 100 SF	Overhead per 100 SF	Profit per 100 SF	Total cost per 100 SF
Adhesive coverage, per gallon of liquid paste									
Ready-mix									
Light weight vinyl (material #60)									
Slow	350	275	7.10	2.86	.81	2.58	1.94	1.23	9.42
Medium	375	250	6.30	4.20	1.46	2.52	2.00	1.12	11.30
Fast	400	225	5.40	5.25	2.19	2.40	1.82	.70	12.36
Heavy weight vinyl (material #61)									
Slow	275	200	7.30	3.64	1.04	3.65	2.58	1.64	12.55
Medium	300	175	6.40	5.25	1.83	3.66	2.63	1.47	14.84
Fast	325	150	5.50	6.46	2.70	3.67	2.37	.91	16.11
Cellulose (material #62)									
Slow	325	250	7.80	3.08	.88	3.12	2.19	1.39	10.66
Medium	350	225	6.90	4.50	1.57	3.07	2.24	1.25	12.63
Fast	375	200	5.90	5.60	2.34	2.95	2.01	.77	13.67
Vinyl to vinyl (material #63)									
Slow	300	225	16.80	3.33	.95	7.47	3.64	2.31	17.70
Medium	325	200	14.70	4.85	1.69	7.35	3.40	1.90	19.19
Fast	350	175	12.60	6.00	2.51	7.20	2.91	1.12	19.74
Powdered cellulose (material #64)									
Slow	425	400	2.70	2.35	.67	.68	1.15	.73	5.58
Medium	450	367	2.40	3.50	1.22	.65	1.32	.74	7.43
Fast	475	333	2.00	4.42	1.85	.60	1.27	.49	8.63
Powdered vinyl (material #65)									
Slow	425	400	3.20	2.35	.67	.80	1.18	.75	5.75
Medium	450	367	2.80	3.50	1.22	.76	1.34	.75	7.57
Fast	475	333	2.40	4.42	1.85	.72	1.29	.50	8.78
Powdered wheat paste (material #66)									
Slow	450	425	2.70	2.22	.63	.64	1.08	.69	5.26
Medium	475	400	2.40	3.32	1.16	.60	1.24	.70	7.02
Fast	500	375	2.00	4.20	1.75	.53	1.20	.46	8.14

These figures are based on gallon quantities of liquid paste. Vinyl to vinyl ready-mix is usually distributed in pint containers which have been converted to gallons. Powdered adhesive coverage is based on water added to powder to prepare gallon quantities. Typically, powdered, ready-mix material is in 2 to 4 ounce packages which will adhere 6 to 12 rolls of wallcovering. See the Adhesive coverage table on the previous page for conversion to rolls and yards. "Slow" work is based on a $10.00 hourly rate, "Medium" work on a $15.75 hourly rate, and "Fast" work on a $21.00 hourly rate.

Wallcovering application, average rates, medium rooms

	Labor rolls per day	Material by others	Labor cost per 10 rolls	Labor burden 10 rolls	Overhead per 10 rolls	Profit per 10 rolls	Total cost per 10 rolls	Total price per roll
Walls								
Slow	17	--	47.06	13.41	18.74	11.88	91.09	9.11
Medium	19	--	66.32	23.12	21.91	12.25	123.60	12.36
Fast	21	--	80.00	33.42	20.98	8.06	142.46	14.24
Ceilings								
Slow	15	--	53.33	15.20	21.24	13.46	103.23	10.32
Medium	17	--	74.12	25.84	24.49	13.69	138.14	13.81
Fast	19	--	88.42	36.94	23.19	8.91	157.46	15.74

The table above assumes that residential rolls are hand pasted. Add surface preparation time on page 420 as needed. For heights above 8 feet, use the High Time Difficulty Factors on page 137. "Slow" work is based on a $10.00 hourly wage, "Medium" work on a $15.75 hourly wage, and "Fast" work on a $21.00 hourly wage.

Wallcovering application, average rates, small rooms

	Labor rolls per day	Material by others	Labor cost per 10 rolls	Labor burden 10 rolls	Overhead per 10 rolls	Profit per 10 rolls	Total cost per 10 rolls	Total price per roll
Walls								
Slow	7	--	114.29	32.57	45.51	28.85	221.22	22.12
Medium	9	--	140.00	48.80	46.25	25.85	260.90	26.09
Fast	11	--	152.73	63.79	40.05	15.39	271.96	27.19
Ceilings								
Slow	6	--	133.33	38.00	53.09	33.65	258.07	25.80
Medium	8	--	157.50	54.90	52.04	29.09	293.53	29.35
Fast	10	--	168.00	70.16	44.06	16.93	299.15	29.91

The table above assumes that residential rolls are hand pasted. Add surface preparation time on page 420 as needed. For heights above 8 feet, use the High Time Difficulty Factors on page 137. "Slow" work is based on a $10.00 hourly wage, "Medium" work on a $15.75 hourly wage, and "Fast" work on a $21.00 hourly wage.

	Labor LF per manhour	Material by others	Labor cost per 100 LF	Labor burden 100 LF	Overhead per 100 LF	Profit per 100 LF	Total cost per 100 LF	Total price per LF
Borders 3" to 8" wide, commercial, machine pasted								
Medium size rooms (10 x 10 range)								
Slow	140	--	6.43	1.82	2.56	1.62	12.43	.12
Medium	158	--	9.02	3.15	3.77	2.39	18.33	.18
Fast	175	--	10.86	4.52	4.77	3.02	23.17	.23

The table above assumes that commercial rolls are machine pasted. Add surface preparation time on page 420 as needed. For heights above 8 feet, use the High Time Difficulty Factors on page 137. "Slow" work is based on a $9.00 hourly wage, "Medium" work on a $14.25 hourly wage, and "Fast" work on a $19.00 hourly wage.

	Labor LF per manhour	Material by others	Labor cost per 100 LF	Labor burden 100 LF	Overhead per 100 LF	Profit per 100 LF	Total cost per 100 LF	Total price per LF
Borders 3" to 8" wide, residential, hand pasted								
Medium size rooms (bedrooms, dining rooms)								
Slow	100	--	10.00	2.85	3.98	2.52	19.35	.19
Medium	113	--	13.94	4.86	4.61	2.58	25.99	.26
Fast	125	--	16.80	7.02	4.40	1.69	29.91	.30

The table above assumes that residential rolls are hand pasted. Add surface preparation time on page 420 as needed. For heights above 8 feet, use the High Time Difficulty Factors on page 137. "Slow" work is based on a $10.00 hourly wage, "Medium" work on a $15.75 hourly wage, and "Fast" work on a $21.00 hourly wage.

	Labor SF per manhour	Material by others	Labor cost per 100 SF	Labor burden 100 SF	Overhead per 100 SF	Profit per 100 SF	Total cost per 100 SF	Total price per SF

Flexible wood sheet and veneer

Wood veneer flexwood (residential or commercial)

Medium rooms (bedrooms, dining rooms, offices, reception areas)

	Labor SF per manhour	Material by others	Labor cost per 100 SF	Labor burden 100 SF	Overhead per 100 SF	Profit per 100 SF	Total cost per 100 SF	Total price per SF
Slow	14	--	67.86	19.36	27.02	17.13	131.37	1.31
Medium	20	--	75.00	26.15	24.78	13.85	139.78	1.40
Fast	26	--	76.92	32.11	20.17	7.75	136.95	1.37

Flexi-wall systems (residential or commercial)

Medium rooms (bedrooms, dining rooms, offices, reception areas)

	Labor SF per manhour	Material by others	Labor cost per 100 SF	Labor burden 100 SF	Overhead per 100 SF	Profit per 100 SF	Total cost per 100 SF	Total price per SF
Slow	12	--	79.17	22.58	31.52	19.98	153.25	1.53
Medium	18	--	83.33	29.07	27.53	15.39	155.32	1.55
Fast	24	--	83.33	34.80	21.85	8.40	148.38	1.48

Flexible wood sheet and veneer appears under section 09980 in the Construction Specifications Institute (CSI) indexing system. For heights above 8 feet, use the High Time Difficulty Factors on page 137. The labor rates in the table above are an average of the residential and commercial rates. Thus, "Slow" work is based on a $9.50 hourly rate, "Medium" work on a $15.00 hourly rate, and "Fast" work on a $20.00 hourly rate.

	Labor SF per manhour	Material by others	Labor cost per 100 SF	Labor burden 100 SF	Overhead per 100 SF	Profit per 100 SF	Total cost per 100 SF	Total price per SF

Surface preparation, wallcovering

Rule of thumb, typical preparation

	Labor SF per manhour	Material by others	Labor cost per 100 SF	Labor burden 100 SF	Overhead per 100 SF	Profit per 100 SF	Total cost per 100 SF	Total price per SF
Slow	100	--	10.00	2.85	2.38	.91	16.14	.16
Medium	125	--	12.60	4.39	3.14	1.21	21.34	.21
Fast	150	--	14.00	5.86	3.67	1.41	24.94	.25

Putty cracks, sand and wash

	Labor SF per manhour	Material by others	Labor cost per 100 SF	Labor burden 100 SF	Overhead per 100 SF	Profit per 100 SF	Total cost per 100 SF	Total price per SF
Slow	120	--	8.33	2.37	1.98	.76	13.44	.13
Medium	135	--	11.67	4.07	2.91	1.12	19.77	.20
Fast	150	--	14.00	5.86	3.67	1.41	24.94	.25

Rule of thumb: For surface preparation and sizing, allow 1 hour per room, maximum. Also see the Preparation operation tables. For heights above 8 feet, use the High Time Difficulty Factors on page 137. "Slow" work is based on a $10.00 hourly rate, "Medium" work on a $15.75 hourly rate, and "Fast" work on a $21.00 hourly rate.

	Labor rolls per day	Material by others	Labor cost per 10 rolls	Labor burden 10 rolls	Overhead per 10 rolls	Profit per 10 rolls	Total cost per 10 rolls	Total price per roll
Vinyl wallcover, commercial, machine pasted, single rolls per day 48" to 54" wide								
Cut-up areas (stair halls, landing areas)								
Walls								
Slow	16	--	45.00	12.80	17.92	11.36	87.08	8.71
Medium	18	--	63.33	22.08	20.92	11.70	118.03	11.80
Fast	20	--	76.00	31.72	19.93	7.66	135.31	13.53
Ceilings								
Slow	13	--	55.38	15.76	22.05	13.98	107.17	10.72
Medium	15	--	76.00	26.50	25.11	14.04	141.65	14.16
Fast	17	--	89.41	37.32	23.45	9.01	159.19	15.92
Small rooms (baths, utility rooms)								
Walls								
Slow	17	--	42.35	12.05	16.86	10.69	81.95	8.20
Medium	20	--	57.00	19.88	18.83	10.53	106.24	10.62
Fast	22	--	69.09	28.83	18.12	6.96	123.00	12.30
Ceilings								
Slow	15	--	48.00	13.65	19.11	12.12	92.88	9.29
Medium	18	--	63.33	22.08	20.92	11.70	118.03	11.80
Fast	20	--	76.00	31.72	19.93	7.66	135.31	13.53
Medium rooms (bedrooms, offices)								
Walls								
Slow	25	--	28.80	8.19	11.47	7.27	55.73	5.57
Medium	30	--	38.00	13.26	12.55	7.02	70.83	7.08
Fast	35	--	43.43	18.13	11.39	4.38	77.33	7.73
Ceilings								
Slow	22	--	32.73	9.30	13.03	8.26	63.32	6.33
Medium	27	--	42.22	14.73	13.95	7.80	78.70	7.87
Fast	32	--	47.50	19.83	12.46	4.79	84.58	8.46
Large rooms (conference rooms)								
Walls								
Slow	30	--	24.00	6.83	9.56	6.06	46.45	4.65
Medium	35	--	32.57	11.37	10.76	6.01	60.71	6.07
Fast	40	--	38.00	15.86	9.97	3.83	67.66	6.77
Ceilings								
Slow	26	--	27.69	7.88	11.03	6.99	53.59	5.36
Medium	31	--	36.77	12.84	12.15	6.79	68.55	6.85
Fast	36	--	42.22	17.62	11.07	4.26	75.17	7.52

	Labor rolls per day	Material by others	Labor cost per 10 rolls	Labor burden 10 rolls	Overhead per 10 rolls	Profit per 10 rolls	Total cost per 10 rolls	Total price per roll
Large wall areas (corridors, long hallways)								
Walls								
Slow	35	--	20.57	5.86	8.19	5.19	39.81	3.98
Medium	40	--	28.50	9.94	9.42	5.26	53.12	5.31
Fast	45	--	33.78	14.10	8.86	3.40	60.14	6.01
Ceilings								
Slow	30	--	24.00	6.83	9.56	6.06	46.45	4.65
Medium	35	--	32.57	11.37	10.76	6.01	60.71	6.07
Fast	40	--	38.00	15.86	9.97	3.83	67.66	6.77
Paper-backed vinyl on medium room walls								
Bedrooms, dining rooms								
Slow	17	--	42.35	12.05	16.86	10.69	81.95	8.20
Medium	21	--	54.29	18.94	17.94	10.03	101.20	10.12
Fast	25	--	60.80	25.38	15.94	6.13	108.25	10.83
Cork wallcovering on medium room walls								
Bedrooms, dining rooms								
Slow	17	--	42.35	12.05	16.86	10.69	81.95	8.20
Medium	21	--	54.29	18.94	17.94	10.03	101.20	10.12
Fast	25	--	60.80	25.38	15.94	6.13	108.25	10.83

Vinyl coated wallcovering appears under section 09955 in the Construction Specifications Institute (CSI) indexing system. Vinyl wallcovering appears in section 09960. Cork wallcovering appears in section 09965. The table above assumes that commercial rolls are machine pasted. Add surface preparation time on page 420 as needed. For heights above 8 feet, use the High Time Difficulty Factors on page 137. "Slow" work is based on a $9.00 hourly wage, "Medium" work on a $14.25 hourly wage, and "Fast" work on a $19.00 hourly wage.

	Labor rolls per day	Material by others	Labor cost per 10 rolls	Labor burden 10 rolls	Overhead per 10 rolls	Profit per 10 rolls	Total cost per 10 rolls	Total price per roll
Vinyl wallcovering, residential, hand pasted, single rolls per day 18" to 27" wide								
Cut-up areas (stair halls, landing areas)								
Walls								
Slow	10	--	80.00	22.80	31.86	20.19	154.85	15.48
Medium	11	--	114.55	39.93	37.85	21.16	213.49	21.35
Fast	12	--	140.00	58.48	36.71	14.11	249.30	24.93
Ceilings								
Slow	7	--	114.29	32.57	45.51	28.85	221.22	22.12
Medium	8	--	157.50	54.90	52.04	29.09	293.53	29.35
Fast	9	--	186.67	77.96	48.95	18.81	332.39	33.24
Small rooms (baths, utility rooms)								
Walls								
Slow	11	--	72.73	20.73	28.96	18.36	140.78	14.07
Medium	12	--	105.00	36.61	34.69	19.39	195.69	19.57
Fast	13	--	129.23	53.97	33.89	13.02	230.11	23.01
Ceilings								
Slow	9	--	88.89	25.33	35.40	22.44	172.06	17.20
Medium	10	--	126.00	43.92	41.63	23.27	234.82	23.48
Fast	11	--	152.73	63.79	40.05	15.39	271.96	27.19
Medium rooms (bedrooms, dining rooms)								
Walls								
Slow	17	--	47.06	13.41	18.74	11.88	91.09	9.11
Medium	19	--	66.32	23.12	21.91	12.25	123.60	12.36
Fast	21	--	80.00	33.42	20.98	8.06	142.46	14.24
Ceilings								
Slow	15	--	53.33	15.20	21.24	13.46	103.23	10.32
Medium	17	--	74.12	25.84	24.49	13.69	138.14	13.81
Fast	19	--	88.42	36.94	23.19	8.91	157.46	15.74
Large rooms (living rooms)								
Walls								
Slow	20	--	40.00	11.40	15.93	10.10	77.43	7.74
Medium	23	--	54.78	19.09	18.10	10.12	102.09	10.21
Fast	26	--	64.62	26.98	16.95	6.51	115.06	11.51
Ceilings								
Slow	18	--	44.44	12.67	17.69	11.22	86.02	8.60
Medium	20	--	63.00	21.96	20.82	11.64	117.42	11.74
Fast	22	--	76.36	31.88	20.02	7.70	135.96	13.60

	Labor rolls per day	Material by others	Labor cost per 10 rolls	Labor burden 10 rolls	Overhead per 10 rolls	Profit per 10 rolls	Total cost per 10 rolls	Total price per roll
Large wall areas (corridors, long hallways)								
Walls								
Slow	24	--	33.33	9.50	13.27	8.41	64.51	6.45
Medium	27	--	46.67	16.26	15.42	8.62	86.97	8.70
Fast	30	--	56.00	23.40	14.69	5.64	99.73	9.97
Ceilings								
Slow	21	--	38.10	10.86	15.17	9.62	73.75	7.37
Medium	23	--	54.78	19.09	18.10	10.12	102.09	10.21
Fast	25	--	67.20	28.06	17.62	6.77	119.65	11.97
Paper-backed vinyl on medium room walls								
Bedrooms, dining rooms								
Slow	8	--	100.00	28.50	39.82	25.24	193.56	19.35
Medium	10	--	126.00	43.92	41.63	23.27	234.82	23.48
Fast	12	--	140.00	58.48	36.71	14.11	249.30	24.93
Cork wallcovering on medium room walls								
Bedrooms, dining rooms								
Slow	8	--	100.00	28.50	39.82	25.24	193.56	19.35
Medium	10	--	126.00	43.92	41.63	23.27	234.82	23.48
Fast	12	--	140.00	58.48	36.71	14.11	249.30	24.93

Vinyl coated wallcovering appears under section 09955 in the Construction Specifications Institute (CSI) indexing system. Vinyl wallcovering appears in section 09960. Cork wallcovering appears in section 09965. The table above assumes that residential rolls are hand pasted. Add surface preparation time on page 420 as needed. For heights above 8 feet, use the High Time Difficulty Factors on page 137. "Slow" work is based on a $10.00 hourly wage, "Medium" work on a $15.75 hourly wage, and "Fast" work on a $21.00 hourly wage. Add for ready-mix paste.

	Labor rolls per day	Material by others	Labor cost per 10 rolls	Labor burden 10 rolls	Overhead per 10 rolls	Profit per 10 rolls	Total cost per 10 rolls	Total price per roll

Wall fabric, commercial, machine pasted, single rolls per day, 48" to 54" wide

	Labor rolls per day	Material by others	Labor cost per 10 rolls	Labor burden 10 rolls	Overhead per 10 rolls	Profit per 10 rolls	Total cost per 10 rolls	Total price per roll
Coated fabric								
Cut-up areas (stair halls, landing areas)								
Walls								
Slow	23	--	31.30	8.91	12.46	7.90	60.57	6.06
Medium	25	--	45.60	15.90	15.07	8.42	84.99	8.50
Fast	27	--	56.30	23.49	14.76	5.67	100.22	10.02
Ceilings								
Slow	14	--	51.43	14.62	20.48	12.98	99.51	9.95
Medium	16	--	71.25	24.85	23.54	13.16	132.80	13.28
Fast	18	--	84.44	35.24	22.14	8.51	150.33	15.03
Small rooms (baths, utility rooms)								
Walls								
Slow	24	--	30.00	8.53	11.95	7.57	58.05	5.81
Medium	26	--	43.85	15.29	14.49	8.10	81.73	8.17
Fast	28	--	54.29	22.65	14.24	5.47	96.65	9.67
Ceilings								
Slow	16	--	45.00	12.80	17.92	11.36	87.08	8.71
Medium	19	--	60.00	20.94	19.82	11.08	111.84	11.18
Fast	22	--	69.09	28.83	18.12	6.96	123.00	12.30
Medium rooms (bedrooms, offices)								
Walls								
Slow	27	--	26.67	7.58	10.62	6.73	51.60	5.16
Medium	32	--	35.63	12.42	11.77	6.58	66.40	6.64
Fast	37	--	41.08	17.14	10.77	4.14	73.13	7.31
Ceilings								
Slow	24	--	30.00	8.53	11.95	7.57	58.05	5.81
Medium	29	--	39.31	13.72	12.99	7.26	73.28	7.33
Fast	34	--	44.71	18.66	11.73	4.51	79.61	7.96
Large rooms (conference rooms)								
Walls								
Slow	34	--	21.18	6.02	8.44	5.35	40.99	4.10
Medium	39	--	29.23	10.19	9.66	5.40	54.48	5.45
Fast	44	--	34.55	14.41	9.06	3.48	61.50	6.15
Ceilings								
Slow	27	--	26.67	7.58	10.62	6.73	51.60	5.16
Medium	32	--	35.63	12.42	11.77	6.58	66.40	6.64
Fast	37	--	41.08	17.14	10.77	4.14	73.13	7.31

	Labor rolls per day	Material by others	Labor cost per 10 rolls	Labor burden 10 rolls	Overhead per 10 rolls	Profit per 10 rolls	Total cost per 10 rolls	Total price per roll
Large wall areas (corridors, long hallways)								
Walls								
Slow	38	--	18.95	5.38	7.55	4.78	36.66	3.67
Medium	44	--	25.91	9.03	8.56	4.79	48.29	4.83
Fast	50	--	30.40	12.69	7.97	3.06	54.12	5.41
Ceilings								
Slow	31	--	23.23	6.61	9.25	5.86	44.95	4.50
Medium	36	--	31.67	11.04	10.46	5.85	59.02	5.90
Fast	41	--	37.07	15.47	9.72	3.74	66.00	6.60

Canvas sheeting: see Wall fabric, residential, hand pasted

Grasscloth: see Wall fabric, residential, hand pasted

Burlap: see Wall fabric, residential, hand pasted

Natural fabric, silk: see Wall fabric, residential, hand pasted

Natural fabric, felt, linen, cotton: see Wall fabric, residential, hand pasted

Wall fabric appears under section 09975 in the Construction Specifications Institute (CSI) indexing system. The table above assumes that commercial rolls are machine pasted. Add surface preparation time on page 420 as needed. For heights above 8 feet, use the High Time Difficulty Factors on page 137. "Slow" work is based on a $9.00 hourly wage "Medium" work on a $14.25 hourly wage, and "Fast" work on a $19.00 hourly wage. Add for ready-mix paste.

	Labor rolls per day	Material by others	Labor cost per 10 rolls	Labor burden 10 rolls	Overhead per 10 rolls	Profit per 10 rolls	Total cost per 10 rolls	Total price per roll
Wall fabric, residential, hand pasted, single rolls per day 18" to 27" wide								
Coated fabrics								
Cut-up areas (stair halls, landing areas)								
Walls								
Slow	12	--	66.67	19.00	26.55	16.83	129.05	12.90
Medium	13	--	96.92	33.79	32.02	17.90	180.63	18.06
Fast	14	--	120.00	50.11	31.47	12.09	213.67	21.37
Ceilings								
Slow	8	--	100.00	28.50	39.82	25.24	193.56	19.35
Medium	9	--	140.00	48.80	46.25	25.85	260.90	26.09
Fast	10	--	168.00	70.16	44.06	16.93	299.15	29.91
Small rooms (baths, utility rooms)								
Walls								
Slow	13	--	61.54	17.54	24.51	15.53	119.12	11.91
Medium	14	--	90.00	31.37	29.74	16.62	167.73	16.77
Fast	15	--	112.00	46.76	29.37	11.29	199.42	19.94
Ceilings								
Slow	10	--	80.00	22.80	31.86	20.19	154.85	15.48
Medium	11	--	114.55	39.93	37.85	21.16	213.49	21.35
Fast	12	--	140.00	58.48	36.71	14.11	249.30	24.93
Medium rooms (bedrooms, dining rooms)								
Walls								
Slow	18	--	44.44	12.67	17.69	11.22	86.02	8.60
Medium	20	--	63.00	21.96	20.82	11.64	117.42	11.74
Fast	22	--	76.36	31.88	20.02	7.70	135.96	13.60
Ceilings								
Slow	16	--	50.00	14.25	19.91	12.62	96.78	9.68
Medium	18	--	70.00	24.39	23.13	12.93	130.45	13.05
Fast	20	--	84.00	35.08	22.03	8.47	149.58	14.96
Large rooms (living rooms)								
Walls								
Slow	22	--	36.36	10.36	14.48	9.18	70.38	7.04
Medium	25	--	50.40	17.57	16.65	9.31	93.93	9.39
Fast	28	--	60.00	25.05	15.73	6.05	106.83	10.68
Ceilings								
Slow	19	--	42.11	12.00	16.77	10.63	81.51	8.15
Medium	21	--	60.00	20.92	19.82	11.08	111.82	11.18
Fast	23	--	73.04	30.50	19.15	7.36	130.05	13.00

	Labor rolls per day	Material by others	Labor cost per 10 rolls	Labor burden 10 rolls	Overhead per 10 rolls	Profit per 10 rolls	Total cost per 10 rolls	Total price per roll
Large wall areas (corridors, long hallways)								
Walls								
Slow	26	--	30.77	8.77	12.25	7.77	59.56	5.95
Medium	29	--	43.45	15.15	14.35	8.02	80.97	8.10
Fast	32	--	52.50	21.93	13.77	5.29	93.49	9.35
Ceilings								
Slow	22	--	36.36	10.36	14.48	9.18	70.38	7.04
Medium	24	--	52.50	18.29	17.35	9.70	97.84	9.79
Fast	26	--	64.62	26.98	16.95	6.51	115.06	11.51
Canvas sheeting								
Medium room walls (bedrooms, dining rooms)								
Walls								
Slow	14	--	57.14	16.28	22.75	14.42	110.59	11.06
Medium	16	--	78.75	27.45	26.02	14.54	146.76	14.68
Fast	17	--	98.82	41.28	25.91	9.96	175.97	17.60
Grasscloth								
Medium room walls (bedrooms, dining rooms)								
Walls								
Slow	15	--	53.33	15.20	21.24	13.46	103.23	10.32
Medium	20	--	63.00	21.96	20.82	11.64	117.42	11.74
Fast	24	--	70.00	29.22	18.36	7.06	124.64	12.47
Burlap								
Medium rooms (bedrooms, dining rooms)								
Walls								
Slow	10	--	80.00	22.80	31.86	20.19	154.85	15.48
Medium	14	--	90.00	31.37	29.74	16.62	167.73	16.77
Fast	18	--	93.33	38.97	24.48	9.41	166.19	16.62

	Labor rolls per day	Material by others	Labor cost per 10 rolls	Labor burden 10 rolls	Overhead per 10 rolls	Profit per 10 rolls	Total cost per 10 rolls	Total price per roll
Natural fabric, silk								
Medium rooms (bedrooms, dining rooms)								
Walls								
Slow	8	--	100.00	28.50	39.82	25.24	193.56	19.35
Medium	12	--	105.00	36.61	34.69	19.39	195.69	19.57
Fast	15	--	112.00	46.76	29.37	11.29	199.42	19.94
Natural fabric, felt, linen, cotton								
Medium rooms (bedrooms, dining rooms)								
Walls								
Slow	7	--	114.29	32.57	45.51	28.85	221.22	22.12
Medium	9	--	140.00	48.80	46.25	25.85	260.90	26.09
Fast	11	--	152.73	63.79	40.05	15.39	271.96	27.19

Wall fabric appears under section 09975 in the Construction Specifications Institute (CSI) indexing system. The table above assumes that residential rolls are hand pasted. Add surface preparation time on page 420 as needed. For heights above 8 feet, use the High Time Difficulty Factors on page 137. "Slow" work is based on a $10.00 hourly wage "Medium" work on a $15.75 hourly wage, and "Fast" work on a $21.00 hourly wage. Add for ready-mix paste.

	Labor rolls per day	Material by others	Labor cost per 10 rolls	Labor burden 10 rolls	Overhead per 10 rolls	Profit per 10 rolls	Total cost per 10 rolls	Total price per roll

Wallpaper, commercial, machine pasted, single rolls per day, 48" to 54" wide

	Labor rolls per day	Material by others	Labor cost per 10 rolls	Labor burden 10 rolls	Overhead per 10 rolls	Profit per 10 rolls	Total cost per 10 rolls	Total price per roll
Blind stock (lining)								
Cut-up areas (stair halls, landing areas)								
Walls								
Slow	19	--	37.89	10.79	15.09	9.56	73.33	7.33
Medium	21	--	54.29	18.94	17.94	10.03	101.20	10.12
Fast	23	--	66.09	27.57	17.33	6.66	117.65	11.77
Ceilings								
Slow	16	--	45.00	12.80	17.92	11.36	87.08	8.71
Medium	18	--	63.33	22.08	20.92	11.70	118.03	11.80
Fast	20	--	76.00	31.72	19.93	7.66	135.31	13.53
Small rooms (baths, utility rooms)								
Walls								
Slow	21	--	34.29	9.75	13.66	8.66	66.36	6.64
Medium	25	--	45.60	15.90	15.07	8.42	84.99	8.50
Fast	29	--	52.41	21.89	13.74	5.28	93.32	9.33
Ceilings								
Slow	17	--	42.35	12.05	16.86	10.69	81.95	8.20
Medium	20	--	57.00	19.88	18.83	10.53	106.24	10.62
Fast	22	--	69.09	28.83	18.12	6.96	123.00	12.30
Medium rooms (bedrooms, offices)								
Walls								
Slow	29	--	24.83	7.06	9.89	6.27	48.05	4.81
Medium	35	--	32.57	11.37	10.76	6.01	60.71	6.07
Fast	41	--	37.07	15.47	9.72	3.74	66.00	6.60
Ceilings								
Slow	25	--	28.80	8.19	11.47	7.27	55.73	5.57
Medium	30	--	38.00	13.26	12.55	7.02	70.83	7.08
Fast	34	--	44.71	18.66	11.73	4.51	79.61	7.96
Large rooms (conference rooms)								
Walls								
Slow	35	--	20.57	5.86	8.19	5.19	39.81	3.98
Medium	41	--	27.80	9.70	9.19	5.13	51.82	5.18
Fast	47	--	32.34	13.49	8.48	3.26	57.57	5.76
Ceilings								
Slow	29	--	24.83	7.06	9.89	6.27	48.05	4.81
Medium	34	--	33.53	11.69	11.08	6.19	62.49	6.25
Fast	38	--	40.00	16.69	10.49	4.03	71.21	7.12

	Labor rolls per day	Material by others	Labor cost per 10 rolls	Labor burden 10 rolls	Overhead per 10 rolls	Profit per 10 rolls	Total cost per 10 rolls	Total price per roll
Large wall areas (corridors, long hallways)								
Walls								
Slow	40	--	18.00	5.12	7.17	4.54	34.83	3.48
Medium	45	--	25.33	8.84	8.37	4.68	47.22	4.72
Fast	50	--	30.40	12.69	7.97	3.06	54.12	5.41
Ceilings								
Slow	33	--	21.82	6.20	8.69	5.51	42.22	4.22
Medium	38	--	30.00	10.46	9.91	5.54	55.91	5.59
Fast	42	--	36.19	15.11	9.49	3.65	64.44	6.44
Ordinary pre-trimmed wallpaper or butt joint work								
Medium room walls (bedrooms, offices)								
Walls								
Slow	26	--	27.69	7.88	11.03	6.99	53.59	5.36
Medium	31	--	36.77	12.84	12.15	6.79	68.55	6.85
Fast	36	--	42.22	17.62	11.07	4.26	75.17	7.52
Ceilings								
Slow	23	--	31.30	8.91	12.46	7.90	60.57	6.06
Medium	28	--	40.71	14.20	13.45	7.52	75.88	7.59
Fast	32	--	47.50	19.83	12.46	4.79	84.58	8.46
Hand-crafted wallpaper								
Medium room walls (bedrooms, offices)								
Walls								
Slow	20	--	36.00	10.24	14.33	9.09	69.66	6.97
Medium	24	--	47.50	16.56	15.69	8.77	88.52	8.85
Fast	28	--	54.29	22.65	14.24	5.47	96.65	9.67
Ceilings								
Slow	18	--	40.00	11.37	15.93	10.10	77.40	7.74
Medium	21	--	54.29	18.94	17.94	10.03	101.20	10.12
Fast	23	--	66.09	27.57	17.33	6.66	117.65	11.77
Flock wallpaper, medium rooms (bedrooms, offices)								
Walls								
Slow	14	--	51.43	14.62	20.48	12.98	99.51	9.95
Medium	17	--	67.06	23.39	22.16	12.38	124.99	12.50
Fast	20	--	76.00	31.72	19.93	7.66	135.31	13.53
Foil wallpaper, medium rooms (bedrooms, offices)								
Walls								
Slow	13	--	55.38	15.76	22.05	13.98	107.17	10.72
Medium	16	--	71.25	24.85	23.54	13.16	132.80	13.28
Fast	19	--	80.00	33.40	20.98	8.06	142.44	14.24

	Labor rolls per day	Material by others	Labor cost per 10 rolls	Labor burden 10 rolls	Overhead per 10 rolls	Profit per 10 rolls	Total cost per 10 rolls	Total price per roll
Canvas wallpaper, medium rooms (bedrooms, offices)								
Walls								
Slow	12	--	60.00	17.07	23.89	15.14	116.10	11.61
Medium	14	--	81.43	28.39	26.90	15.04	151.76	15.18
Fast	16	--	95.00	39.65	24.91	9.57	169.13	16.91
Scenic wallpaper, medium rooms (bedrooms, offices)								
Walls								
Slow	16	--	45.00	12.80	17.92	11.36	87.08	8.71
Medium	18	--	63.33	22.08	20.92	11.70	118.03	11.80
Fast	20	--	76.00	31.72	19.93	7.66	135.31	13.53

Wallpaper appears under section 09970 in the Construction Specifications Institute (CSI) indexing system. The table above assumes that commercial rolls are machine pasted. Add surface preparation time on page 420 as needed. For heights above 8 feet, use the High Time Difficulty Factors on page 137. "Slow" work is based on a $9.00 hourly wage, "Medium" work on a $14.25 hourly wage, and "Fast" work on a $19.00 hourly wage. Add for ready-mix paste.

National Estimator Quick Start

Use the Quick Start on the next ten pages to get familiar with National Estimator. In less than an hour you'll be printing labor and material cost estimates for your jobs.

To install National Estimator, put the National Estimator disk in a disk drive (such as A:). Start Windows™.

In Windows 3.1 or 3.11
go to the Program Manager.

1. Click on File.
2. Click on Run . . .
3. Type A:SETUP
4. Press Enter.

In Windows 95,

1. Click on Start
2. Click on Run . . .
3. Type A:SETUP
4. Press Enter.

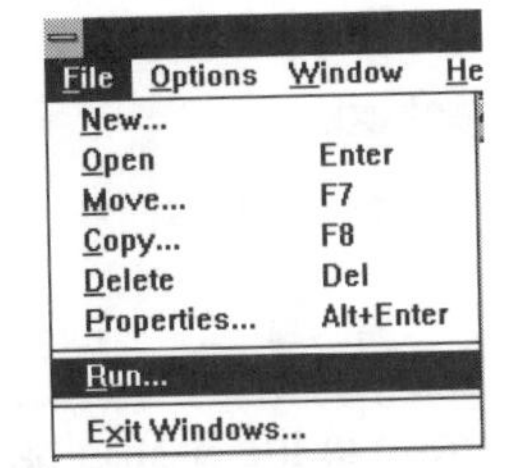

Click on File. Click on Run.

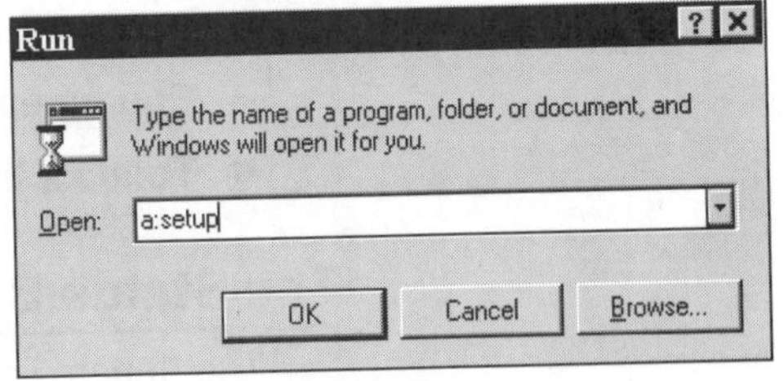

Type A:SETUP in Windows 95.

National Estimator Icon

Then follow instructions on the screen. Installation should take about two minutes. We recommend installing all National Estimator files to the NATIONAL directory. If you have trouble installing National Estimator, call 619-438-7828.

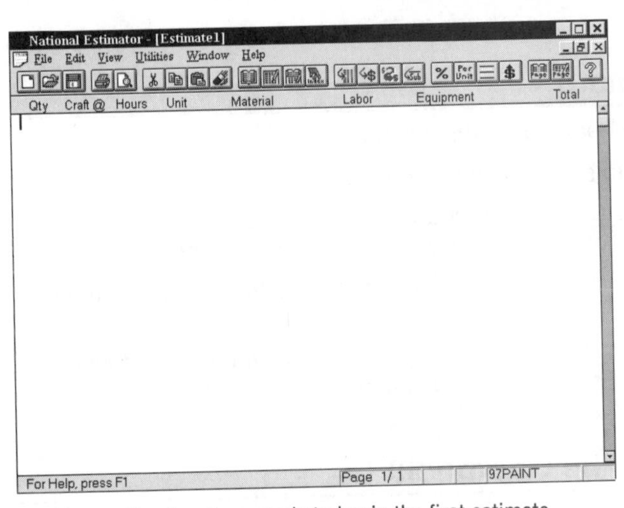

Your estimating form ready to begin the first estimate.

The installation program will create a Construction Estimating group on Program Manager and put the National Estimator Icon in that group. When installation is complete, click on OK to begin using National Estimator. In a few seconds your electronic estimating form will be ready to begin the first estimate.

On the title bar at the top of the screen you see the program name, National Estimator, and [Estimate1], showing that Estimate1 is on the screen. Let's take a closer look at other information at the top of your screen.

The Menu Bar

Below the title bar you see the menu bar: Every option in National Estimator is available on the menu bar. Click with your left mouse button on any item on the menu to open a list of available commands.

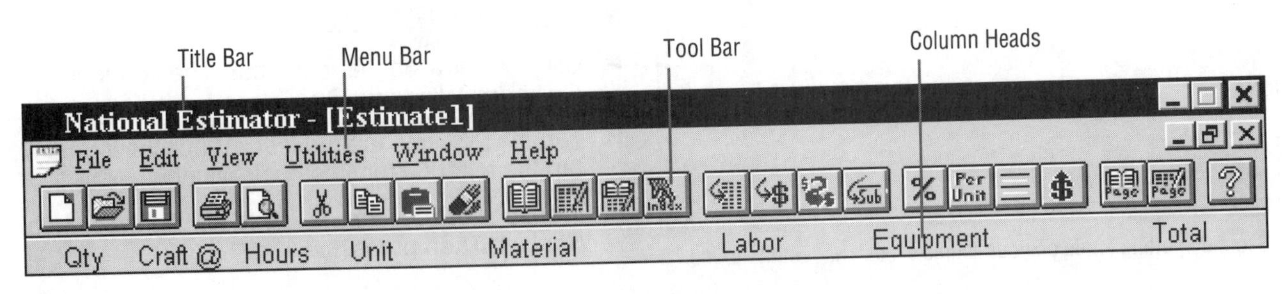

Buttons on the Tool Bar

Below the menu bar you see 24 buttons that make up the tool bar. The options you use most in National Estimator are only a mouse click away on the tool bar.

Column Headings

Below the tool bar you'll see column headings for your estimate form:

- **Qty** for quantity
- **Craft@Hours** for craft (the crew doing the work) and manhours (to complete the task)
- **Unit** for unit of measure, such as per linear foot or per square foot
- **Material** for material cost
- **Labor** for labor cost
- **Equipment** for equipment cost
- **Total** for the total of all cost columns

The Status Bar

The bottom line on your screen is the status bar. Here you'll find helpful information about the choices available. Notice "Page 1/ 1" near the right end of the status line. That's a clue that you're looking at page 1 of a one-page estimate.

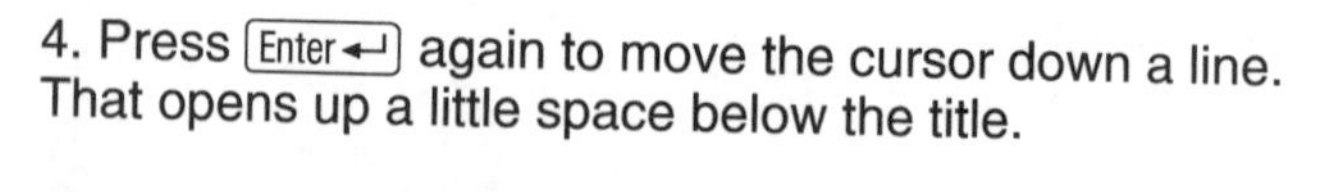

Check the status bar occasionally for helpful tips and explanations of what you see on screen.

Beginning an Estimate

The Blinking Cursor (insert point)

Mouse Pointer

Mouse Pointer

Let's start by putting a heading on this estimate.

1. Press [Enter↵] once to space down one line.
2. Press [Tab] four times (or hold the space bar down) to move the Blinking Cursor (the insert point) near the middle of the line.
3. Type "Estimate One" and press [Enter↵]. That's the title of this estimate, "Estimate One."
4. Press [Enter↵] again to move the cursor down a line. That opens up a little space below the title.

Begin by putting a title on your estimate, such as "Estimate One."

The Costbook

Let's leave your estimating form for a moment to take a look at estimates stored in the costbook. To switch to the costbook, either:

- Click on the [costbook] button, -Or-
- Click on View on the menu bar. Then click on Costbook Window, -Or-
- Tap the [Alt] key, tap the letter V (for View) and tap the letter C (for Costbook Window), -Or-
- Press [Esc]. Press the [↓] key to highlight Costbook Window. Then press [Enter↵].

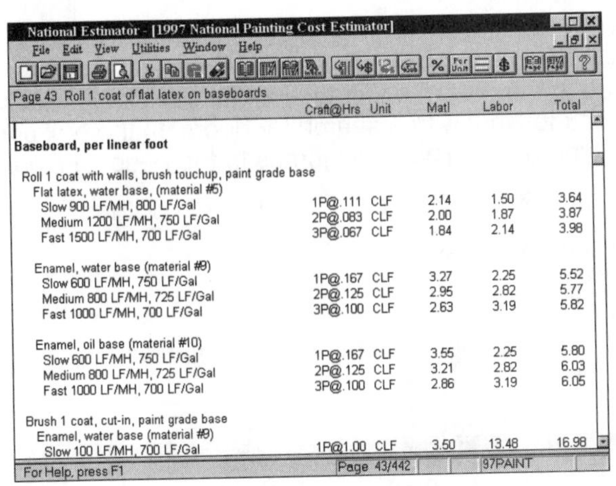

The costbook window has the entire National Painting Cost Estimator.

The entire 1997 National Painting Cost Estimator is available in the Costbook Window. Notice the words *Page 43 Roll 1 coat of flat latex on baseboards* at the left side of the screen just below the tool bar. That's your clue that the baseboard section of page 43 is on the screen. To turn to the next page, either:

- Press PgDn (with Num Lock off), -Or-
- Click on the lower half of the scroll bar at the right edge of the screen.

To move down one line at a time, either:

- Press the ↓ key (with Num Lock off), -Or-
- Click on the arrow on the down scroll bar at the lower right corner of the screen.

Press PgDn about 1,400 times and you'll page through the entire National Painting Cost Estimator. Obviously, there's a better way. To turn quickly to any page, either:

- Click on the Page button near the right end of the tool bar, -Or-
- Click on View on the menu bar. Then click on Turn to Costbook Page, -Or-
- Tap the Alt key, tap the letter V (for View) and tap the letter T (for Turn to Costbook page), -Or-
- Press Esc. Press the ↓ key to highlight Turn to Costbook Page. Then press Enter↵.

Type the page number you want to see.

Type the number of the page you want to see and press Enter↵. National Estimator will turn to the top of the page you requested.

An Even Better Way

Find the small square in the slide bar at the right side of the Costbook Window. Click and hold on that square while rolling the mouse up or down. Keep dragging the square until you see the page you want in the Page: box. Release the mouse button to turn to the top of that page.

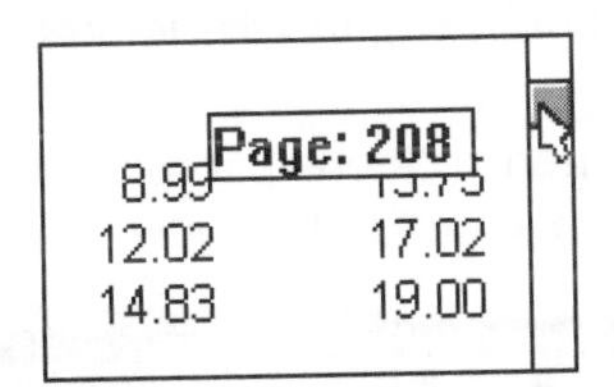

Drag the square to see any page.

A Still Better Way: Keyword Search

To find any cost estimate in seconds, search by keyword in the index. To go to the index, either:

- Click on the Index button near the center of the tool bar, -Or-
- Press Esc. Press Enter↵, -Or-
- Tap the Alt key, tap the letter V (for View) and press Enter↵, -Or-
- Click on View on the menu bar. Then press Enter↵.

Notice that the cursor is blinking in the Enter Keyword to Locate box at the right of the screen. Obviously, the index is ready to begin a search.

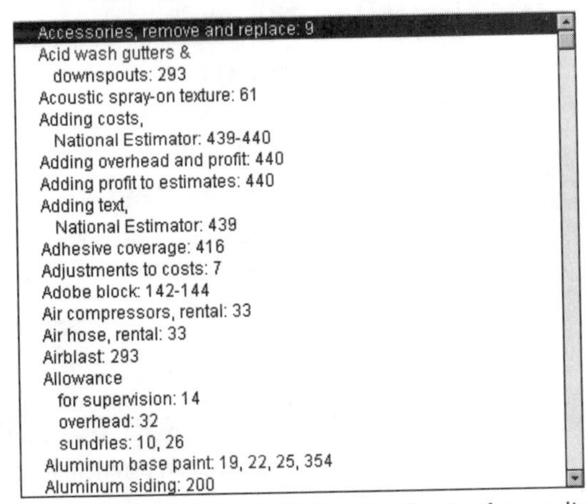

Use the electronic index to find cost estimates for any item.

Your First Estimate

Enter Keyword to Locate:

doors

Enter keyword to locate

Suppose we're estimating the cost of painting eight interior doors with brush and roller. Let's put the index to work with a search for doors. In the box under Enter Keyword to Locate, type *doors*. The index jumps to the heading *Doors*.

The sixth item under Doors is *interior: 101, 107, 109, 111*. Either:

- Click once on that line and press [Enter←], -Or-
- Double click on that line, -Or-
- Press [Tab] and the [↓] key to move the highlight to *interior: 101, 107, 109, 111*. Then press [Enter←].

The index jumps to *Doors*.

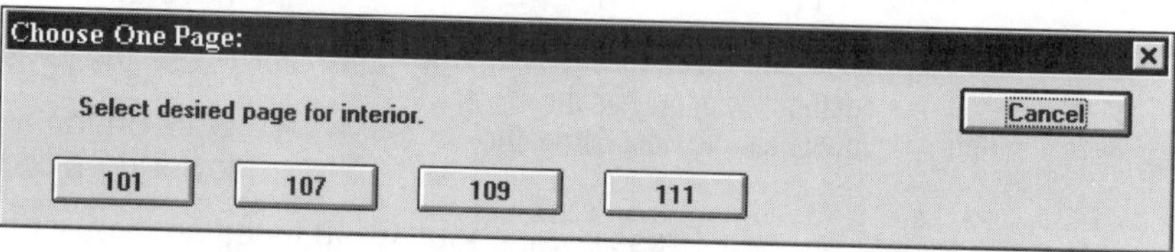

If costs appear on several pages, click on the page you prefer.

To select the page you want to see (page 107 in this case), either:

- Click on number 107, -Or-
- Press [Tab] to highlight 107. Then press [Enter←].

National Estimator turns to the top of page 107. See the example at the left below. Notice that estimates at the top of this page are for painting interior flush doors.

Splitting the Screen

Most of the time you'll want to see what's in both the costbook and your estimate. To split the screen into two halves, either:

- Click on the [button] button near the center of the tool bar, -Or-
- Press [Esc] and the [↓] key to move the selection bar to Split Window, -Or-
- Tap the [Alt] key, tap the letter V (for View), tap the letter S (for Split Window), -Or-
- Click on View on the menu bar. Then click on split window and your screen should look like the example at the right below.

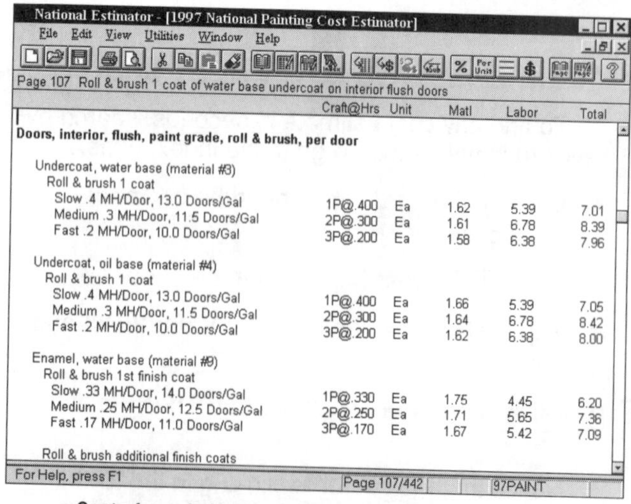

Costs for painting interior flush doors on page 107.

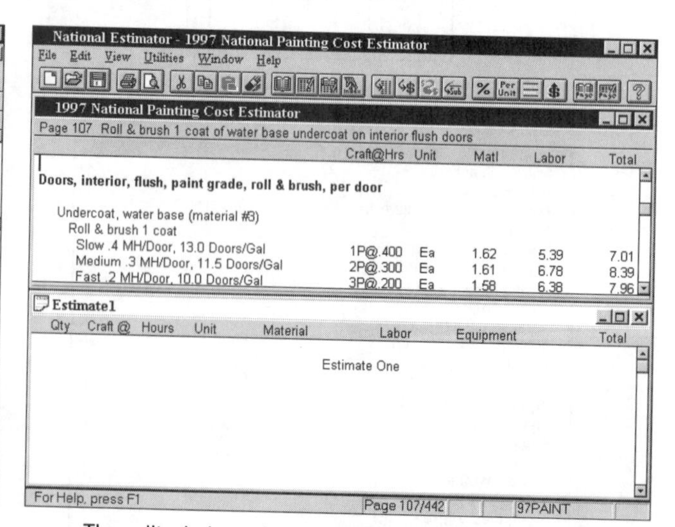

The split window: Costbook above and estimate below.

Notice that eight lines of the costbook are at the top of the screen and your estimate is at the bottom. You should recognize "Estimate One." It's your title for this estimate. Column headings are at the top of the costbook and across the middle of the screen (for your estimate).

To Switch from Window to Window

- Click in the window of your choice, -Or-
- Hold the Ctrl key down and press Tab.

Notice that a window title bar turns dark when that window is selected. The selected window is where keystrokes appear as you type.

Copying Costs to Your Estimate

Next, we'll estimate the cost of painting eight doors. Click on the [button] button on the tool bar to be sure you're in the split window. Click anywhere in the costbook (the top half of your screen). Then press the ↓ key until the cursor is on the line:

Medium .3 MH/Door, 11.5 Doors/Gal	2P@.300	Ea	1.61	6.78	8.39

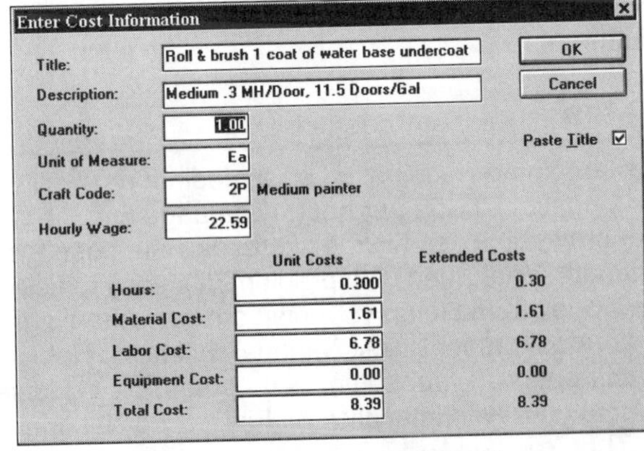

Use the Enter Cost Information dialog box to copy or change costs.

Enter Cost Information

Title:	Roll & brush 1 coat of water base undercoat c	
Description:	Medium .3 MH/Door, 11.5 Doors/Gal	
Quantity:	8.00	
Unit of Measure:	Ea	
Craft Code:	2P Medium painter	
Hourly Wage:	22.59	
	Unit Costs	**Extended Costs**
Hours:	0.300	2.40
Material Cost:	1.61	12.88
Labor Cost:	6.78	54.24
Equipment Cost:	0.00	0.00
Total Cost:	8.39	67.12

OK / Cancel / Paste Title

Costs for the eight doors (extended costs) are on the right.

To copy this line to your estimate:

1. Click on the line.
2. Click on the [button] button.
3. Click on the [button] button to open the Enter Cost Information dialog box.

Notice that the blinking cursor is in the Quantity box:

1. Type a quantity of 8 because the job has eight interior doors.
2. Press Tab and check the estimate for accuracy.
3. Notice that the column headed Unit Costs shows costs per unit, per "Ea" (each) in this case.
4. The column headed Extended Costs shows costs for the entire eight doors.
5. The lines opposite Title and Description show what's getting painted and how you plan to do the work. You can change the words in either of these boxes. Just click on what you want to change and start typing or deleting.
6. You can also change any numbers in the Unit Cost column. Just click and start typing.
7. When the words and costs are exactly right, press Enter↵ or click on OK to copy these figures to the end of your estimate.

Roll & brush 1 coat of water base undercoat on interior flush doors
Medium .3 MH/Door, 11.5 Doors/Gal

8.00	2P@2.400	Ea	12.88	54.24	0.00	67.12

Extended costs for eight doors as they appear on your estimate form.

The new line at the bottom of your estimate shows:

8.00 is the quantity of doors
2P is the recommended crew, a "medium" productivity painter
@2.40 shows the manhours required for the work
Ea is the unit of measure, each in this case
12.88 is the material cost (the paint)
54.24 is the labor cost for the job
0.00 shows there is no equipment cost
67.12 is the total of material, labor and equipment columns

Copy Anything to Anywhere in Your Estimate

Anything in the costbook can be copied to your estimate. Just click on the line (or select the words) you want to copy and press the F8 key. It's copied to the last line of your estimating form. If your selection includes costs you'll have a chance to enter the quantity. To copy to the middle of your estimate:

1. Select what you want to copy.
2. Click on the button.
3. Click in the estimate where you want to paste.
4. Click on the button.

Your Own Wage Rates

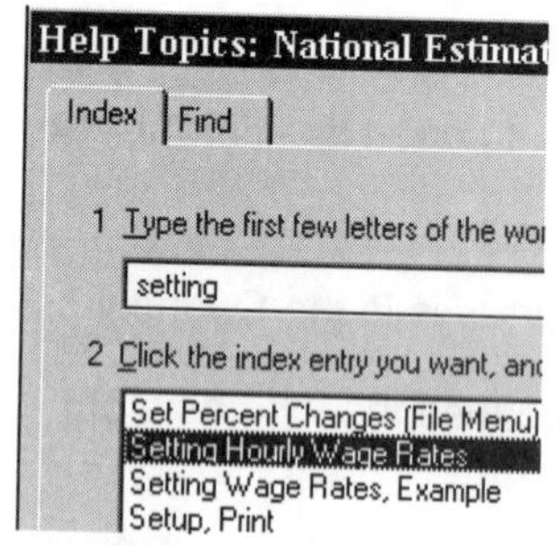

Search for information on setting wage rates.

The labor cost in the example above is based on a medium productivity painter working at an hourly cost of $22.59 per hour. (See page 6 in the National Painting Cost Estimator for crew rates used in the costbook.) Suppose $22.59 per hour isn't right for your estimate. What then? No problem! It's easy to use your own wage rate for any crew or even make up your own crew codes. To get more information on setting wage rates, press F1. At National Estimator Help Contents, click on the Search button. Type "setting" and press Enter to see how to set wage rates. To return to your estimate, click on File on the National Estimator Help menu bar. Then click on Exit.

Changing Cost Estimates

With Num Lock off, use the ↑ or ↓ key to move the cursor to the line you want to change (or click on that line). In this case, move to the line that begins with a quantity of 8. To open the Enter Cost Information Dialog box, either:

- Press Enter, -Or-
- Click on the button on the tool bar.

	Unit Costs
Hours:	0.300
Material Cost:	2.00
Labor Cost:	6.78
Equipment Cost:	0.00
Total Cost:	8.78

Change the material cost to two dollars.

To make a change, either

- Click on what you want to change, -Or-
- Press Tab until the cursor advances to what you want to change.

Then type the correct figure. In this case, change the material cost to 2.00 (two dollars).

Press Tab and check the Extended Costs column. If it looks OK, press Enter and the change is made on your estimating form.

Changing Text (Descriptions)

Click on the button on the tool bar to be sure you're in the estimate. With Num Lock off, use the ↑ or ↓ key or click the mouse button to put the cursor where you want to make a change. In this case, we're going to make a change on the line that begins "Medium . . ."

To make a change, click where the change is needed. Then either:

- Press the Del or ←Bksp key to erase what needs deleting, -Or-
- Select what needs deleting and click on the button on the tool bar.
- Type what needs to be added.

Roll & brush 1 coat of water base undercoat on
Medium .3 MH/Door, 11.5 Doors/Gal
8.00 2P@2.400 Ea 16.00

To select, click and hold the mouse button while dragging the mouse.

In this case, click just before the word "Medium." Then hold the left mouse button down and drag the mouse to the right until you've put a dark background behind the word "Medium." The dark background shows that this word is selected and ready for editing.

Press the Del key (or click on the button on the tool bar), and the selection is cut from the estimate. If that's not what you wanted, click on the button and word "Medium" is back again.

Adding Text (Descriptions)

Some of your estimates will require descriptions (text) and costs that can't be found in the National Painting Cost Estimator. What then? With National Estimator it's easy to add descriptions and costs of your choice anywhere in the estimate. For practice, let's add an estimate for four hinge brackets to Estimate One.

Roll & brush 1 coat of water base undercoat on int
Medium .3 MH/Door, 11.5 Doors/Gal
8.00 2P@2.400 Ea 16.00
Hinge brackets

Total Manhours, Material, Labor, and Equipment:

Adding "Hinge brackets."

Click on the button to be sure the estimate window is maximized. We can add lines anywhere on the estimate. But in this case, let's make the addition at the end. Press the ↓ key to move the cursor down until it's just above the horizontal line that separates estimate detail lines from estimate totals. To open a blank line, either:

- Press Enter↵, -Or-
- Click on the button on the tool bar, -Or-
- Click on Edit on the menu bar. Then click on Insert a Text Line.

Type "Hinge brackets" and press Enter↵.

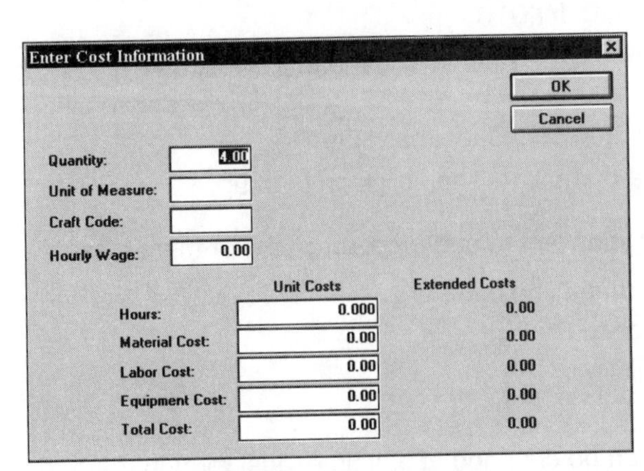

Adding a cost line with the Enter Cost Information dialog box.

Adding a Cost Estimate Line

Now let's add a cost for "Hinge brackets" to your estimate. Begin by opening the Enter Cost Information dialog box. Either:

- Click on button on the tool bar, -Or-
- Click on Edit on the menu bar. Then click on Insert a Cost Line.

1. The cursor is in the Quantity box. Type the number of units (4 in this case) and press Tab.

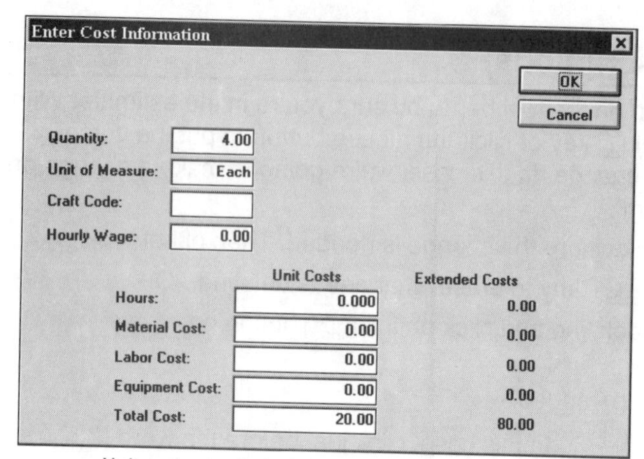

Unit and extended costs for four hinge brackets.

2. The cursor moves to the next box, Unit of Measure.
3. In the Unit of Measure box, type *Each* and press [Tab].
4. Press [Tab] twice to leave the Craft Code blank and Hourly Wage at zero.
5. Since these brackets will be installed by supplier, there's no material, labor or equipment cost. So press [Tab] four times to skip over the Hours, Material Cost, Labor Cost and Equipment Cost boxes.
6. In the Total Cost box, type 20.00. That's the cost per bracket quoted by your supplier.
7. Press [Tab] once more to advance to OK.
8. Press [Enter←] and the cost of four hinge brackets is written to your estimate.

Note: The sum of material, labor, and equipment costs appears automatically in the Total Cost box. If there's no cost entered in the Material Cost, Labor Cost or Equipment Cost boxes (such as for a subcontracted item), you can enter any figure in the Total Cost box.

Adding Lines to the Costbook

Add lines or make changes in the costbook the same way you add lines or make changes in an estimate. The additions and changes you make become part of the user costbook. For more information on user costbooks, press [F1]. Click on Search. Type "user" and press [Enter←].

Enter Tax Rates

Tax on Materials: 7.25 %
Tax on Labor: 0. %
Tax on Equipment: 0. %
Tax on Total Only Costs: 0. %
Tax on the Contract Price: 0. %

OK Cancel

Type the tax rate that applies.

Adding Tax

To include sales tax in your estimate:

1. Click on Edit.
2. Click on Tax Rates.
3. Type the tax rate in the appropriate box.
4. Press [Tab] to advance to the next box.
5. Press [Enter←] or click on OK when done.

In this case, the tax rate is 7.25% on materials only. Tax will appear as the last line of the estimate.

Adding Overhead and Profit

Set markup percentages in the Add for Overhead & Profit dialog box. To open the box, either:

- Click on the [$] button on the tool bar, -Or-
- Click on Edit on the menu bar. Then click on Markup.

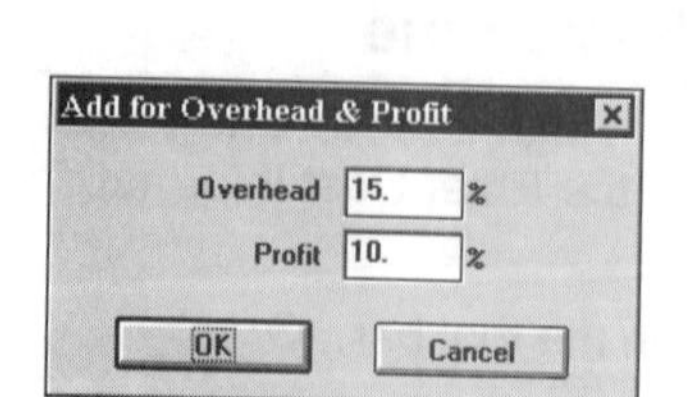

Adding overhead & profit.

Type the percentages you want to add for overhead. For this estimate:

1. Type 15 on the Overhead line.
2. Press [Tab] to advance to Profit.
3. Type 10 on the Profit line.
4. Press [Enter←].

Markup percentages can be changed at any time. Just reopen the Add for Overhead & Profit Dialog box and type the correct figure.

File Name: Estimate 1 — Construction Estimate — Page 1

Qty	Craft@Hours	Unit	Material	Labor	Equipment	Total
			Estimate One			
Roll & brush 1 coat of water base undercoat on interior flush doors						
Medium .3 MH/Door, 11.5 Doors/Gal						
8.00	2P@2.400	Ea	16.00	54.24	0.00	70.24
Hinge brackets						
4.00	--@ .0000	--	0.00	0.00	0.00	80.00
Total Manhours, Material, Labor, and Equipment:						
	2.4		16.00	54.24	0.00	70.24
Total Only (Subcontract) Costs:						80.00
			Subtotal:			150.24
			15.00% Overhead:			22.54
			10.00% Profit:			17.28
			Estimate Total:			190.06
			Tax on Materials:			1.16
			Grand Total:			191.22

A preview of Estimate One.

Preview Your Estimate

You can display an estimate on screen just the way it will look when printed on paper. To preview your estimate, either:

- Click on the [Print Preview] button on the tool bar, -Or-
- Click on File on the menu bar. Then click on Print Preview.

In print preview:

- Click on Next Page or Prev. Page to turn pages.
- Click on Two Page to see two estimate pages side by side.
- Click on Zoom In to get a closer look.
- Click on Close when you've seen enough.

Use buttons on Print Preview to see your estimate as it will look when printed.

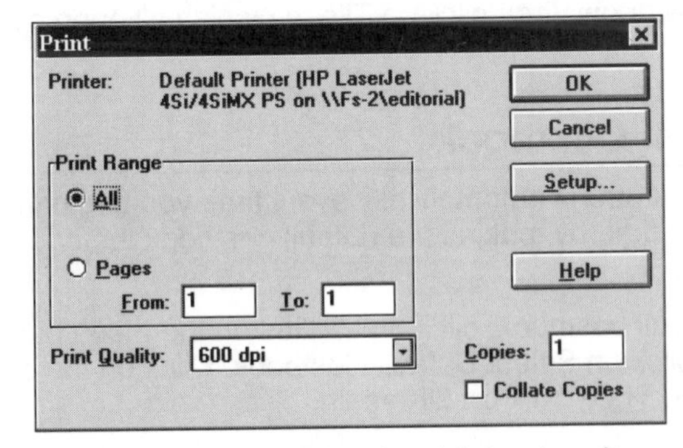

Options available depend on the printer you're using.

Printing Your Estimate

When you're ready to print the estimate, either:

- Click on the [Print] button on the tool bar, -Or-
- Click on File on the menu bar. Then click on Print, -Or-
- Hold the Ctrl key down and type the letter P.

Press Enter or click on OK to begin printing.

Save Your Estimate to Disk

To store your estimate on the hard disk where it can be re-opened and changed at any time, either:

- Click on the [Save] button on the tool bar, -Or-
- Click on File on the menu bar. Then click on Save, -Or-
- Hold the Ctrl key down and type the letter S.

The cursor is in the File Name box. Type the name you want to give this estimate, such as FIRST. The name can be up to eight letters and numbers, but don't use symbols or spaces. Press Enter or click on OK and the estimate is written to disk.

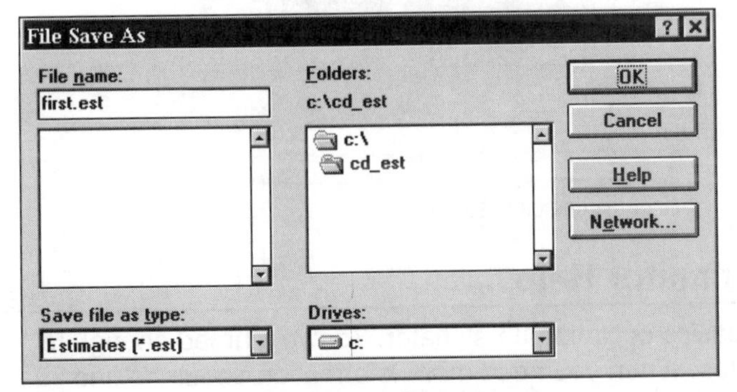

Type the estimate name in the File Name box to assign a file name.

Opening Other Costbooks

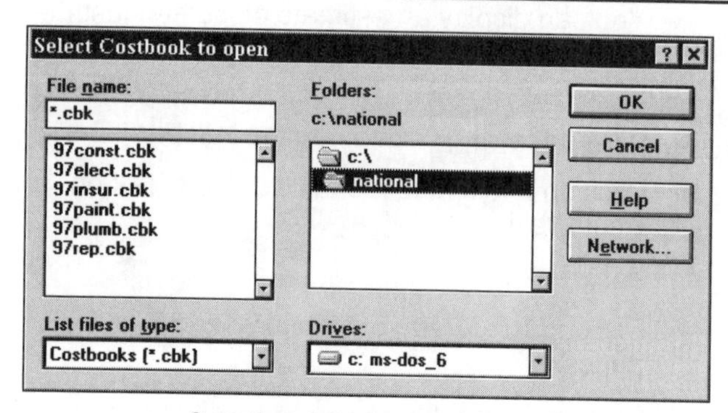

Open the costbook of your choice.

Many construction cost estimating databases are available for the National Estimator program. For example, a National Estimator costbook comes with each of these estimating manuals:

- *National Construction Estimator*
- *National Repair & Remodeling Estimator*
- *National Electrical Estimator*
- *National Plumbing & HVAC Estimator*
- *National Painting Cost Estimator*
- *National Renovation & Insurance Repair Estimator*

To open any of the other costbooks on your computer:

- Click on File
- Click on Open Costbook
- Be sure the drive and directory are correct, usually c:\national.
- Double click on the costbook of your choice.

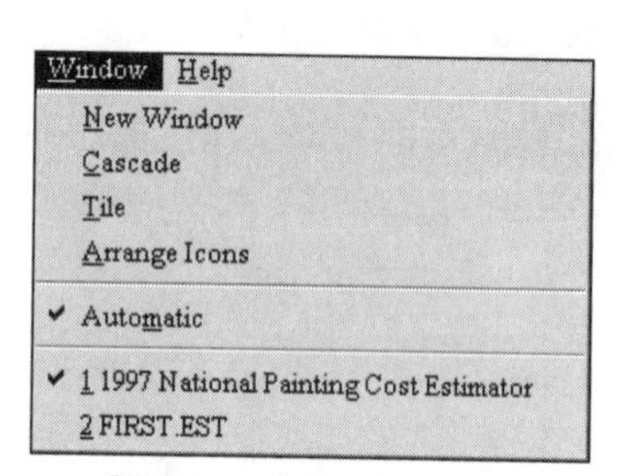

Click to switch costbooks.

To see a list of the costbooks open, click on Window. The name of the current costbook will be checked. Click on any other costbook name to display that costbook. Click on Window, then click on Tile to display all open costbooks and estimates.

Select Your Default Costbook

Your default costbook opens automatically every time you begin using National Estimator. Save time by making the default costbook the one you use most.

To change your default costbook, click on Utilities on the menu bar. Then click on Options. Next, click on Select Default Costbook. Click on the costbook of your choice. Click on OK. Then click on OK again.

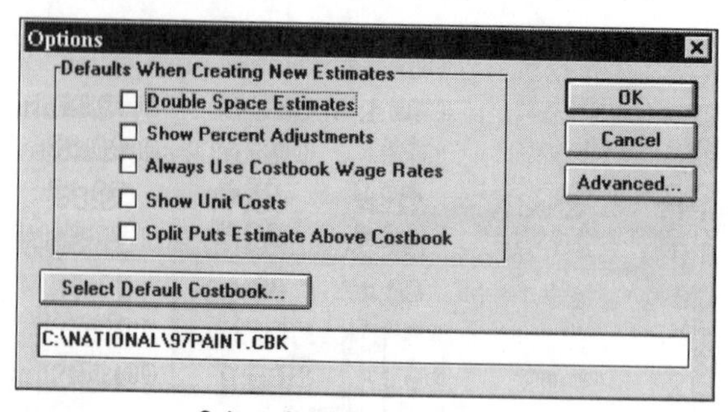

Select the default costbook.

Use National Estimator Help

Click on Print All Topics to print the entire Guide to National Estimator (27 pages).

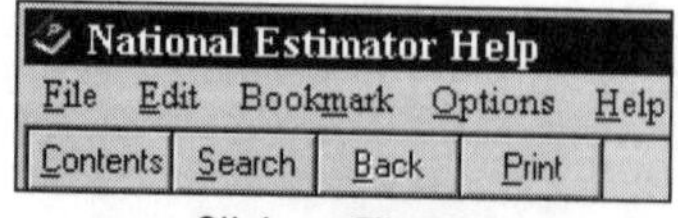

Click on File and then Print Topic.

That completes the basics of National Estimator. You've learned enough to complete most estimates. When you need more information about the fine points, use National Estimator Help. Click on [?] to see Help Contents. Then click on the menu selection of your choice. To print 27 pages of instructions for National Estimator, go to Help Contents. Click on Print All Topics (at the bottom of Help Contents). Click on File on the Help menu bar. Then click on Print Topic. Click on OK.

Index

Other Practical References

Builder's Guide to Room Additions

How to tackle problems that are unique to additions, such as requirements for basement conversions, reinforcing ceiling joists for second-story conversions, handling problems in attic conversions, what's required for footings, foundations, and slabs, how to design the best bathroom for the space, and much more. Besides actual construction, you'll even find help in designing, planning and estimating your room addition jobs. **352 pages, 8½ x 11, $27.25**

Profits in Building Spec Homes

If you've ever wanted to make big profits in building spec homes yet were held back by the risks involved, you should have this book. Here you'll learn how to do a market study and feasibility analysis to make sure your finished home will sell quickly, and for a good profit. You'll find tips that can save you thousands in negotiating for land, learn how to impress bankers and get the financing package you want, how to nail down cost estimating, schedule realistically, work effectively yet harmoniously with subcontractors so they'll come back for your next home, and finally, what to look for in the agent you choose to sell your finished home. Includes forms, checklists, worksheets, and step-by-step instructions. **208 pages, 8½ x 11, $27.25**

How to Succeed With Your Own Construction Business

Everything you need to start your own construction business: setting up the paperwork, finding the work, advertising, using contracts, dealing with lenders, estimating, scheduling, finding and keeping good employees, keeping the books, and coping with success. If you're considering starting your own construction business, all the knowledge, tips, and blank forms you need are here.
336 pages, 8½ x 11, $24.25

Finish Carpenter's Manual

Everything you need to know to be a finish carpenter: assessing a job before you begin, and tricks of the trade from a master finish carpenter. Easy-to-follow instructions for installing doors and windows, ceiling treatments (including fancy beams, corbels, cornices and moldings), wall treatments (including wainscoting and sheet paneling), and the finishing touches of chair, picture, and plate rails. Specialized interior work includes cabinetry and built-ins, stair finish work, and closets. Also covers exterior trims and porches. Includes manhour tables for finish work, and hundreds of illustrations and photos. **208 pages, 8½ x 11, $22.50**

Handbook of Construction Contracting

Volume 1: Everything you need to know to start and run your construction business; the pros and cons of each type of contracting, the records you'll need to keep, and how to read and understand house plans and specs so you find any problems before the actual work begins. All aspects of construction are covered in detail, including all-weather wood foundations, practical math for the job site, and elementary surveying. **416 pages, 8½ x 11, $28.75**

Volume 2: Everything you need to know to keep your construction business profitable; different methods of estimating, keeping and controlling costs, estimating excavation, concrete, masonry, rough carpentry, roof covering, insulation, doors and windows, exterior finishes, specialty finishes, scheduling work flow, managing workers, advertising and sales, spec building and land development, and selecting the best legal structure for your business.
320 pages, 8½ x 11, $30.75

CD Estimator

If your computer has *Windows*™ and a CD-ROM drive, CD Estimator puts at your fingertips 85,000 construction costs for new construction, remodeling, renovation & insurance repair, electrical, plumbing, HVAC and painting. You'll also have the National Estimator program — a stand-alone estimating program for *Windows* that Remodeling magazine called a "computer wiz." Included is a communications program you can use to download cost updates on Contractor's Bulletin Board. To help you create professional-looking estimates, the disk includes over 40 construction estimating and bidding forms in a format that's perfect for nearly any word processing or spreadsheet program for *Windows*. And to top it off, a 70-minute interactive video teaches you how to use this CD-ROM to estimate construction costs. **CD Estimator is $59.00**

Video: Paint Contractor's Manual 1

How to run a paint contracting business, set up jobs for good production, and prepare surfaces for painting. Topics include planning, scheduling, using job work orders, setting production targets, advising on color selection, patching, textures, and the most common reasons for paint failure. **40 minutes, VHS, $24.75**

Video: Paint Contractor's Manual 2

How to select the right tools for the best results, get the most from spray paint equipment, lay down a glass-like finish in enamel, paint doors, jambs, windows, and cabinets quickly, and much more.
40 minutes, VHS, $24.75

Manual of Professional Remodeling

The practical manual of professional remodeling that shows how to evaluate a job so you avoid 30-minute jobs that take all day, what to fix and what to leave alone, and what to watch for in dealing with subcontractors. Includes how to calculate space requirements; repair structural defects; remodel kitchens, baths, walls, ceilings, doors, windows, floors and roofs; install fireplaces and chimneys (including built-ins), skylights, and exterior siding. Includes blank forms, checklists, sample contracts, and proposals you can copy and use. **400 pages, 8½ x 11, $23.75**

Contractor's Year-Round Tax Guide Revised

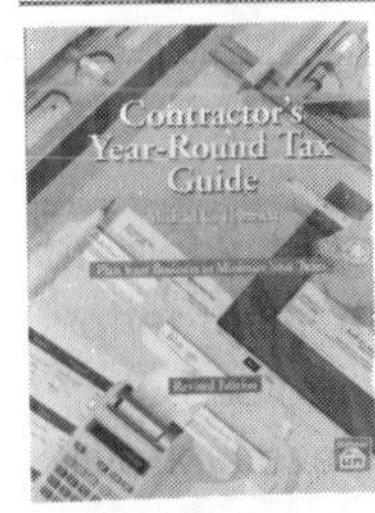

How to set up and run your construction business to minimize taxes: corporate tax strategy and how to use it to your advantage, and what you should be aware of in contracts with others. Covers tax shelters for builders, write-offs and investments that will reduce your taxes, accounting methods that are best for contractors, and what the I.R.S. allows and what it often questions. **192 pages, 8½ x 11, $26.50**

Renovating & Restyling Vintage Homes

Any builder can turn a run-down old house into a showcase of perfection — if the customer has unlimited funds to spend. Unfortunately, most customers are on a tight budget. They usually want more improvements than they can afford — and they expect you to deliver. This book shows how to add economical improvements that can increase the property value by two, five or even ten times the cost of the remodel. Sound impossible? Here you'll find the secrets of a builder who has been putting these techniques to work on Victorian and Craftsman-style houses for twenty years. You'll see what to repair, what to replace and what to leave, so you can remodel or restyle older homes for the least amount of money and the greatest increase in value. **416 pages, 8½ x 11, $33.50**

Craftsman Book Company
6058 Corte del Cedro, P.O. Box 6500
Carlsbad, CA 92018

24 hour order line
1-800-829-8123 Fax (619) 438-0398

Order online: http://www.craftsman-book.com

In A Hurry? We accept phone orders charged to your Visa, MasterCard, Discover or American Express

Name ______________________

Company ______________________

Address ______________________

City/State/Zip ______________________

Total enclosed______________________(In California add 7.25% tax)

We pay shipping when your check covers your order in full.
If you prefer, use your ❑Visa ❑MasterCard ❑Discover or ❑American Express

Card#______________________

Expiration date______________________Initials______________________

10-Day Money Back Guarantee

- ❑27.25 Builder's Guide to Room Additions
- ❑59.00 CD Estimator
- ❑39.75 Construction Forms & Contracts with a 3½" HD disk. Add $15.00 if you need ❑Macintosh disks.
- ❑39.00 Contractor's Guide to Building Code Revised
- ❑26.50 Contractor's Year-Round Tax Guide Revised
- ❑18.25 Drywall Contracting
- ❑22.50 Finish Carpenter's Manual
- ❑28.75 Handbook of Construction Contracting Volume 1
- ❑30.75 Handbook of Construction Contracting Volume 2
- ❑24.25 How to Succeed w/Your Own Construction Business
- ❑23.75 Manual of Professional Remodeling
- ❑38.50 National Repair & Remodeling Estimator with FREE stand-alone *Windows* estimating program on a 3½" HD disk.
- ❑24.00 Paint Contractor's Manual
- ❑21.25 Painter's Handbook
- ❑27.25 Profits in Building Spec Homes
- ❑33.50 Renovating & Restyling Vintage Homes
- ❑24.75 Video: Paint Contractor's Manual 1
- ❑24.75 Video: Paint Contractor's Manual 2
- ❑38.00 National Painting Cost Estimator with FREE stand-alone *Windows* estimating program on a 3½" HD disk.
- ❑FREE Full Color Catalog

Receive Fresh Cost Data Every Year – Automatically

Join Craftsman's Standing Order Club and automatically receive special membership discounts on annual cost books!

___ Qty. National Building Cost Manual
Standing Order price $19.55

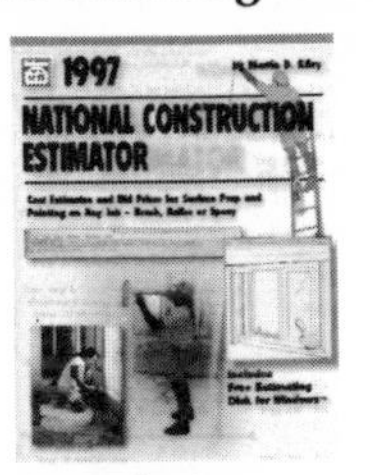

___ Qty. National Construction Estimator
Standing Order price $31.88

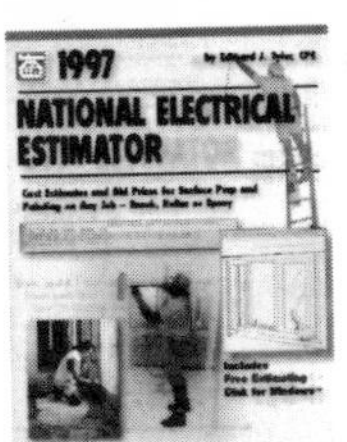

___ Qty. National Electrical Estimator
Standing Order price $32.09

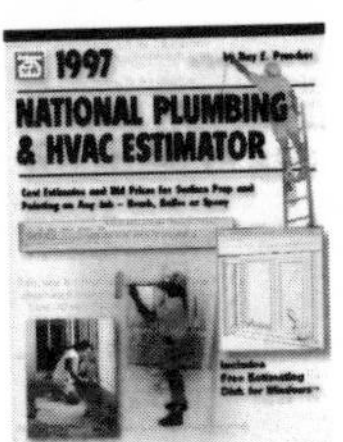

___ Qty. National Plumbing & HVAC Estimator
Standing Order price $32.52

___ Qty. National Painting Cost Estimator
Standing Order price $32.30

___ Qty. National Renovation & Insurance Repair Est.
Standing Order price $33.58

___ Qty. National Repair & Remodeling Est.
Standing Order price $32.73

How many times have you missed one of Craftsman's pre-publication discounts, or procrastinated about buying the updated book and ended up bidding your jobs using obsolete cost data that you "updated" on your own? As a Standing Order Member you never have to worry about ordering every year. Instead, you will receive a confirmation of your standing order each September. If the order is correct, do nothing, and your order will be shipped and billed as listed on the confirmation card. If you wish to change your address, or your order, you can do so at that time. Your standing order for the books you selected will be shipped as soon as the publications are printed. You'll automatically receive any pre-publication discount price.

Bill to:
Name ______________________
Company ______________________
Address ______________________
City/State/Zip ______________________

Ship to:
Name ______________________
Company ______________________
Address ______________________
City/State/Zip ______________________

Purchase order # ______________________
Phone # (   ) ______________________
Signature ______________________

Mail This Card Today for a Free Full Color Catalog

Over 100 books, videos, and audios at your fingertips with information that can save you time and money. Here you'll find information on carpentry, contracting, estimating, remodeling, electrical work, and plumbing.

All items come with an unconditional 10-day money-back guarantee.

If they don't save you money, mail them back for a full refund.

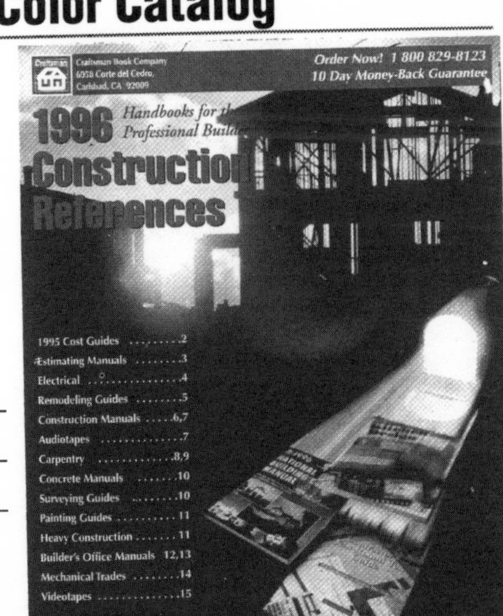

Name / Company ______________________

Address ______________________

City / State / Zip ______________________

Craftsman Book Company / 6058 Corte del Cedro / P.O. Box 6500 / Carlsbad, CA 92018

NO POSTAGE
NECESSARY
IF MAILED
IN THE
UNITED STATES

BUSINESS REPLY MAIL
FIRST CLASS MAIL PERMIT NO. 271 CARLSBAD CA

POSTAGE WILL BE PAID BY ADDRESSEE

Craftsman Book Company
6058 Corte del Cedro
P.O. Box 6500
Carlsbad CA 92018-9974

NO POSTAGE
NECESSARY
IF MAILED
IN THE
UNITED STATES

BUSINESS REPLY MAIL
FIRST CLASS MAIL PERMIT NO. 271 CARLSBAD CA

POSTAGE WILL BE PAID BY ADDRESSEE

Craftsman Book Company
6058 Corte del Cedro
P.O. Box 6500
Carlsbad CA 92018-9974

NO POSTAGE
NECESSARY
IF MAILED
IN THE
UNITED STATES

BUSINESS REPLY MAIL
FIRST CLASS MAIL PERMIT NO. 271 CARLSBAD CA

POSTAGE WILL BE PAID BY ADDRESSEE

Craftsman Book Company
6058 Corte del Cedro
P.O. Box 6500
Carlsbad CA 92018-9974